AF832702

Cardiovascular Molecular Morphogenesis

Series Editor

Roger R. Markwald
Medical University of South Carolina

Editorial Advisory Board

Paul Barton
National Heart and Lung Institute
London, United Kingdom

Clayton Buck
University of Pennsylvania

María V. de la Cruz
Hospital Infantil de Mexico
"Federico Gómez," Mexico

Mark Fishman
Massachusetts General Hospital

Adriana Gittenberger-de Groot
University of Leiden
The Netherlands

Julie Korenberg
Cedars-Sinai Medical Center
Los Angeles

Wout H. Lamers
University of Amsterdam
The Netherlands

Kersti Linask
University of Medicine and Dentistry of New Jersey

Charles D. Little
Medical University of South Carolina

John Lough
Medical College of Wisconsin

Takashi Mikawa
Cornell Medical College

Jeffrey Robbins
Children's Hospital Medical Center, Cincinnati

Thomas Rosenquist
University of Nebraska Medical School

Raymond B. Runyan
University of Arizona

Robert Schwartz
Baylor College of Medicine

Books in the Series

Vascular Morphogenesis: In vivo, in vitro, in mente — *0-8176-3920-9*
C.D. Little, V.A. Mironov, and E.H. Sage, Editors

Living Morphogenesis of the Heart — *0-8176-4037-1*
M.V. de la Cruz and R.R. Markwald

Formation of the Heart and Its Regulation — *0-8176-4216-1*
R.J. Tomanek and R.B. Runyan, Editors

Vascular Morphogenesis in the Female Reproductive System — *0-8176-4221-8*
H. Augustin, L. Iruela-Arispe, P. Rogers, and S.K. Smith, Editors

Vascular Morphogenesis in the Female Reproductive System

Hellmut G. Augustin
M. Luisa Iruela-Arispe
Peter A.W. Rogers
Stephen K. Smith
Editors

With 45 Figures, Including 5 Color Plates

Birkhäuser
Boston • Basel • Berlin

Hellmut G. Augustin
Department of Vascular Biology and Angiogenesis Research
Tumor Biology Center
Freiburg D-79106
Germany

M. Luisa Iruela-Arispe
Molecular Biology Institute
University of California
Los Angeles, CA 90095-1570
USA

Peter A.W. Rogers
Department of Obstetrics and Gynaecology
Monash Medical Centre
Victoria 3168
Australia

Stephen K. Smith
Department of Obstetrics and Gynaecology
Rosie Maternity Hospital
Cambridge CB2 2SW
England

Cover illustration: Adapted from Figure 15.7, page 293, and Figure 15.8, page 294. See Chapter 15, "Placental Functional Morphology," by Graham J. Burton for additional details.

Library of Congress Cataloging-in-Publication Data
Vascular morphogenesis in the female reproductive system / edited by
Hellmut Augustin ... [et al.]. p. cm.—(Cardiovascular molecular morphogenesis)
Includes bibliographical references and index.
ISBN 0-8176-4221-8 (alk. paper)
1. Generative organs, Female—Blood-vessels—Growth. 2. Neovascularization.
I. Augustin, Hellmut. II. Series.
QP259 .V37 2001
612.6′2—dc21 00-065098

Printed on acid-free paper.
© 2001 Birkhäuser Boston
Birkhäuser ®

All rights reserved. This work may not be translated or copied in whole or in part without the written permission of the publisher (Birkhäuser Boston, c/o Springer-Verlag New York, Inc., 175 Fifth Avenue, New York, NY 10010, USA), except for brief excerpts in connection with reviews or scholarly analysis. Use in connection with any form of information storage and retrieval, electronic adaptation, computer software, or by similar or dissimilar methodology now known or hereafter developed is forbidden.
The use of general descriptive names, trade names, trademarks, etc., in this publication, even if the former are not especially identified, is not to be taken as a sign that such names, as understood by the Trade Marks and Merchandise Marks Act, may accordingly be used freely by anyone.

ISBN 0-8176-4221-8
ISBN 3-7643-4221-8 SPIN 10835847

Production managed by Louise Farkas; manufacturing supervised by Jerome Basma.
Typeset by Best-set Typesetter Ltd., Hong Kong.
Printed and bound by Maple-Vail Book Manufacturing Group, York, PA.
Printed in the United States of America.

9 8 7 6 5 4 3 2 1

Contents

Contributors

Kari Alitalo, Molecular/Cancer Biology Laboratory, Haartman Institute, University of Helsinki, SF-00014 Helsinki, Finland

Maria Asuncion Ortega, Molecular Biology Institute, University of California Los Angeles, Los Angeles, CA 90095-1570, USA

Hellmut G. Augustin, Department of Vascular Biology and Angiogenesis Research, Tumor Biology Center, Freiburg D-79106, Germany

Graham J. Burton, Department of Anatomy, University of Cambridge, CB2 3DY Cambridge, England

Alicia Collado-Hidalgo, Division of Gynecologic Oncology, UCLA School of Medicine, Los Angeles, CA 90095-1740, USA

D. Stephen Charnock-Jones, Department of Obstetrics and Gynaecology, Rosie Maternity Hospital, Cambridge CB2 2SW, England

Caroline H. Damsky, University of San Francisco Cancer Center, San Francisco, CA 94143, USA

Robert B. Dickson, Lombardi Cancer Center, Georgetown University, Washington, DC 20007, USA

Napoleone Ferrara, Genentech Inc., San Francisco, CA 94080-4990, USA

Susan Fisher, University of San Francisco Cancer Center, San Francisco, CA 94143, USA

Stephen B. Fox, Department of Anatomical Pathology, Christchurch School of Medicine, Christchurch, New Zealand

Ian S. Fraser, Department of Obstetrics and Gynaecology, Sydney Centre for Reproductive Health Research, The University of Sydney, Sydney NSW 2006, Australia

Caroline E. Gargett, Department of Obstetrics and Gynaecology, Monash Medical Centre, Victoria, 3168 Australia

Anna T. Grazul-Bilska, Department of Animal and Range Sciences, North Dakota State University, Fargo, ND 58105-5727, USA

M. Luisa Iruela-Arispe, Molecular Biology Institute, University of California Los Angeles, Los Angeles, CA 90095-1570, USA

Marika J. Karkkainen, Molecular/Cancer Biology Laboratory, Haartman Institute, University of Helsinki, SF-00014 Helsinki, Finland

Elise C. Kohn, Laboratory of Pathology, National Cancer Institute, Bethesda, MD 20892, USA

Timothy F. Lane, Division of Gynecologic Oncology, UCLA School of Medicine, Los Angeles, CA 90095-1740, USA

Steven K. Libutti, Laboratory of Pathology, National Cancer Institute, Bethesda, MD 20892, USA

Sandra W. McLeskey, School of Nursing and Greenebaum Cancer Center, University of Maryland, Baltimore, MD 21201, USA

Roberto Montesano, Department of Morphology, University Medical Center, Geneva, Switzerland

Michael D. Mueller, Department of Obstetrics, Gynecology, and Reproductive Sciences, University of California, San Francisco, CA 94143-0556, USA

Sarah Oikemus, Molecular Biology Institute, University of California Los Angeles, Los Angeles, CA 90095-1570, USA

Michael S. Pepper, Molecular Biology Institute, University of California Los Angeles, Los Angeles, CA 90095-1570, USA

M. Graziella Persico, International Institute of Genetics and Biophysics, CNR, I-80125 Naples, Italy

Dale A. Redmer, Department of Animal and Range Sciences, North Dakota State University, Fargo, ND 58105-5727, USA

Lawrence P. Reynolds, Department of Animal and Range Sciences, North Dakota State University, Fargo, ND 58105-5727, USA

Peter A.W. Rogers, Department of Obstetrics and Gynaecology, Monash Medical Centre, Victoria, 3168, Australia

Constance E. Runyan, Department of Pediatrics, Northwestern University Medical School, Chicago, IL 60611-3008, USA

Anne Saaristo, Molecular/Cancer Biology Laboratory, Haartman Institute, University of Helsinki, SF-00014 Helsinki, Finland

H. William Schnaper, Department of Pediatrics, Northwestern University Medical School, Chicago, IL 60611-3008, USA

Stephen K. Smith, Department of Obstetrics and Gynaecology, Rosie Maternity Hospital, Cambridge CB2 2SW, England

Robert N. Taylor, Department of Obstetrics, Gynecology, and Reproductive Sciences, University of California, San Francisco, CA 94143-0556, USA

Series Preface

The overall scope of this new series will be to evolve an understanding of the genetic basis of (1) how early mesoderm commits to cells of a heart lineage that progressively and irreversibly assemble into a segmented, primary heart tube that can be remodeled into a four-chambered organ, and (2) how blood vessels are derived and assembled both in the heart and in the body. Our central aim is to establish a four-dimensional, spatiotemporal foundation for the heart and blood vessels that can be genetically dissected for function and mechanism.

Since Robert DeHaan's seminal chapter "Morphogenesis of the Vertebrate Heart" published in *Organogenesis* (Holt Rinehart & Winston, NY) in 1965, there have been surprisingly few books devoted to the subject of cardiovascular morphogenesis, despite the enormous growth of interest that occurred nationally and internationally. Most writings on the subject have been scholarly compilations of the proceedings of major national or international symposia or multiauthored volumes, often without a specific theme. What is missing are the unifying concepts that can make sense out of a burgeoning database of facts. The Editorial Board of this new series believes the time has come for a book series dedicated to cardiovascular morphogenesis that will serve not only as an important archival and didactic reference source for those who have recently come into the field but also as a guide to the evolution of a field that is clearly coming of age. The advances in heart and vessel morphogenesis are not only serving to reveal general basic mechanisms of embryogenesis but are also now influencing clinical thinking in pediatric and adult cardiology.

Undoubtedly, the Human Genome Project and other genetic approaches will continue to reveal new genes or groups of genes that may be involved in heart development. A central goal of this series will be to extend the identification of these and other genes into their functional role at the molecular, cellular, and organ levels. The major issues in morphogenesis highlighted in the series will be the local (heart or vessel) regulation of cell growth and death, cell adhesion and migration, and gene expression responsible for the cardiovascular cellular phenotypes.

Specific topics will include the following:

- The roles of extracardiac populations of cells in heart development.
- Coronary angiogenesis.

- Vasculogenesis.
- Breaking symmetry, laterality genes, and patterning.
- Formation and integration of the conduction cell phenotypes.
- Growth factors and inductive interactions in cardiogenesis and vasculogenesis.
- Morphogenetic role of the extracellular matrix.
- Genetic regulation of heart development.
- Application of developmental principles to cardiovascular tissue engineering.

Cardiovascular Developmental Biology Center
Medical University of South Carolina
Charleston, South Carolina

Roger R. Markwald

Preface

Angiogenesis is a downregulated process in the healthy adult that is primarily associated with pathological conditions such as tumor growth, inflammation, ischemia, and retinopathies. Given the preeminent role of pathological angiogenesis for tumor growth and metastasis, angiogenesis research during the last 25 years was primarily aimed at deciphering the molecular mechanisms of tumor angiogenesis in order to develop antiangiogenic therapies as a novel antitumorigenic strategy. As a result, an increasing list of antiangiogenic substances is in different phases of clinical trials, and antiangiogenic therapies are on the verge of entering the clinic.

In contrast to the intense efforts aimed at understanding tumor angiogenesis, physiological angiogenic processes in the healthy adult have long been ignored. Most of the early studies of female reproductive angiogenesis were aimed at analyzing these processes paradigmatically for tumor angiogenic processes. More recently, however, the importance of female reproductive vascular morphogenesis for proper reproductive function has increasingly been recognized and has helped to shape the field of reproductive angiogenesis research. Inhibition of angiogenesis has been shown to completely block ovarian function. Ovarian hyperstimulation is associated with an activation of the vascular system. Endometriosis involves distinct angiogenic processes. Placental growth is critically dependent on an adequate formation of the vascular system. Likewise, an inability of cytotrophoblastic cells to acquire a vascular phenotype has been identified as the underlying cause of the complex cascade of events that leads to the manifestation of preeclampsia. Much has been learned about the ability of reproductive hormones to directly act on the vascular system and to affect the expression of angiogenic cytokines in other cells. Lastly, the association of angiogenesis with reproductive tumors of the mammary gland, the ovary, and the uterus has also been a driving force for reproductive biologists and gynecologists to better understand reproductive angiogenesis.

Given these rapid developments, we felt the need for an authoritative reference work that systematically summarizes the significant scientific progress that has been generated in the last few years in fields as diverse as cellular and molecular biology, endocrinology, clinical gynecology, and tumor biology. We thus applied an interdisciplinary and integrative approach to bring together the specialists in the field in order to synergistically summarize and review this large body of timely and diverse information. As a result, the book is formatted into four parts reflecting the four female reproductive organs (mammary gland, ovary, uterus, placenta). Each part is covered by three to five individual chapters. The first chapter of each part is intended to provide an introductory overview of the subject. This introductory chapter is supplemented by more specialized chapters that focus on individual aspects of vascular morphogenetic events in the female reproductive organs.

We sincerely hope that the book will be a useful reference for students and vascular biologists with an interest in the female reproductive system as well as for clinical gynecologists with a scientific interest in the vascular system. We also hope that the book will be thought provoking and stimulate further research in the basic sciences and in clinical disciplines. Future work will continue to shape the field of reproductive vascular biology for the benefit of women's health.

Freiburg, Germany *Hellmut G. Augustin*
Los Angeles, California *M. Luisa Iruela-Arispe*
Victoria, Australia *Peter A.W. Rogers*
Cambridge, United Kingdom *Stephen K. Smith*

PART I

Vascular Morphogenesis in the Mammary Gland

CHAPTER 1

Vascular Morphogenesis in the Mammary Gland: Introduction and Overview

M. Luisa Iruela-Arispe, Maria Asuncion Ortega, Sarah Oikemus, and Michael S. Pepper

The mammary gland is unique, it undergoes most of its development well after birth. Although organ immaturity is a constant feature of most tissues after birth, no other structure is associated with the significant changes in size, shape, and function that occur in the breast during puberty, pregnancy, lactation, and involution.

Much of mammary gland research has focused on understanding the growth, differentiation and apoptosis of epithelial cells, which compose the parenchyma of the gland. The contribution of the stroma has been acknowledged as an important modulator of epithelial growth and differentiation. In contrast, the vascular component of the mammary gland has received little attention. The mammary gland offers an important platform for studying blood vessel function including permeability, modulation of angiogenesis, and regulation of vascular regression following weaning. Chapters 1 to 5 illustrate several intrinsic aspects of mammary gland biology, summarize the current knowledge of vascular regulation associated with the development, function, and pathology of this organ, report the available data related to the vascular component of the mammary gland, and emphasize the importance of organ physiology in the regulation of vascular function.

Because of the importance of the angiogenic switch for tumor growth, Chapters 3 to 5 are devoted to this issue. Chapters 3 and 4 provide a summary of the current knowledge of angiogenesis in breast cancer and discuss some of the discrepancies present in the field. Chapter 5 summarizes the current mouse models used to study mammary carcinogenesis and discusses the information gathered related to the development of a vascular supply in these models.

The mammary gland consists of two primary components: the parenchyma, which forms a system of branching ducts from which alveolar secretory units develop, and the adipose stroma, which provides a substrate within which the parenchyma expands and differentiates.

Given the importance of lactation for nurturing offspring and the high incidence of breast cancer, the mammary gland has been the focus of intense research. Most of the interest has concentrated on the epithelial components. In some cases, the stroma, in particular the role of the mesenchyme in the development of the mammary gland, has also been assessed. However, little attention has been devoted to the vascular network. This chapter summarizes the basic information related to the biology of the murine mammary gland in concert with the growth and regression of vessels in this organ.

Recently, several investigators in the vascular field have concentrated their attention on the mammary gland, and the results from these efforts are beginning to emerge. There is increasing awareness of certain experimentally useful features of this organ:

Late development and accessibility: The mammary gland parenchyma undergoes most of its growth during puberty. Thus, experiments on development can be performed in the adolescent or adult, and there are opportunities for experimental manipulation that are not as available in other organs.

Understanding of the biology of the organ: Much is now understood about the development of the parenchyma and the physiology of the gland. This knowledge can be used to understand the influence of the parenchyma and stroma on the regulation of vascular growth and involution.

Existence of well-developed animal models: Several tumor models have been generated using the mammary gland. In addition, the utilization of "mammary-specific" promoters (including the mammary tumor virus and the whey acidic protein promoters) to target the mammary epithelium offers the opportunity to generate transgenic animal models for overexpression of a variety of gene products. This is discussed at length in Chapter 5. In addition, a large number of techniques, including transplantation, in vitro growth and manipulation of parenchyma, and retroviral and adenoviral infections, have been well established in the literature.

Postembryonic angiogenesis: The mammary gland is one of the few organs that undergoes extensive physiologic angiogenesis in the adult. In addition, events related to alterations in vascular permeability during lactation make this an excellent site to study differentiation and endothelial function including vascular permeability.

Physiological remodeling: During involution (following lactation), a great number of vessels undergo regression, making this organ an excellent site to evaluate remodeling and apoptosis of endothelial cells. The effects of vascular remodeling can also be extended to other cells, particularly pericytes and smooth muscle cells, which contribute to the morphogenesis of blood vessels.

DEVELOPMENT OF THE DUCTAL EPITHELIAL TREE

The development of the mammary gland can be divided into four phases: fetal, postnatal, pubertal, and adult. During fetal life, mammary gland development mirrors the morphogenesis of other glands. The incipient mammary bud grows as an invagination of epithelial cells from the overlying ectoderm. In the mouse fetus this occurs at days 10 to 11 and in the human fetus between weeks 4 and 5. At that time, the initial indications of mammary gland primordia are two parallel lines of ectodermal thickening that extend from the anterior to the posterior limb bud on each side of the embryo (Turner and Gomez, 1933). In the mouse, these lines, referred to as mammary streaks, subdivide into five pairs of mammary gland primordia at day 12, three of which are thoracic and two inguinal. These lens-shaped epithelial primordia grow into bulb-shaped structures with a narrow neck by day 14, a time when the sexual phenotype in the mammary gland is determined. By day 16 in female embryos, the bud elongates and sprouts into the fatty stroma. Further growth and branching continues until each gland develops 15 to 20 branching ducts at the time of birth. In male mice, further development of the mammary primordia is suppressed by fetal androgens around day 14. The process involves detachment of the mammary bud from the surface epidermis by rupture of the epithelial stalk. This is mediated by testosterone receptor–positive stromal cells that in response to this hormone grow around the epithelial stalk and promote detachment (reviewed in Sakakura, 1987).

In newborn female mice, the parenchyma is organized into small cords of epithelial cells that branch from the nipple (Figure 1.1A). The mammary ducts invade the subdermal mammary fat pad and are highly vascularized by a profuse rete of capillaries that surrounds the epithelial branches. The postnatal phase is characterized by elongation of mammary ducts and further ramification. Growth, however, is slow and is dependent on the presence of the ovary since it can be suppressed by castration even though sexual maturation is not reached until 4 to 5 weeks in the mouse (reviewed by Knight and Peaker, 1982; Daniel and Silberstein, 1987).

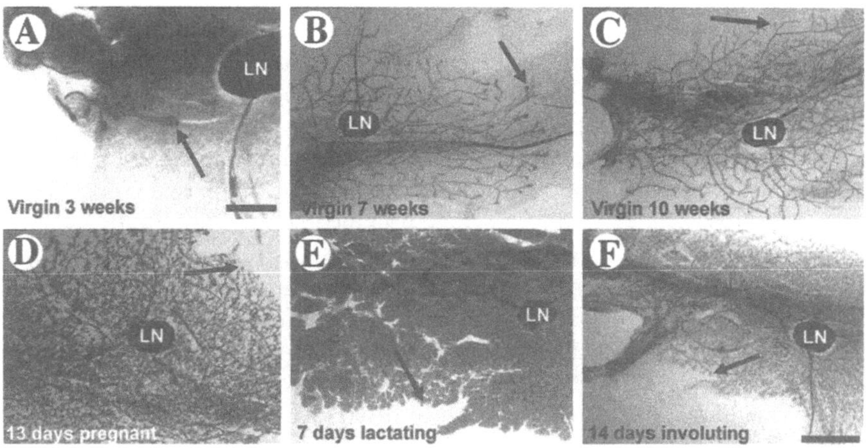

FIGURE 1.1. Growth of murine mammary gland parenchyma. At the specified times, the fourth inguinal murine mammary gland was removed and mounted on a microscope slide, briefly dried, and fixed overnight in 3.7% formaldehyde. Fat was removed by several washes in acetone. Hydration was achieved by serial washes in decreasing concentrations of ethanol. The glands were then stained overnight with carmine solution, washed, dehydrated, and mounted in permount. Carmine strongly stains the parenchyma. The lymphonode (LN) is clearly detected and is a landmark of this gland (fourth pair). A: Ductal invasion into the stroma can be detected by 3 weeks. Note the distended appearance of the terminal end bud (arrow). B: Growth continues during the pubertal stage (7 weeks) by proliferation of the terminal epithelial end-buds (arrow). C: By week 10, parenchymal growth has ceased and the distended end-buds are no longer visible (arrow). D: Additional and profuse epithelial branching (arrow) and growth occurs during pregnancy. E: The terminal alveolar units are distended and occupy most of the available space leaving little room for the fatty stroma. F: Significant remodeling and reduction in the epithelial tree (arrow) are associated with involution of the gland. Bar in A = 2 mm; Bar in B–F = 6 mm.

At about 4 weeks in mice, the ductal tips begin to proliferate, elongate, and invade the mammary fat pad. This is associated with and dependent on the presence of sexual steroid hormones, and marks the pubertal phase. In mice, the extent of epithelial invasion of the fat pad is genetically controlled and varies from strain to strain. Full expansion of the gland is completed between 8 and 10 weeks, again depending on genetic background.

In the parenchyma, the terminal end bud, a distended and highly cellular bulb-like structure (arrows in Figure 1.1A,B) is responsible for morphogenesis of the epithelial tree. The bud is the source of ductal and myoepithelial cells. It has been demonstrated that the end bud is controlled by systemic hormones and by local environmental signals that provide the direction of growth and also lead to regression and involution of the epithelial tree (Faulkin and DeOme, 1960). The final pattern of the mammary ducts is established by an interaction between epithelial and stromal factors.

As the invasion of the ductal tree progressess, blood vessels surround the terminal bud and provide nutrients and oxygen to the rapidly dividing epithelial cells. (A more detailed discussion of vascular growth is provided below.) Concomitant with epithelial invasion there is intense stromal reorganization near the end bud. This area becomes more cellular. Fibroblasts orient themselves along the growing ducts and fill the interstitial spaces with fibrillar collagen (Williams and Daniel, 1983). Secretion of glycosaminoglycans also takes place. In fact, these extracellular matrix components have been shown to be important in the regulation of branching morphogenesis and are likely to be determinants of mammary gland patterning (Bernfield et al, 1984).

Complete growth and maturation is achieved at the point when the mammary parenchyma has grown to the limits of the fat pad and the gland reaches "stasis." Subsequently, during pregnancy, the ductal system increases in complexity by further growth of side branches that sprout from preexisting ducts (Figure 1.1D) (Daniel and Silberstein, 1987). The process can be divided into two stages: the early stage is characterized by rapid growth and branching, resulting in the neoformation of terminal structures called acini or alveolar units. This is clearly seen by day 5 of pregnancy in the mouse. The epithelial cells continue to grow in number and size mainly because of cytoplasmic enlargement. In the second stage, by day 16 to 18 of pregnancy, the structure of the acini becomes more complex with evident myoepithelial cells, distended lumina, and accumulation of secretory material. During lactation, the acini become fully distended as acinar cells secrete milk (Figure 1.1E).

Although the development of the murine, human, and rat glands are very similar, there are some important differences that should be considered when studying each individual model. Figure 1.2 shows the normal histologic alterations in the rat mammary gland during pregnancy, lactation, and involution. Similar alterations are seen in human and mouse glands, with an important exception being that the myoepithelial cells undergo apoptosis in the rat but not in the human or murine mammary gland (Zena Werb, personal communication).

THE MAMMARY GLAND STROMA

As previously alluded to, the mammary gland grows into and is maintained by the underlying and surrounding mesenchyme. Epithelial-stromal interactions appear to be important both for the growth of the mammary epithelium during the

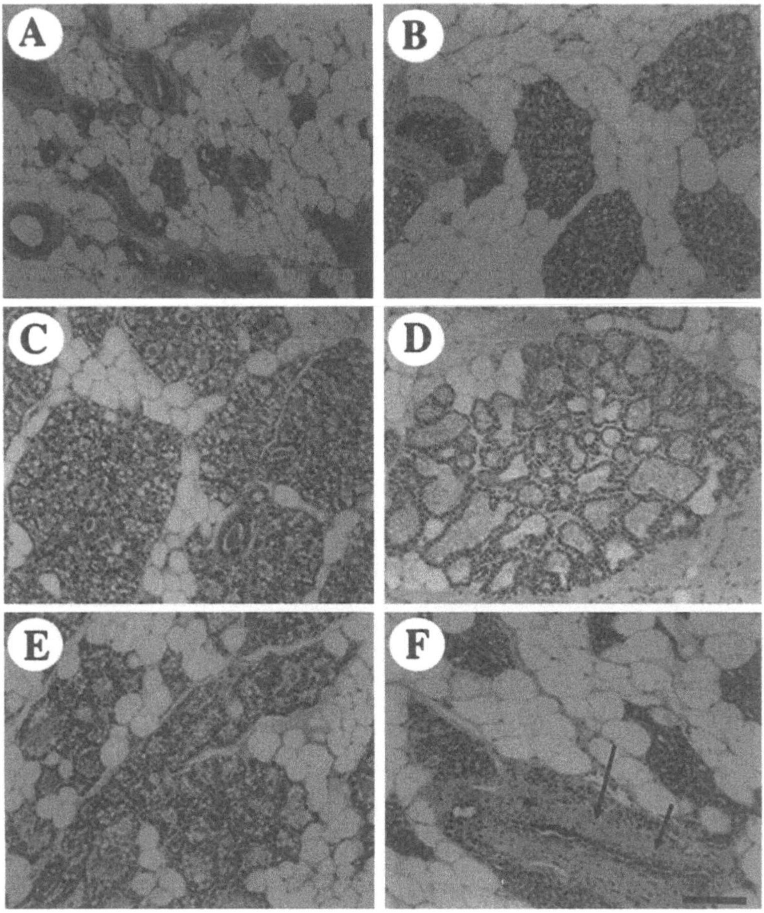

FIGURE 1.2. Histologic changes in rat mammary gland parenchyma and stroma. Rat mammary gland during pregnancy, lactation, and involution. Hematoxylin- and eosin-stained paraffin-embedded sections of (A) 3-month virgin, (B) 12-day pregnant, (C) 18-day pregnant, (D) 2-day lactating (E) 7-day postlactating, and (F) 21-day postlactating mammary gland. Bar = 100 μm.

embryo and for further growth and differentiation in the adult animal. It has also been recognized that the nature of the stroma is important. If mammary epithelium is separated from the surrounding tissue and recombined with mesenchyme derived from other organs, the mammary epithelium makes a typical mammary gland with elongated ducts and end bud formation only if fat pad precursors are present (Sakakura et al, 1982). During development, growth of the adipose stroma occurs normally in the absence of parenchyma. However, mammary epithelial cells cannot grow in the absence of adipose stroma. Furthermore, the stroma can fully support subsequent mammary parenchyma development. Studies by Kratochwil (1987), Sakakura (1991), Lin and Bissel (1993), and Zangani et al (1999a,b) suggest that the stroma plays an inductive role in mammary gland development and function.

Genetic evidence for the influence of the stroma in parenchymal morphogenesis has come from the inhibin null mouse (Robinson and Hennighausen, 1997). These animals are unable to nurse their pups due to aberrant and incomplete development of the mammary epithelium. The results clearly demonstrate that ductal elongation and epithelial cell differentiation during puberty require activin/inhibin signaling from the stroma.

Additional important genetic evidence that underscores the relevance of stromal-epithelial interactions has been elucidated through the use of tissue recombinants and transplantation experiments (Wiesen et al, 1999). Ductal growth and branching morphogenesis was normal in transplants of mammary epithelium from Egfr –/– mice into wild-type fat pads. However, epithelial development (growth and branching) was impaired in tissue recombinants prepared with Egfr –/– stroma regardless of the genomic nature of the epithelium (stroma –/– Epi^{wt}, stroma –/– Epi –/–) (Wiesen et al, 1999).

The complete pattern of reciprocal stromal-epithelial interactions is still far from being elucidated, but a few steps are known. First, mammary mesenchyme "determines" mammary epithelium and fixes the ability of embryonic mammary epithelium to interact with the fatty stroma (Propper, 1970; Sakakura et al, 1982; Zangani et al, 1999a). Second, the mammary epithelium induces androgen receptors in the mammary mesenchyme by direct cell-cell contact and thereby controls the development of androgen responsiveness in this tissue. If androgen is present at this stage, the mammary mesenchyme responds to the hormone and condenses around the mammary epithelium, causing detachment of the mammary bud from the ectodermal surface and subsequent suppression of epithelial growth (Kratochwil, 1971). From these studies, it is apparent that the mammary mesenchyme is both the target and the mediator of the morphogenetic effect of androgens on mammary epithelium during development. Interestingly, only mesenchyme from the mammary gland is able to respond to androgens and regulate epithelial morphogenesis. Fibroblasts from other organs were unable to condensate around the epithelium in response to androgens. It is clear that fibroblasts from mammary glands have androgen receptors and that there is a strong degree of specificity as to the ability of other cells to affect mammary epithelium.

Third, a very important contribution of the stroma is the synthesis and secretion of growth factors, such as transforming growth factors α and β (TGF-α, -β), among many others (Daniel and Robinson, 1992) as well as extracellular matrix proteins and proteases that clearly modulate epithelial function.

Through a series of very elegant studies, Bissell and colleagues (Howlett and Bissell, 1993; Boudreau et al, 1995; Pujuguet et al, 2000) have provided proof that specific extracellular matrix components are essential for the induction of mammary-specific genes, maintenance of the epithelial differentiated phenotype, and prevention of apoptosis.

An equally important contribution of the stroma is the expression of matrix metalloproteinases and their inhibitors. A clear example has been provided by experiments in which overexpression of stromelysin-1, also known as matrix metalloproteinase-3 (MMP3) was targeted to the mammary epithelium. In these animals, the glands develop more branches, show early differentiation of alveoli, and undergo precocious apoptosis during pregnancy. This is also associated with formation of an abnormal stroma and development of neoplasia (Sympson et al, 1994; Sternlicht et al, 1999). In summary, these data support the notion that disruption of the basement membrane can regulate branching morphogenesis during development, apoptosis, stromal organization, as well as induction and progression of cancer.

KINETICS OF VASCULAR GROWTH

Early studies indicated that the mammary gland vasculature is a dynamic and highly specialized structure that undergoes alterations associated with the physiologic state of the gland (Wahl, 1915; Turner and Gomez, 1933; Soemarwoto and Bern, 1958). For example, functional changes in permeability have been reported in late pregnancy and during lactation (Matsumoto et al, 1992a,b, 1994a, 1995a,b; Yasugi et al, 1989). However, with the exception of these studies, little has been documented on the growth and involution of capillaries in the mammary gland, despite a large amount of interest in angiogenesis in mammary tumors (see Chapters 3 and 4).

Our interest in understanding the physiologic changes in mammary vasculature was born from the need to interpret the effect of overexpression of angiogenesis inhitors in this organ (Iruela-Arispe et al, 1999a,b; Ortega and Iruela-Arispe, 2000). While comparing the vascular changes in the transgenic animals, we performed a systematic morphologic and quantitative analysis of the vasculature during fetal and pubertal growth, pregnancy, lactation, and involution, which we summarize below.

During midembryonic stages in the mouse (day 10), differentiation of vascular structures with progressive angiogenic growth occurs in concert with the development of the mammary fat pad stroma and prior to the differentiation of the parenchyma. As epithelial buds invade the underlying mesenchymal tissue (embryonic day 14), capillary sprouts grow toward and surround the incipient epithelial structures. A second wave of angiogenesis progresses in concert with epithelial expansion during puberty by recruitment of new vessels from the highly vascularized fat pad (Figure 1.3A,B). Although interconnected, parenchymal vessels are subject to greater changes (angiogenesis and regression) than those in the fatty stroma. These vessels are more stable, as revealed by proliferation (data not shown) and apoptosis assays (discussed below).

The association between parenchyma and new vessels becomes clearly apparent in whole-mount preparations using specific labeling to visualize the vasculature. We found that whole-mount analysis is a very effective method for examining the extent of parenchymal-vascular interactions. For example, the topologic distribution of new vessels to the ingrowing alveoli during pregnancy (Figure 1.3C), as well as the embracement of epithelial ducts by capillaries, has only become apparent through this technique. Also variations in the capillary plexus, such as distention in the absence of proliferation, can be observed in whole-mount preparations of lactating glands. Note the distended thin capillaries in Figure 1.3D.

As a complement for visualization of epithelial-capillary interactions, a more detailed cellular resolution can be achieved by confocal microscopy. This technique provides, in a single specimen, cytologic evaluation within a visual three-dimensional framework (Figure 1.4; see color insert). It is also prone to quantitative applications through computer programs that enable assessment of several morphometric parameters. Using this approach, we have undertaken a systematic study of mammary gland vasculature and have gathered morphometric information related to the changes associated with growth and involution of the gland. As presented in Table 1.1, expansion of the vascular component follows the expansion of epithelial structures during puberty and pregnancy. To reveal vascular architecture, animals were injected with fluorescein isothiocyanate (FITC)

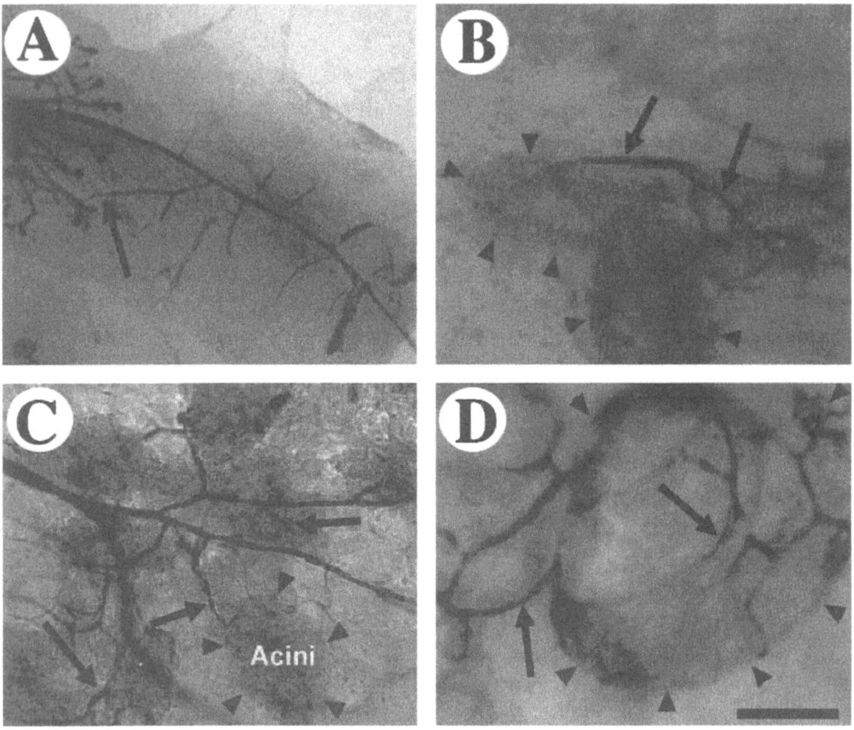

FIGURE 1.3. Capillary growth and expansion in the murine mammary gland. A: During pubertal growth (week 6) ingrowing parenchyma recruits vessels (arrow) from the previously vascularized fatty stroma as can be seen in this carmine-stained whole-mount preparation. B: High magnification of a terminal end bud (arrowheads) at week 6 of postnatal development. Note the intimal association of epithelial cells with the newly angiogenic vessels (arrows). C: By midpregnancy, significant angiogenesis has occurred and results in vascularization (arrows) of the differentiating lobuloalveolar units (acini, arrowheads). D: In the 2-day lactating gland, distended capillaries (arrows) embraced the milk-producing alveolar units (arrowheads). Bar in B and D = 50 μm.

conjugated *Lycopersicon esculentum* lectin followed by perfusion fixation. The *L. esculentum* lectin has been shown to bind to the luminal surface of murine endothelial cells and it binds equivalently to all categories of normal vessels (arterial, venous, and capillaries) (Thurston et al, 1998). Precautions were taken to allow fair comparison between different animals and mammary gland stages. The following criteria were used for analysis: (1) evaluations were carried out using the same scanning parameters (magnification, thickness, laser intensity, reconstruction analysis); (2) for pregnant and lactating animals, only females pregnant with eight embryos and suckling offspring were included; (3) for involuting glands the females were allowed to feed eight pups for 7 days, at which time the animals were removed (time of removal = day 0 of involution); (4) several females (three to five) were evaluated per stage to ensure appropriate sampling and to enable statistical analysis; (5) the area scanned was maintained as X (width) = 2.5 mm by Y (length) = 2.5 mm by Z (depth) = 250 μm; and (6) four areas were evaluated from each gland; the location used to perform the scan was exactly 500 μm from the lymph node (north, south, east, and west). The gland studied was always the fourth inguinal left gland. Volume measurements were obtained using ImagePro software algo-

TABLE 1.1. Morphometric changes of parenchymal and capillary structures during mammary gland growth and involution.

Stage	Vascular-epithelial volume (mm³)	Parenchymal volume (mm³)	Vascular-stromal volume (mm³)	Intercapillary volume (mm³)
Virgin				
7 weeks	53 (± 7)	354 (± 12)	194 (± 21)	87 (± 7)
10 weeks	297 (± 58)	427 (± 21)	173 (± 19)	98 (± 3.5)
13 weeks	288 (± 39)	431 (± 19)	187 (± 17)	85 (± 8.2)
Pregnant				
2 days	284 (± 28)	529 (± 22)	199 (± 25)	99 (± 10.5)
4 days	527 (± 59)	927 (± 31)	252 (± 22)	52 (± 15)
6 days	993 (± 108)	1,227 (± 47)	203 (± 38)	39 (± 11)
12 days	1,127 (± 87)	2,096 (± 52)	195 (± 19)	43 (± 7)
18 days	1,183 (± 77)	2,772 (± 51)	184 (± 20)	74 (± 12)
Lactating				
2 days	1,034 (± 83)	3,455 (± 47)	174 (± 23)	201 (± 25)
4 days	1,187 (± 128)	6,470 (± 42)	168 (± 27)	527 (± 45)
7 days	1,274 (± 144)	7,273 (± 59)	162 (± 19)	558 (± 25)
Involuting				
2 days	1,179 (± 98)	4,274 (± 39)	177 (± 21)	374 (± 42)
4 days	1,034 (± 92)	2,478 (± 29)	172 (± 17)	281 (± 38)
7 days	743 (± 34)	1,567 (± 33)	194 (± 19)	134 (± 25)
14 days	493 (± 52)	674 (± 21)	187 (± 22)	157 (± 19)

Evaluation of volume was performed by scanning fluorescently labeled specimens (Figure 1.4) and following specific parameters (described in text). Numbers represent the average of three independent specimens (± SD). Volume obtained from one specimen is the mean of four scans, as indicated in text.

rithms. Quantitation was performed at two levels within the gland: (1) at the same level as the parenchyma, and (2) in the fatty stoma located above the parenchyma. These evaluations provided the first indication that the vascular component in the stroma varies little in comparison to the vessels associated with the parenchyma (Table 1.1).

Capillary growth is associated with the invasion of parenchyma during the pubertal phase. A significant increase in capillary volume is detected by week 10 in comparison to week 7. After week 10, there are no significant alterations in volume. At this time, the vasculature reaches "stasis." A second wave of postnatal angiogenesis occurs during pregnancy and can be detected as early as day 4. Little to no expansion was observed late in pregnancy or during lactation. No statistically significant vascular expansion was seen after day 12 postconception.

Evaluation of intercapillary distance was also revealing. In early pregnancy, capillaries are closer to one another (Figure 1.4B and Table 1.1). During lactation, however, a 10-fold or greater intercapillary distance can be observed (Figure 1.3D and Table 1.1). This is most likely due to the distention associated with alveolar secretion. Variations in intercapillary distance are an important consideration, since quantitation of vascular profiles by histologic means alone would most likely indicate a reduction in the vascular density during this time, which is artifactual. No apoptosis in vessels was detected during this stage (data not shown).

An important conclusion also reached from these analyses is that regression of capillaries during involution of the mammary gland occurs subsequent to the remodeling events in the parenchyma. While parenchymal volume decreased significantly by days 2 to 7, vascular volume was only reduced from 7 to 14 days postweaning (Table 1.1). Vascular regression is discussed in detail below.

Studies using mammary-derived endothelial cell cultures are scant (Gumkowski et al, 1987; Hewett et al, 1993). However a clear protocol for their isolation has been reported (Hewett et al, 1993). The heterogeneous response of parenchymal versus stromal-associated vessels is intriguing, and further understanding of the molecular basis of this diversity is warranted.

GROWTH REGULATION IN THE MAMMARY GLAND

Systemic Growth Regulators of the Mammary Gland

The initial understanding of the stimuli responsible for mammary gland growth resulted from classic endocrine ablation and replacement studies (Lyons, 1958; Nandi, 1958). More recently, the use of homologous recombination and the generation of null mice that lack hormone receptors have confirmed and expanded these original observations (Lubahn et al, 1993; Lydon et al, 1995), but fundamentally the major conclusions reached at that time stand.

Early endocrine studies performed with ovariectomized and hypophysectomized rodents showed that treatment with estrogen and growth hormone stimulated vigorous end bud growth in areas where ducts faced noninvaded stroma (Lyons, 1958). Interestingly, increased branching was not seen, and little to no effect was detected if mature glands were used (Nandi, 1958). These findings support the concept that local environmental clues are in place to further regulate the growth of the gland independently of systemic signals.

Estrogen

Administration of estrogen to ovariectomized females leads to expansion and branching of the ductal parenchymal system. Nonetheless, in vitro studies using mammary epithelium have mostly failed to demonstrate a direct proliferative effect with estradiol (Yang et al, 1980a,b; Imagawa et al, 1982). However, mixed cultures of mammary epithelial cells and stromal cells have resulted in increased epithelial proliferation in the presence of estradiol (McGrath, 1983; Haslam and Levely, 1985; Haslam, 1986). Together these results indicate that the influence of estrogens on mammary gland morphogenesis is likely to be indirect, and dependent on stromal cells.

The generation of animals that lack the estrogen α-receptor support the essential role of estrogen signaling for mammary epithelial expansion (Lubahn et al, 1993). Although ductal invasion and proliferation occurred during fetal development, estrogen receptor null mice showed poor ductal expansion during puberty and upon exogenous administration of estrogen and progesterone.

Endothelial cells express the estrogen receptor (Kim-Schulze et al, 1996). The effects of estradiol on endothelial cells include increased angiogenesis, nitric oxide production, and suppression of apoptosis (Morales et al, 1995; Rubanyi et al, 1997; Spyridopoulos et al, 1997). Although no systematic studies have been performed on mammary vasculature in estrogen receptor (ER)-null mouse, the compromised nature of the epithelium is likely to be a caveat in the interpretation of results. The generation of cell-specific ER-null animals (particularly endothelial cells with no ER) through the cre-lox system could add important information as to the role of estrogen signaling in mammary gland angiogenesis, particularly during pregnancy.

Progesterone

Until recently the contribution of progesterone to mammary gland development has been difficult to ascertain, with studies supporting both suppressive and proliferative roles. However, the phenotype displayed by the progesterone receptor knockout (PRKO) mouse has recently shed light on this controversial issue (Lydon et al, 1995). Although there is no significant difference in ductal development between wild-type and PRKO mice at 6 weeks, hormonally treated glands were markedly different. Treated progesterone receptor (PR)-null animals showed poor ductal side-branching and lack of lobuloalveolar development in contrast to control littermates (Lydon et al, 1995; Humphreys et al, 1997; Brisken et al, 1998). These findings provided unquestionable proof that progesterone is required for complete mammary gland development, and it is likely that the effects of progesterone are indirect (Brisken et al, 1998). To this end, it has been demonstrated that the lack of branching displayed by the PRKO animals can be overcome by ectopic expression of the proto-oncogene *Wnt-1*. It has been proposed that Wnt proteins function as paracrine factors that operate downstream of progesterone to mediate the process of side-branching. In addition, Wnt-4 has an essential role in branching during early pregnancy. Progesterone has been shown to induce Wnt-4 expression during pregnancy. Thus, it appears that Wnt signaling is essential in mediating progesterone function during mammary gland morphogenesis (Brisken et al, 2000).

Endothelial cells also express the progesterone receptor (Vazquez et al, 1999). Treatment of culture endothelial cells with progesterone results in cell cycle arrest in early G1 (Vazquez et al, 1999). It is likely that this hormone suppresses angiogenesis, given its effects on endothelial growth suppression, yet these experiments have not been performed in vivo. As with estrogen, the role of progesterone in the regulation of mammary gland vasculature is unknown.

Prolactin

Mammary gland development is severely compromised in mice lacking prolactin (Brisken et al, 1999). Although growth of epithelial structures appears normal up to puberty, further expansion and alveolar differentiation is suppressed. In addition, heterozygous females showed significantly reduced mammary gland development during the first pregnancy (Ormandy et al, 1997). This was associated with almost complete failure to lactate. The effects of prolactin on mammary development during puberty appear to be indirect, since transplantation of epithelium from null mice into fat pads of wild-type animals showed normal developmental expansion and differentiation. In contrast, the effects on lobuloalveolar differentiation appear to be directly mediated by this hormone (Ormandy et al, 1997).

Interestingly, a 16 kD fragment of prolactin has been shown to suppress the mitogenic response of endothelial cells to vascular endothelial growth factor (VEGF) and fibroblast growth factor-2 (FGF-2) and to inhibit angiogenesis in vivo

(Clapp and Weiner, 1992; Clapp et al, 1993; Duenas et al, 1999). Although the specific mechanism of action of prolactin on endothelial cells is unclear, it appears that its effects are not mediated by the classic prolactin receptor. Treatment of endothelial cell cultures with 16 kD prolactin suppresses VEGF-induced Ras activation (D'Angelo et al, 1999). In addition, prolactin has been shown to stimulate plasminogen activator inhibitor-1 and suppress urokinase activity, which can also contribute to its antiangiogenic effects (Lee et al, 1998).

Local Regulators of Mammary Gland Growth

Epidermal Growth Factor

The lack of a more profound growth effect in vivo in hormonal experiments with mature mammary glands, or even with end buds in vitro, suggested that other factors act in conjunction with systemic hormones to achieve a well-balanced growth regulation. Using end buds implanted in collagen gels in vitro, Yang et al (1980b) demonstrated that epidermal growth factor (EGF), as well as agents that increase intracellular levels of cyclic adenosine monophosphate (cAMP), strongly stimulated mammary epithelial proliferation. Similar experiments performed later by Nandi et al (1984) demonstrated that a combination of insulin, EGF, transferrin, bovine serum albumin, and cholera toxin stimulated cellular proliferation up to 10-fold within 10 days. Additional experiments demonstrated that EGF at low concentrations (1 ng/mL) could synergize with prolactin and progesterone to stimulate growth (Imagawa et al, 1985).

Evaluation of EGF receptor activity by autoradiography in a hormonally intact, 5-week-old animal showed a heavy concentration of receptors in the stroma immediately adjacent to the end bud, with less dense labeling in stroma surrounding the cap region. Few, if any, receptors were seen in the luminal epithelium. These results were also supported by quantitative analysis of EGF receptor in vitro using fibroblast and epithelial cell lines from mammary glands. The evaluation revealed about a 100-fold more receptors in the fibroblast cell line (Danielson et al, 1984).

More mechanistic experiments showing the participation of EGF signaling were recently performed using EGF receptor (EGFR) null mammary tissue in combination with transplantation assays. Growth of neonatal EGFR null mammary glands under the renal capsule of wild-type animals showed that the participation of EGF is essential for mammary ductal growth and branching morphogenesis, but not for mammary lobuloalveolar development. The impairment in ductal growth is mediated by the stroma and not the epithelial components (Wiesen et al, 1999). These experiments provide additional support to the role of stromal-epithelial interactions in mammary mophogenesis.

Transforming growth factor-α is related to and mimics EGF by eliciting its effects through binding to the EGF receptor. TGF-α is known to be present in rapidly growing normal tissues including the mammary gland. Transgenic animals overexpressing TGF-α exhibit mammary hyperplasia and increased incidence of mammary carcinogenesis, and a shorter latency period for mouse mammary tumor formation after carcinogen treatment (Jhappan et al, 1990; Matsui et al, 1990; Sandgren et al, 1990).

Transforming growth factor-β has been demonstrated by several investigators to play an important role in mammary gland biology. Slow release pellets containing TGF-β can inhibit ductal growth (Silberstein and Daniel, 1987). The other two mammalian isoforms, TGF-β_2 and -β_3, have also been examined and shown to cause disappearance of the proliferating mammary stem cell layer, involution of ductal end buds, and suppression of glandular growth (Daniel and Robinson, 1992). These results are consistent with growth inhibitory effects of TGF-β on mammary epithelial cells in vitro (Ethier and Van de Velde, 1990; Knabbe et al, 1987). Overexpression of TGF-β in transgenic mice elicits general mammary hypoplasia (Jhappan et al, 1993; Pierce et al, 1993). Reduction of mammary ductal branching was evident at 7 weeks of age and even more pronounced at 13 weeks (Jhappan et al, 1993). Lactation is not possible in TGF-β transgenic mice due to failure of lobuloalveolar development and suppression of milk protein production (Pierce et al, 1993).

Angiogenic Growth Factors

Little is known about the mechanisms that regulate growth and remodeling of mammary gland vasculature. Recently we have reported the expression of VEGF and VEGF-C, as well as VEGFR1, -2, and -3 during the cycle of pregnancy-associated mammary growth, lactation, and involution (Pepper et al, 2000). Angiogenic growth factors, and in particular the VEGF family of cytokines, are discussed in depth in Chapter 2.

Inhibitors of Angiogenesis

The 16kD form of prolactin has already been discussed, although the role of this fragment in the regulation of the vasculature in the mammary gland is not known.

Another inhibitor of angiogenesis, thrombospondin-1 (TSP-1) (Iruela-Arispe et al, 1991, 1999b; Tolsma et al, 1993), has been more thoroughly studied in the mammary gland. TSP-1 messenger RNA (mRNA) is easily detected in the virgin postpubertal gland, yet it is rapidly downregulated during pregnancy and lactation. Increased transcript levels were detected in regressing glands, particularly 4 days after the removal of pups (Ortega and Iruela-Arispe, 2001). Overexpression of TSP-1 in the mammary epithelium using the mammary mouse tumor virus promoter results in a dramatic suppression of parenchymal-associated vessels during the first pregnancy (Ortega and Iruela-Arispe, 2001). Interestingly, mammary glands of TSP-1–null animals showed defective vascular regression during involution. In both cases, the vascular phenotypes were also associated with epithelial abnormalities. In the null animals, excessive epithelial branching was detected, while hypoplasia was noted in the TSP-1–overexpressing mice. Although direct effects of TSP-1 were also detected on mammary epithelial cells in culture, these results raise important questions related to the effects of abnormal vascularization on parenchymal growth and differentiation. A recent report utilizing an in vitro model that allows heterotypic cell interaction (endothelial and breast cancer cell tissue) has demonstrated that actively proliferating endothelial cells are essential for ductal-alveolar morphogenesis of preneoplastic epithelial cells (Shekhar et al, 2000). It is hoped that cell-targeting knockout models will enable us to discriminate between direct and indirect effects of TSP-1 in mammary epithelium.

VASCULAR REGRESSION DURING APOPTOSIS

The involuting mammary gland offers one of the most impressive examples of tissue remodeling, during which epithelial cells engaged in the production of milk die by apoptosis and the entire organ is restructured. The process appears to involve two phases (Lund et al, 1996):

1. Apoptotic phase, which is characterized by rapid induction of proapoptotic genes within the epithelium (days 1 to 5 postweaning).
2. Remodeling phase, which is characterized by the induction of genes encoding proteases within stromal cells that results in the remodeling of the gland (days 4 to 21 postweaning).

These phases, particularly the apoptotic phase, are significantly enhanced by loss or removal of the suckled offspring. Natural weaning differs greatly from forced weaning, as autocrine inhibition of milk secretion is likely to prevent the dramatic milk stasis that occurs if the pups are abruptly removed. During natural weaning, there is a progressive reduction in the milk secretion rate (controlled by galactopoietic hormones), which matches the lowering milking demand. Apoptosis is thus less abrupt and more spread out during late lactation stages. (For a review on epithelial mammary gland involution see the *Journal of Mammary Gland Biology and Neoplasia* 4(2) 1999.) The initial signals that lead to apoptosis in the mammary gland have not been completely defined, but major candidates include alterations in survival molecules such as insulin-like growth factor-1 (IGF-1) and glucocorticoids, availability and distribution of active metalloproteases, and presence of high levels of milk, which contains apoptosis-inducing factors (Strange et al, 1992; Feng et al, 1995; Lund et al, 1996; Tonner et al, 1997; Sudhakaran et al, 1999; and reviewed by Streuli et al, 1997; Furth, 1999).

Recently much of the intracellular pathway that leads to apoptosis has been elucidated, and it is known that the cysteine protease family of caspases partakes in the processes that result in program cell death. In addition, several members of the Bcl-2 family of proteins participate in the fine-tune regulation of the apoptotic process (Farrow and Brown, 1996; Jacobson, 1997; Metcalfe and Streuli, 1997; Green and Kroemer, 1998; Colitti et al, 1999). Members of the Bcl-2 family include proapoptotic proteins such as Bax and Bak, and death-suppressors such as Bcl-x and Bcl-w (Oltavai et al, 1993; Farrow et al, 1995; Gibson et al, 1996). All of these proteins have been found to regulate epithelial apoptosis during the involution of the mammary gland (Metcalfe et al, 1999). Although the specific mode of action of these proteins has not been completely elucidated, they appear to act as switches that determine the apoptotic cell fate. They form homo- and heterodimers with each other and are able to act in independent pathways to promote or repress apoptosis (Gross et al, 1998; Knudson and Korsmeyer, 1997). In a recent paper, Metcalfe and colleagues (1999) demonstrated that mammary parenchymal apoptosis, triggered during weaning, is regulated by de novo expression of Bak and Bad and repression of Bcl-x and Bcl-2. The timing was associated with maximal epithelial apoptosis during days 1 to 3 postweaning. This was also associated with increased levels of p53, another important player during mammary gland apoptosis (Jerry et al, 1999).

Currently there is no information in the literature regarding the remodeling of the vascular bed upon regression of mammary parenchyma. Recently, we have been able to reproduce the results published by Metcalfe and shed some light on capillary regression and postweaning apoptosis. Our findings indicate that regression of the capillary network occurs secondarily to that of the mammary epithelium (Figure 1.5; see color insert). These findings were initially obtained by morphometric analysis during mammary gland involution (Table 1.1) and were subsequently reinforced by gene expression analysis (Figure 1.6).

To determine the progression of apoptosis in the capillary network independent from that of epithelial structures, we devised a rapid isolation procedure that separates endothelial from epithelial structures. The technique involves digestion with collagenase and differential filtration (Figure 1.5C). The preparations are

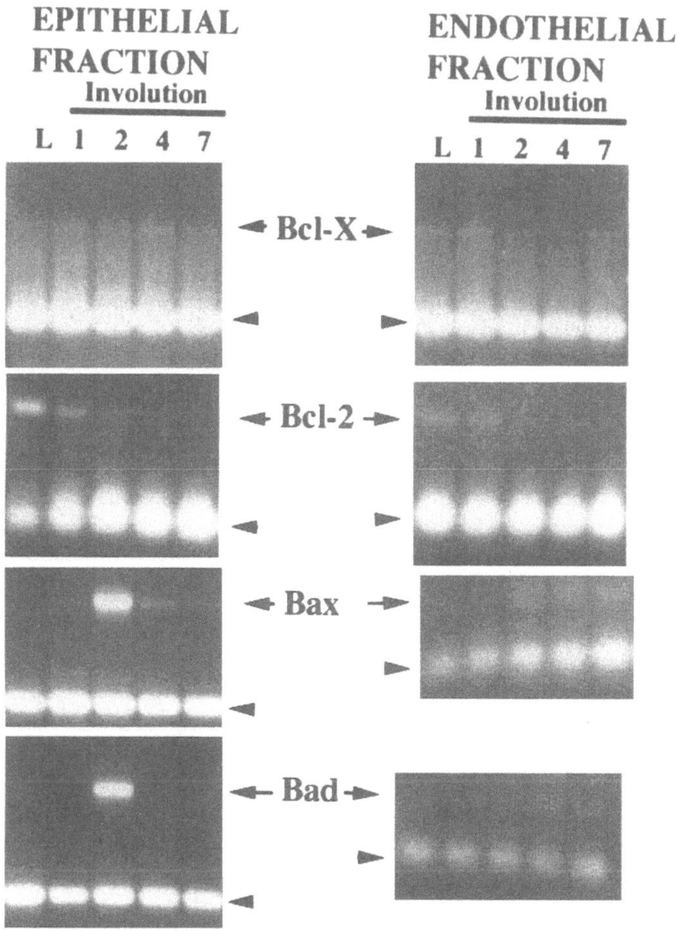

FIGURE 1.6. Evaluation of Bcl-2 and other family members during murine mammary gland apoptosis. RNA from parenchymal (epithelial) and vascular (endothelial) fractions (Fig. 1.5D,E) was subjected to RT-PCR for assessment of several members of the Bcl-2 family as indicated. L, lactation; 1, 2, 4, and 7 days after removal of pups. Arrowhead in each gel represents migration of L-32 product used as an internal control.

about 85% pure capillaries or epithelial alveoli and some ducts. Most of the fibroblasts are contained in the capillary fraction. The purity of the samples has been assessed by reverse-transcriptase polymerase chain reaction (RT-PCR) using keratin 18 as a marker for epithelial cells and CD-31 as indicator of endothelial cells (Figure 1.5D,E).

Initial assessment of apoptosis was performed by Southern blot analysis (Figure 1.5F,G). While epithelial apoptosis peaked at day 2, apoptosis of endothelial cells peaked at days 4 to 5. Another important distinction is that while most alveolar epithelial cells died postweaning by apoptosis, the extent of endothelial cell death did not appear to be as significant, as revealed by Southern blot (Figure 1.5F,G) and terminal transferase UTP nick end labeling (TUNEL) analysis (data not shown).

A more refined study of epithelial apoptosis was performed by RT-PCR analysis for expression of Bcl-2 family members; specifically, we examined Bcl-x, Bcl-2, Bac, and Bad (Figure 1.6). Our results confirmed the recent publications on epithelial apoptosis (Figure 1.6-epithelial fraction). We found constant levels of Bcl-x, reduction in Bcl-2, and upregulation of Bak and Bad. The Bak and Bad peak was seen 2 days after weaning, in contrast to day 1 in the published report (Metcalfe et al, 1999). Nonetheless, our animals were 129-BALB/c mice while the Metcalfe study used ICR mice. Both studies concur on the point that both Bak and Bad appear to play a significant role in triggering epithelial apoptosis.

The same evaluation for levels of Bcl-2 family members was applied to the isolated endothelial cells. Here the levels of Bak and Bad were only increased at days 4 to 5 during involution of the gland. Furthermore, Bak appeared to be more prevalent than Bad. In contrast to epithelial cells, the decreased levels of Bcl-2 were not as dramatic.

CONCLUSION

Little is understood about the biology of vascular alterations associated with the mammary gland. Yet this organ offers a significant opportunity to answer questions related to the mechanisms of angiogenic expansion and remodeling in a whole-organ system. Of great importance are the differences in the response of the parenchymal versus the stromal-associated vessels. It is clear that these two populations, although interconnected, are regulated by different mechanisms, some of which might be epithelial in nature. Additional and equally important questions include: What are the factors that mediate angiogenic growth and regression? Is there a direct role for estrogen and progesterone in vascular growth regulation? What is the participation of vascular mural cells (pericytes and smooth muscle) in the mammary vascular cycle?

The increasing number of animal models (including gene deletion studies), combined with the accessibility of the organ, should enable extensive mechanistic experimentation. It is clear that additional and more molecular studies are required to advance our current knowledge in this important area of vascular biology.

ACKNOWLEDGMENTS

The authors thank Ivy Panlilio for assistance with the manuscript. This work was supported by the National Cancer Institute (grants CA 65624-05 and

CA-86256-01), the U.S. Department of Defense (contract DAMD 17-94-J-4346), and the Swiss National Science Foundation (grant 3100-043364.95).

REFERENCES

Bernfield, M., Banerjee, S. D., Koda, J. E., and Rapraeger, A. C. 1984. Remodeling of the basement membrane as a mechanism for morphogenetic tissue interactions. In: Robert, L. Trelstad, ed., *The Role of the Extracellular Matrix in Development*. Alan, R. Liss, New York, pp. 545–572.

Boudreau, N., Sympson, C. J., Werb, Z., and Bissell, M. J. 1995. Suppression of ICE and apoptosis in mammary epithelial cells by extracellular matrix. *Science* 267:891–893.

Brisken, C., Heineman, A., Chavarria, T., Elenbaas, B., Tan, J., Dey, S. K., McMahon, J. A., McMahon, A. P., and Weinberg, R. A. 2000. Essential function of Wnt-r in mammary gland development downstream of progesterone signaling. *Genes Dev.* 14:650–654.

Brisken, C., Kaur, S., Chavarria, T., Binart, N., Sutherland, R. L., Weinberg, R. A., Kelly, P. A., and Ormandy, C. J. 1999. Prolactin controls mammary gland development via direct and indirect mechanisms. *Dev. Biol.* 210:96–106.

Brisken, C., Park, S., Vass, T., Lydon, J., O'Malley, B., and Weinberg, R. 1998. A paracrine role for the epithelial progesterone receptor in mammary gland development. *Proc. Natl. Acad. Sci. USA* 95:5076–5081.

Clapp, C., Martial, J. A., Guzman, R. C., Rentier-Delure, F., and Weiner, R. I. 1993. The 16-kilodalton N-terminal fragment of human prolactin is a potent inhibitor of angiogenesis. *Endocrinology* 133:1292–1299.

Clapp, C., and Weiner, R. I. 1992. A specific, high affinity, saturable binding site for the 16-kilodalton fragment of prolactin on capillary endothelial cells. *Endocrinology* 130:1380–1386.

Colitti, M., Stefanon, B., and Wilde, C. J. 1999. Apoptotic cell death, bax and bcl-2 expression during sheep mammary gland involution. *Anat. Histol. Embryol.* 28:257–264.

D'Angelo, G., Martini, J. F., Iiri, T., Fantl, W. J., Martial, J., and Weiner, R. I. 1999. 16K human prolactin inhibits vascular endothelial growth factor-induced activation of Ras in capillary endothelial cells. *Mol. Endocrinol.* 13:692–704.

Daniel, C. W., and Robinson, S. D. 1992. Regulation of mammary growth and function by TGF-beta. *Mol. Reprod. Dev.* 32:145–151.

Daniel, C. W., and Silberstein, G. B. 1987. Postnatal development of the rodent mammary gland. In: Nevile, M. C., and Daniel, C. W., eds. *The Mammary Gland: Development, Regulation, and Function.* Plenum Press, New York, pp. 3–36.

Danielson, K. G., Oborn, C. J., Durban, E. M., Butel, J. S., and Medina, D. 1984. An epithelial mouse mammary cell line exhibiting normal morphogenesis in vivo and functional differentiation in vitro. *Proc. Natl. Acad. Sci. USA* 81:3756–3760.

Duenas, Z., Torner, L., Corbacho, A. M., Ochoa, A., Gutierrez-Ospina, G., Lopez-Barrera, F., Barrios, F. A., Berger, P., Martinez de la Escalera, G., and Clapp, C. 1999. Inhibition of rat cornal angiogenesis by 16-kDa prolactin and by endogenous prolactin-like molecules. *Invest. Ophthalmol. Vis. Sci.* 40:2498–2505.

Ethier, S. P., and Van de Velde, R. M. 1990. Secretion of a TGF-β-like growth inhibitor by normal rat mammary epithelial cells in vitro. *J. Cell. Physiol.* 142:15–20.

Farrow, S. N., and Brown, R. 1996. New members of the Bcl-2 family and their protein partners. *Curr. Opin. Genet. Dev.* 6:45–49.

Farrow, S. N., White, J. H. M., Marinou, I., Raven, T., Pun, K. T., Grinham, C. J., Marinou, J. C., and Brown, R. 1995. Cloning of a Bcl-2 homolog by interaction with adenovirus E1b 19k. *Nature* 374:731–733.

Faulkin, L. J., Jr., and DeOme, K. B. 1960. Regulation of growth and spacing of gland elements in the mammary fat pad of the C3H mouse. *J. Natl. Cancer Inst.* 24:953–969.

Feng, Z. W., Marti, A., Jehn, B., Altermatt, H. J., Chicaiza, G., and Jaggi, R. 1995. Glucocorticoid and progesterone inhibit involution and programmed cell-death in the mouse mammary-gland. *J. Cell Biol.* 131:1095–1103.

Furth, P. A. 1999. Introduction: Mammary gland involution and apoptosis of mammary epithelial cells. *J. Mammary Gland Biol. Neoplasia* 4:123–127.

Gibson, L., Holmgreen, S. P., Huang, D. C. S., Bernand, O., Copeland, N. G., Jenkins, N. A., Sutherland, G. R., Baker, E., Adams, J. M., and Cory, S. 1996. Bcl-W, a novel member of the Bcl-2 family, promotes cell-survival. *Oncogene* 13:665–675.

Green, D., and Kroemer, G. 1998. The central executioners of apoptosis: caspases or mitochondria? *Trends Cell Biol.* 8:267–271.

Gross, A., Jockel, J., Wei, M. C., and Korsmeyer, S. J. 1998. Enforced dimerization of Bax results in its translocation, mitochondrial dysfunction and apoptosis. *EMBO J.* 17:3878–3885.

Gumkowski, F., Kaminska, G., Kaminski, M., Morrissey, L. W., and Auerbach, R. 1987. Heterogeneity of mouse vascular endothelium. In vitro studies of lymphatic, large blood vessel and microvascular endothelial cells. *Blood Vessels* 24:11–23.

Haslam, S. Z. 1986. Mammary fibroblast influence on normal mouse mammary epithelial cell responses to estrogen in vitro. *Cancer Res.* 46:310–316.

Haslam, S. Z., and Levely, M. L. 1985. Estradiol responsiveness of normal mouse mammary cells in primary cell culture: association of mammary fibroblasts with estradiol regulation of progesterone receptors. *Endocrinology* 116:1835–1841.

Hewett, P. W., Murray, J. C., Price, E. A., Watts, M. E., and Woodcock, M. 1993. Isolation and characterization of microvessel endothelial cells from human mammary adipose tissue. *In Vitro Cell Dev. Biol. Anim.* 29A:325–331.

Howlett, A. R., and Bissell, M. J. 1993. The influence of tissue microenvironment (stroma and extracellular matrix) on the development and function of mammary epithelium. *Epithelial Cell Biol.* 2(2):79–89.

Humphreys, R., Lydon, J., O'Malley, B., and Rosen, J. 1997. Mammary gland development is mediated by both stromal and epithelial progesterone receptors. *Mol. Endocrinol.* 11:801–811.

Imagawa, W., Tomooka, Y., Hamamoto, S., and Nandi, S. 1985. Stimulation of mammary epithelial cell growth in vitro: interaction of epidermal growth factor and mammogenic hormones. *Endocrinology* 116:1514–1524.

Imagawa, W., Tomooka, Y., and Nandi, S. 1982. Serum-free growth of normal and tumor mouse mammary epithelial cells in primary culture. *Proc. Natl. Acad. Sci. USA* 79:4074–4077.

Iruela-Arispe, M. L., Bornstein, P., Sage, H. 1991. Thrombospondin exerts an antiangiogenic effect on cord formation by endothelial cells in vitro. *Proc. Natl. Acad. Sci. USA* 88:5026–5030.

Iruela-Arispe, M. L., Lombardo, M., Krutzsch, H. C., Lawler, J., and Roberts, D. D. 1999a. Inhibition of angiogenesis by thrombospondin-1 is mediated by 2 independent regions within the type 1 repeats. *Circulation* 100:1423–1431.

Iruela-Arispe, M. L., Ortega, M. A., and Vazquez, F. 1999b. Anti-angiogenic domain of thrombospondin. In: Rubany, G., ed. *Angiogenesis in Health and Disease: Basic Mechanisms and Clinical Applications.* pages:349–357.

Jacobson, M. D. 1997. Apoptosis: Bcl-2-related proteins get connected. *Curr. Biol.* 7:R277–R281.

Jerry, D. J., Pinkas, J., Kuperwasser, C., Dickinson, E. S., and Naber, S. P. 1999. Regulation of p53 and its targets during involution of the mammary gland. *J. Mammary Gland Biol. Neoplasia* 4:177–181.

Jhappan, C., Geiser, A. G., Kordon, E. C., Bagheri, D., Hennighausen, L., Roberts, A. B., Smith, G. H., and Merlino, G. 1993. Targeting expression of a transforming growth factor beta 1 transgene to the pregnant mammary gland inhibits alveolar development and lactation. *EMBO J.* 12:1835–1845.

Jhappan, C., Stahle, C., Harkins, R. N., Fausto, N., Smith, G. H., and Merlino, G. T. 1990. TGFα overexpression in transgenic mice induces liver neoplasia and abnormal development of the mammary gland and pancreas. *Cell* 61:1137–1146.

Kim-Schulze, S., McGowan, L. A., Hubchak, S. C., Cid, M. C., Martin, M. B., Kleinman, H. K., Greene, G. L., and Schnaper, H. W. 1996. Expression of an estrogen receptor by human coronary artery and umbilical vein endothelial cells. *Circulation* 94:1402–1407.

Knabbe, C., Lippman, M. E., Wakefield, L. M., Flanders, K. C., Kasid, A., Derynck, R., and Dickson, R. B. 1987. Evidence that transforming growth factor-beta is a hormonally regulated negative growth factor in human breast cancer cells. *Cell* 48:417–428.

Knight, C. H., and Peaker, M. 1982. Development of the mammary gland. *J. Reprod. Fertil.* 65:521–536.

Knudson, C. M., and Korsmeyer, S. J. 1997. Bcl-2 and Bax function independently to regulate cell death. *Nat. Genet.* 16:358–363.

Kratochwil, K. 1971. In vitro analysis of the hormonal basis for the sexual dimorphism in the embryonic development of the mouse mammary gland. *J. Embryol. Exp. Morphol.* 25:141–153.

Kratochwil, K. 1987. Epithelium-mesenchyme interactions in the fetal mammary gland. In: Medina, D., Kidwell, W., Heppner, G., and Anderson, E. eds. *Cellular and Molecular Biology of Mammary Cancer.* Plenum Press, New York, pp. 67–80.

Lee, H., Struman, I., Clapp, C., Martial, J., and Weiner, R. I. 1998. Inhibition of urokinase activity by the antiangiogenic factor 16K prolactin: activation of plasminogen activator inhibitor 1 expression. *Endocrinology* 139:3696–3703.

Lin, C. Q., and Bissel, M. J. 1993. Multi-faceted regulation of cell differentiation by extracellular matrix. *FASEB J.* 7:737–743.

Lubahn, D. B., Moyer, J. S., Golding, T. S., Couse, J. F., Korach, K. S., and Smithies, O. 1993. Alteration of reproductive function but not prenatal sexual development after insertional disruption of the mouse estrogen receptor gene. *Proc. Natl. Acad. Sci. USA* 90:11162–11166.

Lund, L. R., Romer, J., Thomasset, N., Solberg, H., Pyke, C., Bissell, M. J., Dano, K., and Werb, Z. 1996. 2 distinct phases of apoptosis in mammary-gland involution—proteinase-independent and proteinase-dependent pathways. *Development* 122:181–193.

Lydon, J. P., DeMayo, F. J., Funk, C. R., Mani, S. K., Hughes, A. R., Montgomery Jr., C. A., Shyamala, G., Conneely, O. M., and O'Malley, B. W. 1995. Mice lacking progesterone receptor exhibit pleiotropic reproductive abnormalities. *Genes Dev.* 9:2266–2278.

Lyons, W. R. 1958. Hormonal synergism in mammary growth. *Proc. R. Soc. (Lond.) [B]* 49:303–325.

Matsui, Y., Halter, S. A., Holt, J. T., Hogan, B. L. M., and Coffey, R. J. 1990. Development of mammary hyperplasia and neoplasia in MMTV-TGFα transgenic mice. *Cell* 61:1147–1155.

Matsumoto, M., Kurohmaru, M., Hayashi, Y., Nishinakagawa, H., and Otsuka, J. 1994. Permeability of mammary gland capillaries to ferritin in mice. *J. Vet. Med. Sci.* 56:65–70.

Matsumoto, M., Nishinakagawa, H., Kurohmaru, M., and Hayashi, Y. 1995a. Effects of estrogen and progesterone on the parenchyma and blood vessels of the mammary gland in ovariectomized adult mice. *J. Vet. Med. Sci.* 57:39–44.

Matsumoto, M., Nishinakagawa, H., Kurohmaru, M., Hayashi, Y., and Awal, M. A. 1995b. Ultrastructural changes in fat cells and blood capillaries of the mammary gland in starved mice. *J. Vet. Med. Sci.* 57:733–736.

Matsumoto, M., Nishinakagawa, H., Kurohmaru, M., Hayashi, Y., and Otsuka, J. 1992a. Effects of estrogen and progesterone on the development of the mammary gland and the associated blood vessels in ovariectomized mice. *J. Vet. Med. Sci.* 54:1117–1124.

Matsumoto, M., Nishinakagawa, H., Kurohmaru, M., Hayashi, Y., and Otsuka, J. 1992b. Pregnancy and lactation affect the microvasculature of the mammary gland in mice. *J. Vet. Med. Sci.* 54:937–943.

McGrath, C. M. 1983. Augmentation of the response of normal mammary epithelial cells to estradiol by mammary stroma. *Cancer Res.* 43:1355–1360.

Metcalfe, A. D., Gilmore, A., Klinowka, T., Oliver, J., Valentijn, A. J., Brown, R., Ross, A., MacGregor, G., Hickman, J. A., and Streuli, C. H. 1999. Developmental regulation of Bcl-2 family protein expression in the involuting mammary gland. *J. Cell Sci.* 112:1771–1783.

Metcalfe, A., and Streuli, C. 1997. Epithelial appoptosis. *BioEssays* 19:711–720.

Morales, D. E., McGowan, K. A., Grant, D. S., Maheshwari, S., Bhartiya, D., Cid, M. C., Kleinman, H. K., and Schnaper, H. W. 1995. Estrogens promote angiogenic activity in human umbilical vein endothelial cells in vitro and in a murine model. *Circulation* 91:755–763.

Nandi, S. 1958. Endocrine control of mammary-gland development and function in the C3H/He Crgl mouse. *J. Natl. Cancer Inst.* 21:1039–1063.

Nandi, S., Imagawa, W., Tomooka, Y., McGrath, M. F., and Edery, M. 1984. Collagen gel culture system and analysis of estrogen effects on mammary carcinogenesis. *Arch. Toxicol.* 55:91–96.

Oltavai, Z. N., Milliman, C. L., and Korsmeyer, S. J. 1993. Bcl-2 heterodimerizes in-vivo with a conserved homolog, Bax, that accelerates programmed cell-death. *Cell* 74:609–619.

Ormandy, C. J., Camus, A., Barra, J., Damotte, D., Lucas, B., Buteau, H., Edery, M., Brousse, N., Babinet, C., Binart, N., and Kelly, P. A. 1997. Null mutation of the prolactin receptor gene produces multiple reproductive defects in the mouse. *Genes Dev.* 11:167–178.

Ortega, M. A., and Iruela-Arispe, M. L. 2001. Thrombospondin-1 is an endogenous regulator of angiogenesis in the mammary gland. *J. Cell Biol.* (submitted).

Pepper, M. S., Baetens, D., Mandriota, S. J., Di Sanza, C., Oikemus, S., Lane, T. F., Soriano, J. V., Montesano, R., and Iruela-Arispe, M. L. 2000. Regulation of VEGF and VEGF receptor expression in the rodent mammary gland during pregnancy, lactation and involution. *Dev. Dyn.* 218:507–524.

Pierce, D. F., Jr., Johnson, M. D., Matsui, Y., Robinson, S. D., Gold, L. I., Purchio, A. F., Daniel, C. W., Hogan, B. L., and Moses, H. L. 1993. Inhibition of mammary duct development but not alveolar outgrowth during pregnancy in transgenic mice expressing active TGF-beta 1. *Genes Dev.* 7:2308–2317.

Propper, A. 1970. Experimental study of the 1st stages of mammary morphogenesis. *Annee Biol.* 9:267–275.

Pujuguet, P., Simian, M., Liaw, J., Timpl, R., Werb, Z., and Bissell, M. J. 2000. Nidogen-1 regulates laminin-1-dependent mammary-specific gene expression. *J. Cell Sci.* 113:849–858.

Robinson, G. W., and Hennighausen, L. 1997. Inhibins and activins regulate mammary epithelial cell differentiation through mesenchymal-epithelial interactions. *Development* 124:2701–2708.

Rubanyi, G. M., Freay, A. D., Kauser, K., Sukovich, D., Burton, G., Lubahn, D. B., Couse, J. F., Curtis, S. W., and Korach, K. S. 1997. Vascular estrogen receptors and endothelium-derived nitric oxide production in the mouse aorta. *J. Clin. Invest.* 99:2429–2437.

Sakakura, T. 1987. Mammary embryogenesis. In: Neville, M. C., and Daniel, C. W., eds. *The Mammary Gland: Development, Regulation, and Function.* Plenum Press, New York, pp. 37–66.

Sakakura, T. 1991. New aspects of stroma-parenchyma relations in mammary gland differentiation. *Int. Rev. Cytol.* 125:165–202.

Sakakura, T., Sakagami, Y., and Nishizuka, Y. 1982. Dual origin of mesenchymal tissues participating in mouse mammary gland embryogenesis. *Dev. Biol.* 91:202–207.

Sandgren, E. P., Luetteke, N. C., Palmiter, R. D., Brinster, R. L., and Lee, D. C. 1990. Overexpression of TGFα in transgenic mice: Induction of epithelial hyperplasia, pancreatic metaplasia, and carcinoma of the breast. *Cell* 61:1121–1135.

Shekhar, M. P., Werdell, J., and Tait, L. 2000. Interaction with endothelial cells is a prerequisite for branching ductal-alveolar morphogenesis and hyperplasia of preneoplastic human breast epithelial cells: regulation by estrogen. *Cancer Res.* 60:439–449.

Silberstein, G. B., and Daniel, C. W. 1987. Reversible inhibition of mammary gland growth by transforming growth factor-beta. *Science* 237:291–293.

Soemarwoto, I. N., and Bern, H. A. 1958. The effects of hormones on the vascular pattern of the mouse mammary gland. *Am. J. Anat.* 103:403–435.

Spyridopoulos, I., Sullivan, A. B., Kearney, M., Isner, J. M., and Losordo, D. W. 1997. Estrogen receptor-mediated inhibition of human endothelial cell apoptosis: estradiol as a survival factor. *Circulation* 95:1505–1514.

Sternlicht, M. D., Lochter, A., Sympson, C. J., Huey, B., Rougier, J., Gray, J. W., Pinkel, D., Bissell, M. J., and Werb, Z. 1999. The stromal proteinase MMP3/stromelysin-1 promotes mammary carcinogenesis. *Cell* 98:137–146.

Strange, R., Li, F., Saurer, S., Burkhardt, A., and Friis, R. R. 1992. Apoptotic cell-death and tissue remodeling during mouse mammary-gland involution. *Development* 115:49–58.

Streuli, C. H., Dive, C., Hickman, J. A., Farrelly, N., and Metcalfe, A. 1997. Control of apoptosis in breast epithelium. *Endocr. Rel. Cancer* 4:45–53.

Sudhakaran, P. R., Ambili, M., and Philip, S. 1999. Matrix metalloproteinase in mammary gland remodeling-modulation by glycosaminoglycans. *Biosci. Rep.* 19:485–490.

Sympson, C. J., Talhouk, R. S., Alexander, C. M., Chin, J. R., Clift, S. M., Bissell, M. J., and Werb, Z. 1994. Targeted expression of stromelysin-1 in mammary gland provides evidence for a role of proteinases in branching morphogenesis and the requirement for an intact basement membrane for tissue-specific gene expression. *J. Cell Biol.* 125:681–693.

Thurston, G., Murphy, T., Baluk, P., Lindsey, J., and McDonald, D. 1998. Angiogenesis in mice with chronic airway inflammation: strain-dependent differences. *Am. J. Pathol.* 153:1099–1112.

Tolsma, S. S., Volpert, O. V., Good, D. J., Frazier, W. A., Polverini, P. J., and Bouck, N. 1993. Peptides derived from 2 separate domains of the matrix protein thrombospondin-1 have anti-angiogenic activity. *J. Cell Biol.* 122:497–511.

Tonner, E., Barber, M. C., Travers, M. T., Logan, A., and Flint, D. J. 1997. Hormonal control of insulin-like growth factor binding protein-5 production in the involuting mammary gland of the rat. *Endocrinology* 138:5101–5107.

Turner, C. W., and Gomez, E. T. 1933. The normal development of the mammary gland of the male and female albino mouse. I. Intrauterine. *Mo. Agric. Exp. Stn. Res. Bull.* 182:3–20.

Vazquez, F., Rodriguez-Manzaneque, J. C., Lydon, J. P., Edwards, D. P., O'Malley, B. W., and Iruela-Arispe, M. L. 1999. Progesterone regulates proliferation of endothelial cells. *J. Biol. Chem.* 274:2185–2192.

Wahl, H. M. 1915. Development of the blood vessels of the mammary gland in the rabbit. *Am. J. Anat.* 18:515–524.

Wiesen, J. F., Young, P., Werb, Z., and Cunha, G. R. 1999. Signaling through the stromal epidermal growth factor receptor is necessary for mammary ductal development. *Development* 126:335–344.

Williams, J. M., and Daniel, C. W. 1983. Mammary ductal elongation: differentiation of myoepithelium and basal lamina during branching morphogenesis. *Dev. Biol.* 97:274–290.

Yang, J., Guzman, R., Richards, J., Imagawa, W., McCormack, K., and Nandi, S. 1980b. Growth factor- and cyclic nucleotide-induced proliferation of normal and malignant mammary epithelial cells in primary culture. *Endocrinology* 107:35–41.

Yang, J., Richards, J., Guzman, R., Imagawa, W., and Nandi, S. 1980a. Sustained growth in primary culture of normal mammary epithelial cells embedded in collagen gels. *Proc. Natl. Acad. Sci. USA* 77:2088–2092.

Yasugi, T., Kaido, T., and Uehara, Y. 1989. Changes in density and architecture of microvessels of the rat mammary gland during pregnancy and lactation. *Arch. Histol. Cytol.* 52:115–122.

Zangani, D., Darcy, K. M., Masso-Welch, P. A., Bellamy, E. S., Desole, M. S., and Ip, M. M. 1999a. Multiple differentiation pathways of rat mammary stromal cells in vitro: acquisition of a fibroblast, adipocyte or endothelial phenotype is dependent on hormonal and extracellular matrix stimulation. *Differentiation* 64:91–101.

Zangani, D., Darcy, K. M., Shoemaker, S., and Ip, M. M. 1999b. Adipocyte-epithelial interactions regulate the in vitro development of normal mammary epithelial cells. *Exp. Cell Res.* 247:399–409.

CHAPTER 2

Growth Factor Regulation of Physiologic Angiogenesis in the Mammary Gland

Anne Saaristo, Marika J. Karkkainen, Kari Alitalo, Roberto Montesano, M. Luisa Iruela-Arispe, and Michael S. Pepper

During embryogenesis, the formation of new blood vessels occurs via two processes: vasculogenesis and angiogenesis. Vasculogenesis involves the de novo differentation of endothelial cells from mesoderm-derived precursors called angioblasts, which then cluster and reorganize to form capillary-like tubes (Risau and Flamme, 1995). Once the primary vascular plexus is formed, new capillaries form by sprouting or by splitting (intussusception) from preexisting capillaries in the processes called sprouting or nonsprouting angiogenesis, respectively (Risau, 1997). In postnatal life, the growth of normal as well as neoplastic tissues depends on angiogenesis. Angiogenesis is particularly important for normal reproductive function, including the cyclical growth of capillaries within the ovary (required for ovulation and corpus luteum formation) and the endometrium (required for regeneration following menstruation). Angiogenesis also occurs following implantation of the blastocyst, and is required for the formation of the placenta (Findlay, 1986). This chapter discusses the role of physiologic angiogenesis and angiogenic growth factors in relation to mammary gland function.

The mammary gland is one of the few organs that undergo hormonally regulated cycles of growth and morphogenesis during adult life. This makes it a particularly attractive model for studying angiogenesis (during pregnancy), vascular permeability (during lactation), and endothelial cell apoptosis (during postweaning involution). However, surprisingly few studies have been done, and thus very little is known about the molecular mechanisms of physiologic angiogenesis in the mammary gland.

Various cytokine families have been implicated in the neovascularization process. These include, but are not limited to, the vascular endothelial growth factor (VEGF) and angiopoietin families. The former is comprised of structurally related members, including VEGF (also known as VEGF-A), placenta growth factor (PIGF), VEGF-B, VEGF-C, VEGF-D and Orf virus VEGFs (also called VEGF-E) (PlGF). VEGFs bind to three transmembrane receptors (VEGFR-1 to -3), which possess intrinsic tyrosine kinase activity and are mainly expressed in the endothelial cells. Two angiopoietins have been well characterized, namely Ang-1 and Ang-2, both of which bind with similar affinity to the Tie-2 (Tek) receptor. The VEGFRs and Tie-2 are expressed almost exclusively by endothelial cells.

While angiogenesis is required for growth of normal and neoplastic tissues during adult life, active lymphangiogenesis has not been detected in tumors or in normal adult tissues. However, the function of lymphatics is crucial both for the normal physiology of the mammary gland as well as for the understanding of the mechanisms of lymphatic metastasis in breast cancer. VEGF-C is a growth factor for both lymphatic and blood vessels, and the expression of its receptor VEGFR-3 becomes restricted mainly to the lymphatic endothelium during development (Kaipainen et al, 1995).

VEGF FAMILY AND RECEPTORS

All VEGF family members possess a homology domain containing eight distinctly spaced cysteine residues. In addition, VEGF-C and VEGF-D have long N- and C-terminal extensions, of which the N-terminal parts do not display sequence homology to any known protein, whereas their C-terminal portions contain motifs similar to the Balbiani ring 3 protein, a major component of silk produced by the midge *Chironomus tentans* (Joukov et al, 1996; Lee et al, 1996; Orlandini et al, 1996; Achen et al, 1998). Each of the VEGFs binds selectively to one or two members of the VEGF receptor family (Table 2.1).

VEGF is considered to be the most important stimulator of vascular endothelial cell proliferation and migration, and it has a role in the maintenance of newly formed, immature vessels (Ferrara and Henzel, 1989; Plouet et al, 1989; Alon et al, 1995; Benjamin and Keshet, 1997). VEGF is expressed as several splice isoforms

TABLE 2.1. Summary of the interactions between vascular endothelial growth factor (VEGF) family ligands and their receptors.

	VEGFR-1	VEGFR-2	VEGFR-3	NP-1	NP-2	Tie-1	Tie-2
VEGF165	×	×		×	×		
VEGF121	×	×					
VEGF145	×	×			×		
VEGF189	×	×					
VEGF206	×	×					
VEGF-B 167	×			×			
VEGF-B 186	×						
VEGF-C		×	×				
VEGF-D		×	×				
VEGF-E (NZ-7 VEGF)		×					
VEGF-E (NZ-2 VEGF)		×		×			
PlGF-1	×						
PlGF-2	×			×	×		
PlGF/VEGF heterodimer		×					
Ang-1							×
Ang-2							×*

* A receptor antagonist.

Ang, angiopoietin; NP-1, neuropilin-1; PlGF, placenta growth factor.

consisting of polypeptides with 121, 145, 165, 189, and 206 amino acid residues. These isoforms differ in their ability to interact with heparan sulfate proteoglycans in the extracellular matrix (reviewed in Neufeld et al, 1999). The VEGF molecule is an antiparallel disulfide-linked dimer and it mediates its effects via endothelial cell–specific tyrosine kinase receptors VEGFR-1 and VEGFR-2. The most abundant and presumably major biologically active form is $VEGF_{165}$, which binds VEGFR-1 and VEGFR-2 with dissociation constants (K_d's) of 10 to 20 pM and 75 to 125 pM, respectively (De Vries et al, 1992; Terman et al, 1992). $VEGF_{165}$ also binds with high affinity to nonreceptor tyrosine kinase neuropilin-1 (NP-1) and NP-2 (Soker et al, 1998; Gluzman-Poltoraz et al, 2000). The crucial role of VEGF-mediated signaling during embryogenesis has been demonstrated in knockout mice, where the inactivation of a single *Vegf* allele results in aberrant vessel formation and lethality *in utero* (Carmeliet et al, 1996; Ferrara et al, 1996). This phenotype appears to be due to decreased gene dosage, and it is the first case where the loss of a single autosomal allele is lethal, indicating that a tight dose-dependent regulation of embryonic vessel development is mediated by VEGF. Among the members of the VEGF family, VEGF is the only one that is strongly upregulated by hypoxia and low glucose, conditions that occur in poorly vascularized tissues, such as tumors. In addition, VEGF is more potent than histamine in increasing capillary permeability to plasma proteins (Dvorak et al, 1995).

Recently, PlGF has also been found to bind to NP-2 (Gluzman-Poltorak et al, 2000) PlGF and VEGF-B are two members of the VEGF family that interact exclusively with VEGFR-1 and NP-1 (Migdal et al, 1998; Olofsson et al, 1998; Soker et al, 1998; Makinen et al, 1999). In normal tissues PlGF is expressed mainly in the placenta, but also in some malignant tumors such as renal cancer, suggesting that in addition to VEGF, it might regulate tumor angiogenesis in some settings (Takahashi et al, 1994). As a result of alternative splicing of the gene, PlGF exists as isoforms of 131, and 152 amino acid residues, the latter containing a stretch of 21 basic residues, which confers heparin- and neuropilin-1–binding ability (Maglione et al, 1991; Hauser and Weich, 1993). VEGF-B resembles PlGF in several aspects: there are two currently known splice isoforms of VEGF-B, and VEGF-B also binds selectively VEGFR-1 and NP-1. Of the splice isoforms, VEGF-B_{167} remains cell associated via C-terminal basic sequences. VEGF-B_{186}, which lacks the highly basic region, is secreted from cells in a freely soluble form. Both forms of VEGF-B are produced as disulfide-linked homodimers, but they can also form heterodimers with VEGF (Olofsson et al, 1996b). VEGF-B–containing conditioned medium was able to stimulate thymidine incorporation into DNA in human umbilical vein endothelial cells (ECs) and bovine capillary endothelial cells, suggesting that it is mitogenic for ECs (Olofsson et al, 1996a). However, so far this has not been reproduced using recombinant VEGF-B, suggesting that the activity could be due to VEGF-B/VEGF heterodimers in transfected cells. VEGF-B is expressed in heart during development, and deletion of the *Vegfb* gene results in vascular dysfunction after coronary occlusion, and impaired recovery from experimentally induced myocardial ischemia (Bellomo et al, 2000).

VEGF-C and VEGF-D are produced as long precursor proteins that undergo proteolytic processing to generate several forms that differ in their receptor binding properties (Lee et al, 1996; Orlandini et al, 1996; Joukov et al, 1997; Yamada et al, 1997; Achen et al, 1998). Partially processed VEGF-C and VEGF-

D are able to bind and activate exclusively VEGFR-3, whereas their processed forms stimulate both VEGFR-2 and VEGFR-3. The mature form of VEGF-C, containing only the VEGF homology domain, binds to VEGFR-2 and VEGFR-3 with K_d's of 410pM and 135pM, respectively. Presumably through its interaction with VEGFR-2, VEGF-C shares the major functions of VEGF, including induction of angiogenesis and vascular permeability. In embryos, targeted disruption of the *Vegfr3* gene leads to failure of remodeling of the primary vascular network, and death after embryonic day 9.5 (Dumont et al, 1998). This indicates that VEGFR-3 has an essential role in the development of the embryonic vasculature before the emergence of lymphatic vessels. However, during further development, the expression of VEGFR-3 becomes restricted mainly to lymphatic vessels (Kaipainen et al, 1995; Kukk et al, 1996; Dumont et al, 1998). Overexpression of VEGF-C in transgenic mice using a basal keratin promoter induced a hyperplastic lymphatic vessel network in the skin of adult mice (Jeltsch et al, 1997). Lymphatic growth, therefore, is thought to be mediated predominantly via VEGFR-3.

Recently, viral homologs of VEGF were found to be encoded by two strains (NZ2 and NZ7) of parapox Orf viruses (Ogawa et al, 1998; Meyer et al, 1999). The resulting polypeptides carry the characteristic dimer cysteine knot motif present in all mammalian VEGFs. These factors are collectively termed VEGF-E, they exclusively activate VEGFR-2, and at least NZ2-VEGF-E can use neuropilin-1 as a co-receptor (Wise et al, 1999). The Orf virus produces a pustular dermatitis in sheep, goats, and humans. The induced lesions show extensive proliferation of vascular endothelial cells, dilation of blood vessels, and dermal swelling.

All the above-mentioned members of the VEGF family transmit angiogenic signals to the vascular endothelium via high-affinity VEGFRs, currently designated VEGFR-1 (originally named *flt*: *fms*-like tyrosine kinase), VEGFR-2 (KDR: kinase insert-domain containing receptor in humans; or *flk-1*: fetal liver kinase-1 in mice), and VEGFR-3 (*FLT4*) (Shibuya et al, 1990; Aprelikova et al, 1992; Galland et al, 1992; Terman et al, 1992). All of these receptors have seven immunoglobulin (Ig) homology domains in their extracellular region, and an intracellular tyrosine kinase signaling domain. In adults, VEGFR-1 and VEGFR-2 are expressed mainly in the blood vascular endothelium, while VEGFR-3 is mostly restricted to the lymphatic endothelium. VEGFR-1 is a 180-kd transmembrane glycoprotein, but due to alternative splicing, a shorter soluble protein consisting of only the six first Ig homology domains is also produced (De Vries et al, 1992; Kendall and Thomas, 1993). VEGFR-2 is a 230-kd protein, with no reported splice variants. In human VEGFR-3, alternative 3′ polyadenylation signals result in a 4.5-kilobase (kb) transcript and a more prevalent 5.8-kb transcript (Pajusola et al, 1993). The latter encodes 65 additional amino acid residues and is the major form detected in tissues. After biosynthesis, the glycosylated 195-kd VEGFR-3 is proteolytically cleaved in the fifth Ig homology domain, but the resulting 125- and 75-kd chains remain linked by a disulfide bond (Pajusola et al, 1993).

Like other receptor tyrosine kinases, the VEGFRs are thought to dimerize and to undergo transautophosphorylation upon ligand binding. Both VEGFR-2 and VEGFR-3 are tyrosine phosphorylated when stimulated with their respective ligands (Waltenberger et al, 1994; Joukov et al, 1996), but VEGFR-1 autophosphorylation is less obvious and has been studied mostly in receptor over-expressing transfected cells (Waltenberger et al, 1994; Seetharam et al, 1995). Phosphorylated tyrosine residues serve to control the kinase activity of the receptor and to create docking sites for cytoplasmic signaling molecules, which provide substrates for the kinase. These molecules, either adapters or enzymes themselves, link VEGFRs to the signaling pathways leading to specific responses like proliferation, migration, survival, tube formation, or increased permeability of target endothelial cells.

ANGIOPOIETINS AND TIE RECEPTORS

Angiopoietins are a second growth factor family with specificity for vascular endothelial cells (Table 2.1). Like VEGFs, angiopoietins have also been shown to play a role in the formation of the blood vascular system. Ang-1 is a ligand for Tie-2 (tyrosine kinase with Ig and epidermal growth factor [EGF] homology domains receptor tyrosine kinase-2, also called Tek: tunica interna endothelial cell kinase), and a related growth factor termed Ang-2 functions as a naturally occurring antagonist for Ang-1 and Tie-2 signaling (Maisonpierre et al, 1997; Suri et al, 1997). The receptors of the Tie family (Tie-1 and Tie-2) are structurally closely related and contain two immunoglobulin homology (IgH) domains and three epidermal growth factor and fibronectin type III homology domains in the extracellular part of the receptor. Trans- and juxtamembrane domains are followed by a tyrosine kinase domain and a carboxyterminal tail (Partanen et al, 1992). No ligand has been identified to date for the Tie-1 receptor. However, Tie-1 deficient mice die in utero between embryo day (E) 13.5 and birth mainly as a consequence of a decrease in vascular integrity (Puri et al, 1995). As a result, edema and localized hemorrhage occur in embryos deficient in this receptor.

Like VEGFRs, Tie-1 and Tie-2 are specifically expressed in endothelial cells, and these receptors have been shown to have distinct roles during vascular development. The phenotype of Tie-2– and Ang-1–deficient mice suggests a role for this ligand-receptor pair in maintaining communication between endothelial cells and the surrounding mesenchyme to establish stable cellular and biochemical interactions between endothelial cells and pericytes/smooth muscle cells (Dumont et al, 1994; Suri et al, 1997). On the other hand, overexpression of Ang-1 in the skin of transgenic mice demonstrated that Ang-1 can reduce vascular leakage due to inflammatory mediators even in the presence of excess VEGF (Thurston et al, 1999). This result suggests that Ang-1 can be used for reducing microvascular leakage in diseases in which the leakage results from chronic inflammation or elevated VEGF, or, alternatively, Ang-1 may be used, in combination with VEGF, for promoting growth of nonleaky vessels. In contrast to Ang-1, the expression of Ang-2 is mainly detected in sites of active angiogenesis (Maisonpierre et al, 1997). Analysis of the Ang-2 function in normal adults suggests that it plays a crucial role in destabilizing quiescent adult vessels, and thus seems to be involved in the initiation of normal vascular remodeling.

DEVELOPMENT AND MORPHOGENESIS OF THE MAMMARY GLAND

The mammary gland is one of the few organs that undergo cycles of growth, morphogenesis, differentiation, functional activity, and involution. It arises from an epidermal thickening on the ventral surface of the body (the milk line) at approximately the sixth week of fetal development. At birth, the mammary gland consists of branching ducts connected to the nipple by a single primary duct.

Development of the mammary gland is incomplete at the time of birth. Although growth and branching of the mammary ducts occur slowly during prepubertal life, during puberty, with the progressive increase in estrogen and progesterone levels, ductal growth and ramification are rapid, resulting in the formation of the mammary gland ductal tree. Following the onset of pregnancy, in response to sustained elevated levels of estrogens and progesterone, ductal elongation and branching resumes, and clusters of alveoli bud off from the growing ducts (Figure 2.1; see color insert). During the second half of gestation, alveolar morphogenesis is followed by the structural and functional differentiation of alveolar epithelial cells in preparation for milk fat and protein secretion, which occurs during lactation (Figure 2.1). After weaning, the mammary gland involutes rapidly, ultimately leaving only a rudimentary ductal system and a few remaining alveoli (reviewed by Daniel and Silberstein, 1987).

ALTERATIONS IN MAMMARY GLAND VASCULATURE DURING PREGNANCY, LACTATION, AND INVOLUTION

The arteries that supply the breast are derived from the thoracic branches of the axillary artery, the internal thoracic artery, and anterior intercostal arteries. Ramifications of these vessels accompany ductal structures, ultimately reaching the capillary network, which surrounds alveoli.

Previous descriptive studies have revealed that the mammary gland vascular bed is a highly dynamic structure that undergoes profound qualitatitive and quantitative alterations during the mammary cycle (Wahl, 1915; Turner and Gomez, 1933; Soemarwoto and Bern, 1958; Yasugi et al, 1989; Matsumoto et al, 1992). Thus, in the virgin mammary gland, the presence of a periductal capillary plexus has been described, which becomes richly developed with advancing pregnancy in association with epithelial growth. This is accompanied by concomitant growth of arterioles and venules. During lactation, there is a progressive increase in capillary size due to dilatation of existing capillaries in association with increased milk secretory function. During involution, the capillary bed progressively disappears, so that relatively thick-walled venules and arterioles appear disproportionately large compared with the capillary field that they supply. Gradually, however, the arterioles and venules also begin to involute, and ultimately the vascular bed resembles that of the virgin mammary gland.

Using a novel whole-mount lectin-based technique to quantitate vascularization in the murine mammary gland, we have observed that the mature virgin mammary gland contains a fully developed capillary system, which increases significantly in density during early pregnancy (Table 2.2.) with vascular networks surrounding the expanding parenchyma and developing alveolar units (Pepper et al, 2000). During lactation, expansion of terminal alveoli distends the capillary networks; however, this stage is not associated with increased vascular growth. In fact, the vascular density appears to decrease slightly. This is probably artifactual and due to alveolar distention during lactation. Upon weaning, regression of parenchyma results in involution of capillary loops. Vessels appear to be thinner and further apart; eventually many of these vessels regress. These events are also associated with loss of pericytes and complete obliteration of the lumen. Thus,

TABLE 2.2. Quantitation of capillary density in the mouse mammary gland during pregnancy, lactation and involution.

Stage	Capillary density (per mm^2) (mean ± SD)
Virgin	
5 weeks	27 ± 4.6
7 weeks	32 ± 8.9
10 weeks	39 ± 4.5
Pregnant	
5 days	67 ± 6.5
14 days	110 ± 3.4
19 days	198 ± 7.6
Lactation	
2 days	119 ± 5.7
4 days	105 ± 7.5
7 days	109 ± 6.5
Involution	
2 days	93 ± 7.6
4 days	72 ± 8.7
7 days	52 ± 5.7
21 days	69 ± 6.4

Quantification was performed using whole-mount preparations stained with *Bandeiraea simplicifonia* lectin (Vector Labs, Burlingame, CA). Images were scanned using a Toshiba cool CCD camera and quantified with ImagePro software analysis program. Four independent animals were evaluated per stage.

vascular density increases progressively during pregnancy and subsequently decreases during lactation, to levels observed during midpregnancy; during involution, there is a further reduction in vessel density. These findings are consistent with a study in the rat using India ink injection (Yasugi et al, 1989), in which vessel density increased progressively during pregnancy and decreased thereafter through late pregnancy, lactation, and involution. Alterations in vascular volume have been discussed in Chapter 1.

VEGF AND VEGFRS IN THE MAMMARY GLAND DURING PREGNANCY, LACTATION, AND INVOLUTION

Despite the long-known and well-characterized morphologic alterations in mammary gland vasculature described above, surprisingly little is known about mammary endothelial cell function or the molecular mechanisms that regulate angiogenesis, permeability, and endothelial cell death in the mammary gland.

Studies in the virgin rodent mammary gland have revealed that VEGF (188, 164, and 120 amino acid isoforms), VEGF-C, and VEGFR-1, -2, and -3 are expressed, with VEGF messenger RNA (mRNA) and protein being localized predominantly in mammary epithelium (Figures 2.2A and 2.3) (Pepper et al, 2000). Low levels of VEGF mRNA have also been observed in terminal duct epithelium of normal

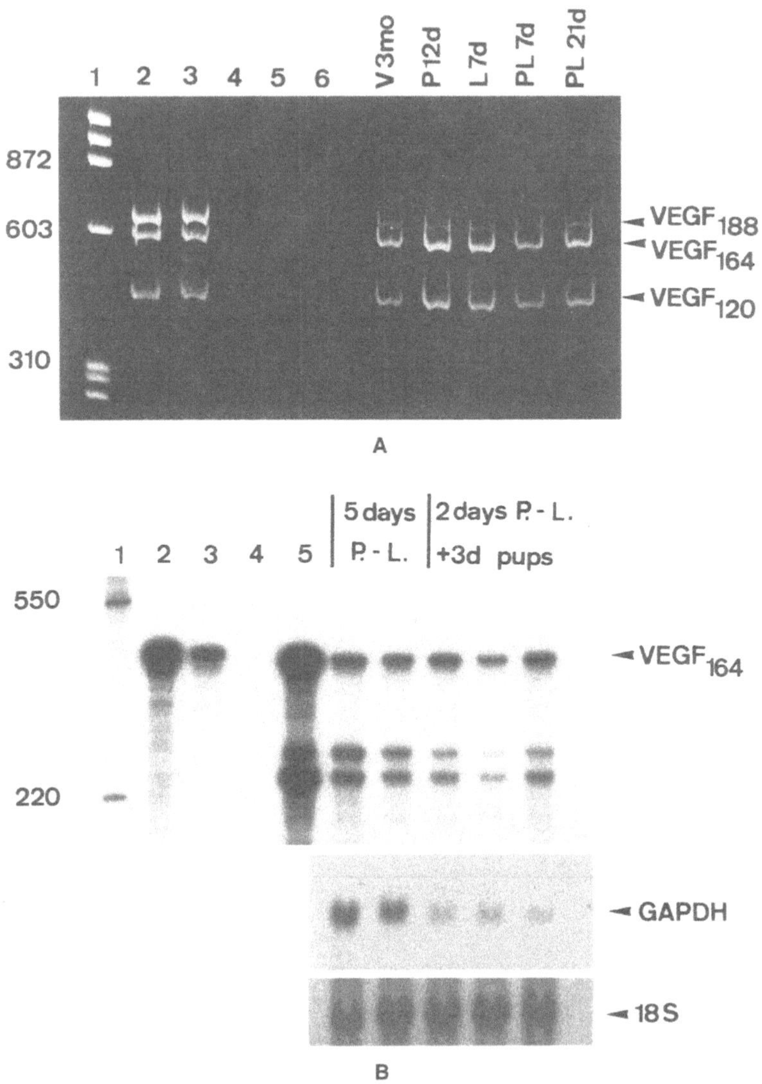

FIGURE 2.2. A: Vascular endothelial growth factor (VEGF) isoforms expressed in rat mammary gland. Reverse transcriptase-polymerase chain reaction (RT-PCR) analysis of VEGF mRNA expression at various stages including 3-month virgin (V 3mo), 12-day pregnant (P 12d), 7-day lactating (L 7d), 7-day postlactating (PL 7d), and 21-day postlactating (PL 21d). Lane 1: molecular size markers; lane 2: mouse lung; lane 3: rat lung; lane 4: H_2O; lane 5: mouse lung—reverse transcriptase; lane 6: rat lung—reverse transcriptase. B: Induction of $VEGF_{164}$ during lactation. Samples from rat mammary glands were analyzed by RNase protection as follows: 5 days postlactation/weaning (5 days P.-L.), two animals; 2 days of weaning/postlactation followed by replacement of pups for a further 3 days (2 days P.L. + 3d pups), three animals. Lane 1: molecular size markers; lane 2: chromosomal RNA (cRNA) probe; lane 3: cRNA probe + hybridization mixture; lane 4: yeast transfer RNA (tRNA); lane 5: rat lung. 18S ribosomal RNA is shown as an indicator of RNA loading and integrity. When normalized to glyceraldehyde-3-phosphate dehydrogenase (GAPDH), $VEGF_{164}$ is increased 4.1-fold during lactation.

human breast tissue (Brown et al, 1995), and VEGF immunoreactivity has been observed in these cells in normal human and nonhuman primate mammary gland (Nakamura et al, 1999, and our own observations). Low levels of VEGF protein, as determined by enzyme-linked immunosorbent assay (ELISA), have been observed in nonneoplastic human breast tissue (Obermair et al, 1997; Greb et al, 1999), and weak VEGF-C immunoreactivity has been reported in normal human mammary epithelium (Valtola et al, 1999). Using a dorsal skinfold chamber model in nude mice, Lichtenbeld et al (1998) have reported that in contrast to tissue samples from breast cancer, samples from normal healthy breasts had no angiogenic activity. The precise role of the VEGF ligand-receptor system in the normal mammary gland is not known.

In the rat, mRNA for $VEGF_{164}$ is increased during pregnancy (5.0-fold increase on day 12) and during lactation (18.5-fold increase on day 7) (Pepper et al, 2000). Levels of VEGF appear to be minimally altered during involution. VEGF 120, 164, and 188 amino acid isoforms are expressed during pregnacy, lactation, and involution, and their abundance relative to one another does not alter during the mammary cycle (Figure 2.2A). In the mouse, there is an increase in VEGF mRNA levels throughout pregnancy (maximum 5.5-fold at 5 days) and a more marked increase during lactation (maximal 9.7-fold at 7 days); levels of VEGF decreased progressively during the phase of involution. These findings point to a causal role for VEGF-VEGFR interactions in the increase in vascularization, which occurs during pregnancy-associated mammary growth, as well as endothelial cell function during lactation. Whether the downregulation of VEGFRs observed following weaning is causally related to involution of the vasculature is not known.

The relationship between increased levels of VEGF mRNA and lactation was confirmed by removing rat pups from their lactating mothers, and after 2 days reintroducing the pups for a further 3 days. In a parallel (control) group, pups were removed for a period of 5 days (Figure 2.2B). Analysis of VEGF mRNA expression, relative to glyceraldehyde-3-phosphate dehydrogenase (GAPDH), revealed a 4.1-fold increase in the experimental group relative to controls, thereby establishing a clear relationship between lactation and increased VEGF expression. The increase in VEGF was in contrast to the decrease in hepatocyte growth factor (HGF) and its receptor c-Met (71% and 86%, respectively, when normalized to GAPDH), previously observed under the same conditions (Pepper et al, 1995).

With respect to VEGFR-1 and -2, a quantitative analysis in the rat mammary gland revealed an increase in VEGFR-2 during pregnancy (1.6-fold at 4 days). During lactation, VEGFR-1 (2.7-fold at 7 days) and VEGFR-2 (3.8-fold at 7 days) were increased. In contrast, VEGFR-2 was decreased in the early phases of involution (45%, 50%, and 34% on days 1, 2, and 3, respectively). In the mouse, VEGFR-2 mRNA levels were moderately increased during pregnancy (maximum 2.7-fold at 19 days), and increased further during lactation (maximum 3.7-fold increase on day 7).

As can be appreciated from the above, the greatest increase in VEGF and VEGFR-1 and -2 was seen during lactation. What is the relevance of these findings? The capillary is the primary site of transport between the blood and alveolar epithelium, and constitutes the so-called blood–milk barrier. In the virgin mammary gland, the capillary wall is usually continuous with rare fenestrations of 30 to 55 nm (Stirling and Chandler, 1976; Matsumoto et al, 1992). During lactation, there is a marked increase in permeability as well as in the number of cytoplasmic vesicles in endothelial cells surrounding alveoli (Matsumoto et al, 1992, 1994). These vesicles tend to fuse with one another to form clusters. These authors also describe a striking increase in the number and length of microvillous processes emanating from the surface of parenchyma-associated capillary endothelial cells. Following weaning, these features progressively disappear, with the endothelium ultimately returning to its prepregnancy state. VEGF, which was first identified as a result of its capacity to increase vascular permeability (reviewed by Dvorak et al, 1995), induces morphologic alterations in endothelial cells, including the appearance of vesiculovacuolar organelles (reviewed by Feng et al, 1999), which are entirely consistent with this function. We suggest that during lactation, the marked increase in VEGF induces functional alterations in endothelial cells, which lead to the increase in vascular permeability.

Mammary tissue is composed of parenchymal (epithelial structures) and stromal (connective tissue, blood and lymphatic vessels, and nerves) compartments. Since fibroblasts, epithelium, endothelial, and smooth muscle cells have been reported to synthesize VEGF in vitro and in several organs in vivo, we evaluated the relative contribution of parenchyma vs. stroma by examining VEGF mRNA expression in cleared murine mammary glands devoid of epithelial components. Quantitation revealed that approximately 25% of VEGF mRNA is synthesized by stroma, indicating that mammary epithelium is the major source of VEGF mRNA. It has previously been demonstrated that mammary gland fat pads devoid of parenchyma fail to develop the vascular pattern associated with pregnancy (Soemarwoto and Bern, 1958). The authors concluded that the various changes observed in the vascular bed occur in association with alterations in the epithelial structures (ducts and alveoli), and that these changes do not occur in the absence of parenchyma. Our in situ and immunohistochemistry data suggest that although VEGF is expressed predominantly by epithelial cells (Figure 2.3; see color insert), there is also a significant contribution from the stroma. VEGF has been detected in early passage mammary fibroblast cultures, in which its expression is dramatically increased by hypoxia (Hlatky et al, 1994). These findings indicate that rapidly growing epithelial cells are the major source of VEGF in the mammary gland. VEGF has also been detected in human milk (Siafakas et al, 1999, Pepper et al, 2000).

What are the factors that regulate VEGF expression during pregnancy and lactation? Using a clonal mouse mammary epithelial cell line (Soriano et al, 1995), we have observed that VEGF is downregulated by HGF. Interestingly, the period of maximal VEGF expression (i.e., lactation) is also the period during which HGF and c-Met are markedly downregulated (Pepper et al, 1995; Yang et al, 1995). It has been demonstrated that HGF, which regulates many epithelial cell functions (Matsumoto and Nakamura, 1996) including the induction of mammary ductal morphogenesis (Soriano et al, 1995), is downregulated during lactation (Pepper et al, 1995; Yang et al, 1995), when the HGF-dependent tubulogenic process is essentially completed. It is not possible at this point to ascertain whether there is causality in the inverse relationship between VEGF and HGF levels. In contrast to HGF, mammotrophic hormones did not directly alter VEGF mRNA levels in cultured mammary epithelial cells (Pepper et al, 2000).

The mouse mammary tumor virus long terminal repeat (MMTV LTR) is widely used as a promoter in mouse models to study the role of different growth factors in mammary gland and in breast cancer. This promoter results in hormonally regulated, mammary gland–specific expression during pregnancy and lactation. To study the effects of VEGF in mammary gland angiogenesis, transgenic mice were generated that express human $VEGF_{165}$ under the MMTV promoter (Karkkainen et al, 1998). Morphologic analysis of the mammary glands in these mice revealed edema and loss of the regular alveolar pattern (Figure 2.4; see color insert). During pregnancy, the mammary glands were hypervascularized, and there were abnormally large and malformed vessels surrounding the lobular structures, indicating that they might have developed in response to the expression of the transgene in the ductal epithelial cells, as detected by in situ hybridization. These remarkable alterations are likely to explain why the multiparous female founder mice had problems in fostering their pups after their third or fourth litters.

LYMPHATIC VESSELS IN THE BREAST

The lymphatics of the breast are especially numerous and well developed. Four principal lymphatic pathways drain the breast: cutaneous, axillary, internal thoracic, and posterior intercostal lymphatics (Haagensen, 1986). Most of the lymphatics (75%) drain chiefly to axillary nodes. Lymphatics play a crucial role in homeostasis of normal breast tissue, and also serve to disseminate tumor cells to regional lymph nodes in metastatic breast cancer.

As indicated previously, the VEGF-C/VEGFR-3 ligand-receptor pair is clearly implicated in the growth of lymphatic vessels (Kaipainen et al, 1995; Jeltsch et al, 1997; Dumont et al, 1998). In contrast to VEGF, the levels of VEGF-C mRNA are only modestly increased in the rat mammary gland during pregnancy (2.8-fold increase on day 4) and lactation (1.9-fold increase on day 2). Similarly, VEGFR-3 is modestly increased during pregnancy (2.2-fold at 4 days) and lactation (1.5-fold at 21 days). In contrast, VEGFR-3 is decreased in the early phases of involution (33%, 21%, and 45% on days 1, 2, and 3, respectively). VEGFR-3, which is shown to be specific for lymphatic vessels in most other tissues, is also expressed in the blood capillary endothelial cells in the resting mammary gland (Valtola et al, 1999).

The role of VEGF-C in the function of the lymphatic network of the mammary gland is still unknown. Recently, VEGFR-3 was found to be expressed in the endothelium of certain fenestrated and discontinuous capillaries in certain tissues such as nasal mucosa (Partanen et al, 2000; Saaristo et al, 2000). These results suggest that VEGFR-3 might play a role in discontinuous endothelia, which are known to display increased permeability to macromolecules. Regulation of vascular permeability is also crucial in the mammary gland during lactation. VEGF-C may therefore function as a growth factor for lymphatic vessels, and might also play a role in modifying the permeability of lymphatics as well as blood vessels during lactation.

CONCLUSION

The existence of a precise sequence of tightly scheduled, hormonally regulated changes in tissue structure and function makes the mammary gland an attractive model for studying the mechanisms of angiogenesis, vascular permeability, and endothelial cell death. Results obtained from our studies point to a causal role for VEGF and its receptors in the increase in vascularization, that occurs during pregnancy. They also support a role for VEGF in increasing vascular permeability during lactation, during which increased transport of molecules from the blood is required for efficient milk protein synthesis.

The relationship between microvessel density and breast cancer progression has previously been reported (Weidner et al, 1991). The controlled increase and subsequent decrease we have observed in VEGF/VEGFR expression in response to the physiologic requirements of the mammary gland contrast sharply with the sustained elevation in VEGF and its receptors reported in breast carcinoma (Brown et al, 1995; Gasparini et al, 1997; Obermair et al, 1997). Breast cancer cells also express VEGF-C, and there is upregulation of its receptor VEGFR-3 in the surrounding angiogenic blood vessels (Valtola et al, 1999). In an animal model VEGF-C stimulates the growth of tumor-associated lymphatic vessels and promotes the intralymphatic spread of breast cancer cells (Karpanen et al, unpublished observation). In the future anti-VEGF-C therapy might be used in preventing the formation of tumor metastasis.

REFERENCES

Achen, M. G., Jeltsch, M., Kukk, E., Makinen, T., Vitali, A., Wilks, A. F., Alitalo, K., and Stacker, S. A. 1998. Vascular endothelial growth factor D (VEGF-D) is a ligand for the tyrosine kinases VEGF receptor 2 (Flk1) and VEGF receptor 3 (Flt4). *Proc. Natl. Acad. Sci. USA.* 95:548–553.

Alon, T., Hemo, I., Itin, A., Pe'er, J., Stone, J., and Keshet, E. 1995. Vascular endothelial growth factor acts as a survival factor for newly formed retinal vessels and has implications for retinopathy of prematurity. *Nat. Med.* 1:1024–1028.

Aprelikova, O., Pajusola, K., Partanen, J., Armstrong, E., Alitalo, R., Bailey, S. K., McMahon, J., Wasmuth, J., Huebner, K., and Alitalo, K. 1992. *FLT4*, a novel class III receptor tyrosine kinase in chromosome 5q33-qter. *Cancer Res.* 52:746–748.

Bellomo, D., Headrick, J. P., Silins, G. U., Paterson, C. A., Thomas, P. S., Gartside, M., Mould, A., Cahill, M. M., Tonks, I. D., Grimmond, S. M., Townson, S., Wells, C., Little, M., Cummings, M. C., Hayward, N. K., and Kay, G. F. 2000. Mice lacking the vascular endothelial growth factor-B gene (Vegfb) have smaller hearts, dysfunctional coronary vasculature, and impaired recovery from cardiac ischemia. *Circ. Res.* 86:E29–E35.

Benjamin, L. E., and Keshet, E. 1997. Conditional switching of vascular endothelial growth factor (VEGF) expression in tumors: induction of endothelial cell shedding and regression of hemangioblastoma-like vessels by VEGF withdrawal. *Proc. Natl. Acad. Sci. USA* 94:8761–8766.

Brown, L. F., Berse, B., Jackman, R. W., Tognazzi, K., Guidi, A. J., Dvorak, H. F., Senger, D. R., Connolly, J. L., and Schnitt, S. J. 1995. Expression of vascular permeability factor (vascular endothelial growth factor) and its receptors in breast cancer. *Hum. Pathol.* 26:86–91.

Carmeliet, P., Ferreira, V., Breier, G., Pollefeyt, S., Kieckens, L., Gertsenstein, M., Fahrig, M., Vandenhoeck, A., Harpal, K., Ebenhardt, C., Declercq, C., Pawling, J., Moons, L., Collen, D., Risau, W., and Nagy, A. 1996. Abnormal blood vessel development and lethality in embryos lacking a single VEGF allele. *Nature* 380:435–439.

Daniel, C. W., and Silberstein, G. B. 1987. Postnatal development of rodent mammary gland. In: Neville, M. C., and Daniel, C. W., eds. *The Mammary Gland.* Plenum Press, New York, pp. 3–36.

De Vries, C., Escobedo, J. A., Ueno, H., Houck, K., Ferrara, N., and Williams, L. T. 1992. The *fms*-like tyrosine kinase, a receptor for vascular endothelial growth factor. *Science* 255:989–991.

Dumont, D. J., Gradwohl, G., Fong, G.-H., Puri, M. C., Gertsenstein, M., Auerbach, A., and Breitman, M. L. 1994. Dominant-negative and targeted null mutations in the endothelial receptor tyrosine kinase, *tek*, reveal a critical role in vasculogenesis of the embryo. *Genes Dev.* 8:1897–1909.

Dumont, D. J., Jussila, L., Taipale, J., Lymboussaki, A., Mustonen, T., Pajusola, K., Breitman, M., and Alitalo, K. 1998. Cardiovascular failure in mouse embryos deficient in VEGF receptor-3. *Science* 282:946–949.

Dvorak, H. F., Brown, L. F., Detmar, M., and Dvorak, A. M. 1995. Vascular permeability factor/vascular endothelial growth factor, microvascular hyperpermeability, and angio genesis. *Am. J. Pathol.* 146:1029–1039.

Feng, D., Nagy, J. A., Pyne, K., Hammel, I., Dvorak, H. F., and Dvorak, A. M. 1999. Pathways of macromolecular extravasation across microvascular endothelium in responce to VPF/VEGF and other vasoactive mediators. *Microcirculation* 6:23–44.

Ferrara, N., Carver-Moore, K., Chen, H., Dowd, M., Lu, L., O'Shea, K. S., Powell-Braxton, L., Hilan, K. J., and Moore, M. W. 1996. Heterozygous embryonic lethality induced by targeted inactivation of the VEGF gene. *Nature* 380:438–442.

Ferrara, N., and Henzel, W. J. 1989. Pituitary follicular cells secrete a novel heparin-binding growth factor specific for vascular endothelial cells. *Biochem. Biophys. Res. Commun.* 161:851–855.

Findlay, J. K. 1986. Angiogenesis in reproductive tissues. *J. Endocrinol.* 111:357–366.

Galland, F., Karamysheva, A., Mattei, M.-G., Rosnet, O., Marchetto, S., and Birnbaum, D. 1992. Chromosomal localization of *FLT4*, a novel receptor-type tyrosine kinase gene. *Genomics* 13:475–478.

Gasparini, G., Toi, M., Gion, M., Verderio, P., Dittadi, R., Hanatani, M., Matsubara, I., Vinante, O., Bonoldi, E., Boracchi, P., Gatti, C., Suzuki, H., and Tominaga, T. 1997. Prognostic significance of vascular endothelial growth factor protein in node-negative breast carcinoma. *J. Natl. Cancer Inst.* 89:139–147.

Gluzman-Poltorak, Z., Cohen, T., Herzog, Y., and Neufeld, G. 2000. Neuropilin-2 and neuropilin-1 are receptors for the 165-amino acid form of vascular endothelial growth factor (VEGF) and of placenta growth factor-2, but only neuropilin-2 functions as a receptor for the 145-amino acid form of VEGF. *J. Biol. Chem.* 275:18040–18045.

Greb, R. R., Maier, I., Wallwiener, D., and Kiesel, L. 1999. Vascular endothelial growth factor A (VEGF-A) mRNA expression levels decrease after menopause in normal breast tissue but not in breast cancer lesions. *Br. J. Cancer.* 81:225–231.

Haagensen, D. C. 1986. *Diseases of the Breast*, 3rd ed. W. B. Saunders, Philadelphia.

Hauser, S., and Weich, H. A. 1993. A heparin-binding form of placenta growth factor (PlGF-2) is expressed in human umbilical vein endothelial cells and in placenta. *Growth Factors* 9:259–268.

Hlatky, L., Tsionou, C., Hahnfeldt, P., and Coleman, C. N. 1994. Mammary fibroblasts may influence breast tumor angiogenesis via hypoksia-induced vascular endothelial growth factor up-regulation and protein expression. *Cancer Res.* 54:6083–6086.

Jeltsch, M., Kaipainen, A., Joukov, V., Meng, X., Lakso, M., Rauvala, H., Swartz, M., Fukumura, D., Jain, R. K., and Alitalo, K. 1997. Hyperplasia of lymphatic vessels in VEGF-C transgenic mice. *Science* 276:1423–1425.

Joukov, V., Pajusola, K., Kaipainen, A., Chilov, D., Lahtinen, I., Kukk, E., Saksela, O., Kalkkinen, N., and Alitalo, K. 1996. A novel vascular endothelial growth factor, VEGF-C, is a ligand for the Flt4 (VEGFR-3) and KDR (VEGFR-2) receptor tyrosine kinases. *EMBO J.* 15:290–298.

Joukov, V., Sorsa, T., Kumar, V., Jeltsch, M., Claesson-Welsh, L., Cao, Y., Saksela, O., Kalkkinen, N., and Alitalo, K. 1997. Proteolytic processing regulates receptor specificity and activity of VEGF-C. *EMBO J.* 16:3898–3911.

Kaipainen, A., Korhonen, J., Mustonen, T., van Hinsbergh, V. M., Fang, G.-H., Dumont, D., Breitman, M., and Alitalo, K. 1995. Expression of the *fms*-like tyrosine kinase FLT4 gene becomes restricted to endothelium of lymphatic vessels during development. *Proc. Natl. Acad. Sci. USA* 92:3566–3570.

Karkkainen M., Valtola, R., Korpelainen E., and Alitalo, K. 1999. VEGF, VEGF-C and their receptors in tumor angigenesis and metastasis. In: Fidler, I. J., Niitsu, Y., Seiki, M., Sugimura, T., and Yokota, J., eds. Extended abstracts of the 29th International Symposium of The Princess Takamatsu Cancer Research Fund. Molecular basis for invasion and metastasis. (Tokyo: Princess Takamatsu Cancer Research Fund), pp. 94–104.

Kendall, R. L., and Thomas, K. A. 1993. Inhibition of vascular endothelial cell growth factor activity by an endogenously encoded soluble receptor. *Proc. Natl. Acad. Sci. USA* 90:10705–10709.

Lee, J., Gray, A., Yuan, J., Louth, S.-M., Avraham, H., and Wood, W. 1996. Vascular endothelial growth factor-related protein: a ligand and specific activator of the tyrosine kinase receptor Flt4. *Proc. Natl. Acad. Sci. USA* 93:1988–1992.

Lichtenbeld, H. C., Barendsz-Janson, A. F., van Essen, H., Struijker, B., Griffioen, A. W., and Hillen, H. F. 1998. Angiogenic potential of malignant and non-malignant human breast tissues in an in vivo angiogenesis model. *Int. J. Cancer* 77:455–459.

Maglione, D., Guerriero, V., Viglietto, G., Delli-Bovi, P., and Persico, M. G. 1991. Isolation of a human placenta cDNA coding for a protein related to the vascular permeability factor. *Proc. Natl. Acad. Sci. USA* 88:9267–9271.

Maisonpierre, P. C., Suri, C., Jones, P. F., Bartunkova, S., Wiegand, S. J., Radziejewski, C., Compton, D., McClain, J., Aldrich, T. H., Papadopoulos, N., Daly, T. J., Davis, S., Sato, T. N., and Yancopoulos, G. D. 1997. Angiopoietin-2, a natural antagonist for Tie2 that disrupts in vivo angiogenesis. *Science* 277:55–60.

Makinen, T., Olofsson, B., Karpanen, T., Hellman, U., Soker, S., Klagsbrun, M., Eriksson, U., and Alitalo, K. 1999. Differential binding of vascular endothelial growth factor B splice and proteolytic isoforms to neuropilin-1. *J. Biol. Chem.* 274:21217–21222.

Matsumoto, K., and Nakamura, T. 1996. Emerging multipotential aspects of hepatocyte growth factor. *J. Biochem.* 119:591–600.

Matsumoto, M., Kurohmaru, M., Hayashi, Y., Nishinakagawa, H., and Otsuka, J. 1994. Permeability of mammary gland capillaries to ferritin in mice. *J. Vet. Med. Sci.* 56:65–70.

Matsumoto, M., Nishinakagawa, H., Kurohmaru, M., Hayashi, Y., and Otsuka, J. 1992. Pregnancy and lactation affect the microvasculature of the mammary gland in mice. *J. Vet. Med. Sci.* 54:937–943.

Meyer, M., Clauss, M., Lepple-Wienhues, A., Waltenberger, J., Augustin, H. G., Ziche, M., Lanz, C., Böttner, M., Rziha, H.-J., and Dehio, C. 1999. A novel vascular endothelial growth factor encoded by Orf virus, VEGF-E, mediates angiogenesis via signaling through VEGFR-2 (KDR) but not VEGFR-1 (Flt-1) receptor tyrosine kinases. *EMBO J.* 18:363–374.

Migdal, M., Huppertz, B., Tessler, S., Comforti, A., Shibuya, M., Reich, R., Baumann, H., and Neufeld, G. 1998. Neuropilin-1 is a placenta growth factor-2 receptor. *J. Biol. Chem.* 273:22272–22278.

Nakamura, J., Lu, Q., Aberdeen, G., Albrecht, E., and Brodie, A. 1999. The effect of estrogen on aromatase and vascular endothelial growth factor messenger ribonucleic acid in the normal nonhuman primate mammary gland. *J. Clin. Endocrinol. Metab.* 84:1432–1437.

Neufeld, G., Cohen, T., Gengrinovitch S., and Poltorak, Z. 1999. Vascular endothelial growth factor (VEGF) and its receptors. *FASEB J.* 13:9–22.

Obermair, A., Kucera, E., Mayerhofer, K., Speicer, P., Seifert, M., Czerwenka, K., Kaider, A., Leodolter, S., Kainz, C., and Zellinger, R. 1997. Vascular endothelial growth factor (VEGF) in human breast cancer: correlation with disease-free survival. *Int. J. Cancer.* 74:455–458.

Ogawa, S., Oku, A., Sawano, A., Yamaguchi, S., Yazaki, Y., and Shibuya, M. 1998. A novel type of vascular endothelial growth factor: VEGF-E (NZ-7 VEGF) preferentially utilizes

KDR/Flk-1 receptor and carries a potent mitotic activity without heparin-binding domain. *J. Biol. Chem.* 273:31273–31282.

Olofsson, B., Korpelainen, E., Pepper, M. S., Mandriota, S. J., Aase, K., Kumar, V., Gunji, Y., Jeltsch, M. M., Shibuya, M., Alitalo, K., and Eriksson, U. 1998. Vascular endothelial growth factor B (VEGF-B) binds to VEGF receptor-1 and regulates plasminogen activator activity in endothelial cells. *Proc. Natl. Acad. Sci. USA* 95:11709–11714.

Olofsson, B., Pajusola, K., Kaipainen, A., Von Euler, G., Joukov, V., Saksela, O., Orpana, A., Pettersson, R. F., Alitalo, K., and Eriksson, U. 1996a. Vascular endothelial growth factor B, a novel growth factor for endothelial cells. *Proc. Natl. Acad. Sci. USA* 93:2576–2581.

Olofsson, B., Pajusola, K., von Euler, G., Chilov, D., Alitalo, K., and Eriksson, U. 1996b. Genomic organization of the mouse and human genes for vascular endothelial growth factor B (VEGF-B) and characterization of a second splice isoform. *J. Biol. Chem.* 271:19310–19317.

Orlandini, M., Marconcini, L., Ferruzzi, R., and Oliviero, S. 1996. Identification of a c-*fos*-induced gene that is related to the platelet-derived growth factor/vascular endothelial growth factor family. *Proc. Natl. Acad. Sci. USA* 93:11675–11680.

Pajusola, K., Aprelikova, O., Armstrong, E., Morris, S., and Alitalo, K. 1993. Two human FLT4 receptor tyrosine kinase isoforms with distinct carboxyterminal tails are produced by alternative processing of primary transcripts. *Oncogene* 8:2931–2937.

Partanen, J., Armstrong, E., Mäkelä, T. P., Korhonen, J., Sandberg, M., Renkonen, R., Knuutila, S., Huebner, K., and Alitalo, K. 1992. A novel endothelial cell surface receptor tyrosine kinase with extracellular epidermal growth factor homology domains. *Mol. Cell. Biol.* 12:1698–1707.

Partanen, T., Arola, J., Saaristo, A., Jussila, L., Ora, A., Miettinen, M., and Alitalo, K. 2000. VEGF-C and VEGF-D expression in neuroendocrine cells and their receptor, VEGFR-3 in fenestrated blood vessels in human tissues. *FASEB J.* 14:2087–2096.

Pepper, M. S., Baetens, D., Mandriota, S., Di Sanza, C., Oikesmus, S., Lane, T. F., Soriano, J. V., Montesano, R., and Iruela-Arispe, L. 2000. Regulation of VEGF and VEGF receptor expression in the mammary gland during pregnancy, lactation and involution. *Dev. Dyn.* 218:507–524.

Pepper, M. S., Soriano, J. V. Menould, P.-A., Sappino A.-P., Orci, L., Montesano, R. 1995. Modulation of hepatocyte growth factor and c-met expression in the rat mammary gland during pregnancy, lactation and involution. *Exp. Cell Res.* 219:204–210.

Plouet, J., Schilling, J., and Gospodarowicz, D. 1989. Isolation and characterization of a newly identified endothelial cell mitogen produced by AtT-20 cells. *EMBO J.* 8:3801–3806.

Puri, M. C., Rossant, J., Alitalo, K., Bernstein, A., and Partanen, J. 1995. The receptor tyrosine kinase TIE is required for integrity and survival of vascular endothelial cells. *EMBO J.* 14:5884–5891.

Risau, W. 1997. Mechanisms of angiogenesis. *Nature* 386:671–674.

Risau, W., and Flamme, I. 1995. Vasculogenesis. *Annu. Rev. Cell Dev. Biol.* 11:73–91.

Saaristo, A., Partanen, T. A., Jussila, L., Arola, J., Hytonen, M., Makitie, A., Vento, S., Kaipainen, A., Malmberg, H., and Alitalo, K. 2000. Vascular endothelial growth factor-C and its receptor VEGFR-3 in nasal mucosa and in nasopharyngeal tumors. *Am. J. Pathol.* 157:7–14.

Seetharam, L., Gotoh, N., Maru, Y., Neufeld, G., Yamaguchi, S., and Shibuya, M. 1995. A unique signal transduction from FLT tyrosine kinase, a receptor for vascular endothelial growth factor VEGF. *Oncogene* 10:135–147.

Shibuya, M., Yamaguchi, S., Yamane, A., Ikeda, T., Tojo, A., Matsushime, H., and Sato, M. 1990. Nucleotide sequence and expression of a novel human receptor type tyrosine kinase gene (*flt*) closely related to the *fms* family. *Oncogene* 5:519–524.

Siafakas, C. G., Anatolitou, F., Fusunyan, R. D., Walker, W. A., and Sanderson, I. R. 1999. Vascular endothelial growth factor (VEGF) is present in human breast milk and its receptor is present on intestinal epithelial cells. *Pediatr. Res.* 45:652–657.

Soemarwoto, I. N., and Bern, H. A. 1958. The effects of hormones on the vascular pattern of the mouse mammary gland. *Am. J. Anat.* 103:403–435.

Soker, S., Takashima, S., Miao, H. Q., Neufeld, G., and Klagsbrun, M. 1998. Neuropilin-1 is expressed by endothelial and tumor cells as an isoform-specific receptor for vascular endothelial growth factor. *Cell* 92:735–745.

Soriano, J. V., Pepper, M. S., Nakamura, T., Orci, L., and Montesano, R. 1995. Hepatocyte growth factor stimulates extensive development of branching dut-like structures by cloned mammary gland epithelial cells. *J. Cell. Sci.* 108:413–430.

Stirling, J. W., and Chandler, J. A. 1976. The fine structure of the normal, resting terminal ductal-lobular unit of the female breast. *Virchows. Arch. [A]* 372:205–226.

Suri, C., Jones, P. F., Patan, S., Bartunkova, S., Maisonpierre, P. C., Davis, S., Sato, T. N., and Yancopoulos, G. D. 1997. Requisite role of Angiopoietin-1, a ligand for the TIE2 receptor, during embryonic angiogenesis. *Cell* 87:1161–1169.

Takahashi, A., Sasaki, H., Kim, S. J., Tobisu, K., Kakizoe, T., Tsukamoto, T., Kumamoto, Y., Sugimura, T., and Terada, M. 1994. Markedly increased amounts of messenger RNAs for vascular endothelial growth factor and placenta growth factor in renal cell carcinoma associated with angiogenesis. *Cancer Res.* 54:4233–4237.

Terman, B. I., Dougher-Vermazen, M., Carrion, M. E., Dimitrov, D., Armellino, D. C., Gospodarowicz, D., and Böhlen, P. 1992. Identification of the KDR tyrosine kinase as a receptor for vascular endothelial cell growth factor. *Biochem. Biophys. Res. Commun.* 187:1579–1586.

Thurston, G., Suri, C., Smith, K., McClain, J., Sato, T. N., Yancopoulos, G. D., and McDonald, D. M. 1999. Leakage-resistant blood vessels in mice transgenically overexpressing angiopoietin-1. *Science* 286:2511–2514.

Turner, C. W., and Gomez, E. T. 1933. The normal development of the mammary gland in the male and female albino mouse. *Res. Bull. Mo. Agric. Exper. Sta.* 182:3–43.

Valtola, R., Salven, P., Heikkilä, P., Taipale, J., Joensuu, H., Rehn, M., Pihlajaniemi, T., Weich, H., de Waal, R., and Alitalo, K. 1999. VEGFR-3 and its ligand VEGF-C are associated with angiogenesis in breast cancer. *Am. J. Pathol.* 154:1381–1390.

Wahl, H. M. 1915. Development of the blood vessels of the mammary gland in the rabbit. *Am. J. Anat.* 18:515–524.

Waltenberger, J., Claesson-Welsh, L., Siegbahn, A., Shibuya, M., and Heldin, C.-H. 1994. Different signal transduction properties of KDR and Flt1, two receptors for vascular endothelial growth factor. *J. Biol. Chem.* 269:26988–26995.

Weidner, N., Semple, J. P., Welch, W. R., and Folkman, J. 1991. Tumor angiogenesis and metastasis-correlation in invasive breast carcinoma. *N. Engl. J. Med.* 324:1–8.

Wise, L. M., Veikkola, T., Mercer, A. A., Savory, L. J., Fleming, S. B., Caesar, C., Vitali, A., Makinen, T., Alitalo, K., and Stacker, S. A. 1999. Vascular endothelial growth factor (VEGF)-like protein from orf virus NZ2 binds to VEGFR2 and neuropilin-1. *Proc. Natl. Acad. Sci. USA* 96:3071–3076.

Yamada, Y., Nezu, J., Shimane, M., and Hirata, Y. 1997. Molecular cloning of a novel vascular endothelial growth factor, VEGF-D. *Genomics* 42:483–488.

Yang, Y., Spitzer, E., Meyer, D., Sachs, M., Niemann, C., Hartmann, G., Weidner, K. M., Birchmeier, C., and Birchmeier, W. 1995. Sequential requirement of hepatocyte growth factor and neuregulin in morphogenesis and differentiation of the mammary gland. *J. Cell. Biol.* 131:215–226.

Yasugi, T., Kaido, T., and Uehara, Y. 1989. Changes in density and architecture of microvessels of the rat mammary gland during pregnancy and lactation. *Arch. Histol. Cytol.* 52:115–122.

CHAPTER 3

The Role of Angiogenesis in Breast Cancer Progression

Sandra W. McLeskey and Robert B. Dickson

Increasingly abundant evidence indicates that the ability of a tumor to command angiogenesis is an important determinant of its phenotype. The data that support this viewpoint were first gained in patients with carcinoma of the breast (Weidner et al, 1991), and confirmatory data have subsequently been obtained in studies conducted in breast and many other solid tumors (see Chapter 4). Additionally, the discovery of various proangiogenic and antiangiogenic molecules has led to the viewpoint that in the normally quiescent vasculature of an adult, the effects of angiogenesis inhibitors predominate for endothelial cells. Therefore, it follows that in angiogenic tumors, the balance between inhibition and stimulation of angiogenesis is disrupted so that endothelial stimulation is favored. Much effort has been directed toward identifying particular angiogenesis stimulators or inhibitors important in neoangiogenesis of breast cancer. Although a multiplicity of candidate stimulators and inhibitors have been investigated, and although vascular endothelial growth factor (VEGF) has been correlated with poor prognosis in breast cancer (Toi et al, 1995a; Gasparini et al, 1997; Relf et al, 1997; Eppenberger et al, 1998), it is becoming clear that we will probably not find any single angiogenic molecule that is responsible for the neovascularization of this disease. It appears more likely that tumors can influence expression of a large number of angiogenesis stimulators and inhibitors, and that this influence extends to molecules expressed by stromal cells as well as those expressed by the tumor cells themselves (reviewed in Pluda, 1997). Recently, increasing attention has been paid to the mechanism(s) whereby tumor cells increase their expression of angiogenic stimulators or decrease their expression of angiogenic inhibitors; this has been termed the angiogenic switch. This chapter reviews the evidence supporting the existence of an angiogenic switch in breast cancer and examines a number of potentially contributory components. Although it covers many areas of interest to breast cancer researchers, it is not intended as an exhaustive review of the multiplicity of putative angiogenesis factors or inhibitors. Areas for further research will be identified.

MECHANISM OF THE ANGIOGENIC SWITCH

Evidence from transplanted tumors first suggested that a sudden phenotypic transformation occurs, during which a tumor becomes able to command neoangiogenesis (Gimbrone et al, 1972; Folkman et al, 1989; Kandel et al, 1991). More recent evidence has come from naturally occurring breast tumors, observed longitudinally in animals (Brem et al, 1977), and from microvessel quantitation or assays of angiogenesis or angiogenic factors in preneoplastic breast lesions in humans (Fregene et al, 1994; Guidi et al, 1994, 1997; Brown et al, 1995; Engels et al, 1997b). These more recent observations suggest that in breast cancer the angiogenic switch may involve a more gradual process, encompassing preneoplastic as well as frankly neoplastic lesions. In other cases, it has been demonstrated that the angiogenic stimulator, inhibitor, or receptor repertoire of a given breast tumor can change over time; it may include multiple angiogenic factors, expressed sequentially or simultaneously (Relf et al, 1997; de Jong et al, 1998; Morelli et al, 1998). A seminal question arising from these data involves the mechanism(s) of the angiogenic switch(es).

REGULATION OF EXPRESSION OF ANGIOGENIC FACTORS IN BREAST NEOPLASMS

Although many of the fibroblast growth factor (FGFs) have the ability to transform various cell types (Dickson et al, 1984; Delli-Bovi et al, 1987; Blam et al, 1988; Rogelj et al, 1988; Zhan et al, 1988; Jouanneau et al, 1991; Forough et al, 1993), and hence qualify as oncogenes in their own right, the genes encoding angiogenesis activators and inhibitors are usually not mutated in malignant cells that express them. Instead, increased expression of angiogenesis activators or decreased expression of inhibitors regulates the increased angiogenesis associated with increased tumor growth, metastasis, or poor prognosis. Evidence is accumulating that shows that the same genetic alterations that produce malignant transformation may also contribute to the regulation of the angiogenic switch. Moreover, these mechanisms are reinforced by the microenvironment of the tumor. Expression of angiogenesis activators or inhibitors could be regulated at either transcriptional or translational levels; there is evidence that both may be involved.

Transcriptional Control of Angiogenesis Stimulators or Inhibitors

It has long been known that breast and other tumors are frequently hypoxic. When hypoxia was initially shown to induce VEGF expression (Shweiki et al, 1992), one rationale for the angiogenic switch was validated. Although a large part of this induction is posttranscriptional (below), hypoxia-inducible factor-1 (HIF-1) has been identified as an important transcription factor that regulates transcription of many hypoxia-induced genes including both VEGF and platelet-derived growth factor (PDGF), another putative angiogenesis factor (Semenza and Wang, 1992; Wang and Semenza, 1993; Levy et al, 1995; Wang et al, 1995; Forsythe et al, 1996; Flamme et al, 1998; as well as reviewed in Semenza et al, 1998). HIF-1 is a dimer of α- and β-subunits (Jiang et al, 1996); it may also interact with the E1A binding protein p300 in mediating hypoxia-induced transcription (Arany et al, 1996). The β-subunit of HIF-1, arylhydrocarbon receptor nuclear translocator (ARNT), is constitutively expressed, but the α-subunit itself is induced by hypoxia (Semenza and Wang, 1992). The identity of the putative "oxygen sensor" that brings this about is unknown. To address this issue, HIF-1α null embryonic stem cells have been studied in comparison with their wild-type controls. Briefly, the HIF-1α null embryonic stem cells have reduced ability to form teratocarcinomas; this appears to be due to their poor vascularization (Ryan et al, 1998). Other studies have shown that HIF-1α null embryonic stem cells form larger tumors; these are also less vascularized and more hypoxic tumors (Carmeliet et al, 1998). HIF-1α has been shown to be inappropriately expressed under normoxic conditions in prostate cancer cells (Zhong et al, 1998), implying that HIF-1α–regulated genes, including VEGF or PDGF, might also be inappropriately expressed in this type of cancer. The importance of this finding for breast cancer is unknown. Moreover, HIF-1α expression is positively regulated by insulin, insulin-like growth factor-1 (IGF-1), and IGF-2 (Feldser et al, 1999). These data imply that IGF signaling, which is thought to be important in breast cancer (reviewed in Surmacz et al, 1998), could upregulate transcription of VEGF or PDGF by upregulating HIF-1α. In addition, a novel dimerization partner of ARNT, termed HIF-1α–like factor (HLF) due to sequence homology to HIF-1α, has been identified in a murine hypothalamus complementary DNA (cDNA) library. The expression pattern of HLF closely parallels that of VEGF (Ema et al, 1997). Its significance for governing VEGF expression in breast cancer remains unknown.

Hypoxia-induced secretion of angiogenic factors is a physiologic response that may be important in multiple processes, including development, formation of the placenta, and wound healing (D'Amore and Shima, 1996). However, it is becoming clear that successful tumors have the additional ability to stimulate angiogenesis as part of the pathologic process of malignant transformation.

Hypoxia has been shown to activate the c-*src* proto-oncogene leading to increased VEGF expression (Mukhopadhyay et al, 1995a,b) in glioma cells and fibroblasts, a physiologic response. However, antisense to c-*src* downregulated VEGF expression in colon cancer cells (Ellis et al, 1998), implicating pathologically activated c-*src* in their increased expression of VEGF (Ellis et al, 1998). Moreover, activated *src*, in the form of v-*src*, has been shown to induce HIF-1α in the absence of hypoxia in hepatoma cells and thereby increase expression of VEGF (Jiang et al, 1997). Although these data were not obtained in breast cancer, activation and/or increased expression of c-*src* has recently been reported in breast cancer (reviewed in Biscardi et al, 1999), implying that similar mechanisms may be operative.

The transcription factor AP-1, itself composed of two proto-oncogene proteins, c-*fos* and c-*jun*, can potentiate the HIF-1 induction of VEGF transcription in glioma cells (Damert et al, 1997). In addition, c-*jun* transfection of MCF-7 breast carcinoma cells increased FGF-2 expression (Lee et al, 1995). Stimulation of growth factor pathways in breast cancer cells resulted in increases of AP-1 activity (Chen et al, 1996a), and progression to an antiestrogen-resistant phenotype was accompanied by increased AP-1 activity (Dumont et al, 1996). However, the possible pathologic role of AP-1 in breast cancer is open to question since breast cancer cells have been shown to have lower AP-1 activity than normal mammary epithelial cells grown in tissue culture (Smith et al, 1997). Thus, the contribution of increased activity of AP-1 to the angiogenic switch in breast cancer requires further clarification.

The von Hippel–Lindau (VHL) tumor suppressor gene product, which functions to suppress transcription elongation (Duan et al, 1995; Kaelin et al, 1998), has been shown to negatively regulate VEGF expression by interacting with the transcription factor Sp1 (Mukhopadhyay et al, 1997). However, the VHL gene product may also affect the stability of VEGF messenger RNA (mRNA) (see below). In any case, VHL gene mutations that produce inactive protein products result in upregulated VEGF expression (Wizigmann-Voos et al, 1995; Iliopoulos et al, 1996; Siemeister et al, 1996; Stratmann et al, 1997). To our knowledge, however, no mutations of the VHL gene have been demonstrated in familial or sporadic breast cancer.

Mutation of the transcription factor p53 is of relevance in Li-Fraumini syndrome (Malkin et al, 1990; Birch et al, 1998), a class of inherited breast cancers, and also in approximately one third of sporadic breast cancers (reviewed in Ozbun and Butel, 1995). The p53 mutation in fibroblast models is associated with downregulation of the angiogenesis inhibitor thrombospondin-1 (Dameron et al, 1994a,b; Volpert et al, 1997) and upregulation of VEGF (Volpert et al, 1997; Takahashi et al, 1998). Supplying an intact *p53* gene to fibroblasts that lack it has the opposite effect (Bouvet et al, 1998; Holmgren et al, 1998). Moreover, increased transcription of VEGF mediated by activation of c-*src* (see above) was prevented by wild-type p53 in colon carcinoma cells (Mukhopadhyay et al, 1995a). Conversely, a mutated murine *p53* induced VEGF mRNA expression after its transient transfection in human embryonic kidney cells. Intact p53 may interact with HIF-1 and the coactivator of nuclear receptors, p300, to inhibit hypoxia-induced transcription of HIF-1 targets (Blagosklonny et al, 1998). However, expression of a nonmutated *p53* gene does not prevent hypoxia-induced upregulation of VEGF in hepatoma cells (Agani et al, 1997). Future studies should address mechanisms of p53 regulation of thrombospondin and VEGF in breast carcinoma cells.

Transcriptional upregulation of FGF-2 has been demonstrated in both melanoma and breast carcinoma cells by the homeobox protein HOXB7(Caré et al, 1996, 1998). This protein is constitutively expressed in some melanomas (Caré et al, 1996, 1998). Moreover, a mutant form of the *HOXB7* gene has been cloned from MCF-7 breast carcinoma cells. This mutant gene is present in clinical breast cancer specimens, as well as normal tissues, and it is able to stimulate transcription (Chariot et al, 1998). The significance of this mutant gene to the mechanism of the angiogenic switch is unclear at this time.

Posttranscriptional Control of Angiogenesis Factors or Inhibitors

There is considerable evidence for posttranscriptional regulation both of VEGF and FGF-2. Although hypoxia does increase the transcription rate of VEGF (see above), it has an even greater effect in prolonging the half-life of VEGF mRNA (Stein et al, 1995; reviewed in Levy et al, 1997). The regulatory elements for this effect of hypoxia seem to reside in the 3′ untranslated region (UTR) of the VEGF mRNA (Levy et al, 1996a). The hypoxia effect appears to involve protein-mRNA interactions and tyrosine kinase activity, and these effects may be negatively modulated by the VHL protein (Gnarra et al, 1996; Levy et al, 1996b). At least one RNA-binding protein, HuR, which may mediate the hypoxia-induced prolongation of VEGF mRNA half-life, has been identified, but its exact role in this activity is not clear (Levy et al, 1998). Moreover, there is evidence implicating multiple such mRNA-binding proteins as determinants of mRNA stability in response to hypoxia (Scandurro and Beckman, 1998).

The translation of angiogenic factor mRNA may also be positively influenced by hypoxia. Under hypoxic conditions, translation is globally decreased (Kraggerud et al, 1995). However, VEGF, FGF-2, and PDGF-B have long, complex 5′ UTRs that contain internal ribosome entry sites that enable cap-independent translation, which may proceed efficiently during hypoxic conditions (Bernstein et al, 1995, 1997; Vagner et al, 1995; Akiri et al, 1998; Stein et al, 1998). Moreover, FGF-2 translation by this mechanism may proceed from an alternative CUG translation start site, which produces an isoform of FGF-2 with distinct activity (see below) (Vagner et al, 1995).

Similar to the effect of hypoxia on transcription, its effects on translation are of physiologic relevance. Again, there is evidence that pathologic alterations of tumors also promotes translational upregulation of angiogenic factors. The eukaryotic initiation factor 4E (eIF4E) is a 25-kd phosphoprotein that forms part of the translation complex that binds to the mRNA cap. The phosphorylation state and activity of eIF4E each may be under partial regulation by c-*ras* (Okada et al, 1998). Specifically, activating mutations or increased c-*ras* signaling such as is found in breast cancer (see below), would upregulate eIF4E activity. Decreasing eIF4E abundance experimentally reverses the effects of c-*ras* transformation in fibroblasts (Rinker-Schaeffer et al, 1993; Graff et al, 1995). Overexpression of eIF4E induces transformation in fibroblasts or HeLa cells (reviewed in Rhoads 1991, 1993), suggesting that this protein has oncoprotein activity. Moreover, overexpression of eIF4E has been found in ductal carcinoma in situ, invasive breast carcinomas (Li et al, 1997; Nathan et al, 1997), and breast carcinoma cell lines (Anthony et al, 1996); it has been correlated with prognosis in breast cancer (Li et al, 1998). Transcription of eIF4E is upregulated by c-*myc* (Jones et al, 1996), another oncoprotein whose expression may be abnormally increased in some breast cancers (Nass and Dickson, 1997). Overexpression of eIF4E does not result in globally increased translation, but preferential translation of mRNAs with long, complex 5′ UTRs, such as those encoding VEGF and FGF-2. Moreover, the long, complex 5′ UTR of FGF-2 contains alternative translation start sites, involving both AUG and CUG codons. Overexpression of eIF4E favors translation of longer isoforms of FGF-2, which initiate from upstream CUG codons (Kevil et al, 1995; Nathan et al, 1997). These longer isoforms of FGF-2, which have different effects than the shorter ones (Bikfalvi, 1995), were found in greater concentrations in conditioned media than in cells, indicating that they are preferentially externalized (Kevil et al, 1995). Likewise, eIF4E-overexpressing Chinese hamster ovary cells preferentially translated increased amounts of VEGF mRNA, relative to glyceraldehyde-3-phosphate dehydrogenase (GAPDH) mRNA, when compared to nonoverexpressing cells (Kevil et al, 1996). Moreover, VEGF levels in breast tumors correlate with eIF4E levels (Scott et al, 1998). Thus, overexpression of eIF4E may be a mediator of the angiogenic switch in breast cancer at the translational level.

Ras-Mediated Influences on the Angiogenic Switch

Although the *ras* family of oncogenes is not usually mutated in breast cancer, it is often overexpressed (Miyakis et al, 1998), and increased signaling through Ras-activated pathways has been documented in human breast cancer (Clark and Der, 1995). Activating mutations in c-*ras*^{H} can increase VEGF expression, either by its increased transcription or by increased mRNA half-life, in fibroblasts, keratinocytes, or intestinal epithelial cells (Rak et al, 1995; Larcher et al, 1996; Arbiser et al, 1997). In addition, Ras signaling has been shown to increase VEGF translation (above). Moreover, hypoxic upregulation of VEGF in astrocytoma cells was shown to be mediated by Ras (Feldkamp et al, 1999). It is not clear which of the downstream effects of Ras produce these effects. However, in immortalized murine pancreatic endothelial cells, Ras-mediated upregulation of VEGF was partially abrogated by an inhibitor of phosphatidylinositol 3-kinase (PI3-kinase) (Arbiser et al, 1997). In transformed NIH-3T3 cells, the PI 3-kinase/Akt signaling pathway was shown to be involved in the HIF-1–induced upregulation of VEGF (Mazure et al, 1997). In addition, activation of p42/p44 mitogen-activated protein (MAP) kinase by Raf-1, one downstream effector of Ras signaling, was shown to induce activity of the VEGF promoter in fibroblasts, in part due to increases in binding of the AP-2 and Sp-1 transcription factors (Milanini et al, 1998). Although none of these data is derived from breast cancer, the positive correlation of VEGF expression with prognosis in this disease (Toi et al, 1995a; Gasparini et al, 1997; Relf et al, 1997; Eppenberger et al, 1998), coupled with the finding of increased expression and signaling of Ras in breast cancer (see above), argues that these mechanisms may contribute to the angiogenic switch in breast cancer.

EXTRATUMORAL INFLUENCES AS MEDIATORS OF THE ANGIOGENIC SWITCH

Although the theory of the angiogenic switch initially presupposed that angiogenic factors were produced by the tumor cells, evidence exists that macrophages or other immune cells in the area may also be associated with increased angiogenesis (Arbiser et al, 1997). These immune cells may be the source of the angiogenic factors (Freeman et al, 1995; Lewis et al, 1995; Takahashi et al, 1996) or of other proangiogenic molecules such as fibrin (Shoji et al, 1998). In one study, the number of tumor-associated macrophages was correlated with both microvessel density and poor prognosis (Leek et al, 1996). Additionally, patients with breast tumors containing associated monocytes that expressed thymidine phosphorylase, a proangiogenic enzyme (below), had significantly poorer prognosis than those whose tumor-associated monocytes were negative for this enzyme (Toi et al, 1999).

Other stromal cells or the extracellular matrix may also contribute to tumor angiogenesis. Mammary fibroblasts isolated from breast tumors showed an increased ability to upregulate VEGF in response to hypoxia, compared with fibroblasts isolated from normal breast (Hlatky et al, 1994). Moreover, evidence in a transgenic mouse model expressing a reporter construct coupled to promoter sequences of VEGF suggests that fibroblasts in the area of a tumor rather than the tumor cells themselves may be responsible for VEGF synthesis (Fukumura et al, 1998). These data are countered by experiments performed in a model of a VEGF null fibrosarcoma implanted into immunosuppressed mice, which did not upregulate stromal VEGF expression. These experiments showed that, in spite of significant tissue hypoxia adjacent to the tumor, lack of tumor-derived VEGF resulted in smaller tumors with fewer, less fenestrated blood vessels (Grunstein et al, 1999). FGFs may be stored in extracellular matrix and activated by enzymes that degrade matrix (Vlodavsky et al, 1996). Alternatively, proteins that bind to the FGF may make it more available to FGF receptors (Czubayko et al, 1997).

Multiple physiologic and pathologic pathways may control expression of angiogenesis stimulators and inhibitors. These pathways may be operative in several cell types within a tumor and may control expression of multiple molecules that may influence tumor-associated angiogenesis. Therefore, it is doubtful that any one angiogenesis stimulator or inhibitor will be identified as being central to angiogenesis associated with breast cancer.

MIMICKING THE ANGIOGENIC SWITCH IN ANIMAL MODELS

Mimicking the angiogenic switch by directed expression of various angiogenic factors in breast carcinoma cells in animal models has produced evidence of increased tumorigenicity or metastatic capability. Conversely, directed expression of particular angiogenesis inhibitors has resulted in decreased tumorigenicity or abrogation of metastasis. These experimental approaches can provide evidence that a particular factor is capable of increasing or decreasing the neoangiogenic capability of experimental breast carcinomas in animals. These approaches also testify to the importance of angiogenesis in breast tumor progression. It is difficult to judge the applicability of findings from animal models to the situation in human breast cancer, especially since many of the studies employ immunodeficient mice whose ability to reject tumors is impaired by definition. Nonetheless, these types of studies cannot be performed in humans, and they have greatly increased our knowledge of the effects of various factors.

Manipulation of Endogenous Angiogenesis Stimulators in Animal Models

Transfection of FGF-4 or FGF-1 into MCF-7 breast carcinoma cells has led to tumor progression in immunosuppressed mice manifested as increased tumorigenicity, estrogen independence and tamoxifen resistance, or increased metastatic capability (Kurebayashi et al, 1993; McLeskey et al, 1993; Zhang et al, 1997; Hajitou et al, 1998). Moreover, transfection with FGF-4 of the nontumorigenic breast epithelial cell line HBL100 resulted in the acquisition of a tumorigenic phenotype (Soutou et al, 1996). However, FGFs are growth factors whose action is not solely specific for endothelial cells, and their effects might be due to autocrine growth stimulation of the tumor cells or to effects on other stromal cells in addition to endothelial cells. Transfection of a dominant negative FGF receptor construct has shown that, in the case of FGF-1 transfection of MCF-7 breast carcinoma cells, paracrine effects, possibly including angiogenesis, are primarily responsible for the tamoxifen-resistant phenotype conferred by transfection of FGF-1 (Zhang et al, 1998).

Since expression of the known VEGF receptors is largely restricted to endothelial cells, one might expect that if increased angiogenesis is responsible for tumor progression, transfection with VEGF might produce a phenotypic change similar to the FGF transfections. VEGF has multiple isoforms that seem to have distinct effects (reviewed in Neufeld et al, 1999), implying that directed expression of a particular isoform might produce results that are different from those observed with another isoform. However, directed expression of either $VEGF_{121}$ or $VEGF_{165}$ in breast carcinoma cells has not produced the same magnitude of phenotypic change as observed with the FGF transfections. Transfection of $VEGF_{121}$ into MCF-7 cells produced a small increase in tumor growth but not estrogen independence (Zhang et al, 1995). In contrast, substantially increased tumor size coupled with tamoxifen resistance but not estrogen independence was observed with transfection of the $VEGF_{165}$ isoform into the same cell line. These transfectants also formed micrometastases, similar to the FGF transfectants (Bullocks et al, 1997). Retroviral transduction of a tetracycline-inducible vector driving expression of an unspecified isofrom of VEGF also has been used in T47D breast carcinoma cells. This experiment produced a clonal cell line, which exhibited a large increase in xenograft tumor size when compared with control transductants, and a somewhat smaller increase in tumor size when compared with wild-type T47D cells (Yoshiji et al, 1997), raising the question of whether the observed increased tumorigenicity of the VEGF transduced cells might be due to clonal variation. In any case, since VEGF has few effects other than increasing angiogenesis, comparison of the effects of directed expression of FGF or VEGF in breast carcinoma cells confirms the importance of multiple paracrine activities of the FGFs in promoting tumor growth.

Thymidine phosphorylase, also known as platelet-derived endothelial cell growth factor (PDECGF), is an enzyme of the pyrimidine salvage pathway; it is highly expressed in breast cancer epithelium (Toi et al, 1995b) and in the surrounding endothelium (Fox et al, 1996). Its expression in the endothelium has been correlated with a ring-like formation of blood vessels around ductal carcinoma in situ lesions (Engels et al, 1997a). Expression of thymidine phosphorylase is correlated with prognosis in breast cancer (Nagaoka et al, 1998; Gasparini et al, 1999) and has been specifically correlated with microvessel density (Toi et al, 1995b; Nagaoka et al, 1998). The most likely explanation is that 2-deoxy-D-ribose, the catabolic product of thymidine produced by thymidine phosphorylase, is an endothelial chemoattractant (reviewed in Brown and Bicknell, 1998). Expression of thymidine phosphorylase is induced by hypoxia and by low pH in the MDA-MB 231 breast carcinoma cell line in vitro. Xenograft tumors, produced by the same cell line, showed upregulated expression of thymidine phosphorylase adjacent to necrotic areas of tumor, or in whole tumors whose blood supply was occluded (Griffiths et al, 1997). Transfection of MCF-7 breast carcinoma cells with an expression vector for thymidine phosphorylase did not affect in vitro growth; however, it increased tumorigenicity (Moghaddam et al, 1995).

In summary, increasing the angiogenic potential of experimental tumors has resulted in increased tumorigenicity in many models. However, of the angiogenic factors reviewed, only the transfection of FGF-1, FGF-4 (McLeskey et al, 1993; Zhang et al, 1997), or $VEGF_{165}$ (Bullocks et al, 1997) increased metastasis as well as tumor size. Moreover, the magnitude of increase in tumor size reported for the FGF transfections in the poorly tumorigenic MCF-7 cell line is larger than for any other angiogenic factor reported. In addition, the tamoxifen resistance reported for these transfections raises the issue of the contribution of angiogenesis to hormonal responsiveness.

Manipulation of Endogenous Angiogenesis Inhibitors in Animal Models

The most studied putative angiogenesis inhibitor, thrombospondin-1, is a very large matrix protein, capable of interactions with multiple receptors on multiple cell types, components of the extracellular matrix, growth factors, and proteases. Therefore, its functions are complex and its effects are often seemingly contradictory (Bornstein, 1995; Roberts, 1996). Thrombospondin added to tissue culture systems has been shown to inhibit endothelial cell proliferation (Taraboletti et al, 1990), sprouting (Canfield and Schor, 1995), migration toward an angiogenic stimulus, and tube formation (Rastinejad et al, 1989; Good et al, 1990; Taraboletti et al, 1990; Iruela-Arispe et al, 1991; Tolsma et al, 1993). Somewhat paradoxically, at high concentrations it is chemotactic for endothelial cells (Taraboletti et al, 1990). Also paradoxically, despite its putative role as an angiogenesis inhibitor, expression of thrombospondin is upregulated in breast cancer, compared to normal breast or to benign breast disease (Tuszynski and Nicosia, 1994; Bertin et al, 1997). Moreover, expression of neither thrombospondin-1 nor -2 is correlated with any known prognostic factor (Bertin et al, 1997). In addition, both breast cancer cells and endothelial cells adhere to thrombospondin (Taraboletti et al, 1990; Incardona et al, 1995; Chen et al, 1996b), and breast cancer cells may adhere to endothelial cells via thrombospondin (Incardona et al, 1995), implying that it has a function in motility of both cell types and in metastasis of cancer cells. Also, in vitro, thrombospondin increases the invasive capacity of breast carcinoma cells into collagen (Wang et al, 1996a). Furthermore, thrombospondin upregulates the expression of members of the plasminogen and metalloproteinase systems (Bagavandoss et al, 1993; Arnoletti et al, 1995; Qian et al, 1997), implying that it has an effect on proteolysis that might mediate the invasion of endothelial cells into matrix during the formation of new vessels. However, transfection of thrombospondin-1 into a metastatic breast carcinoma line inhibited both tumor growth and metastasis, with a corresponding decrease in angiogenesis (Weinstat-Saslow et al, 1994). Moreover, an antithrombospondin receptor antibody prevented metastasis of a xenograft animal model of breast cancer (Wang et al, 1996b). Thrombospondin expression is negatively regulated by p53 (Dameron et al, 1994a,b; Stellmach et al, 1996), and loss of expression of thrombospondin has been shown to be associated with an angiogenic, metastatic phenotype to experimentally induced tumors (Volpert et al, 1997). Also, purified thrombospondin or thrombospondin-derived peptides were shown to inhibit in vivo angiogenesis in the cornea model that was stimulated by FGF-2 (Rastinejad et al, 1989; Good et al, 1990; Tolsma et al, 1993). Thus, the complex effects of thrombospondin make it difficult to predict what effect it might have on tumor progression or metastasis.

Recently, two novel proteins have been described with homology to the above-mentioned, antiangiogenic, thrombospondin-derived peptides. These proteins, METH-1 and METH-2, contain metalloproteinase and disintegrin domains in addition to areas of homology to thrombospondin. They were shown to inhibit VEGF-induced angiogenesis in a chick chorioallantoic membrane (CAM) assay, and to inhibit FGF-2–stimulated corneal angiogenesis (Vazquez et al, 1999). Data concerning their in vivo function in animal models of breast or other tumors have not been published to date.

The endogenous angiogenesis inhibitors angiostatin and endostatin were discovered in animal tumor systems in which the primary tumor seemed to inhibit the growth of metastases. These molecules are cleavage products of plasminogen and collagen XVIII, respectively (O'Reilly et al, 1994, 1997). Multiple proteases can generate angiostatin and/or endostatin from their larger precursors (Gately et al, 1996; Dong et al, 1997; Gately et al, 1997; Patterson and Sang, 1997; Stathakis et al, 1997; Cornelius et al, 1998; O'Mahony et al, 1998b). Moreover, the formation of angiostatin may be regulated (O'Mahony et al, 1998a). Unfortunately, since they are identical to parts of larger molecules, detection of their presence in normal tissues or in disease states presents a challenge. Therefore, their normal role in the angiogenesis of wound healing, development, or cancer is unclear. Nonetheless, when administered exogenously or expressed, they have been shown to inhibit tumor growth. These experiments have utilized angiostatin or endostatin alone (O'Reilly et al, 1994, 1996, 1997; Boehm et al, 1997; Lannutti et al, 1997; Sim et al, 1997; Wu et al, 1997; Cao et al, 1998; Griscelli et al, 1998; Tanaka et al, 1998; Chen et al, 1999; Dhanabal et al, 1999; Yamaguchi et al, 1999) or in combination with radiation (Gorski et al, 1998; Mauceri et al, 1998) in a variety of animal models, including xenografts of human MDA-MB 231 (Griscelli et al, 1998) or MDA-MB 435 (Chen et al, 1999) breast carcinoma cells. Their activity may be due to inhibition of endothelial cell proliferation (O'Reilly et al, 1994, 1997; Griscelli et al, 1998). However, angiostatin also inhibits endothelial cell migration (Claesson-Welsh et al, 1998; Ji et al, 1998) or endothelial cell tube formation (Claesson-Welsh et al, 1998), and induces endothelial cell apoptosis (Claesson-Welsh et al, 1998; Lucas et al, 1998). Endostatin also inhibits endothelial cell migration (Yamaguchi et al, 1999). Currently, efforts are under way to identify structural components of these molecules that produce their effects (Boehm et al, 1998; Hohenester et al, 1998; Ji et al, 1998). In this regard, angiostatin has been shown to bind to cell surface adenosine triphosphate (ATP) synthase on endothelial cells; its ability to inhibit endothelial cell proliferation was antagonized by an antibody to the α-subunit of this protein (Moser et al, 1999). Collectively, these data raise the possibility of the development of small molecule congeners of angiostatin.

In summary, increasing expression of angiogenesis inhibitors can abrogate tumor growth and may shrink tumors. In one study, using the murine Lewis lung carcinoma, such a strategy did not induce the drug resistance (Boehm et al, 1997) commonly seen with chemotherapeutic agents that target the tumor cells themselves. Although such findings need to be extended to animal models of human breast cancer and to humans, it is intriguing to speculate that such therapies will be effective over long periods of time and in a wide variety of human tumors, including breast.

CONCLUSION

Although multiple mechanisms for increase in expression of angiogenic stimulators or decrease in expression of angiogenic inhibitors have been elucidated in many tumor types, their relevance to breast cancer is still under investigation. Nonetheless, animal models provide abundant data that the angiogenic switch is an important mechanism in breast cancer progression. The data implicating angiogenesis in tamoxifen resistance are particularly intriguing, but need further confirmation and clarification.

REFERENCES

Agani, F., Kirsch, D. G., Friedman, S. L., Kastan, M. B., and Semenza, G. L. 1997. p53 does not repress hypoxia induced transcription of the vascular endothelial growth factor gene. *Cancer Res.* 57:4474–4477.

Akiri, G., Nahari, D., Finkelstein, Y., Le, S.-Y., Elroy-Stein, O., and Levi, B.-Z. 1998. Regulation of vascular endothelial growth factor (VEGF) expression is mediated by internal initiation of translation and alternative initiation of transcription. *Oncogene* 17:227–236.

Anthony, B., Carter, P., and de Benedetti, A. 1996. Overexpression of the proto-oncogene/translation factor 4E in breast-carcinoma cell lines. *Int. J. Cancer* 65:858–863.

Arany, Z., Huang, L. E., Eckner, R., Bhattacharya, S., Jiang, C., Goldberg, M. A., Bunn, H. F., and Livingston, D. M. 1996. An essential role for p300/CBP in the cellular response to hypoxia. *Proc. Natl. Acad. Sci. USA* 93:12969–12973.

Arbiser, J. L., Moses, M. A., Fernandez, C. A., Ghiso, N., Cao, Y., Klauber, N., Frank, D., Brownlee, M., Flynn, E., Parangi, S., Byers, H. R., and Folkman, J. 1997. Oncogenic H-ras stimulates tumor angiogenesis by two distinct pathways. *Proc. Natl. Acad. Sci. USA* 94:861–866.

Arnoletti, J. P., Albo, D., Granick, M. S., Solomon, M. P., Gastiglioni, A., Rothman, V. L., and Tuszynski, G. P. 1995. Thrombospondin and transforming growth factor-beta 1 increase expression of urokinase-type plasminogen activator and plasminogen activator inhibitor-1 in human MDA-MB-231 cells. *Cancer* 76:998–1005.

Bagavandoss, P., Kaytes, P., Vogeli, G., Wells, P. A., and Wilks, J. W. 1993. Recombinant truncated thrombospondin-1 monomer modulates endothelial cell plasminogen activator inhibitor 1 accumulation and proliferation in vitro. *Biochem. Biophys. Res. Commun.* 192:325–332.

Bernstein, J., Sella, O., Le, S.-Y., and Elroy-Stein, O. 1997. PDGF2/c-*sis* mRNA leader contains a differentiation-linked internal ribosomal entry site (D-IRES). *J. Biol. Chem.* 272:9356–9362.

Bernstein, J., Shefler, I., and Elroy-Stein, O. 1995. The translational repression mediated by the platelet-derived growth factor 2/c-*sis* mRNA leader is relieved during megakaryocytic differentiation. *J. Biol. Chem.* 270:10559–10565.

Bertin, N., Clezardin, P., Kubiak, R., and Frappart, L. 1997. Thrombospondin-1 and -2 messenger RNA expression in normal, benign, and neoplastic human breast tissues: correlation with prognostic factors, tumor angiogenesis, and fibroblastic desmoplasia. *Cancer Res.* 57:396–399.

Bikfalvi, A. 1995. Significance of angiogenesis in tumour progression and metastasis. *Eur. J. Cancer* 31A:1101–1104.

Birch, J. M., Blair, V., Kelsey, A. M., Evans, D. G., Harris, M., Tricker, K. J., and Varley, J. M. 1998. Cancer phenotype correlates with constitutional TP53 genotype in families with the Li-Fraumeni syndrome. *Oncogene* 17:1061–1068.

Biscardi, J. S., Tice, D. A., and Parsons, S. J. 1999. c-Src, receptor tyrosine kinases, and human cancer. *Adv. Cancer Res.* 76:61–119.

Blagosklonny, M. V., An, W. G., Romanova, L. Y., Trepel, J., Fojo, T., and Neckers, L. 1998. p53 inhibits hypoxia-inducible factor-stimulated transcription. *J. Biol. Chem.* 273:11995–11998.

Blam, S. B., Mitchell, R., Tischer, E., Rubin, J. S., Silva, M., Silver, S., Fiddes, J. C., Abraham, J. A., and Aaronson, S. A. 1988. Addition of growth hormone secretion signal to basic fibroblast growth factor results in cell transformation and secretion of aberrant forms of the protein. *Oncogene* 3:129–136.

Boehm, T., Folkman, J., Browder, T., and O'Reilly, M. S. 1997. Antiangiogenic therapy of experimental cancer does not induce acquired drug resistance. *Nature* 390:404–407.

Boehm, T., O'Reilly, M. S., Keough, K., Shiloach, J., Shapiro, R., and Folkman, J. 1998. Zinc-binding of endostatin is essential for its antiangiogenic activity. *Biochem. Biophys. Res. Commun.* 252:190–194.

Bornstein, P. 1995. Diversity of function is inherent in matricellular proteins: an appraisal of thrombospondin 1. *J. Cell Biol.* 130:503–506.

Bouvet, M., Ellis, L. M., Nishizaki, M., Fujiwara, T., Liu, W., Bucana, C. D., Fang, B., Lee, J. J., and Roth, J. A. 1998. Adenovirus-mediated wild-type p53 gene transfer down-regulates vascular endothelial growth factor expression and inhibits angiogenesis in human colon cancer. *Cancer Res.* 58:2288–2292.

Brem, S. S., Gullino, P. M., and Medina, D. 1977. Angiogenesis: a marker for neoplastic transformation of papillary hyperplasia. *Science* 195:880–881.

Brown, L. F., Berse, B., Jackman, R. W., Tognazzi, K., Guidi, A. J., Dvorak, H. F., Senger, D. R., Connolly, J. L., and Schnitt, S. J. 1995. Expression of vascular permeability factor (vascular endothelial growth factor) and its receptors in breast cancer. *Hum. Pathol.* 26:86–91.

Brown, N. S., and Bicknell, R. 1998. Thymidine phosphorylase, 2-deoxy-ribose and angiogenesis. *Biochem. J.* 334:1–8.

Bullocks, J., Zhang, L., Ding, I. Y. F., McLeskey, S. W., Tobias, C. A., Miller, D. L., and Kern, F. G. 1997. Overexpression of vascular endothelial cell growth factor (VEGF) in MCF-7 breast carcinoma cells facilitates growth in tamoxifen-treated nude mice and tumor cell dissemination. Proc. *Am. Assoc. Cancer Res.* 38:3521.

Canfield, A. E., and Schor, A. M. 1995. Evidence that tenascin and thrombospondin-1 modulate sprouting of endothelial cells. *J. Cell Sci.* 108:797–809.

Cao, Y., O'Reilly, M. S., Marshall, B., Flynn, E., Ji, R.-W., and Folkman, J. 1998. Expression of angiostatin cDNA in a murine fibrosarcoma suppresses primary tumor growth and produces long-term dormancy of metastases. *J. Clin. Invest.* 101:1055–1063.

Caré, A., Silvani, A., Meccia, E., Mattia, G., Peschle, C., and Colombo, M. P. 1998. Transduction of the SkBr3 breast carcinoma cell line with the HOXB7 gene induces bFGF expression, increases cell proliferation and reduces growth factor dependence. *Oncogene* 16:3285–3289.

Caré, A., Silvani, A., Meccia, E., Mattia, G., Stoppacciaro, A., Parmiani, G., Peschle, C., and Colombo, M. P. 1996. HOXB7 constitutively activates basic fibroblast growth factor in melanomas. *Mol. Cell. Biol.* 16:4842–4851.

Carmeliet, P., Dor, Y., Herbert, J. M., et al. 1998. Role of HIF-1alpha in hypoxia-mediated apoptosis, cell proliferation and tumour angiogenesis. *Nature* 394:485–490.

Chariot, A., Senterre-Lesenfants, S., Sobel, M. E., and Castronovo, V. 1998. Molecular cloning of a mutated HOXB7 cDNA encoding a truncated transactivating homeodomain-containing protein. *J. Cell Biochem.* 71:46–54.

Chen, Q. R., Kumar, D., Stass, S. A., and Mixson, A. J. 1999. Liposomes complexed to plasmids encoding angiostatin and endostatin inhibit breast cancer in nude mice. *Cancer Res.* 59:3308–3312.

Chen, T. K., Smith, L. M., Gebhardt, D. K., Birrer, M. J., and Brown, P. H. 1996a. Activation and inhibition of the AP-1 complex in human breast cancer cells. *Mol. Carcinog.* 15:215–226.

Chen, Z.-S., Pohl, J., Lawley, T. J., and Swerlick, R. A. 1996b. Human microvascular endothelial cells adhere to thrombospondin-1 via an RGD/CSVTCG domain independent mechanism. *J. Invest. Dermatol.* 106:215–220.

Claesson-Welsh, L., Welsh, M., Ito, N., Anand-Apte, B., Soker, S., Zetter, B., O'Reilly, M., and Folkman, J. 1998. Angiostatin induces endothelial cell apoptosis and activation of

focal adhesion kinase independently of the integrin-binding motif RGD. *Proc. Natl. Acad. Sci. USA* 95:5579–5583.
Clark, G. J., and Der, C. J. 1995. Aberrant function of the Ras signal transduction pathway in human breast cancer. *Breast Cancer Res. Treat.* 35:133–144.
Cornelius, L. A., Nehring, L. C., Harding, E., Bolanowski, M., Welgus, H. G., Kobayashi, D. K., Pierce, R. A., and Shapiro, S. D. 1998. Matrix metalloproteinases generate angiostatin: effects on neovascularization. *J. Immunol.* 161:6845–6852.
Czubayko, F., Liaudet-Coopman, E. D. E., Aigner, A., Tuveson, A. T., Berchem, G. J., and Wellstein, A. 1997. A secreted FGF-binding protein can serve as the angiogenic switch in human cancer. *Nat. Med.* 10:1137–1140.
D'Amore, P. A., and Shima, D. T. 1996. Tumor angiogenesis: a physiological process or genetically determined? *Cancer Metastasis Rev.* 15:205–212.
Dameron, K. M., Volpert, O. V., Tainsky, M. A., and Bouck, N. 1994a. The *p53* tumor suppressor gene inhibits angiogenesis by stimulating the production of thrombospondin. *Cold Spring Harb. Symp. Quant. Biol.* 59:483–489.
Dameron, K. M., Volpert, O. V., Tainsky, M. A., and Bouck, N. 1994b. Control of angiogenesis in fibroblasts by p53 regulation of thrombospondin-1. *Science* 265:1582–1584.
Damert, A., Ikeda, E., and Risau, W. 1997. Activator-protein-1 binding potentiates the hypoxia-inducible factor-1-mediated hypoxia-induced transcriptional activation of vascular-endothelial growth factor expression in C6 glioma cells. *Biochem. J.* 327:419–423.
de Jong, J. S., van Diest, P. J., van der Valk, P., and Baak, J. P. A. 1998. Expression of growth factors, growth inhibiting factors, and their receptors in invasive breast cancer. I: An inventory in search of autocrine and paracrine loops. *J. Pathol.* 184:44–52.
Delli-Bovi, P., Curatola, A. M., Kern, F. G., Greco, A., Ittmann, M., and Basilico, C. 1987. An oncogene isolated by transfection of Kaposi's sarcoma DNA encodes a growth factor that is member of the FGF family. *Cell* 50:729–737.
Dhanabal, M., Ramchandran, R., Volk, R., Stillman, I. E., Lombardo, M., Iruela-Arispe, M. L., Simons, M., and Sukhatme, V. P. 1999. Endostatin: yeast production, mutants, and antitumor effect in renal cell carcinoma. *Cancer Res.* 59:189–197.
Dickson, C., Smith, R., Brookes, S., and Peters, G. 1984. Tumorigenesis by mouse mammary tumor virus: proviral activation of a cellular gene in the common integration region int-2. *Cell* 37:529–536.
Dong, Z., Kumar, R., Yang, X., and Fidler, I. J. 1997. Macrophage-derived metalloelastase is responsible for the generation of angiostatin in Lewis lung carcinoma. *Cell* 88:801–810.
Duan, D. R., Pause, A., Burgess, W. H., Aso, T., Chen, D. Y. T., Garrett, K. P., Conaway, R. C., Conaway, J. W., Linehan, W. M., and Klausner, R. D. 1995. Inhibition of transcription elongation by the VHL tumor suppressor protein. *Science* 269:1402–1406.
Dumont, J. A., Bitonti, A. J., Wallace, C. D., Baumann, R. J., Cashman, E. A., Gross-Doersen, D. E., and Cross-Doersen, D. E. 1996. Progression of MCF-7 breast cancer cells to antiestrogen-resistant phenotype is accompanied by elevated levels of AP-1 DNA binding activity. *Cell Growth Differ.* 7:351–359.
Ellis, L. M., Staley, C. A., Liu, W., Fleming, R. Y. D., Parikh, N. U., Bucana, C. D., and Gallick, G. E. 1998. Down-regulation of vascular endothelial growth factor in a human colon carcinoma cell line transfected with an antisense expression vector specific for c-*src*. *J. Biol. Chem.* 273:1052–1057.
Ema, M., Taya, S., Yokotani, N., Sogawa, K., Matsuda, Y., and Fujii-Kuriyama, Y. 1997. A novel bHLH-PAS factor with close sequence similarity to hypoxia-inducible factor 1 alpha regulates the VEGF expression and is potentially involved in lung and vascular development. *Proc. Natl. Acad. Sci. USA* 94:4273–4278.
Engels, K., Fox, S. B., Whitehouse, R. M., Gatter, K. C., and Harris, A. L. 1997a. Up-regulation of thymidine phosphorylase expression is associated with a discrete pattern of angiogenesis in ductal carcinomas in situ of the breast. *J. Pathol.* 182:414–420.

Engels, K., Fox, S. B., Whitehouse, R. M., Gatter, K. C., and Harris, A. L. 1997b. Distinct angiogenic patterns are associated with high-grade *in situ* ductal carcinomas of the breast. *J. Pathol.* 181:207–212.

Eppenberger, U., Kueng, W., Schlaeppi, J. M., Roesel, J. L., Benz, C., Mueller, H., Matter, A., Zuber, M., Luescher, K., Litschgi, M., Schmitt, M., Foekens, J. A., and Eppenberger-Castori, S. 1998. Markers of tumor angiogenesis and proteolysis independently define high- and low-risk subsets of node-negative breast cancer patients. *J. Clin. Oncol.* 16:3129–3136.

Feldkamp, M. M., Lau, N., Rak, J., Kerbel, R. S., and Guha, A. 1999. Normoxic and hypoxic regulation of vascular endothelial growth factor (VEGF) by astrocytoma cells is mediated by Ras. *Int. J. Cancer* 81:118–124.

Feldser, D., Agani, F., Iyer, N. V., Pak, B., Ferreira, G., and Semenza, G. L. 1999. Reciprocal positive regulation of hypoxia-inducible factor 1 alpha and insulin-like growth factor 2. *Cancer Res.* 59:3915–3918.

Flamme, I., Krieg, M., and Plate, K. H. 1998. Up-regulation of vascular endothelial growth factor in stromal cells of hemangioblastomas is correlated with up-regulation of the transcription factor HRF/HIF-2α. *Am. J. Pathol.* 153:25–29.

Folkman, J., Watson, K., Ingber, D., and Hanahan, D. 1989. Induction of angiogenesis during the transition from hyperplasia to neoplasia. *Nature* 339:58–61.

Forough, R., Zhan, X., MacPhee, M., Friedman, S., Engleka, K. A., Sayers, T., Wiltrout, R. H., and Maciag, T. 1993. Differential transforming abilities of non-secreted and secreted forms of human fibroblast growth factor-1. *J. Biol. Chem.* 268:2960–2968.

Forsythe, J. A., Jiang, B.-H., Iyer, N. V., Agani, F., Leung, S. W., Koos, R. D., and Semenza, G. L. 1996. Activation of vascular endothelial growth factor gene transcription by hypoxia-inducible factor 1. *Mol. Cell. Biol.* 16:4604–4613.

Fox, S. B., Westwood, M., Moghaddam, A., Comley, M., Turley, H., Whitehouse, R. M., Bicknell, R., Gatter, K. C., and Harris, A. L. 1996. The angiogenic factor platelet-derived endothelial cell growth factor/thymidine phosphorylase is upregulated in breast cancer epithelium and endothelium. *Br. J. Cancer* 73:275–280.

Freeman, M. R., Schneck, F. X., Gagnon, M. L., Corless, C., Soker, S., Niknejad, K., Peoples, G. E., and Klagsbrun, M. 1995. Peripheral blood T lymphocytes and lymphocytes infiltrating human cancers express vascular endothelial growth factor: a potential role for T cells in angiogenesis. *Cancer Res.* 55:4140–4145.

Fregene, T. A., Kellogg, C. M., and Pienta, K. J. 1994. Microvessel quantification as a measure of angiogenic activity in benign breast tissues lesions: a marker for precancerous disease? *Int. J. Oncol.* 4:1199–1202.

Fukumura, D., Xavier, R., Sugiura, T., Chen, Y., Park, E. C., Lu, N., Selig, M., Nielsen, G., Taksir, T., Jain, R. K., and Seed, B. 1998. Tumor induction of VEGF promoter activity in stromal cells. *Cell* 94:715–725.

Gasparini, G., Toi, M., Gion, M., Verderio, P., Dittadi, R., Hanatani, M., Matsubara, I., Vinante, O., Bonoldi, E., Boracchi, P., Gatti, C., Suzuki, H., and Tominaga, T. 1997. Prognostic significance of vascular endothelial growth factor protein in node-negative breast carcinoma. *J. Natl. Cancer Inst.* 89:139–147.

Gasparini, G., Toi, M., Miceli, R., Vermeulen, P. B., Dittadi, R., Biganzoli, E., Morabito, A., Fanelli, M., Gatti, C., Suzuki, H., Tominaga, T., Dirix, L. Y., and Gion, M. 1999. Clinical relevance of vascular endothelial growth factor and thymidine phosphorylase in patients with node-positive breast cancer treated with either adjuvant chemotherapy or hormone therapy. *Cancer J. Sci. Am.* 5:101–111.

Gately, S., Twardowski, P., Stack, M. S., Patrick, M., Boggio, L., Cundiff, D. L., Schnaper, H. W., Madison, L., Volpert, O., Bouck, N., Enghild, J., Kwaan, H. C., and Soff, G. A. 1996. Human prostate carcinoma cells express enzymatic activity that converts human plasminogen to the angiogenesis inhibitor, angiostatin. *Cancer Res.* 56:4887–4890.

Gately, S., Twardowski, P., Stack, M. S., Cundiff, D. L., Grella, D., Castellino, F. J., Enghild, J., Kwaan, H. C., Lee, F., Kramer, R. A., Volpert, O., Bouck, N., and Soff, G. A. 1997. The mechanism of cancer-mediated conversion of plasminogen to the angiogenesis inhibitor angiostatin. *Proc. Natl. Acad. Sci. USA* 94:10868–10872.

Gimbrone, M., Leapman, S. B., Cotran, R. S., and Folkman, J. 1972. Tumor dormancy in vivo by prevention of neovascularization. *J. Exp. Med.* 136:261–276.

Gnarra, J. R., Zhou, S., Merrill, M. J., Wagner, J. R., Krumm, A., Papavassiliou, E., Oldfield, E. H., Klausner, R. D., and Linehan, W. M. 1996. Post-transcriptional regulation of vascular endothelial growth factor mRNA by the product of the *VHL* tumor suppressor gene. *Proc. Natl. Acad. Sci. USA* 93:10589–10594.

Good, D. J., Polverini, P. J., Rastinejad, F., Le Beau, M. M., Lemons, R. S., Frazier, W. A., and Bouck, N. P. 1990. A tumor suppressor-dependent inhibitor of angiogenesis is immunologically and functionally indistinguishable from a fragment of thrombospondin. *Proc. Natl. Acad. Sci. USA* 87:6624–6628.

Gorski, D. H., Mauceri, H. J., Salloum, R. M., Gately, S., Hellman, S., Beckett, M. A., Sukhatme, V. P., Soff, G. A., Kufe, D. W., and Weichselbaum, R. R. 1998. Potentiation of the antitumor effect of ionizing radiation by brief concomitant exposures to angiostatin. *Cancer Res.* 58:5686–5689.

Graff, J. R., Boghaert, E. R., de Benedetti, A., Tudor, D. L., Zimmer, C. C., Chan, S. K., and Zimmer, S. G. 1995. Reduction of translation initiation factor 4E decreases the malignancy of *ras*-transformed cloned rat embryo fibroblasts. *Int. J. Cancer* 60:255–263.

Griffiths, L., Dachs, G. U., Bicknell, R., Harris, A. L., and Stratford, I. J. 1997. The influence of oxygen tension and pH on the expression of platelet-derived endothelial cell growth factor/thymidine phosphorylase in human breast tumor cells grown in vitro and in vivo. *Cancer Res.* 57:570–572.

Griscelli, F., Li, H., Bennaceur-Griscelli, A., Soria, J., Opolon, P., Soria, C., Perricaudet, M., Yeh, P., and Lu, H. 1998. Angiostatin gene transfer: inhibition of tumor growth *in vivo* by blockage of endothelial cell proliferation associated with a mitosis arrest. *Proc. Natl. Acad. Sci. USA* 95:6367–6372.

Grunstein, J., Roberts, W. G., Mathieu-Costello, O., Hanahan, D., and Johnson, R. S. 1999. Tumor-derived expression of vascular endothelial growth factor is a critical factor in tumor expansion and vascular function. *Cancer Res.* 59:1592–1598.

Guidi, A. J., Fischer, L., Harris, J. R., and Schnitt, S. J. 1994. Microvessel density and distribution in ductal carcinoma in situ of the breast. *J. Natl. Cancer Inst.* 86:614–619.

Guidi, A. J., Schnitt, S. J., Fischer, L., Tognazzi, K., Harris, J. R., Dvorak, H. F., and Brown, L. F. 1997. Vascular permeability factor (vascular endothelial growth factor) expression and angiogenesis in patients with ductal carcinoma in situ of the breast. *Cancer* 80:1945–1953.

Hajitou, A., Deroanne, C. F., Noël, A., Colette, J., Nusgens, B. V., Foidart, J. M., and Calberg-Bacq, C.-M. 1998. FGF-3 increases tumorigenicity in MCF7 breast cancer cells but FGF-4 effect is more pronounced and associated with VEGF upregulation. *Proc. Am. Assoc. Cancer Res.* 39:34.

Hlatky, L., Tsionou, C., Hahnfeldt, P., and Coleman, C. N. 1994. Mammary fibroblasts may influence breast tumor angiogenesis via hypoxia-induced vascular endothelial growth factor up-regulation and protein expression. *Cancer Res.* 54:6083–6086.

Hohenester, E., Sasaki, T., Olsen, B. R., and Timpl, R. 1998. Crystal structure of the angiogenesis inhibitor endostatin at 1.5 A resolution. *EMBO J.* 17:1656–1664.

Holmgren, L., Jackson, G., and Arbiser, J. 1998. p53 induces angiogenesis-restricted dormancy in a mouse fibrosarcoma. *Oncogene* 17:819–824.

Iliopoulos, O., Levy, A. P., Jiang, C., Kaelin Jr., W. G., and Goldberg, M. A. 1996. Negative regulation of hypoxia-inducible genes by the von Hippel-Lindau protein. *Proc. Natl. Acad. Sci. USA* 93:10595–10599.

Incardona, F., Lewalle, J. M., Morandi, V., Lambert, S., Legrand, Y., Foidart, J. M., and Legrand, C. 1995. Thrombospondin modulates human breast adenocarcinoma cell adhesion in human vascular endothelial cells. *Cancer Res.* 55:166–173.

Iruela-Arispe, M. L., Bornstein, P., and Sage, H. 1991. Thrombospondin exerts an antiangiogenic effect on cord formation by endothelial cells in vitro. *Proc. Natl. Acad. Sci. USA* 88:5026–5030.

Ji, W. R., Castellino, F. J., Chang, Y., Deford, M. E., Gray, H., Villarreal, X., Kondri, M. E., Marti, D. N., Llinas, M., Schaller, J., Kramer, R. A., and Trail, P. A. 1998. Characterization of kringle domains of angiostatin as antagonists of endothelial cell migration, an important process in angiogenesis. *FASEB J.* 12:1731–1738.

Jiang, B.-H., Agani, F., Passaniti, A., and Semenza, G. L. 1997. V-SRC induces expression of hypoxia-inducible factor 1 (HIF-1) and transcription of genes encoding vascular endothelial growth factor and enolase 1: involvement of HIF-1 in tumor progression. *Cancer Res.* 57:5328–5335.

Jiang, B.-H., Rue, E., Wang, G. L., Roe, R., and Semenza, G. L. 1996. Dimerization, DNA binding, and transactivation properties of hypoxia-inducible factor 1. *J. Biol. Chem.* 271:17771–17778.

Jones, R. M., Branda, J., Johnston, K. A., Polymenis, M., Gadd, M., Rustgi, A., Callanan, L., and Schmidt, E. V. 1996. An essential E box in the promoter of the gene encoding the mRNA cap-binding protein (eukaryotic initiation factor 4E) is a target for activation by c-myc. *Mol. Cell. Biol.* 16:4754–4764.

Jouanneau, J., Gavrilovic, J., Caruelle, D., Jaye, M., Moens, G., Caruelle, J. P., and Thiery, J. P. 1991. Secreted or nonsecreted forms of acidic fibroblast growth factor produced by transfected epithelial cells influence cell morphology, motility, and invasive potential. *Proc. Natl. Acad. Sci. USA* 88:2893–2897.

Kaelin, W. G., Iliopoulos, O., Lonergan, M., and Ohh, M. 1998. Functions of the von Hippel-Lindau tumor suppressor protein. *J. Intern. Med.* 243:535–539.

Kandel, J., Bossy-Wetzel, E., Radvanyi, F., Klagsbrun, M., Folkman, J., and Hanahan, D. 1991. Neovascularization is associated with a switch to the export of bFGF in the multistep development of fibrosarcoma. *Cell* 66:1095–1104.

Kevil, C. G., de Benedetti, A., Payne, D. K., Coe, L. L., Laroux, F. S., and Alexander, J. S. 1996. Translational regulation of vascular permeability factor by eukaryotic intitiation factor 4E: implications for tumor angiogenesis. *Int. J. Cancer* 65:785–790.

Kevil, C., Carter, P., Hu, B., and de Benedetti, A. 1995. Translational enhancement of FGF-2 by eIF-4 factors, and alternate utilization of CUG and AUG codons for translation initiation. *Oncogene* 11:2339–2348.

Kraggerud, S. M., Sandvik, J. A., and Pettersen, E. O. 1995. Regulation of protein synthesis in human cells exposed to extreme hypoxia. *Anticancer Res.* 15:683–686.

Kurebayashi, J., McLeskey, S. W., Johnson, M. D., Lippman, M. E., Dickson, R. B., and Kern, F. G. 1993. Quantitative demonstration of spontaneous metastasis by MCF-7 human breast cancer cells cotransfected with fibroblast growth factor 4 and *LacZ*. *Cancer Res.* 53:2178–2187.

Lannutti, B. J., Gately, S. T., Quevedo, M. E., Soff, G. A., and Paller, A. S. 1997. Human angiostatin inhibits murine hemangioendothelioma tumor growth in vivo. *Cancer Res.* 57:5277–5280.

Larcher, F., Robles, A. I., Duran, H., Murillas, R., Quintanilla, M., Cano, A., Conti, C. J., and Jorcano, J. L. 1996. Up-regulation of vascular endothelial growth factor/vascular permeability factor in mouse skin carcinogenesis correlates with malignant progression state and activated H-*ras* expression levels. *Cancer Res.* 56:5391–5396.

Lee, Y. J., Galoforo, S. S., Berns, C. M., Erdos, G., Gupta, A. K., Ways, D. K., and Corry, P. M. 1995. Effect of ionizing radiation on AP-1 binding activity and basic fibroblast growth factor gene expression in drug-sensitive human breast carcinoma MCF-7 and multidrug-resistant MCF-7/ADR cells. *J. Biol. Chem.* 270:28790–28796.

Leek, R. D., Lewis, C. E., Whitehouse, R., Greenall, M., Clarke, J., and Harris, A. L. 1996. Association of macrophage infiltration with angiogenesis and prognosis in invasive breast carcinoma. *Cancer Res.* 56:4625–4629.

Levy, A. P., Levy, N. S., and Goldberg, M. A. 1996a. Post-transcriptional regulation of vascular endothelial growth factor by hypoxia. *J. Biol. Chem.* 271:2746–2753.

Levy, A. P., Levy, N. S., and Goldberg, M. A. 1996b. Hypoxia-inducible protein binding to vascular endothelial growth factor mRNA and its modulation by the von Hippel-Lindau protein. *J. Biol. Chem.* 271:25492–25497.

Levy, A. P., Levy, N. S., Iliopoulos, O., Jiang, C., Kaelin Jr., W. G., and Goldberg, M. A. 1997. Regulation of vascular endothelial growth factor by hypoxia and its modulation by the von Hippel-Lindau tumor suppressor gene. *Kidney Int.* 51:575–578.

Levy, A. P., Levy, N. S., Wegner, S., and Goldberg, M. A. 1995. Transcriptional regulation of the rat vascular endothelial growth factor gene by hypoxia. *J. Biol. Chem.* 270:1333–1340.

Levy, N. S., Chung, S., Furneaux, H., and Levy, A. P. 1998. Hypoxic stabilization of vascular endothelial growth factor mRNA by the RNA-binding protein HuR. *J. Biol. Chem.* 273:6417–6423.

Lewis, C. E., Leek, R., Harris, A., and McGee, J. O. 1995. Cytokine regulation of angiogenesis in breast cancer: the role of tumor-associated macrophages. *J. Leukoc. Biol.* 57:747–751.

Li, B. D. L., Liu, L., Dawson, M., and de Benedetti, A. 1997. Overexpression of eukaryotic initiation factor 4E (eIF4E) in breast carcinoma. *Cancer* 79:2385–2390.

Li, B. D., McDonald, J. C., Nassar, R., and de Benedetti, A. 1998. Clinical outcome in stage I to III breast carcinoma and eIF4E overexpression. *Ann. Surg.* 227:756–761; discussion 761–763.

Lucas, R., Holmgren, L., Garcia, I., Jimenez, B., Mandriota, S. J., Borlat, F., Sim, B. K., Wu, Z., Grau, G. E., Shing, Y., Soff, G. A., Bouck, N., and Pepper, M. S. 1998. Multiple forms of angiostatin induce apoptosis in endothelial cells. *Blood* 92:4730–4741.

Malkin, D., Li, F. P., Strong, L. C., Fraumeni, J. F., Jr., Nelson, C. E., Kim, D. H., Kassel, J., Gryka, M. A., Bischoff, F. Z., and Tainsky, M. A. 1990. Germ line p53 mutations in a familial syndrome of breast cancer, sarcomas, and other neoplasms. *Science* 250:1233–1238.

Mauceri, H. J., Hanna, N. N., Beckett, M. A., Gorski, D. H., Staba, M. J., Stellato, K. A., Bigelow, K., Heimann, R., Gately, S., Dhanabal, M., Soff, G. A., Sukhatme, V. P., Kufe, D. W., and Weichselbaum, R. R. 1998. Combined effects of angiostatin and ionizing radiation in antitumour therapy. *Nature* 394:287–291.

Mazure, N. M., Chen, E. Y., Laderoute, K. R., and Giaccia, A. J. 1997. Induction of vascular endothelial growth factor by hypoxia is modulated by a phosphatidylinositol 3-kinase/Akt signaling pathway in Ha-ras-transformed cells through a hypoxia inducible factor-1 transcriptional element. *Blood* 90:3322–3331.

McLeskey, S. W., Kurebayashi, J., Honig, S. F., Zwiebel, J., Lippman, M. E., Dickson, R. B., and Kern, F. G. 1993. Fibroblast growth factor 4 transfection of MCF-7 cells produces cell lines that are tumorigenic and metastatic in ovariectomized or tamoxifen-treated athymic nude mice. *Cancer Res.* 53:2168–2177.

Milanini, J., Vinals, F., Pouyssegur, J., and Pages, G. 1998. p42/p44 MAP kinase module plays a key role in the transcriptional regulation of the vascular endothelial growth factor gene in fibroblasts. *J. Biol. Chem.* 273:18165–18172.

Miyakis, S., Sourvinos, G., and Spandidos, D. A. 1998. Differential expression and mutation of the ras family genes in human breast cancer. *Biochem. Biophys. Res. Commun.* 251:609–612.

Moghaddam, A., Zhang, H.-T., Fan, T.-P. D., Hu, D.-E., Lees, V. C., Turley, H., Fox, S. B., Gatter, K. C., Harris, A. L., and Bicknell, R. 1995. Thymidine phosphorylase is angiogenic and promotes tumor growth. *Proc. Natl. Acad. Sci. USA* 92:998–1002.

Morelli, D., Lazzerini, D., Cazzaniga, S., Squicciarini, P., Bignami, P., Maier, J. A. M., Sfondrini, L., Ménard, S., Colnaghi, M. I., and Balsari, A. 1998. Evaluation of the balance between angiogenic and antiangiogenic circulating factors in patients with breast and gastrointestinal cancers. *Clin. Cancer Res.* 4:1221–1225.

Moser, T. L., Stack, M. S., Asplin, I., Enghild, J. J., Hojrup, P., Everitt, L., Hubchak, S., Schnaper, H. W., and Pizzo, S. V. 1999. Angiostatin binds ATP synthase on the surface of human endothelial cells. *Proc. Natl. Acad. Sci. USA* 96:2811–2816.

Mukhopadhyay, D., Knebelmann, B., Cohen, H. T., Ananth, S., and Sukhatme, V. P. 1997. The von Hippel-Lindau tumor suppressor gene product interacts with Sp1 to repress vascular endothelial growth factor promoter activity. *Mol. Cell. Biol.* 17:5629–5639.

Mukhopadhyay, D., Tsiokas, L., and Sukhatme, V. P. 1995a. Wild-type p53 and v-src exert opposing influences on human vascular endothelial growth factor gene expression. *Cancer Res.* 55:6161–6165.

Mukhopadhyay, D., Tsiokas, L., Zhou, X.-M., Foster, D., Brugge, J. S., and Sukhatme, V. P. 1995b. Hypoxic induction of human vascular endothelial growth factor expression through c-Src activation. *Nature* 375:577–581.

Nagaoka, H., Iino, Y., Takei, H., and Morishita, Y. 1998. Platelet-derived endothelial cell growth factor/thymidine phosphorylase expression in macrophages correlates with tumor angiogenesis and prognosis in invasive breast cancer. *Int. J. Oncol.* 13:449–454.

Nass, S. J., and Dickson, R. B. 1997. Defining a role for c-Myc in breast tumorigenesis. *Breast Cancer Res. Treat.* 44:1–22.

Nathan, C.-A., Carter, P., Liu, L., Li, B. D. L., Abreo, F., Tudor, A., Zimmer, S. G., and de Benedetti, A. 1997. Elevated expression of eIF4E and FGF-2 isoforms during vascularization of breast carcinomas. *Oncogene* 15:1087–1094.

Neufeld, G., Cohen, T., Gengrinovitch, S., and Poltorak, Z. 1999. Vascular endothelial growth factor (VEGF) and its receptors. *FASEB J.* 13:9–22.

O'Mahony, C. A., Albo, D., Tuszynski, G. P., and Berger, D. H. 1998a. Transforming growth factor-beta 1 inhibits generation of angiostatin by human pancreatic cancer cells. *Surgery* 124:388–393.

O'Mahony, C. A., Seidel, A., Albo, D., Chang, H., Tuszynski, G. P., and Berger, D. H. 1998b. Angiostatin generation by human pancreatic cancer. *J. Surg. Res.* 77:55–58.

O'Reilly, M. S., Boehm, T., Shing, Y., Fukai, N., Vasios, G., Lane, W. S., Flynn, E., Birkhead, J. R., Olsen, B. R., and Folkman, J. 1997. Endostatin: an endogenous inhibitor of angiogenesis and tumor growth. *Cell* 88:277–285.

O'Reilly, M. S., Holmgren, L., Chen, C., and Folkman, J. 1996. Angiostatin induces and sustains dormancy of human primary tumors in mice. *Nat. Med.* 2:689–692.

O'Reilly, M. S., Holmgren, L., Shing, Y., Chen, C., Rosenthal, R. A., Moses, M., Lane, W. S., Cao, Y., Sage, E. H., and Folkman, J. 1994. Angiostatin: a novel angiogenesis inhibitor that mediates the suppression of metastases by a Lewis lung carcinoma. *Cell* 79:315–328.

Okada, F., Rak, J. W., St. Croix, B., Lieubeau, B., Kaya, M., Roncari, L., Shirasawa, S., Sasazuki, T., and Kerbel, R. S. 1998. Impact of oncogenes in tumor angiogenesis: mutant K-*ras* up-regulation of vascular endothelial growth factor/vascular permeability factor is necessary, but not sufficient for tumorigenicity of human colorectal carcinoma cells. *Proc. Natl. Acad. Sci. USA* 95:3609–3614.

Ozbun, M. A., and Butel, J. S. 1995. Tumor suppressor p53 mutations and breast cancer: a critical analysis. *Adv. Cancer Res.* 66:71–141.

Patterson, B. C., and Sang, Q. A. 1997. Angiostatin-converting enzyme activities of human matrilysin (MMP-7) and gelatinase B/type IV collagenase (MMP-9). *J. Biol. Chem.* 272: 28823–28825.

Pluda, J. M. 1997. Tumor-associated angiogenesis: mechanisms, clinical implications, and therapeutic strategies. *Semin. Oncol.* 24:203–218.

Qian, X., Wang, T. N., Rothman, V. L., Nicosia, R. F., and Tuszynski, G. P. 1997. Thrombospondin-1 modulates angiogenesis in vitro by up-regulation of matrix metalloproteinase-9 in endothelial cells. *Exp. Cell Res.* 235:403–412.

Rak, J., Mitsuhashi, Y., Bayko, L., Filmus, J., Shirasawa, S., Sasazuki, T., and Kerbel, R.S. 1995. Mutant *ras* oncogenes upregulate VEGF/VPF expression: implications for induction and inhibition of tumor angiogenesis. *Cancer Res.* 55:4575–4580.

Rastinejad, F., Polverini, P. J., and Bouck, N. P. 1989. Regulation of the activity of a new inhibitor of angiogenesis by a cancer suppressor gene. *Cell* 56:345–355.

Relf, M., LeJeune, S., Scott, P. A. E., Fox, S., Smith, K., Leek, R., Moghaddam, A., Whitehouse, R., Bicknell, R., and Harris, A. L. 1997. Expression of the angiogenic factors vascular endothelial cell growth factor, acidic and basic fibroblast growth factor, tumor growth factor β-1, platelet-derived endothelial cell growth factor, placenta growth factor and pleiotrophin in human primary breast cancer and its relation to angiogenesis. *Cancer Res.* 57:963–969.

Rhoads, R. E. 1991. Protein synthesis, cell growth and oncogenesis. *Curr. Opin. Cell Biol.* 3:1019–1024.

Rhoads, R. E. 1993. Regulation of eukaryotic protein synthesis by initiation factors. *J. Biol. Chem.* 268:3017–3020.

Rinker-Schaeffer, C. W., Graff, J. R., de Benedetti, A., Zimmer, S. G., and Rhoads, R. E. 1993. Decreasing the level of translation initiation factor 4E with antisense RNA causes reversal of *ras*-mediated transformation and tumorigenesis of cloned rat embryo fibroblasts. *Int. J. Cancer* 55:841–847.

Roberts, D. D. 1996. Regulation of tumor growth and metastasis by thrombospondin-1. *FASEB J.* 10:1183–1191.

Rogelj, S., Weinberg, A., Fanning, P., and Klagsbrun, M. 1988. Basic fibroblast growth factor fused to a signal peptide transforms cells. *Nature* 331:173–175.

Ryan, H. E., Lo, J., and Johnson, R. S. 1998. HIF-1 alpha is required for solid tumor formation and embryonic vascularization. *EMBO J.* 17:3005–3015.

Scandurro, A. B., and Beckman, B. S. 1998. Common proteins bind mRNAs encoding erythropoietin, tyrosine hydroxylase, and vascular endothelial growth factor. *Biochem. Biophys. Res. Commun.* 246:436–440.

Scott, P. A. E., Smith, K., Poulsom, R., de Benedetti, A., Bicknell, R., and Harris, A. L. 1998. Differential expression of vascular endothelial growth factor mRNA vs protein isoform expression in human breast cancer and relationship to eIF-4E. *Br. J. Cancer* 77:2120–2128.

Semenza, G. L., Agani, F., Iyer, N., Jiang, B.-H., Leung, S., Wiener, C., and Yu, A. 1998. Hypoxia-inducible factor 1: from molecular biology to cardiopulmonary physiology. *Chest* 114:40S–45S.

Semenza, G. L., and Wang, G. L. 1992. A nuclear factor induced by hypoxia via de novo protein synthesis binds to the human erythropoietin gene enhancer at a site required for transcriptional activation. *Mol. Cell. Biol.* 12:5447–5454.

Shoji, M., Hancock, W. W., Abe, K., et al. 1998. Activation of coagulation and angiogenesis in cancer. *Am. J. Pathol.* 152:399–411.

Shweiki, D., Itin, A., Soffer, D., and Keshet, E. 1992. Vascular endothelial growth factor induced by hypoxia may mediate hypoxia-initiated angiogenesis. *Nature* 359:843–848.

Siemeister, G., Weindel, K., Mohrs, K., Barleon, B., Mariny-Baron, G., and Marmé, D. 1996. Reversion of deregulated expression of vascular endothelial growth factor in human renal carcinoma cells by von Hippel-Lindau tumor suppressor protein. *Cancer Res.* 56:2299–2301.

Sim, B. K. L., O'Reilly, M. S., Liang, H., Fortier, A. H., He, W., Madsen, J. W., Lapcevich, R., and Nacy, C. A. 1997. A recombinant human angiostatin protein inhibits experimental primary and metastatic cancer. *Cancer Res.* 57:1329–1334.
Smith, L. M., Birrer, M. J., Stampfer, M. R., and Brown, P. H. 1997. Breast cancer cells have lower activating protein 1 transcription factor activity than normal mammary epithelial cells. *Cancer Res.* 57:3046–3054.
Soutou, B., Gamby, C., Crepin, M., and Hamelin, R. 1996. Tumoral progression of human breast epithelial cells secreting FGF2 and FGF4. *Int. J. Cancer* 68:675–681.
Stathakis, P., Fitzgerald, M., Matthias, L. J., Chesterman, C. N., and Hogg, P. J. 1997. Generation of angiostatin by reduction and proteolysis of plasmin. Catalysis by a plasmin reductase secreted by cultured cells. *J. Biol. Chem.* 272:20641–20645.
Stein, I., Itin, A., Einat, P., Skaliter, R., Grossman, Z., and Keshet, E. 1998. Translation of vascular endothelial growth factor mRNA by internal ribosome entry: Implications for translation under hypoxia. *Mol. Cell. Biol.* 18:3112–3149.
Stein, I., Neeman, M., Shweiki, D., Itin, A., and Keshet, E. 1995. Stabilization of vascular endothelial growth factor mRNA by hypoxia and hypoglycemia and coregulation witih other ischemia-induced genes. *Mol. Cell. Biol.* 15:5363–5368.
Stellmach, V., Volpert, O. V., Crawford, S. E., Lawler, J., Hynes, R. O., and Bouck, N. 1996. Tumour suppressor genes and angiogenesis: the role of TP53 in fibroblasts. *Eur. J. Cancer* 32A:2394–2400.
Stratmann, R., Krieg, M., Haas, R., and Plate, K. H. 1997. Putative control of angiogenesis in hemangioblastomas by the *von Hipple-Lindau* tumor suppressor gene. *J. Neuropathol. Exp. Neurol.* 56:1242–1252.
Surmacz, E., Guvakova, M. A., Nolan, M. K., Nicosia, R. F., and Sciacca, L. 1998. Type I insulin-like growth factor receptor function in breast cancer. *Breast Cancer Res. Treat.* 47:255–267.
Takahashi, Y., Bucana, C. D., Cleary, K. R., and Ellis, L. M. 1998. p53, vessel count and vascular endothelial growth factor expression in human colon cancer. *Int. J. Cancer* 79:34–38.
Takahashi, Y., Cleary, K. R., Mai, M., Kitadai, Y., Bucana, C. D., and Ellis, L. M. 1996. Significance of vessel count and vascular endothelial growth factor and its receptor (KDR) in intestinal-type gastric cancer. *Clin. Cancer Res.* 2:1679–1684.
Tanaka, T., Cao, Y., Folkman, J., and Fine, H. A. 1998. Viral vector-targeted antiangiogenic gene therapy utilizing an angiostatin complementary DNA. *Cancer Res.* 58:3362–3369.
Taraboletti, G., Roberts, D., Liotta, L. A., and Giavazzi, R. 1990. Platelet thrombospondin modulates endothelial cell adhesion, motility, and growth: a potential angiogenesis regulatory factor. *J. Cell Biol.* 111:765–772.
Toi, M., Hoshina, S., Taniguchi, T., Yamamoto, Y., Ishitsuka, H., and Tominaga, T. 1995b. Expression of platelet-derived endothelial cell growth factor/thymidine phosphorylase in human breast cancer. *Int. J. Cancer* 64:79–82.
Toi, M., Inada, K., Suzuki, H., and Tominaga, T. 1995a. Tumor angiogenesis in breast cancer: its importance as a prognostic indicator and the association with vascular endothelial growth factor expression. *Breast Cancer Res. Treat.* 36:193–204.
Toi, M., Ueno, T., Matsumoto, H., Saji, H., Funata, N., Koike, M., and Tominaga, T. 1999. Significance of thymidine phosphorylase as a marker of protumor monocytes in breast cancer. *Clin. Cancer Res.* 5:1131–1137.
Tolsma, S. S., Volpert, O. V., Good, D. J., Frazier, W. A., Polverini, P. J., and Bouck, N. 1993. Peptides derived from two separate domains of the matrix protein thrombospondin-1 have anti-angiogenic activity. *J. Cell Biol.* 122:497–511.
Tuszynski, G. P., and Nicosia, R. F. 1994. Localization of thrombospondin and its cysteine-serine-valine-threonine-cysteine-glycine-specific receptor in human breast carcinoma. *Lab. Invest.* 70:228–233.

Vagner, S., Gensac, M.-C., Maret, A., Bayard, F., Amalric, F., Prats, H., and Prats, A.-C. 1995. Alternative translation of human fibroblast growth factor 2 mRNA occurs by internal entry of ribosomes. *Mol. Cell. Biol.* 15:35–44.

Vazquez, F., Hastings, G., Ortega, M. A., Lane, T. F., Oikemus, S., Lombardo, M., and Iruela-Arispe, M. L. 1999. METH-1, a human ortholog of ADAMTS-1, and METH-2 are members of a new family of proteins with angio-inhibitory activity. *J. Biol. Chem.* 274:23349–23357.

Vlodavsky, I., Miao, H.-Q., Medalion, B., Danagher, P., and Ron, D. 1996. Involvement of heparan sulfate and related molecules in sequestration and growth promoting activity of fibroblast growth factor. *Cancer Metastasis Rev.* 15:177–186.

Volpert, O. V., Dameron, K. M., and Bouck, N. 1997. Sequential development of an angiogenic phenotype by human fibroblasts progressing to tumorigenicity. *Oncogene* 14:1495–1502.

Wang, G. L., Jiang, B.-H., Rue, E. A., and Semenza, G. L. 1995. Hypoxia-inducible factor 1 is a basic-helix-loop-helix-PAS heterodimer regulated by cellular O_2 tension. *Proc. Natl. Acad. Sci. USA* 92:5510–5514.

Wang, G. L., and Semenza, G. L. 1993. Characterization of hypoxia-inducible factor 1 and regulation of DNA binding activity by hypoxia. *J. Biol. Chem.* 268:21513–21518.

Wang, T. N., Qian, X.-H., Granick, M. S., Solomon, M. P., Rothman, V. L., Berger, D. H., and Tuszynski, G. P. 1996a. Thrombospondin-1 (TSP-1) promotes in the invasive properties of human breast cancer. *J. Surg. Res.* 63:39–43.

Wang, T. N., Qian, X.-H., Granick, M. S., Solomon, M. P., Rothman, V. L., Berger, D. H., and Tuszynski, G. P. 1996b. Inhibition of breast cancer progression by an antibody to a thrombospondin-1 receptor. *Surgery* 120:449–454.

Weidner, N., Semple, J. P., Welch, W. R., and Folkman, J. 1991. Tumor angiogenesis and metastasis—correlation in invasive breast carcinoma. *N. Engl. J. Med.* 324:1–8.

Weinstat-Saslow, D., Zabrenetzky, S., VanHoutte, K., Frazier, W. A., Roberts, D. D., and Steeg, P. S. 1994. Transfection of thrombospondin 1 complementary DNA into a human breast carcinoma cell line reduces primary tumor growth, metastatic potential, and angiogenesis. *Cancer Res.* 54:6504–6511.

Wizigmann-Voos, S., Breier, G., Risau, W., and Plate, K. H. 1995. Up-regulation of vascular endothelial growth factor and its receptors in von Hippel-Lindau disease-associated and sporadic hemangioblastomas. *Cancer Res.* 55:1358–1364.

Wu, Z., O'Reilly, M. S., Folkman, J., and Shing, Y. 1997. Suppression of tumor growth with recombinant murine angiostatin. *Biochem. Biophys. Res. Commun.* 236:651–654.

Yamaguchi, N., Anand-Apte, B., Lee, M., Sasaki, T., Fukai, N., Shapiro, R., Que, I., Lowik, C., Timpl, R., and Olsen, B. R. 1999. Endostatin inhibits VEGF-induced endothelial cell migration and tumor growth independently of zinc binding. *EMBO J.* 18:4414–4423.

Yoshiji, H., Harris, S. R., and Thorgeirsson, U. P. 1997. Vascular endothelial growth factor is essential for initial but not continued *in vivo* growth of human breast carcinoma cells. *Cancer Res.* 57:3924–3928.

Zhan, X., Bates, B., Hu, X., and Goldfarb, M. 1988. The human FGF-5 oncogene encodes a novel protein related to fibroblast growth factor. *Mol. Cell. Biol.* 8:3487–3495.

Zhang, H.-T., Craft, P., Scott, P. A. E., Ziche, M., Weich, H. A., Harris, A. L., and Bicknell, R. 1995. Enhancement of tumor growth and vascular density by transfection of vascular endothelial cell growth factor into MCF-7 human breast carcinoma cells. *J. Natl. Cancer Inst.* 87:213–219.

Zhang, L., Kharbanda, S., Chen, D., Bullocks, J., Miller, D. L., Ding, I. Y. F., Hanfelt, J., McLeskey, S. W., and Kern, F. G. 1997. MCF-7 breast carcinoma cells overexpressing FGF-1 form vascularized metastatic tumors in ovariectomized or tamoxifen-treated nude mice. *Oncogene* 15:2093–2108.

Zhang, L., Kharbanda, S., Hanfelt, J., and Kern, F. G. 1998. Both autocrine and paracrine effects of transfected acidic fibroblast growth factor are involved in the estrogen-independent and antiestrogen-resistant growth of MCF-7 breast cancer cells. *Cancer Res.* 58:352–361.
Zhong, H., Agani, F., Baccala, A. A., Laughner, E., Rioseco-Camacho, N., Isaacs, W. B., Simons, J. W., and Semenza, G. L. 1998. Increased expression of hypoxia inducible factor-1alpha in rat and human prostate cancer. *Cancer Res.* 58:5280–5284.

CHAPTER 4

Angiogenesis in Mammary Tumors: Quantitation and Relation to Prognosis

Stephen B. Fox

Angiogenesis is the generation of new vessels from the existing vasculature and is essential for tumor growth and metastasis. This requisite increase in tumor vascularity has been recognized for many centuries. Indeed the tumor vasculature was described posthumously by John Hunter in 1794 who noted, "In a disease in which there is an increase in the part, as in tumors, the increase in the vessels is conspicuous." However, it is only since Folkman's (1971) hypothesis on antiangiogenesis that a quantitative method for measuring angiogenesis has been pursued.

This chapter reviews the timing of the angiogenic switch in breast cancer development, discusses some of the difficulties in identifying this transition in the multitude of precursor lesions that have been described, outlines candidate pathways of angiogenesis in in situ disease, presents the published data on the relationship between quantitative angiogenesis and prognosis, discusses some of the difficulties that contribute to the discrepant findings, and describes some alternate measures of angiogenesis that may improve or supersede the present methods.

ANGIOGENIC SWITCH

To initiate neovascularization, the tumor must switch to an angiogenic phenotype. This may be a result of genetic change but also a response to local stresses such as hypoxia. As discussed in depth in Chapter 3, several oncogenes and tumor suppressor genes including *ras*, *src*, von Hippel–Lindau, and *p53*, have been cited as potential mediators of the angiogenic switch. It is likely, however, that different tumor types use different genetic mechanisms to elicit a blood supply. This is not surprising because each vascular bed has unique characteristics and phenotypes that distinguish each organ. This may be reflected in the angiogenic phenotype of different tumors at different stages of development (Fox et al, 1996; O'Brien et al, 1996; Engels et al, 1997b).

Evidence mostly from transgenic models with reproducible temporally and histologically distinct tumor stages has allowed angiogenesis in individual tumor stages to be studied. Three particular models of neoplastic change—the islet cell carcinoma, the dermal fibrosarcoma, and cutaneous squamous cell carcinoma—have enabled the timing of the angiogenic switch to be identified. Evidence from these models has demonstrated that the acquisition of an angiogenic phenotype occurs in the preneoplastic stage (as early as hyperplasia), that the switch appears to be a separate but complementary change in the multistep pathway of malignant change, and that angiogenesis is a rate-limiting step in tumor development (Hanahan and Folkman, 1996).

Data derived from cell lines and both animal and human mammary tissues support these observations. The ability of cultured human mammary epithelial cells to stimulate angiogenesis in a rabbit cornea test preceded neoplastic transformation (Chaudhury et al, 1980; Ziche and Gullino, 1982). In addition, biopsies from normal or hyperplastic mouse or rat breast tissues evoked an angiogenic response albeit with a lesser frequency than carcinomas (Gimbrone and Gullino, 1976a,b; Maiorana and Gullino, 1978). Similar findings have been observed with human tissue explants; tissue fragments from normal, hyperplastic, in situ, and invasive carcinomas of human breast induced angiogenesis in approximately 5%, 30%, 66%, and 65%, respectively (Brem et al, 1978). Furthermore, in rodent models, angiogenesis was observed to occur more frequently in hyperplastic epithelium associated with neoplastic transformation, findings paralleled in human tissues where angiogenic normal tissues had a higher rate of neoplastic transformation than tissues without an angiogenic phenotype (Gimbrone and Gullino, 1976b; Maiorana and Gullino, 1978; Jensen et al, 1982). Thus, testing the angiogenic capacity of a preneoplastic lesion might be a useful screen for the malignant potential of a tissue.

VASCULAR CHANGES IN NORMAL, PROLIFERATIVE AND NEOPLASTIC BREAST TISSUES

General Changes in Blood Vessel Morphology, Architecture, and Blood Flow in Tumors

The abnormal vascular structure, physiology, and blood flow properties of carcinomas are now well recognized. New vessels are often dilated, saccular, sinusoidal, and tortuous, and contain multiple bifurcations, loops, and blind ending sprouts (Warren, 1979; Lin et al, 1984; Skinner et al, 1995). The new capillaries have wide endothelial cell junctions, numerous transendothelial channels, and a fenestrated or a discontinuous lining; vessels are often lined with tumor cells (Warren, 1979). Indeed, in ocular melanomas it has recently been shown that the tumor cells demonstrate vascular mimicry forming a neovasculature without endothelium (Maniotis et al, 1999). The vessels have an irregular basement membrane composed of abnormal ratios of fibronectin, laminin, and collagen (Paku and Paweletz, 1991). However, even after the establishment of vessels there is continued vascular remodeling (Fox et al, 1997a, 2000a).

Owing to a combination of the above structural defects and the influence of permeability factors such as vascular endothelial growth factor (VEGF), which enhances vesicovacuolar organelle formation, tumor vessels are also hyperpermeable (Dvorak et al, 1988). These hyperpermeable vessels, an expanding tumor mass, and an absence of tumor lymphatics cause an elevated interstitial pressure, compressing the vasculature, which, together with altered rheology (which enhances erythrocyte sludging and platelet aggregation) increases blood viscosity, compromising blood flow. Moreover, as the tumor enlarges, the relative endothelial surface area is reduced, leading to a decrease in tumor perfusion. These changes together with arteriovenous shunting of blood often result in lower average perfusion rates for carcinomas than that observed for its normal tissue counterpart and are likely responsible for tumour necrosis.

Breast Carcinogenesis

Unlike the adenoma-carcinoma sequence and the known associated accumulation of genetic alterations in colorectal tumors, the multistep model for breast carcinogenesis is less defined. Problems in the morphologic classification of the large number of heterogeneous precursor lesions and problems in identifying their corresponding genetic changes have confounded the process. Nevertheless, the transition of lobular hyperplasia through lobular carcinoma in situ and invasive lobular carcinoma with consistent genetic changes in the E cadherin gene is emerging as one candidate pathway (Lakhani, 1999). For other histologic types of breast carcinomas, the sequence is less clear. However, it has been suggested that tumors such as tubular, cribriform, and well-differentiated ductal carcinomas arise from low-grade ductal carcinoma in situ (DCIS), while high-grade DCIS leads to poorly differentiated medullary and ductal carcinomas (van Diest, 1999).

Vessel Changes in Benign Breast Disease

In parallel with the above genetic studies, several groups have begun to describe the vascular changes in normal, benign, and malignant breast diseases associated with the angiogenic switch. Using immunohistochemistry, both size (diameter and perimeter) and density of vessels has been reported to be increased in hyperplastic breast lesions when compared to normal breast lobules (Ottinetti and Sapino, 1988; Porter et al, 1993; Fregene et al, 1994; Guinebretiere et al, 1994; Heffelfinger et al, 1996). Indeed, in accordance with the above, where the ability of tissue biopsies to evoke angiogenic response was associated with malignant change, an increase risk of developing breast cancer was observed in biopsies from patients with benign proliferative breast disease with high microvessel density (Guinebretiere et al, 1994).

Angiogenic Pathways in Ductal Carcinoma In Situ

Changes in both vessel number and arrangement have been described during the transition from nonneoplastic precursor lesions to DCIS. Two predominant patterns in DCIS have been delineated: type 1, where there is an increase in stromal vascularity between involved ducts; and type 2, where there is a dense rim closely opposed to the basement membrane of individual effected ducts. These patterns have been reported in 43% to 57% (Guidi et al, 1994; Engels et al, 1997a) and 23% to 62% (Weidner et al, 1991; Guidi et al, 1994; Heffelfinger et al, 1996; Engels et al, 1997a; Lee et al, 1997a,b) for patterns 1 and 2, respectively, with a significant relationship between the two being described (Engels et al, 1997a). These patterns imply, akin to bladder and cervical cancer, that at least two angiogenic pathways are present—one mediated by angiogenic factors released directly by tumor cells, resulting in the rim or necklace of vessels, and another generated by recruitment of accessory cells such as inflammatory cells, which themselves release angiogenic factors. Indeed, there is some evidence to support this hypothesis. Expression of the angiogenic factor thymidine phosphorylase (TP) (previously known as platelet-derived endothelial cell growth factor) in DCIS was significantly associated with the rim but not stromal vascular phenotype (Engels et al, 1997b), whereas the latter but not former vascular pattern was significantly associated with clusters of inflammatory cells (Lee et al, 1997a,b). Furthermore, a significant increase in both the stromal and vascular rim was observed in high-grade DCIS (the type most likely to progress to invasive carcinoma) (Engels et al, 1997a). The changes in stromal vascularity in high-grade DCIS lesions were also significantly correlated with both a higher relapse rate and a higher risk in developing an invasive tumor, possibly through the enlarged vascular window provided by enhanced angiogenesis.

Relationship of DCIS Vascular Pattern and Clinical Behavior

The increase in stromal neovascularization in high-grade DCIS was also associated with stromal desmoplasia (Guidi et al, 1994). The induction of a neoplastic stroma (desmoplasia) by complex interplay between tumor, endothelial, stromal, and inflammatory cells is essential for solid tumor growth. For DCIS and invasive breast cancers, a particular profile of growth factors and structural and adhesive stromal proteins has been reported that promotes angiogenesis and metastasis. High levels of VEGF, thrombospondin-1, collagen type I, decorin, versican, and total and ED-A+ isoforms of fibronectin have been reported that may facilitate cell adhesion, migration, and invasion, the suggestion being that the alteration in stroma paves the way for tumor cell invasion (Brown et al, 1999).

ANGIOGENESIS AS A PROGNOSTIC MARKER

Measuring Tissue Vascularity

Despite the abnormal haemodynamic and often lower perfusion characteristics in tumors, it was recognized that quantitation of the tumor vasculature might play an important a role in patient management. The first objective measure of angiogenesis was developed by Folkman and colleagues. They developed a microscopic angiogenesis grading system (MAGS score) in tinctorially stained tissue sections. The quantitation was based on vessel number, the degree of endothelial cell hyperplasia, and their cytology (Brem et al, 1972). It was hoped that this first objective scoring system of tumor angiogenesis would yield important information on the relationships to other clinicopathologic tumor characteristics and help in testing antiangiogenic therapies. Nevertheless, although large differences in vascularity could be discerned (endothelial poor versus rich), the technical limitations of sample selection, inter- and intraobserver variation, and conceptual biologic problems were such that the technique could not be easily applied. Interest in measuring tumor angiogenesis was rekindled with the advent of nonspecific endothelial markers such as alkaline phosphatase and lectins (Mlynek et al, 1985; Svrivastava et al, 1988; Porschen et al, 1994), but it has been only in the last 10 years with the generation of specific endothelial markers that quantitation of tumor angiogenesis has been performed.

Most of these studies have employed the method of Weidner et al (1992), who demonstrated that the number microvessels, highlighted by antibodies to factor VIII–related antigen, in the most vascular area of breast cancers gave independent prognostic information. The most vascular area containing the maximum number of discrete microvessels was identified by scanning the tumor at low power, and individual microvessels were counted at high-power. Any immunoreactive endothelial cell that was separate from adjacent microvessels was considered a countable vessel. Vascular lumina were not a requirement to be included in the count, and the vessels within any central sclerotic area of the tumor were not included. Since these initial reports in breast cancer, numerous studies in many tumor types have demonstrated that quantitation by microvessel as a measure of angiogenesis is a powerful prognostic tool (Fox and Harris, 2001) (Table 4.1).

Nevertheless, despite the initial confirmatory publications, numerous reports are now appearing in the literature contradicting the positive association between increasing tumor vascularity and reduced patient survival (Table 4.1), and caution as to the clinical utility of tumor angiogenesis is being urged (Page and Dupont, 1992; Page and Jensen, 1995). However, many of these negative studies are due to significant deviation from the method of Weidner et al (1991, 1992), and as originally recognized by Brem et al (1972) with the MAGS score, quantitation of tumor angiogenesis is still limited by the methods used for capillary identification and quantitation. There are several major considerations to take into account when quantifying tumor angiogenesis in histologic sections: (1) the selection of antibody to highlight the capillaries; (2) the vascular parameter quantified (i.e., microvessel number, area, perimeter, etc.); (3) the area of tumor to be assessed; (4) the measuring technique; (5) the experience of the observer; and (6) the statistical analysis used (reviewed in Fox and Harris, 1997b, 2001).

Variables in the Quantitation of Tumor Angiogenesis

Selection of Antibody

Studies vary in the antibody used to highlight tumor capillaries. Early studies used nonspecific endothelial cell markers such as alkaline phosphatase (Mlynek et al, 1985), vimentin (Wakui et al, 1992), antilectin antibodies (Svrivastava et al, 1988; Carnochan et al, 1991; Barnhill et al, 1992b; Porschen et al, 1994), and type IV collagen (Visscher et al, 1993; Vesalainen et al, 1994). Since these mark the basement membrane, only a proportion of the total tumor vasculature is identified. These nonspecific endothelial cell markers also suffer from their lack of specificity and a sensitivity that is below that of the newer endothelial cell markers such as factor VIII–related antigen, CD31 (platelet/endothelial cell adhesion molecule (PECAM)), and CD34. However, these markers also have their own deficiencies (Horak et al, 1992; Lee et al, 1997a; Schor et al, 1998), and heterogeneity of endothelia in different vascular beds may be the ultimate barrier in generating the ideal marker (Schlingemann et al, 1991; Kuzu et al, 1992; Page et al, 1992).

Nevertheless, where different antibodies directed to factor VIII–related antigen, CD31, and CD34 have been compared, many have provided prognostic information (Horak et al, 1992; Toi et al, 1993; Siitonen et al, 1995; Tanigawa et al, 1997). Furthermore, there is a significant correlation between the mean and median

TABLE 4.1. Summary of published series reporting the relationship between quantitative tumor angiogenesis and prognosis in breast cancer (modified from Fox, SB. Tumour angiogenesis and prognosis. Histopathology (1998): 30; 294–301).

Author	Year	*n*	Follow-up (months)	Magnification	Field area (mm^2)	Antibody
For						
Weidner et al	1992	165	51	×200	0.7386	FVIII
Horak et al	1992	103	30	×400	0.196	CD31
Bosari et al	1992	120	108 min	×200	—	FVIII
Toi et al	1993	125	62	×200	—	FVIII/CD31
Visscher et al*	1993	58	—	×50	—	collagen IV
Fox et al	1994	109 (N–)	25	×250	0.196	CD31
Lipponen et al	1994	173	**120**	×40	0.49	collagen IV
Fregene et al	1994	316	**94**	×400	0.19	FVIII
Toi et al•	1995	328	—	×200	—	FVIII
Fox et al	1995	211	42	×250	0.196	CD31
Bevilacqua et al	1995	211 (N–)	78	×200	0.74	CD31
Ogawa et al	1995	155	82	×200	0.785	FVIII
Obermair et al	1995	230	**74**	×200	0.25	FVIII
Barbareschi et al*	1995	91	66.3	×200	0.74	CD31
Kohlberger et al*	1996	60	79	×200	0.25	FVIII
Simpson et al*	1996	178	71	×20	0.74	CD34
Heinmann et al	1996	167	184	×400	0.145	CD34
Martin et al	1997	144	>96	×200	0.68	CD34
Karelia et al	1997	51	>60	×200	0.74	FVIII
Acenero et al	1998	112	>60	×200	0.74	FVIII
Narita et al	1998	100	88	×200	0.25	FVIII
Gasaparini et al	1998	531	76	×200	0.74	CD31
Chu et al	1998	163	**57**	×200	0.74	CD31
Kato et al[$]	1999	109	168	×200	—	FVIII
Kumar et al	1999	106	60	×400	—	CD105
Jitsuiki et al	1999	147	39	×200	0.38	CD31
Against						
Hall et al	1992	87	—	×400	0.126	FVIII
van Hoef et al	1993	93 (N–)	151	×200	0.123	FVIII
Siitonen et al	1995	77 (N–)	84 max	×200	0.739	CD34/FVIII
Axelsson et al	1995	220	136	×250	0.37	FVIII
Goulding et al	1995	141	144 min	×250	0.37	CD31
Costello et al	1995	87	114 max	×200	0.22	FVIII
Morphopoulos et al	1996	160	61.2	×200	0.785	FVIII
Kato et al	1997	109	168	×200	—	FVIII
Tan et al	1997	66	200	×200	—	FVIII
Clahsen et al	1998	346 (N–)	49	×250	0.384	CD31

The number of vessels/mm and follow-up times are median; bold indicates mean; maximal max; minimal min; NS-not significant; * image analysis; –not stated in paper; ∧ per field area; • an update of the 1993 paper; CC-Chalkley count (a CC of 6 equates to a 100 vessels/mm^2 at ×250); N–/+ node negative/positive; # lobular carcinomas; ∧ advanced stage: RFS relapse free survival OS overall survival; [$] update of Kato et al 1997.

TABLE 4.1. *Continued*

Cutoff	No. of fields	Vessels/ mm^2	Univariate analysis RFS	Univariate analysis OS	Multivariate analysis RFS	Multivariate analysis OS
<34; 34–67; 68–100; >100	1	76	$p < .001$	$p < .001$	$p < .001$	$p < .001$
Median	3	100		$p = .04$		
Mean node and recurrence	3	**84∧**	$p < .004$	$p < .008$	$p < .028$	
100/150	3	**99∧**	$p < .01$		$p = .0226$	
—	All tumors	**19**	$p = .001$			
Median	3	5.3 (CC)	$p = .01$	$p = .028$	$p = .039$	$p = .019$
40	6	**86.3**	NS	$p = .0054$	NS	$p = .009$
<25; 25–75; >75 centiles	1	—	$p = .02$		Significant in some models	
100	3	84	$p = .00001$		$p = .0000011$	
Thirds	3	5.7 (CC)				
Continuous variable	3	—	$p = .0001$	$p = .018$	$p = .0033$	$p = .044$
Mean	3	**67.5**	$p < .025$	$p < .01$	$p = .002$	$p = .001$
40	1	**51**	$p < .0001$		$p = .004$	
66	3	84	$p = .002$			
Median	4	52.5	$p = .024$			
Vessel area and number	1	**142**	NS	$p = .023$	$p = .002$ (N+)	$p = .002$ (N+)
<15 per field	3 (<10)	138	$p = .018$		$p = .04$	
75th centile	3	121		$p < .05$		$p < .003$
>34	3	34	$p < .02$	$p < .001$		
Highest: >48	3	54	$p < .05$		$p < .05$	
120	5	—	$p < .0001$	$p < .0001$	$p < .05$	$p < .05$
Continuous variable	3	80	$p < .00001$	$p < .00001$	$p < .00001$	$p < .00001$
<34; 34–67; 68–100; >100	1	61	$p = .001$	$p = .0008$	$p = .016$	$p = .016$
41	11–63	**39**		$p < .05$		$p = .048$
Continuous/ quartiles	4	3.7 (CC)	$p < .0431$	$p < .0017$	$p < .001$	$p < .001$
>46	1	46			$p < .05$	
—	5	13			$p > .05$	
<68; 68–100; >100	4	**245**	$p = .92$	$p = .99$		
Highest/upper tertile	3	**44/64**	NS			
Median	3	184 (N–/N+)			$p = .65$	$p = .67$
Median	3	64	NS	NS	NS	NS
Median	3	186	$p = .6$	$p = .5$		
<75: 76–100: >101–125: >126	3	**139**	NS	NS		
Highest/64	3	70		$p = .03$		$p = .46$
Median	3	50		$p = .4$		$p = .12$
Highest/75	3		$p = .1$			

microvessel counts (Horak et al, 1992; Weidner et al, 1992; Toi et al, 1993; Fox et al, 1994; Martin et al, 1997), suggesting that the negative findings in some studies is mostly due to the differences in methodology for measuring tumor vascularity.

Several groups have tried an alternative approach using antibodies; this approach does not highlight all endothelium but selectively identifies only the tumor-associated vasculature (Kaczmarek et al, 1994; Wang et al, 1994). This strategy not only might more accurately measure tumor angiogenesis, but also might be useful as a objective for antiangiogenesis and/or vascular targeting strategies (Burrows and Thorpe, 1994). Several antibodies have been generated that identify epitopes that are reported to be expressed by tumor-associated endothelium and show highly restricted expression in normal tissues. One such is E-9, an antiendoglin (CD-105) antibody, that is reported to be upregulated on tumor-associated endothelium and has recently been shown to provide prognostic information (Kumar et al, 1999). Similarly, other restricted epitopes such as a fetal fibronectin splice variant is highly expressed on tumor associated vessels (in up to 78% of breast cancers) compared to normal adult tissues (Kaczmarek et al, 1994). There is some evidence to suggest that optimizing what should otherwise be a standard immunohistochemical approach with the above antibodies can also significantly affect microvessel counts (Schor et al, 1998).

The Selection of the Representative Tumor Area

Studies examining different tumor areas have shown significant heterogeneity among different tumors regions. The edge of the tumor is the most active area of endothelial cell proliferation, which is usually mirrored by microvessel density (Fox et al, 1993a, 1995a; Jitsuiki et al, 1999). Thus, the particular tumor area that is measured will have profound effects on the data derived. It has been shown that the biologic variation within a tumor is a major contributor to differences in microvessel quantitation (Hansen et al, 1998). However, it is still remarkable that assessment of only one representative section that is only 1/1,000 of the tumor volume can give a useful measure of tumor angiogenesis.

Many tumors demonstrate significant heterogeneity even within the tumor periphery with the presence of so-called vascular hot spots. Although studies have measured random fields (Protopapa et al, 1993), average vascularity (Svrivastava et al, 1988; Carnochan et al, 1991), and even the entire tumor section (Visscher et al, 1993), most assess hot spots, the justification being that these are the areas that are likely to be biologically important and therefore grow and metastase (Weidner et al, 1991, 1992). However, there is little agreement as to the optimal number of hot spots to assess, which currently ranges from one to five (Weidner et al, 1991; Barnhill et al, 1992a,b; Hall et al, 1992; Sahin et al, 1992; Weidner et al, 1992; van Hoef et al, 1993; Sightler et al, 1994). This number has important bearing on the efficacy of the method, since tumors have a limited number of definable hot spots. Thus, counting too many fields will tend to diminish the power of the hot spot technique, since there will be a tendency to reduce the average tumor vascularity the more fields that are assessed. Results may be further confounded by using the mean or the highest microvessel counts as the vessel value for individual cases in the analysis.

The tumor field area, determined by the microscopic magnification and particular objectives used, will also significantly affect the vascular index. A higher magnification gives an increased resolution, which enables more microvessels to be identified, but leads to the disadvantage that all fields at too higher magnification become an angiogenic hot spot (Horak et al, 1992). Conversely, a low magnification with its lower resolution will identify a smaller number of vessels and will dilute out the hot spot. We have demonstrated that measurements within magnification of ×200 to ×400 over a range of 0.12 to 0.74 mm^2 give comparable information (Fox et al, 1995b).

The Vascular Parameter and the Counting Technique

Most studies quantifying tumor angiogenesis have measured microvessel density as a measure of tumor angiogenesis. However, the vascular perimeter might be a better measure of angiogenesis than microvessel density since it may reflect the endothelial surface area available for interaction with the tumor, thereby increasing the likelihood of dissemination and metastasis. There is some evidence to show that high angiogenesis in the primary tumor leads to an increase in shedding of cancer cells and the presence of micrometastasis in bone marrow (McCulloch et al, 1995; Fox et al, 1997b). Similarly, the vascular surface area might give an indication of the blood volume that the tumor can derive its nutrients from in exchange for its waste products. We have examined the vascular parameters, microvessel density, luminal area, and perimeter (Fox et al, 1995b), and have observed significant correlations between all, suggesting that these indices are equivalent for quantifying tumor vascularity.

In addition to the conceptual difficulties of microvessel density, there are several practical considerations making it unsuitable for a diagnostic pathology service. Microvessel counting is time-consuming, particularly when the tumor is of high vascularity, and there is significant inter- and intraobserver variation, even when accounting for possible variations in microvessel criteria. In a confirmatory study, two pathologists trained by Weidner assessed microvessel density in 220 breast carcinoma patients with an extended period of follow-up (Axelsson et al, 1995). No correlation between microvessel density and clinicopathologic variables including patient survival was observed. These findings are probably due to the experience of the observers. It has been demonstrated that experienced observers differ significantly from trainee observers in quantifying tumor angiogenesis in breast tumors (Barbareschi et al, 1995). Nevertheless, in our experience highly trained observers also occasionally disagreed on counts, even with strict adherence to guidelines.

Due to the above problems, together with the additional conceptual difficulty of adjacent microvessels with significantly different vascular parameters assuming equal importance (Weidner et al, 1991, 1992), we and others have explored the use of a number of different techniques with the aim of developing a rapid and objective method for quantifying tumor angiogenesis.

In an attempt to automate the procedure, improve reproducibility, and overcome the variations associated with manual counting, computer image analysis systems have been assessed (Svrivastava et al, 1988; Wakui et al, 1992; Bigler et al, 1993; Visscher et al, 1993; Wesseling et al, 1993; Brawer et al, 1994; Furusato et al,

1994; Simpson et al, 1996; Williams et al, 1994; Barbareschi et al, 1995; Charpin et al, 1995; Fox et al, 1995b; Belien et al, 1999). Using similar endothelial markers and vascular parameters, these reports have confirmed the findings of the manual studies, and shown that quantitative angiogenesis gives independent prognostic information. However, in a manner similar to manual counting, these computer systems have their own general and specific problems, above their capital and running costs. An endothelial marker that will give a high signal-to-noise ratio is essential. This is important since the software employed to analyze the staining, unlike the human eye, is unable to distinguish specific from nonspecific immunoreactivity. The systems are not usually automated, require a high degree of operator interaction, and will reproduce what the operator observes. To date, no software is available for identifying hot spots, although partially automated systems with area and shape filters using defined color tolerances are available, which require reduced human control. In our experience a computer image analysis system is more costly, time-consuming, unsuited to routine diagnostic practice, and no more accurate than a trained observer (Fox et al, 1995b).

To overcome the time and observer variation associated with microvessel density, we have assessed a method based on a point counting method. The tumor section is scanned at low power (×10 to ×100) to identify the three most vascular fields before, at high power (×250) over each hot spot, a 25-point Chalkley eyepiece graticule is placed (Chalkley, 1943). The eyepiece graticule is oriented so the maximum number of points are on or within areas of highlighted vessels, thereby bypassing interpretation difficulties of microvessel counting and reducing observer bias to the selection of the tumor area. This technique has less observer variation than microvessel density (Hansen et al, 1998) and gives independent prognostic information for breast (Fox et al, 1994, 1995b) and bladder carcinomas (Dickinson et al, 1994).

To simplify the procedure further, and analogous to assessing tumor differentiation, we have explored the use of a vascular grading system based on the subjective appraisal by trained observers over a conference microscope. This method is significantly correlated with both microvessel density and Chalkley count, and although it may not be prognostic (Fox et al, 2000b), high vascular grade is significantly related to the presence of bone marrow micrometastasis in breast cancer patients (Fox et al, 1997b).

Statistical Analysis

There is considerable variation in cutoffs and types of analyses performed on angiogenesis data (Table 4.1). This alone can determine whether angiogenesis will give prognostic information. Studies have used varied stratification indices, including the mean, the mean count in node-negative patients with recurrence (Bosari et al, 1992), and variable cutoffs given as a function of tumor area (Weidner et al, 1992; van Hoef et al, 1993) or microscope magnification (Sahin et al, 1992; Protopapa et al, 1993; Toi et al, 1993; Bundred et al, 1994; Guinebretiere et al, 1994; Olivarez et al, 1994). Although there is some loss of information, we and others have used the median (Horak et al, 1992; Dickinson et al, 1994; Fox et al, 1994; Costello et al, 1995) or tertile groups (Fox et al, 1995b), since these stratifications avoid making any assumption regarding the relationship between tumor vascularity and other variables, including survival.

Proposed Consensus Methodology

To overcome the above problems and to help determine the reliability of the technique, a discussion paper with a consensus methodology has been proposed (Vermeulen et al, 1996). This proposes using formalin-fixed paraffin-embedded tissue sections, antigen retrieval, and an anti-CD31 antibody to highlight the tumor vasculature. A manual method for identifying hot spots at low power only in viable tissue or immediately adjacent stroma (within one field diameter at ×200 magnification). A Chalkley point graticule with a field size of the mean size of a vascular hot spot (tumor type dependent) is used to determine vascularity. The process is performed sequentially by two observers, with a third observer used if >10% discrepancy exists between the initial observers.

There are some unresolved issues such as the number of tissue blocks to be examined, but a prospective controlled trial is the only reliable method to assess the prognostic value of measuring angiogenesis in breast carcinomas.

CLINICAL USES OF QUANTITATIVE ANGIOGENESIS

In addition to a role in providing prognostic information, it has also been suggested that angiogenesis quantitation might be useful in other clinical areas. As indicated above, the identification of the angiogenic switch not only may indicate the likelihood of the progression of individual proliferative or neoplastic breast lesions, but also with complementary investigations may help in breast screening programs, increasing the ability to detect disease.

Despite appropriate surgical therapy, many breast cancer patients with limited disease will die from metastasis undetected by conventional methods at presentation. Since the degree of angiogenesis of the primary tumor may determine bone metastasis (Fox et al, 1997b), quantitative angiogenesis might be a surrogate for advanced disease and useful in staging patients and helping to stratify patients who would benefit from adjuvant chemotherapy. Indeed, it has been suggested that the varied prognoses of patients in different stages of the menstrual cycle are partly due to angiogenic factors levels at the time of surgery (Ludwig et al, 1998).

Quantitative tumor angiogenesis might be useful for treatment stratification and disease monitoring. It may potentially identify a subgroup of patients who have a poor prognosis and might benefit from adjuvant therapy (Macaulay et al, 1995). Patients with estrogen receptor–positive tumors that are highly angiogenic have a significant reduction in relapse-free survival despite tamoxifen treatment (Gasparini et al, 1996). These patients could be offered more aggressive therapeutic options. At present this is limited to high-dose chemotherapy and stem cell rescue, but as antiangiogenic and vascular targeting agents become available, the angiogenic activity of tumors might be used to stratify patients for these novel anticancer therapies. Indeed, particular end points using angiogenic measures should be incorporated into clinical trials of antiangiogenic or vascular targeting therapies since they would be more sensitive and specific than conventional criteria.

Furthermore, there is some evidence to suggest that quantitative angiogenesis might also be useful in predicting patient response to chemotherapy. Although there is somewhat contradictory evidence, high vascular counts have been predictive of response chemotherapy in breast cancers (Protopapa et al, 1993; Gasparini et al, 1996, 1999; Paulsen et al, 1997; Clahsen et al, 1998; Honkoop et al, 1998; Jacquemier et al, 1998; Viens et al, 1999), and a significantly lower microvessel count was observed in tumors after neoadjuvant therapy (Makris et al, 1999). We have also shown that tumor expression of TP is associated with a significant increase in survival in node-positive patients treated with cyclophosphamide, methotrexate, and 5-fluorouracil (5-FU) (Fox et al, 1997a). This might be due to the enzymatic action of this angiogenic factor converting 5-FU to its active metabolite, in addition to preventing thymidine salvage of the methotrexate block (Fox et al, 1997a). Thus, TP, in addition to being used as a predictive marker, might also be helpful in stratifying patients for chemotherapy.

ALTERNATE METHODS FOR QUANTIFYING TUMOR ANGIOGENESIS

Angiogenesis is a complex dynamic process involving endothelial cell migration, proliferation, tube formation, and anastomosis—events controlled by the net effect of both promoting and inhibiting angiogenic factors released by the tumor (Paweletz and Knierim, 1989) (endothelial cells may also vascularize tumors by intussusception).

The target capillary responds in this proangiogenic environment by releasing plasmin (Pepper and Montesano, 1990) and matrix metalloproteinases (Fisher et al, 1994), which proteolytically degrade the basement membrane and stroma around the capillary, enabling endothelial cells to proliferate (Fox et al, 1993a), migrate, and form sprouts. During these processes cell adhesion molecules (CAMs) of the notch (Zimrin et al, 1996), selectin (Bischoff, 1997), integrin (Brooks, 1996), and immunoglobulin families (DeLisser et al, 1997) play an important role not only in the physical interactions between endothelial cells, matrix, and accessory cells, but also in signal transduction. The sprouts eventually fuse to form a new vascular network in which blood flow can be initiated (Paweletz and Knierim, 1989).

Due to limitations of the current methods used to quantitate tumor angiogenesis, several groups have investigated using many of the above molecules that regulate neovascularization in breast cancer as alternative measures of angiogenesis. Thus, angiogenic factors (Toi et al, 1996b; Colomer et al, 1997; Mortensen et al, 1999), proteases (Fox et al, 1993b), and CAMs (Fox et al, 1995a; Schadendorf et al, 1995) in patient tumor (various elements) and/or fluid samples (Kondo et al, 1994; Nguyen, 1997; Morelli et al, 1998; Kraft et al, 1999; Salven et al, 1999) have been measured to assess their clinical utility. Although some studies have demonstrated a significant relationship between microvessel number and VEGF (Toi et al, 1996a; Obermair et al, 1997; Relf et al, 1997) and/or thymidine phosphorylase (Toi et al, 1995) in breast cancer, not all studies have shown a relationship with survival (Fox et al, 1996; Toi et al, 1996a; Gasparini et al, 1997; Obermair et al, 1997; Linderholm et al, 1998). Nevertheless, by analyzing numerous angiogenic factors that have been shown to be important in animal models, including VEGF, acidic and basic fibroblast growth factor (FGF), transforming growth factor-β_1, placenta growth factor, thymidine phosphorylase, and pleiotrophin in series of primary breast cancer, it was observed that although all tumors variably expressed each factor and that levels varied by at least 10-fold, only VEGF related to disease relapse (Relf et al, 1997). These data show that these studies are valuable not only in identifying the most important factors involved in human cancer but also in patient management. We have attempted to measure the degree of continued remodeling in breast cancer and have shown that a stable vasculature is associated with a poorer prognosis, suggesting that vascular remodeling might give an indication of the functionality of the tumor vasculature (Fox et al, 2000a).

Several noninvasive imaging techniques, including color Doppler, thermography (Sterns et al, 1996), scintimammography (Omar et al, 1997), magnetic resonance imaging (MRI), and positron emission spectroscopy, are also being assessed (Passe et al, 1997; Huber, 1998). Varied findings depending on the imaging system and tumor vascularity have been reported. An increase in blood flow as measured by color Doppler has been reported in breast tumors when compared to benign lesions (Dixon et al, 1992; Cosgrove et al, 1993; Peters-Engl et al, 1998), and thus might be useful in their diagnosis (Kedar et al, 1995) and prognosis (by virtue of the association with advanced stage (Cheng et al, 1999). Color Doppler, as a measure of angiogenesis, may also be useful in monitoring or predicting the response of breast cancer patients unsuitable for surgery (Kedar et al, 1994). The corollary of these findings has been the ability of Doppler to identify hypoxic tumor regions (Evans et al, 1997). Nevertheless, a correlation between Doppler and microvessel density has not been consistently observed (Lee et al, 1995; Sterns et al, 1996), and although microvessel density influences MRI in in situ and invasive breast carcinomas, it is not accurate (Frouge et al, 1994; Gilles et al, 1995; Hulka et al, 1995, 1997; Buadu et al, 1996; Buckley et al, 1997; Stomper et al, 1997). New developments in MRI that enable quantitation with "blood oxygenation level dependent" BOLD image contrast or targeting with specific angiogenic CAMs might improve the assessment of angiogenesis (Sipkins et al, 1998; van Zijl et al, 1998). Currently these should be considered research tools to investigate tumor biology rather than clinical methods of selection of poor-risk patients.

CONCLUSION

Although many of the studies have shown significant relationships between a variety of measures of tumor angiogenesis and patient survival, and approximately 5,500 patients have now been assessed for microvessel quantitation, none of the current techniques has been validated as sensitive, specific, and reproducible enough to use in a clinical setting. The complexity of tumor angiogenesis and the likelihood that different tumor types use different angiogenic factors at different stages in their development make it unlikely that an accurate assessment of the angiogenic activity of a tumor will be obtained by measuring one aspect of angiogenesis. More likely is that information derived from a variety of assays will give a clinically useful angiogenic index.

REFERENCES

Acenero, M. J., Gallego, M. G., et al. 1998. Vascular density as a prognostic indicator for invasive ductal breast carcinoma. *Virchows Arch.* 432(2):113–117.

Axelsson, K., Ljung, B. M., Moore, D. H., Thor, A. D., Chew, K. L., Edgerton, S. M., Smith, H. S., and Mayall, B. H. 1995. Tumor angiogenesis as a prognostic assay for invasive ductal breast carcinoma. *J. Natl. Cancer Inst.* 87:997–1008.

Barbareschi, M., Weidner, N., Gasparini, G., Morelli, L., Forti, S., Eccher, C., Fina, P., Caffo, O., Leonardi, E., Mauri, F., Bevilacqua, P., and Dalla Palma, P. 1995. Microvessel quantitation in breast carcinomas. *Appl. Immunochem.* 3:75–84.

Barnhill, R. L., K. Fandrey, K., Levy, M. A., Mihm, M. C., and Hyman, B. 1992a. Angiogenesis and tumor progression of melanoma. Quantification of vascularity in melanocytic nevi and cutaneous malignant melanoma. *Lab. Invest.* 67:331–337.

Barnhill, R. L., Mihm, M. J., and Ceballos, P. I. 1992b. Angiogenesis and regressing cutaneous malignant melanoma. *Lancet* 339:991–992.

Belien, J. A., Somi, S., de Jong, J. S., van Diest, P. J., and Baak, J. P. 1999. Fully automated microvessel counting and hot spot selection by image processing of whole tumour sections in invasive breast cancer. *J. Clin. Pathol.* 52:184–192.

Bevilacqua, P., Barbareschi, M., et al. 1995. Prognostic value of intratumoral microvessel density, a measure of tumor angiogenesis, in node-negative breast carcinoma—results of a multiparametric study. *Breast Cancer Res. Treat.* 36(2):205–217.

Bigler, S., Deering, R., and Brawer, M. 1993. Comparisons of microscopic vascularity in benign and malignant prostate tissue. *Hum. Pathol.* 24:220–226.

Bischoff, J. 1997. Cell adhesion and angiogenesis. *J. Clin. Invest.* 100:37–39.

Bosari, S., Lee, A. K., DeLellis, R. A., Wiley, B. D., Heatley, G. J., and Silverman, M. L. 1992. Microvessel quantitation and prognosis in invasive breast carcinoma. *Hum. Pathol.* 23:755–761.

Brawer, M. K., Deering, R. E., Brown, M., Preston, S. D., and Bigler, S. A. 1994. Predictors of pathologic stage in prostatic carcinoma. The role of neovascularity. *Cancer* 73:678–687.

Brem, S., Cotran, R., and Folkman, J. 1972. Tumor angiogenesis: a quantitative method for histological grading. *J. Natl. Cancer Inst.* 48:347–356.

Brem, S. S., Jensen, H. M., and Gullino, P. M. 1978. Angiogenesis as a marker of preneoplastic lesions of the human breast. *Cancer* 41:239–244.

Brooks, P. C. 1996. Role of integrins in angiogenesis. *Eur. J. Cancer* 32A:2423–2429.

Brown, L. F., Guidi, A. J., Schnitt, S. J., Van De Water, L., Iruela-Arispe, M. L., Yeo, T. K., Tognazzi, K., and Dvorak, H. F. 1999. Vascular stroma formation in carcinoma in situ, invasive carcinoma, and metastatic carcinoma of the breast. *Clin. Cancer Res.* 5:1041–1056.

Buadu, L., Murakami, J., Murayama, S., Hashiguchi, N., Sakai, S., Masuda, K., Toyoshima, S., Kuroki, S., and Ohno, S. 1996. Breast lesions: correlation of contrast medium enhancement patterns on MR images with histopathological findings and tumor angiogenesis. *Radiology* 200:639–649.

Buckley, D. L., Drew, P. J., Mussurakis, S., Monson, J. R., and Horsman, A. 1997. Microvessel density of invasive breast cancer assessed by dynamic Gd-DTPA enhanced MRI. *J. Magn. Reson. Imaging* 7:461–464.

Bundred, N., Bowcott, M., Walls, J., Faragher, E., and Knox, F. 1994. Angiogenesis in breast cancer predicts node metastasis and survival. *Br. J. Surg.* 81:768.

Burrows, F. J., and Thorpe, P. E. 1994. Vascular targeting—a new approach to the therapy of solid tumors. *Pharmacol. Ther.* 64:155–174.

Carnochan, P., Briggs, J. C., Westbury, G., and Davies, A. J. 1991. The vascularity of cutaneous melanoma: a quantitative histological study of lesions 0.85–1.25 mm in thickness. *Br. J. Cancer* 64:102–107.

Chalkley, H. 1943. Method for the quantative morphological analysis of tissues. *J. Natl. Cancer Inst.* 4:47–53.

Charpin, C., Devictor, B., Bergeret, D., Andrac, L., Boulat, J., Horschowski, N., Lavaut, M. N., and Piana, L. 1995. CD31 quantitative immunocytochemical assays in breast carcinomas. *Am. J. Clin. Pathol.* 103:443–448.

Chaudhury, T. K., Lerner, M. P., and Nordquist, R. E. 1980. Angiogenesis by human melanoma and breast cancer cells. *Cancer Lett.* 11:43–49.

Cheng, W. F., Lee, C. N., Chu, J. S., Chen, C. A., Chen, T. M., Shau, W. Y., Hsieh, C. Y., and Hsieh, F. J. 1999. Vascularity index as a novel parameter for the in vivo assessment of angiogenesis in patients with cervical carcinoma. *Cancer* 85:651–657.

Chu, J. S., Huang, C. S., et al. 1998. The prognostic significance of tumor angiogenesis in Taiwanese patients with invasive ductal breast carcinomas. *Cancer Lett.* 134(1):7–14.

Clahsen, P. C., van de Velde, C. J., Duval, C., Pallud, C., Mandard, A. M., Delobelle-Deroide, A., van den Broek, L., Sahmoud, T. M., and van de Vijver, M. J. 1998. p53 protein accumulation and response to adjuvant chemotherapy in premenopausal women with node-negative early breast cancer. *J. Clin. Oncol.* 16:470–479.

Colomer, R., Aparicio, J., Montero, S., Guzman, C., Larrodera, L., and Cortes-Funes, H. 1997. Low levels of basic fibroblast growth factor (bFGF) are associated with a poor prognosis in human breast carcinoma. *Br. J. Cancer* 76:1215–1220.

Cosgrove, D. O., Kedar, R. P., Bamber, J. C., al-Murrani, B., Davey, J. B. Fisher, C., McKinna, J. A., Svensson, W. E., Tohno, E., and Vagios, E. 1993. Breast diseases: color Doppler US in differential diagnosis. *Radiology* 189:99–104.

Costello, P., McCann, A., Carney, D. N., and Dervan, P. A. 1995. Prognostic significance of microvessel density in lymph node negative breast carcinoma. *Hum. Pathol.* 26:1181–1184.

DeLisser, H. M., Christofidou-Solomidou, M., Strieter, R. M., Burdick, M. D., Robinson, C. S., Wexler, R. S., Kerr, J. S., Garlanda, C., Merwin, J. R., Madri, J. A., and Albelda, S. M. 1997. Involvement of endothelial PECAM-1/CD31 in angiogenesis. *Am. J. Pathol.* 151:671–677.

Dickinson, A. J., Fox, S. B., Persad, R. A., Hollyer, J., Sibley, G. N., and Harris, A. L. 1994. Quantification of angiogenesis as an independent predictor of prognosis in invasive bladder carcinomas. *Br. J. Urol.* 74:762–766.

Dixon, J. M., Walsh, J., Paterson, D., and Chetty, U. 1992. Colour Doppler ultrasonography studies of benign and malignant breast lesions. *Br. J. Surg.* 79:259–260.

Dvorak, H. F., Nagy, J. A., Dvorak, J. T., and Dvorak, A. M. 1988. Identification and characterisation of the blood vessels of solid tumours that are leaky to circulating macromolecules. *Am. J. Pathol.* 133:95–109.

Engels, K., Fox, S. B., Whitehouse, R. M., Gatter, K. C., and Harris, A. L. 1997a. Distinct angiogenic patterns are associated with high-grade in situ ductal carcinomas of the breast. *J. Pathol.* 181:207–212.

Engels, K., Fox, S. B., Whitehouse, R. M., Gatter, K. C., and Harris, A. L. 1997b. Up-regulation of thymidine phosphorylase expression is associated with a discrete pattern of angiogenesis in ductal carcinomas in situ of the breast. *J. Pathol.* 182:414–420.

Evans, S. M., Laughlin, K. M., Pugh, C. R., Sehgal, C. M., and Saunders, H. M. 1997. Use of power Doppler ultrasound-guided biopsies to locate regions of tumour hypoxia. *Br. J. Cancer* 76:1308–1314.

Fisher, C., Gilbertson, B. S., Powers, E. A., Petzold, G., Poorman, R., and Mitchell, M. A. 1994. Interstitial collagenase is required for angiogenesis in vitro. *Dev. Biol.* 162:499–510.

Folkman, J. 1971. Tumor angiogenesis: therapeutic implications. *N. Engl. J. Med.* 285:82–86.

Fox, S. B., Engels, K., Comley, M., Whitehouse, R., Turley, H., Gatter, K., and Harris, A. 1997a. Relationship of elevated tumour thymidine phosphorylase in node positive breast carcinomas to the effects of adjuvant CMF. *Ann. Oncol.* 8:271–275.

Fox, S. B., Gatter, K., Bicknell, R., Going, J., Stanton, P., Cooke, T., and Harris, A. 1993a. Relationship of endothelial cell proliferation to tumor vascularity in human breast cancer. *Cancer Res.* 53:9161–9163.

Fox, S. B., Gatter, K. C., Leek, R. D., Harris, A. L., Chew, K., Mayall, B., and Moore, D. H. 2nd. 2000b. Tumor angiogenesis as a prognostic assay for invasive ductal breast carcinoma. *J. Natl. Cancer Inst.* 92:151–152.

Fox, S. B., and Harris, A. L. 2001. Diagnostic and prognostic significance of tumour angiogenesis. In: Fan, T., and Auerbach, R. eds. *New Angiotherapy*. Humana Press, Cambridge, MA.

Fox, S. B., and Harris, A. L. 1997. Markers of tumor angiogenesis: clinical applications in prognosis and anti-angiogenic therapy. *Invest. New Drugs* 15:15–28.

Fox, S. B., Kakolyris, S., Taylor, M., Gatter, K. C., and Harris, A. L. 2000a. Vascular remodelling in breast cancer angiogenesis assessed by LH39 basement membrane antigen and CD31. *Br. J. Cancer* 82:844–851.

Fox, S. B., Leek, R. D., Bliss, J., Mansi, J. L., Gusterson, B., Gatter, K. C., and Harris, A. L. 1997b. Association of tumor angiogenesis with bone marrow micrometastases in breast cancer patients. *J. Natl. Cancer Inst.* 89:1044–1049.

Fox, S. B., Leek, R., Smith, K., Hollyer, J., Greenall, M., and Harris, A. 1994. Tumor angiogenesis in node negative breast carcinomas—relationship to epidermal growth factor receptor and survival. *Breast Cancer Res. Treat.* 29:109–116.

Fox, S. B., Leek, R. D., Weekes, M. P., Whitehouse, R. M., Gatter, K. C., and Harris, A. L. 1995b. Quantitation and prognostic value of breast cancer angiogenesis: comparison of microvessel density, Chalkley count, and computer image analysis. *J. Pathol.* 177:275–283.

Fox, S. B., Stuart, N., Smith, K., Brunner, N., and Harris, A. L. 1993b. High levels of uPA and PAI-1 are associated with highly angiogenic breast carcinomas. *J. Pathol.* 170:388–398.

Fox, S. B., Turner, G., Gatter, K., and Harris, A. 1995a. The increased expression of adhesion molecules ICAM-3, E and P selectin on breast cancer endothelium. *J. Pathol.* 177:369–376.

Fox, S. B., Westwood, M., Moghaddam, A., Comley, M., Turley, H., Whitehouse, R. M., Bicknell, R., Gatter, K. C., and Harris, A. L. 1996. The angiogenic factor platelet-derived endothelial cell growth factor/thymidine phosphorylase is up-regulated in breast cancer epithelium and endothelium. *Br. J. Cancer* 73:275–280.

Fregene, T. A., Kellog, C., and Pienta, K. J. 1994. Microvessel quantitation as a measure of angiogenic activity in benign breast tissues lesions: a marker for precancerous disease? *Int. J. Oncol.* 4:1199–1202.

Frouge, C., Guinbretiere, J., Contesso, G., di Paola, R., and Blery, M. 1994. Correlation between contrast enhancement in dynamic magnetic resonance imaging of the breast and tumor angiogenesis. *Invest. Radiol.* 29:1043–1049.

Furusato, M., Wakui, S., Sasaki, H., Ito, K., and Ushigome, S. 1994. Tumour angiogenesis in latent prostatic carcinoma. *Br. J. Cancer* 70:1244–1246.

Gasparini, G., Fox, S. B., Verderio, P., Bonoldi, E., Bevilacqua, P., Borrachi, P., Dante, S., Marubini, E., and Harris, A. 1996. Determination of angiogenesis adds information to estrogen receptor status in predicting the efficacy of adjuvant tamoxifen in node-positive breast cancer patients. *Clin. Cancer Res.* 2:1191–1198.

Gasparini, G., Toi, M., Gion, M., Verderio, P., Dittadi, R., Hanatani, M., Matsubara, I., Vinante, O., Bonold, E., Boracchi, P., Gatti, C., Suzuki, H., and Tominaga, T. 1997. Prognostic-significance of vascular endothelial growth-factor protein in node-negative breast-carcinoma. *J. Natl. Cancer Inst.* 89:139–147.

Gasparini, G., Toi, M., et al. 1998. Prognostic significance of p53, angiogenesis, and other conventional features in operable breast cancer: Subanalysis in node-positive and node-negative patients. *Int. J. Oncol.* 12(5):1117–1125.

Gasparini, G., Toi, M., Miceli, R., Vermeulen, P. B., Dittadi, R., Biganzoli, E., Morabito, A., Fanelli, M., Gatti, C., Suzuki, H., Tominaga, T., Dirix, L. Y., and Gion, M. 1999. Clinical relevance of vascular endothelial growth factor and thymidine phosphorylase in patients with node-positive breast cancer treated with either adjuvant chemotherapy or hormone therapy. *Cancer J. Sci. Am.* 5:101–111.

Gilles, R., Zafrani, B., Guinebretiere, J. M., Meunier, M., Lucidarme, O., Tardivon, A. A., Rochard, F., Vanel, D., Neuenschwander, S., and Arriagada, R. 1995. Ductal carcinoma in situ: MR imaging-histopathologic correlation. *Radiology* 196:415–419.

Gimbrone, M. A., Jr., and Gullino, P. M. 1976a. Neovascularization induced by intraocular xenografts of normal, preneoplastic, and neoplastic mouse mammary tissues. *J. Natl. Cancer Inst.* 56:305–318.

Gimbrone, M. A. J., and Gullino, P. M. 1976b. Angiogenic capacity of preneoplastic lesions of the murine mammary gland as a marker of neoplastic transformation. *Cancer Res.* 36:2611–2620.

Goulding, H., Abdul, R. N., et al. 1995. Assessment of angiogenesis in breast carcinoma: an important factor in prognosis? *Hum. Pathol.* 26(11):1196–1200.

Guidi, A., Fischer, L., Harris, J., and Schnitt, S. 1994. Microvessel density and distribution in ductal carcinoma in situ of the breast. *J. Natl. Cancer Inst.* 86:614–619.

Guinebretiere, J. M., Le, M. G., Gavoille, A., Bahi, J., and Contesso, G. 1994. Angiogenesis and risk of breast cancer in women with fibrocystic disease. *J. Natl. Cancer Inst.* 86:635–636.

Hall, N. R., Fish, D. E., Hunt, N., Goldin, R. D., Guillou, P. J., and Monson, J. R. 1992. Is the relationship between angiogenesis and metastasis in breast cancer real? *Surg. Oncol.* 1:223–229.

Hanahan, D., and Folkman, J. 1996. Patterns and emerging mechanisms of the angiogenic switch during tumorigenesis. *Cell* 86:353–364.

Hansen, S., Grabau, D. A., Rose, C., Bak, M., and Sorensen, F. B. 1998. Angiogenesis in breast cancer: a comparative study of the observer variability of methods for determining microvessel density. *Lab. Invest.* 78:1563–1573.

Heffelfinger, S., Yassin, R., Miller, M., and Lower, E. 1996. Vascularity of proliferative breast disease and carcinoma in situ correlates with histological features. *Clin. Cancer Res.* 2:1873–1878.

Heinmann, R., Ferguson, D., et al. 1996. Angiogenesis as a predictor of long term survival for patients with node-negative breast cancer. *J. Natl. Cancer Inst.* 88:1764–1769.

Honkoop, A. H., van Diest, P. J., de Jong, J. S., Linn, S. C., Giaccone, G., Hoekman, K., Wagstaff, J., and Pinedo, H. M. 1998. Prognostic role of clinical, pathological and biological characteristics in patients with locally advanced breast cancer. *Br. J. Cancer* 77:621–626.

Horak, E. R., Leek, R., Klenk, N., LeJeune, S., Smith, K., Stuart, N., Greenall, M., Stepniewska, K., and Harris, A. L. 1992. Angiogenesis, assessed by platelet/endothelial cell adhesion molecule antibodies, as indicator of node metastases and survival in breast cancer. *Lancet* 340:1120–1124.

Huber, S. 1998. Doppler ultrasound in the diagnosis of breast lesions. *Anticancer Res.* 18:2147–2150.

Hulka, C. A., Edmister, W. B., Smith, B. L., Tan, L., Sgroi, D. C., Campbell, T., Kopans, D. B., and Weisskoff, R. M. 1997. Dynamic echo-planar imaging of the breast: experience in diagnosing breast carcinoma and correlation with tumor angiogenesis. *Radiology* 205:837–842.

Hulka, C. A., Smith, B. L., Sgroi, D. C., Tan, L., Edmister, W. B., Semple, J. P., Campbell, T., Kopans, D. B., Brady, T. J., and Weisskoff, R. M. 1995. Benign and malignant breast lesions: differentiation with echo-planar MR imaging. *Radiology* 197:33–38.

Jacquemier, J. D., Penault-Llorca, F. M., Bertucci, F., Sun, Z. Z., Houvenaeghel, G. F., Geneix, J. A., Puig, B. D., Bardou, V. J., Hassoun, J. A., Birnbaum, D., and Viens, P. J. 1998. Angiogenesis as a prognostic marker in breast carcinoma with conventional adjuvant chemotherapy: a multiparametric and immunohistochemical analysis. *J. Pathol.* 184:130–135.

Jensen, H. M., Chen, I., De, V. M., and Lewis, A. E. 1982. Angiogenesis induced by normal human breast tissue: a probable marker for precancer. *Science* 218:293–295.

Jitsuiki, Y., Hasebe, T., Tsuda, H., Imoto, S., Tsubono, Y., Sasaki, S., and Mukai, K. 1999. Optimizing microvessel counts according to tumor zone in invasive ductal carcinoma of the breast. *Mod. Pathol.* 12:492–498.

Kaczmarek, J., Castellani, P., Nicolo, G., Spina, B., Allemanni, G., and Zardi, L. 1994. Distribution of oncofetal fibronectin isoforms in normal, hyperplastic and neoplastic human breast tissues. *Int. J. Cancer* 58:11–16.

Karelia, N. H., Patel, D. D., et al. 1997. Prognostic significance of tumor angiogenesis in advanced breast carcinoma: An Indian experience. *Neoplasma* 44(3):163–166.

Kato, T., Kimura, T., et al. 1997. Clinicopathologic study of angiogenesis in Japanese patients with breast cancer. *World Journal of Surgery* 21(1):49–56.

Kato, T., Kimura, T., et al. 1999. The methodology of quantitation of microvessel density and prognostic value of neovascularization associated with long-term survival in Japanese patients with breast cancer. *Breast Cancer Res. Treat.* 53(1):19–31.

Kedar, R. P., Cosgrove, D. O., Bamber, J. C., and Bell, D. S. 1995. Automated quantification of color Doppler signals: a preliminary study in breast tumors. *Radiology* 197:39–43.

Kedar, R. P., Cosgrove, D. O., Smith, I. E., Mansi, J. L., and Bamber, J. C. 1994. Breast carcinoma: measurement of tumor response to primary medical therapy with color Doppler flow imaging. *Radiology* 190:825–830.

Kohlberger, P., Obermair Sliutz, G., et al. 1996. Quantitative immunohistochemistry of factor VIII-related antigen in breast carcinoma. *Am. J. Clin. Pathol.* 105:705–710.

Kondo, S., Asano, M., Matsuo, K., Ohmori, I., and Suzuki, H. 1994. Vascular endothelial growth factor/vascular permeability factor is detectable in the sera of tumor-bearing mice and cancer patients. *Biochem. Biophys. Acta* 1221:211–214.

Kraft, A., Weindel, K., Ochs, A., Marth, C., Zmija, J., Schumacher, P., Unger, C., Marmé, D., and Gastl, G. 1999. Vascular endothelial growth factor in the sera and effusions of patients with malignant and nonmalignant disease. *Cancer* 85:178–187.

Kumar, S., Ghellal, A., Li, C., Byrne, G., Haboubi, N., Wang, J. M., and Bundred, N. 1999. Breast carcinoma: vascular density determined using CD105 antibody correlates with tumor prognosis. *Cancer Res.* 59:856–861.

Kuzu, I., Bicknell, R., Harris, A. L., Jones, M., Gatter, K. C., and Mason, D. Y. 1992. Heterogeneity of vascular endothelial cells with relevance to diagnosis of vascular tumours. *J. Clin. Pathol.* 45:143–148.

Lakhani, S. R. 1999. The transition from hyperplasia to invasive carcinoma of the breast. *J. Pathol.* 187:272–278.

Lee, A., Happerfield, L., and Bobrow, L. 1997a. Comparison of four endothelial markers for assessing angiogenesis in carcinoma of the breast. *J. Cell Pathol.* 2:67–73.

Lee, A. H., Happerfield, L. C., Bobrow, L. G., and Millis, R. R. 1997b. Angiogenesis and inflammation in ductal carcinoma in situ of the breast. *J. Pathol.* 181:200–206.

Lee, W. J., Chu, J. S., Houng, S. J., Chung, M. F., Wang, S. M., and Chen, K. M. 1995. Breast cancer angiogenesis: a quantitative morphologic and Doppler imaging study. *Ann. Surg. Oncol.* 2:246–251.

Lin, G., Lunderquist, A., Hagerstrand, I., and Boijsen, E. 1984. Postmortem examination of the blood supply and vascular pattern of small liver metastases in man. *Surgery* 96:517–526.

Linderholm, B., Tavelin, B., Grankvist, K., and Henriksson, R. 1998. Vascular endothelial growth factor is of high prognostic value in node-negative breast carcinoma. *J. Clin. Oncol.* 16:3121–3128.

Lipponen, P., Ji, H., et al. 1994. Tumour vascularity and basement membrane structure in breast cancer as related to tumour histology and prognosis. *J. Cancer Res. Clin. Oncol.* 120(11):645–650.

Ludwig, C. U., Aebersold, D. M., and Rothen, C. 1998. Perioperative angiogenesis and outcome in premenopausal women with breast cancer. *Lancet* 352:147.

Macaulay, V. M., Fox, S. B., Zhang, H., Whitehouse, R. M., Leek, R. D., Gatter, K. C., and Harris, A. L. 1995. Breast cancer angiogenesis and tamoxifen resistence. *Endocr. Rel. Cancer* 2:1–8.

Maiorana, A., and Gullino, P. M. 1978. Acquisition of angiogenic capacity and neoplastic transformation in the rat mammary gland. *Cancer Res.* 38:4409–4414.

Makris, A., Powles, T. J., Kakolyris, S., Dowsett, M., Ashley, S. E., and Harris, A. L. 1999. Reduction in angiogenesis after neoadjuvant chemoendocrine therapy in patients with operable breast carcinoma. *Cancer* 85:1996–2000.

Maniotis, A. J., Folberg, R., Hess, A., Seftor, E. A., Gardner, L. M., Pe'er, J., Trent, J. M., Meltzer, P. S., and Hendrix, M. J. 1999. Vascular channel formation by human melanoma cells in vivo and in vitro: vasculogenic mimicry. *Am. J. Pathol.* 155:739–752.

Martin, L., Green, B., Renshaw, C., Lowe, D., Rudland, P., Leinster, S. J., and Winstanley, J. 1997. Examining the technique of angiogenesis assessment in invasive breast cancer. *Br. J. Cancer* 76:1046–1054.

McCulloch, P., Choy, A., and Martin, L. 1995. Association between tumour angiogenesis and tumour cell shedding into effluent venous blood during breast cancer surgery. *Lancet* 346:1334–1335.

Mlynek, M., van Beunigen, D., Leder, L.-D., and Streffer, C. 1985. Measurement of the grade of vascularisation in histological tumour tissue sections. *Br. J. Cancer* 52:945–948.

Morelli, D., Lazzerini, D., Cazzaniga, S., Squicciarini, P., Bignami, P., Maier, J. A., Sfondrini, L., Menard, S., Colnaghi, M. I., and Balsari, A. 1998. Evaluation of the balance between angiogenic and antiangiogenic circulating factors in patients with breast and gastrointestinal cancers. *Clin. Cancer Res.* 4:1221–1225.

Morphopoulos, G., Pearson, M., et al. 1996. Tumour angiogenesis as a prognostic marker in infiltrating lobular carcinoma of the breast. *J. Pathol.* 180(1):44–49.

Mortensen, K., Holck, S., Christensen, I. J., Skouv, J., Hougaard, D. M., Blom, J., and Larsson, L. I. 1999. Endothelial cell nitric oxide synthase in peritumoral microvessels is a favorable prognostic indicator in premenopausal breast cancer patients. *Clin. Cancer Res.* 5:1093–1097.

Narita, M., Nakao, K., et al. 1998. Independent prognostic factors in breast cancer patients. *Am. J. Surg.* 175(1):73–75.

Nguyen, M. 1997. Angiogenic factors as tumor markers. *Invest. New Drugs* 15:29–37.

O'Brien, T., Fox, S. B., Dickinson, A., Turley, H., Westwood, M., Moghaddam, A., Gatter, K., Bicknell, R., and Harris, A. 1996. Expression of the angiogenic factor thymidine phosphorylase/platelet derived endothelial cell growth factor in primary bladder cancers. *Cancer Res.* 56:4799–4804.

Obermair, A., Kucera, E., Mayerhofer, K., Speiser, P., Seifert, M., Czerwenka, K., Kaider, A., Leodolter, S., Kainz, C., and Zeillinger, R. 1997. Vascular endothelial growth factor (VEGF) in human breast cancer: correlation with disease-free survival. *Int. J. Cancer* 74:455–458.

Ogawa, Y., Chung, Y., et al. 1995. Microvessel quantitation in invasive breast cancer by staining for factor VIII-related antigen. *Br. J. Cancer* 71:1297–1301.

Olivarez, D., Ulbright, T., DeRiese, W., Foster, R., Reister, T., Einhorn, L., and Sledge, G. 1994. Neovascularization in clinical stage A testicular germ cell tumor: prediction of metastatic disease. *Cancer Res.* 54:2800–2802.

Omar, W. S., Eissa, S., Moustafa, H., Farag, H., Ezzat, I., and Abdel-Dayem, H. M. 1997. Role of thallium-201 chloride and Tc-99m methoxy-isobutyl-isonitrite (sestaMIBI) in evaluation of breast masses: correlation with the immunohistochemical characteristic parameters (Ki-67, PCNA, Bcl, and angiogenesis) in malignant lesions. *Anticancer Res.* 17:1639–1644.

Ottinetti, A., and Sapino, A. 1988. Morphometric evaluation of microvessels surrounding hyperplastic and neoplastic mammary lesions. *Breast Cancer Res. Treat.* 11:241–248.

Page, C., Rose, M., Yacoub, M., and Pigott, R. 1992. Antigenic heterogeneity of vascular endothelium. *Am. J. Pathol.* 141:673–683.

Page, D. L., and Dupont, W. D. 1992. Breast cancer angio-histogenesis: through a narrow window. *J. Natl. Cancer Inst.* 84:1850–1851.
Page, D. L., and Jensen, R. 1995. Angiogenesis in human breast carcinoma. *Hum. Pathol.* 26:1173–1174.
Paku, S., and Paweletz, N. 1991. First steps of tumor-related angiogenesis. *Lab. Invest.* 65:334–346.
Passe, T. J., Bluemke, D. A., and Siegelman, S. S. 1997. Tumor angiogenesis: tutorial on implications for imaging. *Radiology* 203:593–600.
Paulsen, T., Aas, T., Borresen, A. L., Varhaug, J. E., Lonning, P. E., and Akslen, L. A. 1997. Angiogenesis does not predict clinical response to doxorubicin monotherapy in patients with locally advanced breast cancer. *Int. J. Cancer* 74:138–140.
Paweletz, N., and Knierim, M. 1989. Tumor-related angiogenesis. *Crit. Rev. Oncol. Hematol.* 9:197–242.
Pepper, M., and Montesano, R. 1990. Proteolytic balance and capillary morphogenesis. *Cell. Differ. Dev.* 32:319–328.
Peters-Engl, C., Medl, M., Mirau, M., Wanner, C., Bilgi, S., Sevelda, P., and Obermair, A. 1998. Color-coded and spectral Doppler flow in breast carcinomas—relationship with the tumor microvasculature. *Breast Cancer Res. Treat.* 47:83–89.
Porschen, R., Classen, S., Piontek, M., and Borchard, F. 1994. Vascularization of carcinomas of the esophagus and its correlation with tumor proliferation. *Cancer Res.* 54:587–591.
Porter, P., Patton, K., Self, S., Gown, A., and Schmidt, R. 1993. A quantitative study of blood vessel size and density innormal and neoplastic breast tissue. *Mod. Pathol.* 6:18, abstract 87.
Protopapa, E., Delides, G. S., and Revesz, L. 1993. Vascular density and the response of breast carcinomas to mastectomy and adjuvant chemotherapy. *Eur. J. Cancer* 29A:1141–1145.
Relf, M., LeJeune, S., Scott, P. A., Fox, S., Smith, K., Leek, R., Moghaddam, A., Whitehouse, R., Bicknell, R., and Harris, A. L. 1997. Expression of the angiogenic factors vascular endothelial cell growth factor, acidic and basic fibroblast growth factor, tumor growth factor beta-1, platelet-derived endothelial cell growth factor, placenta growth factor, and pleiotrophin in human primary breast cancer and its relation to angiogenesis. *Cancer Res.* 57:963–969.
Sahin, A., Sneige, N., Singletary, E., and Ayala, A. 1992. Tumor angiogenesis detected by factor-VIII immunostaining in node-negative breast carcinoma (NNBC): a possible predictor of distant metastasis. *Mod. Pathol.* 5:17A.
Salven, P., Perhoniemi, V., Tykka, H., Maenpaa, H., and Joensuu, H. 1999. Serum VEGF levels in women with a benign breast tumor or breast cancer. *Breast Cancer Res. Treat.* 53:161–166.
Schadendorf, D., Heidel, J., Gawlik, C., Suter, L., and Czarnetzki, B. M. 1995. Association with clinical outcome of expression of VLA-4 in primary cutaneous malignant melanoma as well as P-selectin and E-selectin on intratumoral vessels. *J. Natl. Cancer Inst.* 87:366–371.
Schlingemann, R. O., Rietveld, F. J., Kwaspen, F., van d. Kerkhof, P. C., de Waal, R. M., and Ruiter, D. J. 1991. Differential expression of markers for endothelial cells, pericytes, and basal lamina in the microvasculature of tumors and granulation tissue. *Am. J. Pathol.* 138:1335–1347.
Schor, A. M., Pendleton, N., Pazouki, S., Smither, R. L., Morris, J., Lessan, K., Heerkens, E., Chandrachud, L. M., Carmichael, G., Adi, M., Chisholm, D. M., and Stevenson, H. 1998. Assessment of vascularity in histological sections: effects of methodology and value as an index of angiogenesis in breast tumours. *Histochem. J.* 30:849–856.
Sightler, H., Borowsky, A., Dupont, W., Page, D., and Jensen, R. 1994. Evaluation of tumor angiogenesis as a prognostic marker in breast cancer. *Lab. Invest.* 70:22A.

Siitonen, S., Haapasalo, H., Rantala, I., Helin, H., and Isola, J. 1995. Comparison of different immunohistochemical methods in the assessment of angiogenesis: lack of prognostic value in a group of 77 selected node-negative breast carcinomas. *Mod. Pathol.* 8:745–752.
Simpson, J., Ahn, C., Battifora, H., and Esteban, J. 1996. Endothelial surface area as a prognostic indicator for invsasive breast carcinoma. *Cancer* 77(10):2077–2085.
Sipkins, D. A., Cheresh, D. A., Kazemi, M. R., Nevin, L. M., Bednarski, M. D., and Li, K. C. 1998. Detection of tumor angiogenesis in vivo by alpha$_V$beta$_3$-targeted magnetic resonance imaging. *Nat. Med.* 4:623–626.
Skinner, S. A., Frydman, G. M., and O'Brien, P. E. 1995. Microvascular structure of benign and malignant tumors of the colon in humans. *Dig. Dis. Sci.* 40:373–384.
Sterns, E., Zee, B., SenGupta, S., and Saunders, F. 1996. Thermography: its relation to pathologic characteristics, vascularity, proliferation rate and survival of patients with invasive ductal carcinoma of the breast. *Cancer* 77:1324–1328.
Stomper, P. C., Winston, J. S., Herman, S., Klippenstein, D. L., Arredondo, M. A., and Blumenson, L. E. 1997. Angiogenesis and dynamic MR imaging gadolinium enhancement of malignant and benign breast lesions. *Breast Cancer Res. Treat.* 45:39–46.
Svrivastava, A., Laidler, P., Davies, R., Horgan, K., and Hughes, L. 1988. The prognostic significance of tumor vascularity in intermediate-thickness (0.76–4.0 mm thick) skin melanoma. *Am. J. Pathol.* 133:419–423.
Tan, P., Cady, B., et al. 1997. The cell cycle inhibitor p27 is an independent prognostic marker in small (T1a,b) invasive breast carcinomas. *Cancer Res.* 57(7):1259–1263.
Tanigawa, N., Matsumura, M., Amaya, H., Kitaoka, A., Shimomatsuya, T., Lu, C., Muraoka, R., and Tanake, T. 1997. Tumor vascularity correlates with the prognosis of patients with esophageal squamous carcinoma. *Cancer* 79:220–225.
Toi, M., Hoshina, S., Taniguchi, T., Yamamoto, Y., Ishitsuka, H., and Tominaga, T. 1995. Expression of platelet derived endothelial cell growth factor/thymidine phosphorylase in human breast cancer. *Int. J. Cancer* 64:79–82.
Toi, M., Kashitani, J., and Tominaga, T. 1993. Tumor angiogenesis is an independent prognostic indicator in primary breast carcinoma. *Int. J. Cancer* 55:371–374.
Toi, M., Kondo, S., Suzuki, H., Yamamoto, Y., Inada, K., Imazawa, T., Taniguchi, T., and Tominaga, T. 1996a. Quantitative analysis of vascular endothelial growth factor in primary breast cancer. *Cancer* 77:1101–1106.
Toi, M., Taniguchi, T., Yamamoto, Y., Kurisaki, T., Suzuki, H., and Tominaga, T. 1996b. Clinical significance of the determination of angiogenic factors. *Eur. J. Cancer* 32A:2513–2519.
van Diest, P. J. 1999. Ductal carcinoma in situ in breast carcinogenesis. *J. Pathol.* 187:383–384.
van Hoef, M. E., Knox, W. F., Dhesi, S. S., Howell, A., and Schor, A. M. 1993. Assessment of tumour vascularity as a prognostic factor in lymph node negative invasive breast cancer. *Eur. J. Cancer* 29A:1141–1145.
van Zijl, P. C., Eleff, S. M., Ulatowski, J. A., Oja, J. M., Ulug, A. M., Traystman, R. J., and Kauppinen, R. A. 1998. Quantitative assessment of blood flow, blood volume and blood oxygenation effects in functional magnetic resonance imaging. *Nat. Med.* 4:159–167.
Vermeulen, P. B., Gasparini, G., Fox, S. B., Toi, M., Martin, L., McCulloch, P., Pezzella, F., Viale, G., Weidner, N., Harris, A. L., and Dirix, L. Y. 1996. Quantification of angiogenesis in solid human tumours: an international consensus on the methodology and criteria of evaluation. *Eur. J. Cancer* 32A:2474–2484.
Vesalainen, S., Lipponen, P., Talja, M., Alhava, E., and Syrjanen, K. 1994. Tumor vascularity and basement membrane structure as prognostic factors in T1–2M0 prostatic adenocarcinoma. *Anticancer Res.* 14:709–714.

Viens, P., Jacquemier, J., Bardou, V. J., Bertucci, F., Penault-Llorca, F., Puig, B., Gravis, G., Oziel-Taieb, S., Resbeut, M., Houvenaeghel, G., Camerlo, J., Birbaum, D., Hassoun, J., and Maraninchi, D. 1999. Association of angiogenesis and poor prognosis in node-positive patients receiving anthracycline-based adjuvant chemotherapy. *Breast Cancer Res. Treat.* 54:205–212.
Visscher, D., Smilanetz, S., Drozdowicz, S., and Wykes, S. 1993. Prognostic significance of image morphometric microvessel enumeration in breast carcinoma. *Anal. Quant. Cytol.* 15:88–92.
Wakui, S., Furusato, M., Itoh, T., Sasaki, H., Akiyama, A., Kinoshita, I., Asano, K., Tokuda, T., Aizawa, S., and Ushigome, S. 1992. Tumour angiogenesis in prostatic carcinoma with and without bone marrow metastases: a morphometric study. *J. Pathol.* 168:257–262.
Wang, J. M., Kumar, S., Pye, D., Haboubi, N., and Al-Nakib, L. 1994. Breast carcinoma: comparative study of tumor vasculature using two endothelial cell markers. *J. Natl. Cancer Inst.* 86:386–388.
Warren, B. 1979. The vascular morphology of tumors. In: Peterson, H. ed. *Tumor Blood Circulation.* CRC Press, Boca Raton, FL, pp. 1–47.
Weidner, N., Folkman, J., Pozza, F., Bevilacqua, P., Allred, E. N., Moore, D. H., Meli, S., and Gasparini, G. 1992. Tumor angiogenesis: a new significant and independent prognostic indicator in early-stage breast carcinoma. *J. Natl. Cancer Inst.* 84:1875–1887.
Weidner, N., Semple, J. P., Welch, W. R., and Folkman, J. 1991. Tumor angiogenesis and metastasis—correlation in invasive breast carcinoma. *N. Engl. J. Med.* 324:1–8.
Wesseling, P., Vandersteenhoven, J. J., Downey, B. T., Ruiter, D. J., and Burger, P. C. 1993. Cellular components of microvascular proliferation in human glial and metastatic brain neoplasms. A light microscopic and immunohistochemical study of formalin-fixed, routinely processed material. *Acta Neuropathol.* 85:508–514.
Williams, J. K., Carlson, G. W., Cohen, C., Derose, P. B., Hunter, S., and Jurkiewicz, M. J. 1994. Tumor angiogenesis as a prognostic factor in oral cavity tumors. *Am. J. Surg.* 168:373–380.
Ziche, M., and Gullino, P. M. 1982. Angiogenesis and neoplastic progression in vitro. *J. Natl. Cancer Inst.* 69:483–487.
Zimrin, A. B., Pepper, M. S., McMahon, G. A., Nguyen, F., Montesano, R., and Maciag, T. 1996. An antisense oligonucleotide to the notch ligand jagged enhances fibroblast growth factor-induced angiogenesis in vitro. *J. Biol. Chem.* 271:32499–32502.

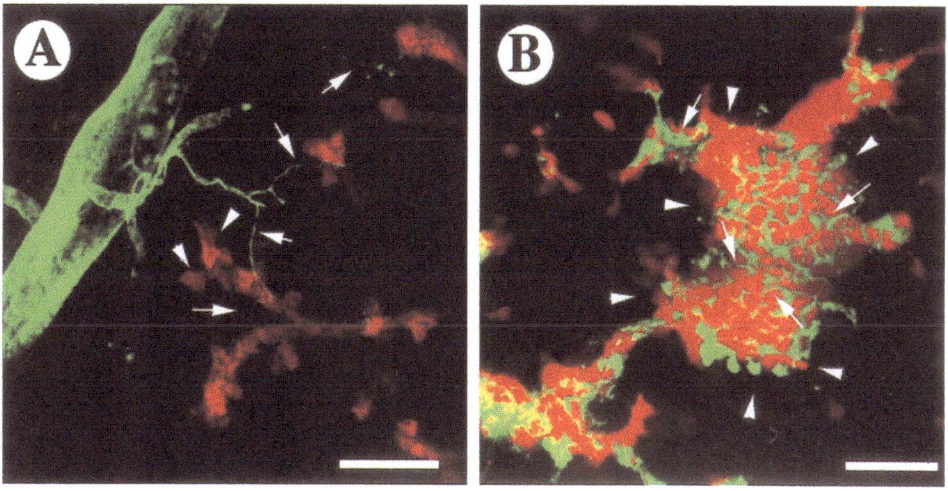

FIGURE 1.4. Confocal visualization of epithelial and endothelial components during murine mammary gland growth. Confocal analysis of whole-mount preparations in which the epithelial components are revealed by pan-keratin antibodies (in red) and the endothelium is visualized by a fluorescein isothiocyanate (FITC)-conjugated lectin (which was perfused through the vasculature prior to the removal of the glands). A: Development of secondary lobuloalveolar units (arrowheads) during early pregnancy (day 5) recruits thin capillary extensions (arrows) from large vessels. B: An elaborate and rich capillary network (arrows) is seen within the lobuloalveolar unit (indicated by arrowheads) late in pregnancy (day 12). At this time, capillaries are in close proximity. Bar in A = 30 μm; Bar in B = 50 μm.

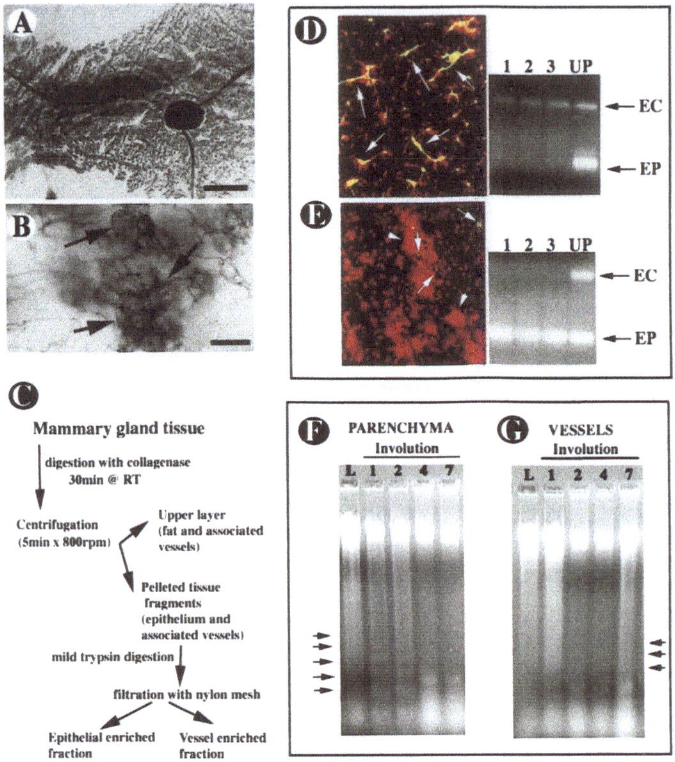

FIGURE 1.5. Epithelial and endothelial apoptosis during murine mammary gland involution. A: Whole-mount preparation of a mammary gland 4 days after removal of pups. The gland has been stained with carmine to reveal the regressing parenchyma. B: High magnification of involuting gland. Vessels associated with the involuting alveolar units were revealed after staining in toto with *Bandeiraea simplicifolia* lectin. C: Experimental procedure to separate epithelial from endothelial components. D: Staining of the vascular fraction after purification. Note enrichment of endothelium (arrows) to epithelium in red. The gel shows a reverse transcriptase–polymerase chain reaction (RT-PCR) that reveals the purity of the preparations. The upper band indicates migration of a CD-31 product. The lower band indicates contaminating epithelial cells, assessed by the levels of keratin 18 PCR product. Lanes 1 to 3, endothelial purifications; lane 4, preparations prior to trypsin digestion (UP, unpurified). E: Visualization of the epithelial fraction after purification with anti–pan-keratin antibodies. Few endothelial cells (in green) can be detected. The RT-PCR shows the purity of the preparation. The lower band represent levels of epithelial cells, as assessed by keratin 18 PCR product. The upper band indicates levels of contaminating endothelial cells. Lanes 1 to 3, epithelilal purifications; lane 4, preparations prior trypsin digestion (UP, unpurified). Southern blots from the (F) epithelial and (G) endothelial fractions during lactation (L) and at several time points during involution are indicated. Note DNA fragmentation at day 1 and 2 in the epithelial fraction (arrows). In the endothelial fraction very low levels of DNA fragmentation were only seen at day 4 and 7 (arrows).

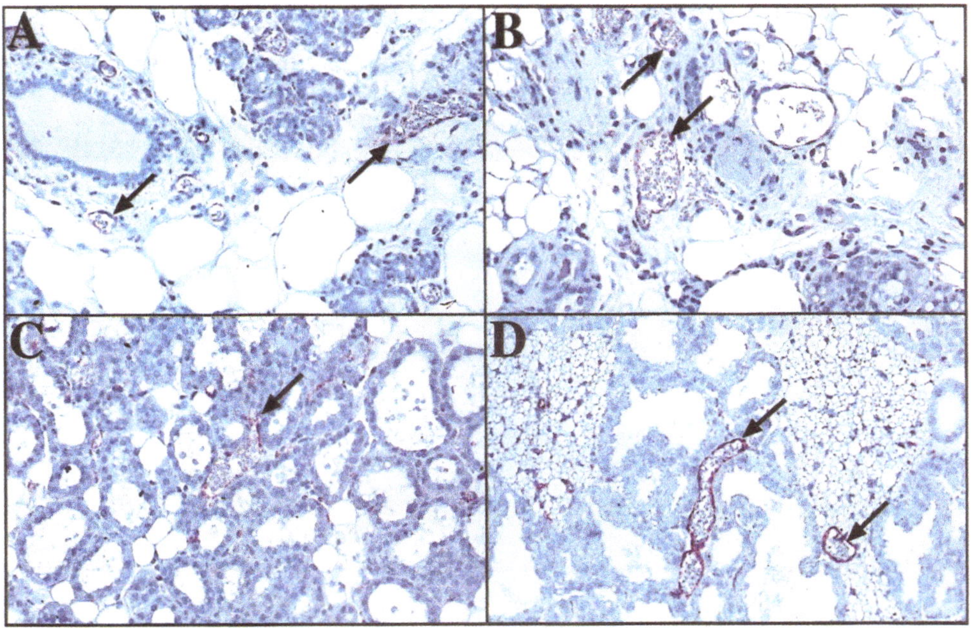

FIGURE 2.1. Mammary gland during pregnancy and lactation in mice. Antiserum against the von Willebrand factor (Dako, code no. A0082) was used to visualize endothelial cells (arrows). A,B: Mammary gland during 12th and 17th day of pregnancy, respectively. Under the influence of estrogens and progesterone, the alveolar duct epithelium proliferates to form secretory alveoli. C,D: Mammary gland during 1st and 12th day of lactation. The breast consists almost entirely of glandular tissue; the alveoli are widely dilated by milk and most of the interlobular connective tissue disappears.

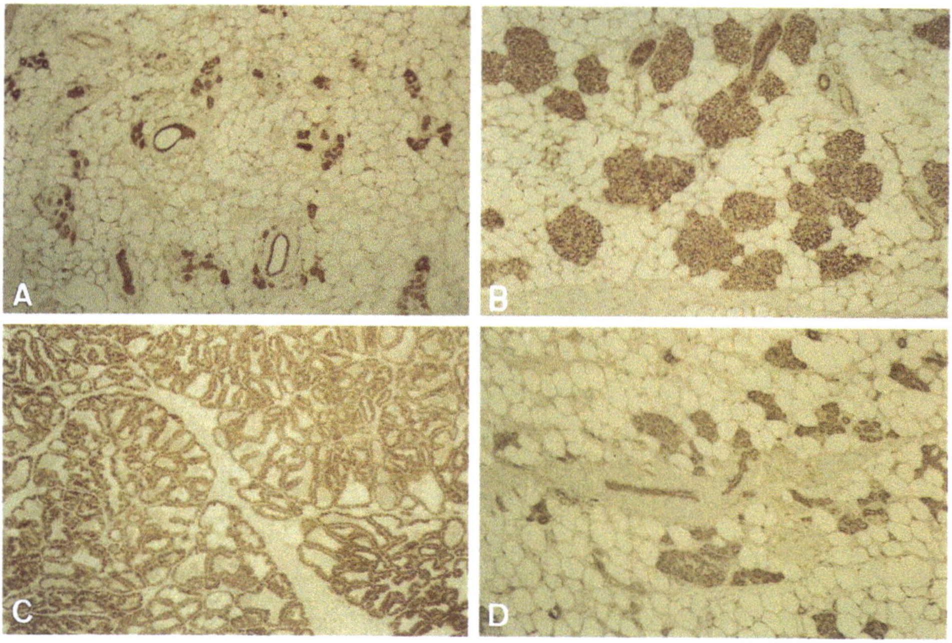

FIGURE 2.3. Immunolocalization of VEGF in rat mammary gland. Bouin-fixed paraffin sections were incubated with an affinity purified rabbit polyclonal antibody prepared against a synthetic peptide corresponding to amino acids 1 to 20 of the amino terminus of human VEGF. A: 3-month virgin. B: 12-day pregnant. C: 7-day lactating. D: 21 days postlactation. Magnification = 60×.

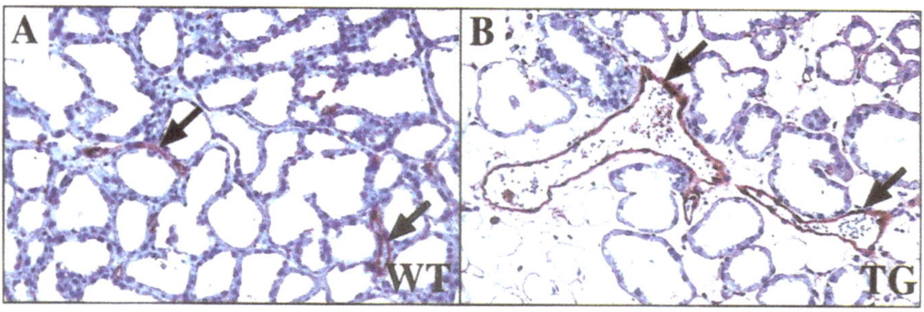

FIGURE 2.4. Blood vessels in lactating mammary glands of mammary tumor virus long terminal repeat (MMTV LTR) VEGF mice. Antiserum against the von Willebrand factor (Dako, code no. A0082) was used to visualize endothelial cells. The mammary glands of the transgenic animals differed markedly from the controls in that the tissue was edematous and the regular alveolar pattern was lost. More numerous, and frequently enlarged vessels (arrows) were seen in the mammary glands of transgenic mice (B) when compared to control mice (A).

CHAPTER 5

Oncogenes, Anti-Oncogenes, and Genetic Regulators of Vascular Development

Timothy F. Lane and Alicia Collado-Hidalgo

The architecture of all tissue represents the cumulative actions of many genes. Similarly, the phenotype of tumors is the product of successive genetic lesions that alter the function and regulation of oncogenes, tumor-suppressors, and modifier genes. A key stage of both tissue development and tumor progression is the recruitment of a vascular network.

From limited experimental and epidemiologic data, it is hypothesized that the response of tumors to specific therapeutic agents may vary considerably (Thompson et al, 1999). In particular, various tumor cell lines produce diverse levels of angiogenic growth factors and respond differently to antiangiogenic therapies (Kim et al, 1993; Takahashi et al, 1996; Yuan et al, 1996). It is therefore critical to understand the contributions of tissue-specific and genetic factors to the vascular response. It is likely that variability in vascular responsiveness will complicate the search for global modulators of angiogenesis. Conversely, such variation may ultimately provide a benefit by increasing the potential for target specificity in distinct disease states.

While there is solid evidence that the progression of solid tumors involves angiogenesis (reviewed in D'Amore and Shima, 1996; Hanahan and Folkman, 1996), it is not immediately clear which features of the tumor vasculature will provide the most appropriate targets for antitumor therapy, or whether these features will vary in different tumor types. This chapter focuses on recent results from studies in mice that begin to define the contributions of specific genes in diverse vascular systems, and the role of genetic diversity in the development of vascular patterning of normal and transformed tissues.

USE OF TRANSGENIC MICE TO STUDY THE IMPACT OF INDIVIDUAL GENES ON CANCER

Heterogeneity of genetic backgrounds, along with developmental age, diet, and a host of other variables, makes direct analysis of gene function a laborious and often imprecise task in humans and in cell lines. For this reason, genetically modified mice provide a unique tool to study the effects of specific genes on tumor growth and on tumor-stromal interactions. Immune-competent mouse models offer the potential to study earlier stages of cancer progression than are possible in humans, and inbred mice offer additional advantages with respect to consistency of phenotype, unlimited access to tissue, and increasingly well-documented parallels to human development and physiology. Importantly, there are rapidly improving resources to identify genetic modifier loci in mice, which will prove critical in understanding genetic heterogeneity and its impact on disease. For specialized treatments of specific topics relevant to the production and analysis of tumor prone mice, several reviews are recommended (Bernstein and Breitman, 1989; Eastin et al, 1998; Macleod and Jacks, 1999; Tallquist et al, 1999; Cardiff et al, 2000).

Transgenic models of cancer provide opportunities to study (1) the effects of single or multiple mutations in all stages of disease progression in an immunocompetent host, (2) the effects of specific therapeutic agents or modifier genes on specific aspects of disease, and (3) the contribution of stromal cells (including endothelial, fibroblast, and immune cells) on tumor progression. While there are some limitations in the direct utility of mouse models for analysis of human genes, they have become an indispensable tool in efforts to improve the overall understanding of molecular mechanisms of disease.

While studies of human cell lines are of indispensable value to molecular studies of disease, the inheritance of multiple uncharacterized mutations in such cells and the limitation of ex vivo analysis make them relatively difficult to interpret in the context of cell behavior in vivo. Clearly, such studies are useful for establishing the effects of treatments on a limited range of cellular responses. Nonetheless, they lack the ability to evaluate the interactions of genes in an intact host and in a neutral genetic background. Eventually, there will be a need for researchers and clinicians to understand the genetic composition of both tumor and host/stromal cells and the interactions of genetic background in disease progression. One promise of the current explosion of information on molecular mechanisms underlying disease is that one may ultimately have access to better markers to define the status and therapeutic potential of specific disease states in individuals. In the case of tumor development, there is a growing appreciation of the diversity of mechanisms that can lead to tumor formation, and a realization that an equally diverse set of treatment options may ultimately be necessary to effectively regulate the process.

Gene targeting in transgenic mice, also referred to as reverse genetics, has progressed rapidly (Bernstein and Breitman, 1989). The ability to derive cells from any tissue that display altered expression of single gene has been a boon for the most advanced studies in every phase of disease. Of tumor-associated gene mutations alone, the opportunity to test specific hypotheses on cells overexpressing ras or epidermal growth factor (EGF) family members (Yuspa et al, 1994; Kim and Muller, 1999), or lacking only p53, *a*deno*p*olyposis coli (APC), or a host of other genes (reviewed in Macleod and Jacks, 1999), has revolutionized our capacity to understand disease pathways. Studies of genetically altered mice have facilitated the first insights into mammalian tumor modifier genes (Parangi et al, 1995; Lifsted et al, 1998). In addition to the methodologies for specifically introducing and removing genes from the entire genome, the use of sequence-specific recombinases (cre and flp) have allowed the targeting of these effects in a manner that was not previously possible (Sauer, 1998).

The use of genetically defined, immunocompetent model mice will ultimately permit the types of rigorous studies needed to define issues related to genetic background and tissue heterogeneity on gene function. This chapter describes some of the resources available, the types of data that have been collected from transgenic mice for dissecting the complex interactions between endothelial cells within the tumor microenvironment, and some recent results on the diversity of angiogenic responses that can be observed within transgenic models of breast cancer in the context of transgenic mice.

THE USE OF TRANSGENIC MICE FOR THE ANALYSIS OF ANGIOGENESIS

Transgenic mice are animals whose DNA has been modified using molecular biology methods and who pass this modification to progeny. The current nomenclature distinguishes transgenic mice generated by random integration, TgN (insertional transgenics), from those generated by targeted integration, TgH (homologous recombination, "knockout," or "knockin"). An extensive discussion of the nomenclature and available lines of mice can be accessed through several Web-based search engines:

1. Genetic nomenclature for mice (http://www.informatics.jax.org/mgihome/nomen/)
2. TBASE (http://tbase.jax.org/)
3. Mouse genome informatics (http://www.informatics.jax.org/)
4. Transgenic database (http://www.biomednet.com/db/mkmd)
5. Internet resources (http://tbase.jax.org/docs/databases.html)
6. More internet resources (http://darwin.ceh.uvic.ca/bigblue/transsites.htm)

A recent survey of these sites identified some 239 publications on mice engineered to overexpress known or putative oncogenes, and 259 publications on mice carrying mutant alleles of specific tumor suppressor loci. While the analysis of angiogenesis in transgenic tumor-prone mice has been relatively neglected, interest in this area is growing rapidly (Table 5.1, part A). This pioneering work has been led by D. Hanahan and colleagues, who have performed extensive analysis of tumor angiogenesis in transgenic models of skin and pancreatic cancer progression (Christofori and Hanahan, 1994; Hanahan and Folkman, 1996). Also evident, are the growing number of targeted deletions of genes thought to regulate vascular development. Some notable examples are included in Table 5.1, part B.

Because of the use of inbred lines, transgenic mice present a homogeneous genetic background in which to assess the involvement of specific genes. One can go on to investigate the interactions of genes by mating lines that contain single mutations and then analyze the phenotype of bigenic offspring. Over 100 different strains of transgenic mice have been generated carrying alterations in genes that affect vascular parameters, and well over 400 have been made involving genes that affect tumor predisposition. Thus, there is considerable opportunity, largely untapped, to study angiogenic parameters and therapeutic strategies in these animals.

While some of the phenotypes listed in Table 5.1 might not seem surprising, many were completely unexpected. The requirement for arylhydrocarbon

TABLE 5.1. Some transgenic models in which the vasculature* has been studied.

Transgene	Allele type	Phenotype	Tumor analysis†	Reference‡
(A) Overexpression phenotypes				
Angiogenin-1	TgN	AS	ND	(Suri et al, 1998)
Angiogenin-2	TgN	Disruption of emb. vasc.	ND	(Maisonpierre et al, 1997)
erbB2/neu	TgN	AS, TP	Yes	(Di Carlo et al, 1999)
FGF-2	TgN	AS	ND	(Fulgham et al, 1999)
fps/fep	TgN	AS, hemangiomas	Yes	(Green et al, 2000)
MMP-3 (activated)	TgN	AS, TP	Yes	(Sternlicht et al, 2000)
Neuropilin	TgN	AS	ND	(Kitsukawa et al, 1995)
v-*src*	TgN	AS, TP	Yes	(Theurillat et al, 1999)
Sv40T antigen	TgN	AS, TP	Yes	(Bergers et al, 1999) (Ganss and Hanahan, 1998)
Thrombospondin-1	TgN	AI		(Iruela-Arispe et al, 1999)
			Yes	(Ortega and Iruela-Arispe, submitted)
VEGF	TgN	AS, TP	Yes	(Larcher et al, 1998) (Zeng et al, 1998) (Detmar et al, 1998)
(B) Loss-of-function phenotypes				
Angiopoitin-1/2	TgH	Def. emb. vasc.	ND	(Suri et al, 1996)
ARNT	TgH	Def. emb. vasc.	ND	(Kozak et al, 1997)
		Def. VEGF expression		(Maltepe et al, 1997)
bmp4	TgH	EC overgrowth	ND	(Lawson et al, 1999)
crebbp	TgH	Def. vasc. remodeling	ND	(Oike et al, 1999)
dHAND	TgH	Def. emb. vasc.	ND	(Yamagishi et al, 2000)
EphB2/B3 Rc	TgH	Def. angiogenesis	ND	(Adams et al, 1999) (Wang et al, 1998a)
endoglin	TgH	Def. emb. vasc. remodeling	ND	(Arthur et al, 2000) (Li et al, 1999)
epo	TgH	Def. cardiac vasc.	ND	(Wu et al, 1999)
factor VII	TgH	Vasc. hemorrhage	ND	(Rosen et al, 1997)
FGF-1/FGF-2	TgH	No vascular defects	ND	(Miller et al, 2000)
Fibronectin	TgH	EC migration defects	ND	(George et al, 1997)
HIF-1α	TgH	Def. hypoxia response	Yes	(Carmeliet et al, 1998)
Integrin α_v	TgH	Brain vasc. def.	ND	(Bader et al, 1998)
Integrin α_4	TgH	Pericardial vasc. def.	ND	(Yang et al, 1995)
Integrin β_1	TgH	Teratomas w/ smaller vessels	Yes	(Block and Poncz, 1995)
Id1/Id3	TgH	Def. angiogenesis	Yes	(Lyden et al, 1999)
Lmo2	TgH	Def. large vessels/ remodeling	ND	(Yamada et al, 2000)
MEF-2C	TgH	Def. vasculature	ND	(Bi et al, 1999)
Mesp	TgH	Def. heart formation	ND	(Saga et al, 1999)
mek-1	TgH	Placental vasc. def.	ND	(Giroux et al, 1999)
myb	TgH	No vasc. def.	ND	(Krause et al, 1998)
Neuropilin-1	TgH	Def. yolk sac and large vessels	ND	(Kawasaki et al, 1999)
eNOS/iNOS/nNOS	TgH	Role for eNOS and nNOS in vasc. tone		(Fagan et al, 1999a)
		Hyperresponsive to hypoxia		(Fagan et al, 1999b)
p53	TgH	No vasc. phenotype	Yes	(Tyner et al, 1999) (Wang et al, 1998b)

TABLE 5.1. *Continued*

Transgene	Allele type	Phenotype	Tumor analysis[†]	Reference[‡]
PDGF-B and Rcβ	TgH	Def. pericytes/ hemorrhage	ND	(Lindahl et al, 1997)
prox1	TgH	Def. lymph angiogenesis	ND	(Wigle and Oliver, 1999)
sapin	TgH	Def. placental and heart EC	ND	(Sapin et al, 1997)
SCL/Tal1	TgH	Def. vasc. remodeling	ND	(Visvader et al, 1998)
smad 5	TgH	Def. vasc.	ND	(Yang, 1999)
TGF-β	TgH	Def. vasc.	ND	(Martin et al, 1995)
TGF-βRc type II	TgH	AS, TP	Yes	(Go et al, 1999)
TSP-1	TgH	Inc. vascularity	ND	(Lawler et al, 1998)
TSP-2	TgH	Inc. vascularity in dermis	ND	(Kyriakides et al, 1998)
Tissue factor (TF)	TgH	Def. emb. vasc.	ND	(Carmeliet et al, 1996)
TFPI	TgH	Emb. vasc. hemorrhage	ND	(Huang et al, 1997)
Tie-1	TgH	Vasc. maturation/ hemorrhage	ND	(Sato et al, 1995)
Tie-2/Tek-1	TgH	Def. vascular remodeling	ND	(Sato et al, 1995) (Dumont et al, 1994)
TNC	TgH	No vascular defect	Yes	(Talts et al, 1999)
VCAM	TgH	Def. emb. vasculature	ND	(Kwee et al, 1995; Terry et al, 1997)
VE-CAD	TgH	Def. vasc. remodeling	ND	(Carmeliet et al, 1999a)
VEGFA	TgH	Def. angioblast development		(Carmeliet et al, 1999b)
		Role in ras mediated tumors	Yes	(Grunstein et al, 1999)
VEGFRc1/flt-1	TgH	Def. emb. vasculature remodeling	ND	(Fong et al, 1995)
VEGFRc2/flk-1	TgH	Def. yolk-sac blood vessels	ND	(Shalaby et al, 1995)
VEGFRc3	TgH	Def. capillary remodeling	ND	(Dumont et al, 1998)
(C) Endothelial promoters characterized in mice				
VE-CAD	TgN	Ubiquitous EC expression		(Gory et al, 1999)
Tie-1	TgN	Heterogeneous EC expression		(Sato et al, 1995) (Korhonen et al, 1995)
Tie-2/Tek-1	TgN	Ubiquitous EC expression		(Schlaeger et al, 1997)
endothelin1	TgN	EC and some SMC expression		(Harats et al, 1995)
vWF	TgN	Heterogeneous EC		(Aird et al, 1995)
ICAM-2	TgN	EC, neutrophils, monocytes		(Cowan et al, 1996)
Flk-1	TgN	EC, angioblasts		(Kappel et al, 1999)

* Genetically manipulated inbred mouse strains with intact immune systems.

† Was the role in tumor development accessed?

‡ See additional references cited by the authors.

AS, angiogenic stimulation; AI, inhibition of angiogenesis; def., defect; EC, endothelial cell; emb., embryonic; ND, not determined; SMC, smooth muscle cell; TgN, transgenic overexpression; TgH, targeted mutation introduced by homologous recombination; TP, tumor promotion; TI, tumor inhibition; vasc., vasculature.

receptor nuclear translocator (ARNT) and its dimerization partner hypoxia-inducible factor-1α (HIF-1α) to mediate hypoxia-induced vascular endothelial growth factor (VEGF) induction (Maltepe et al, 1997; Carmeliet et al, 1998) is a prime example of a pathway defined through gene ablation, and has clear potential as a therapeutic target. Also surprising, yet informative, are the unexpected vascular phenotypes of transcription factors such as prox1, Id1, Id3, dHAND, and the lack of significant vascular phenotypes in fibroblast growth factor-1 (FGF-1), FGF-2, and p53 knockout mice. Of particular value, has been the growing appreciation for the diversity of genes that regulate vascular patterning. Loss of transcription factor prox1 ablates lymph-angiogenesis specifically (Wigle and Oliver, 1999), while the growth factor angiopoietin-1 (Suri et al, 1996) and its receptor Tie-2 (Dumont et al, 1994; Sato et al, 1995) are required for hierarchical vascular structures and vascular remodeling, but not for earlier establishment of the embryonic vasculature. One clear take-home message from these studies is the potential resource that transgenic animals provide in isolating cell types, such as endothelial and smooth muscle cells, that lack specific gene products. These cells can be used to define biochemical pathways such as the requirement of urokiuase-type plasminosen activator (uPA) in activation of L-transforming growth factor-β_1 (L-TGF-β_1) (Herbert and Carmeliet, 1997). Although the effects of many of the induced mutations have yet to be analyzed in the context of tumor promotion, there is a clear expectation that such analysis could provide insight into pathways that hold promise for therapeutic manipulation. In those cases in which null mutations lead to embryonic phenotypes, such analysis will have to await the development of tissue-specific or inducible deletions of the genes in question, which could facilitate investigations of adult phenotypes. A large literature has developed on such systems using sequence-specific recombinases such as cre (Sauer, 1998) and flp (Vooijs et al, 1998). By linking expression of cre-recombinase to endothelial-specific promoters (Table 5.1, part C), several genes have already been ablated from specific vascular cells. These include VEGF (Grunstein et al, 1999) and vascular cell adhesion molecule-1 (VCAM-1) (Terry et al, 1997). Drug-inducible forms of cre and flp seem to be particularly powerful tools for analyzing the role of essential genes in many physiologic states that are not currently accessible by standard targeting strategies (Brocard et al, 1998; Sawicki et al, 1998).

GENETIC MODIFIERS OF ANGIOGENESIS AND VESSEL MORPHOLOGY

There is very little information on classic modifier loci that directly influence angiogenesis; however, major inroads into this potentially important area are beginning to be made. Strain-specific susceptibility to cancer formation in inbred mice has been used to identify loci that affect a variety of tumor phenotypes, such as susceptibility to colon cancer (Cormier et al, 1997; Dove et al, 1998) and retinoblastoma (Griep et al, 1998). However, the identity of modifiers in tumor angiogenesis in these models is currently unknown.

That genetic variants play a role in vascular phenotypes is no longer in doubt. Recently, D'Amato and colleagues have demonstrated large genetic contributions to the response of mice to the angiogenic stimulators FGF-2 and VEGF (Rohan et al, 2000). They demonstrated a 10-fold difference in the response of 129/SvImJ mice and C57BL/6J to FGF-2 in corneal pocket assays. Furthermore, they also found strong evidence of genetic modifiers of the antiangiogenic factors thalidomide and TNP-470. While the D'Amato study is the clearest evidence that genetic modifiers are involved in the magnitude of the angiogenic response, other studies on the genetic susceptibility of mice to tumor progression may also uncover modifiers of this class. Studies that have mapped tumor susceptibility in lung cancer have shown that different modifiers affect tumor initiation and progression. Coussens and colleagues (1996) showed extensive strain differences to tumor progression of squamous carcinomas that develop in the K14-HPV16 transgenic mice. Finally, there is evidence of alleles of a locus on chromosome 4 that predispose to large tumor size in a lung tumor model (Manenti et al, 1997). If modifiers were identified that affect the vascularization of tumors, it is likely that they would be of a class that influences tumor progression as opposed to initiation. Thus, one could potentially devise screens to isolate such variants. With rapid improvement in the tools available for genetic analysis in mice (Macleod and Jacks, 1999), it is very likely that steady gains will be made toward identifying these and other genetic modifiers of angiogenesis.

BLOOD VESSEL RECRUITMENT AND THE ROLE OF MOUSE MODELS IN GENETIC ANALYSIS OF TUMOR ANGIOGENESIS

The mechanism of blood vessel recruitment in tissues has been extensively investigated in mouse models. From analysis of knockout mice, it is clear that VEGF and angiopoietin-1 (Ang-1) act as principal positive comodulators of angiogenesis (see Chapter 2). Conversely, Ang-2 and thrombospondin-1 (TSP-1), are emerging as potent negative regulators (Iruela-Arispe and Dvorak, 1997; Maisonpierre et al, 1997). Targeted gene inactivation studies in mice have shown that VEGF is necessary for very early stages of vascular development, whereas Ang-1 is required for vascular remodeling at later stages (Suri et al, 1996; Carmeliet and Collen, 1999). In addition, mouse models have provided extensive support for the angiogenic-switch hypothesis, which suggests the solid tumors progress from avascular, to vascular stages, in discrete genetic steps (Folkman and Hanahan, 1991; Hanahan and Folkman, 1996). New evidence suggests a potential role for other additional mechanisms of vascular recruitment in tumors such as "co-opting" of existing blood vessels (Holash et al, 1999a,b). Such alternative mechanisms have yet to be demonstrated in model systems, and will undoubtedly benefit from more extensive molecular characterization.

Angiogenic regulators include a variety of primary oncogenes and tumor suppressor genes as well as genetic modifier genes that affect general angiogenic responsiveness within an individual. Growth factors such as TGF-α, EGF, platelet-derived growth factor-BB (PDGF-BB), TGF-β, and several FGF family members (HSTF1, int2) are primary oncogenes that also enhance the recruitment of vascular endothelial cells (Folkman and Klagsbrun, 1987). TSP-1 is currently the most important of the classic tumor suppressor genes known to have direct antiangiogenic activity in tumors (Good et al, 1990; Streit et al, 1999). In addition, it is becoming increasingly clear that a large number of tumor genes regulate the production of angiogenic factors (Rak et al, 1995; Ravi et al, 2000), and thus regulate angiogenesis directly or indirectly. Many oncogenic signals must now be considered as part of the machinery that controls capillary recruitment and maintenance.

As a primary example of this phenomenon, there is growing evidence that VEGF expression is variable in tumors (Brown et al, 1995; Kvanta et al, 1996) and is dependent on both hypoxia (Shweiki et al, 1992) and the expression of particular oncogenes. VEGF expression is highly upregulated in cells transformed by mutated *ras* (Rak et al, 1995; White et al, 1997), v-*src*, and mutated *p53* (Mukhopadhyay et al, 1995), but is not particularly abundant in cells transformed with ornithine decarboxylase (Auvinen et al, 1997). Conversely, by using mouse cells engineered to lack VEGF expression, it has been shown that oncogenic *ras* requires VEGF for transformation. Indeed, VEGF deficient embryonic stem cells expressing *ras* formed small poorly vascularized tumors, providing evidence that VEGF induction is crucial for effective tumor growth in vivo (Shi and Ferrara, 1999). It has also been demonstrated that loss of VEGF expression results in decreased vascular density and permeability, and leads to tumor cell apoptosis (Grunstein et al, 1999), which corroborates the role of VEGF in tumor angiogenesis. Mutations in the tumor suppressor gene *p53* are commonly found in a wide variety of human tumors. Interestingly, there is some evidence to suggest that wild-type *p53* plays a role in angiogenesis by suppressing basal VEGF production (Mukhopadhyay et al, 1995). It has also been shown that transfection of mutant *p53* is able to induce expression of VEGF messenger RNA (mRNA) (Kieser et al, 1994), suggesting that *p53* gene may have a role, directly or indirectly, in regulating VEGF expression.

ANALYSIS OF ANGIOGENIC REGULATORS IN TRANSGENIC TUMOR MODELS IN THE MAMMARY GLAND

While an extensive literature has developed on the ability of primary oncogenes to stimulate breast cancer progression in transgenic mice (Cardiff et al, 2000), few studies have specifically addressed angiogenic parameters within these tumors, or the response of the tumor vasculature to diverse stimuli. Thus, mammary tumors in transgenic mice represent a relatively untapped resource for understanding the vascular biology of mammary tumors.

Mice carrying *erbB2* transgenes e.g., mouse mammary tumor virus (MMTV)-driven *neu* have been widely used because of the short latency and high penetrance of mammary tumor formation. Treatment of the MMTV-*neu* mice with interleukin-12 (IL-12) appeared to inhibit early stages of tumor formation while having less effect on late-stage, more developed, tumors (Boggio et al, 2000). The authors tested the effects of delivering similar doses of IL-12 spaced over different time courses and saw similar effects in mice from two different genetic backgrounds carrying different activating mutations of the *erbB2/neu* oncogene.

In a second example, the effects of metalloproteinase inhibitors has also been analyzed due to the expectation that neoangiogenesis relies heavily on matrix remodeling factors. Coexpression of TIMP-1 (tissue inhibitor of metalloproteinases) has now been shown to inhibit tumor progression in two independent transgenic models. The Khokha lab has shown that regulation of TIMP-1 has reci-

TABLE 5.2. Analysis of angiogenic effectors in tumor-prone transgenic mice.

Transgene	Agent	Dosage	Effect: Tumor growth	Effect: Angiogenesis	Reference
MMTV-TGF-α	TGF-β	TgN	Inc.TFS, dec.TM	ND	(Pierce et al, 1995)
Insulin-Sv40T	TNP-40*/IFN	var.	Dec.TV	Dec.	(Bergers et al, 1999; Parangi et al, 1996)
MMTV-neuT	thbs1	TgN	Inc.TFS, dec.TM	Dec.	(Iruela-Arispe et al, 1999)
MMTV-TGF-β	L-744,832†	Var.	Inc.TFS, dec.TM	ND	(Norgaard et al, 1999)
H2-Sv40T	TIMP1	TgN	Dec.TV	Dec.	(Martin et al, 1999)
MMTV-neuT	IL12	var.	Inc.TFS, dec.TM	Dec.	(Boggio et al, 2000)
MMTV-neuN	IL12	var.	Inc.TFS, dec.TM	Dec.	(Boggio et al, 2000)
C3(1)-Sv40T	IL12/IL2	var.	Dec.TV	Dec.	(Green et al, 2000)

* Fumagillin analog.
† Inhibitor of farnesyl-protein transferase.
var., various; inc., increase; dec., decrease; TgN, inhibitor was delivered as a transgene; TFS, tumor free survival; TM, tumor multiplicity; TV, tumor volume.

procal effects on hepatic tumor progression (Martin et al, 1999) and that this effect is, at least partially, related to inhibition of angiogenic parameters in the tissues. Stromelysin-1 (MMP-3), a target for TIMP-1, can induce mammary carcinogenesis when overexpressed in mammary epithelial cells (Sympson et al, 1995; Witty et al, 1995; Thomasset et al, 1998). Coexpression of TIMP-1 and MMP-3 in double transgenic animals blocked tumor progression (Sternlicht et al, 2000).

Though the transformed phenotype is recognized to be complex in all transgenic tumor prone mice, it is clear that use of such animals will have enormous value for assessing the contribution of angiogenesis, and other phenotypes, in tumor progression. Table 5.2 lists a few additional examples of the types of transgenic systems that have been used to study angiogenic recruitment in genetically defined tumor systems.

EVIDENCE THAT TUMOR VASCULATURE IS MODULATED BY THE INITIATING ONCOGENIC EVENTS IN THE MAMMARY GLAND

Although little is known about the regulators of tumor morphology, a considerable amount of research has been focused on establishing a functional link between genetic alteration(s) and acquisition of angiogenic properties. To examine whether tumor genetics could be predictive of vascular patterning, we examined the tumor vasculature of a variety of breast cancer prone transgenic mice.

The studies were performed on a series of transgenic mice that mimic aspects of human breast cancer progression by overexpression of oncogenes (*erbB2/neu*, *wnt10b*), by underexpression of specific tumors suppressor genes (*Brca1* or *p53*), or by overexpression of tumor suppressor genes (*TGF*-β) implicated in this disease. The lines that we will discuss (Table 5.3) include animals that (1) overexpress an activated form of *erbB2/neu* under control of the MMTV promoter (NK) (Muller et al, 1988); (2) overexpress the oncogenic growth factor *wnt10b* under

TABLE 5.3. Transgenic models used for the analysis of vascular morphology in mammary tumor development.

Transgene	Line	Phenotype	Reference
MMTV-neuT	NK	Rapid transformation	(Muller et al, 1988)
p53 null	BB	Genomic instability	(Donehower et al, 1992)
MMTV-wnt10b	WA	Rapid transformation	(Lane and Leder, 1997)
MMTV-Brca1 antisense	BAS	Genomic instability	(Lane and Leder, in preparation)
MMTV-TGF-β	TB	Tumor suppression	(Pierce et al, 1995)

control of the MMTV promoter (WA) (Lane and Leder, 1997); (3) carry a null mutation in the p53 locus (BB) (Donehower et al, 1992); (4) have reduced levels of *Brca-1* in mammary tissues resulting from expression of *Brca1* antisense RNAs (BAS) (Lane and Leder, in preparation); or (5) overexpress the *TGF-β* tumor suppressor gene in the context of *wnt10b* tumors (Pierce et al, 1995).

The transgenic strains have been created in the FVB/n strain or back-crossed for >12 generations into FVB/n. FVB/n is an inbred strain of mice that provide a powerful tool for genetic analysis as they have good parenting behavior, large litter sizes of 10 to 12, and display good sensitivity to a variety of oncogenic stimuli (Taketo et al, 1991). A recent study comparing the response of several strains to angiogenic stimulation by FGF-2 and VEGF placed FVB/n as intermediate in response to both factors (Rohan et al, 2000).

While dissecting tumors from these mice, it became evident that some tumors appeared more reddish and/or bled more profusely than others. These features were strongly correlated with the initiating genetic mutation, but appeared independent of size, time of onset, or whether the tumors were primary, metastastic, or had been transplanted as single cell suspensions into syngeneic hosts.

Histologic examination of tumors showed distinct, oncogene-specific, patterns of tumor cell organization, as has been described previously (Cardiff et al, 1991). Vascular differences were also evident from these preparations (Figure 5.1), particularly as to whether the tumors showed thin capillaries irradiating into a compact mass of cells (e.g., NK or WA alone), or distended vessels and lacunae (e.g., WA × TGF-β).

As seen in Figure 5.1, three-dimensional (3D) reconstruction of tumor vasculature demonstrates significant differences in density and fine structure in different transgenic tumors. Precautions were taken to allow fair comparison between lines: (1) Confocal micrographic analysis was carried out using the same scanning parameters (magnification, thickness, laser intensity, reconstruction analysis). (2) Since tumors varied in the time of incidence, age of animals could not be matched; we chose to normalize our analysis to tumors of 0.8 to 1.2 cm^3. (3) Several animals (three to nine) were evaluated from each tumor model to ensure reproducibility of the pattern. (4) The area scanned was identical. (5) The area used to perform the scan was exactly 100 μm below the capsule of the tumor (Figure 5.1 inset).

From the samples imaged in Figure 5.1, vascular density (VD) and vascular volume (VV) were calculated using ImagePro software algorithms. VV is a measure of vessel capacity, which can be very informative. For instance, although WA × TGF-β tumors had lower vascular count (VD), the overall capacity of this vascular bed exceeds that of tumors expressing WA alone by 2.8-fold. Therefore, we

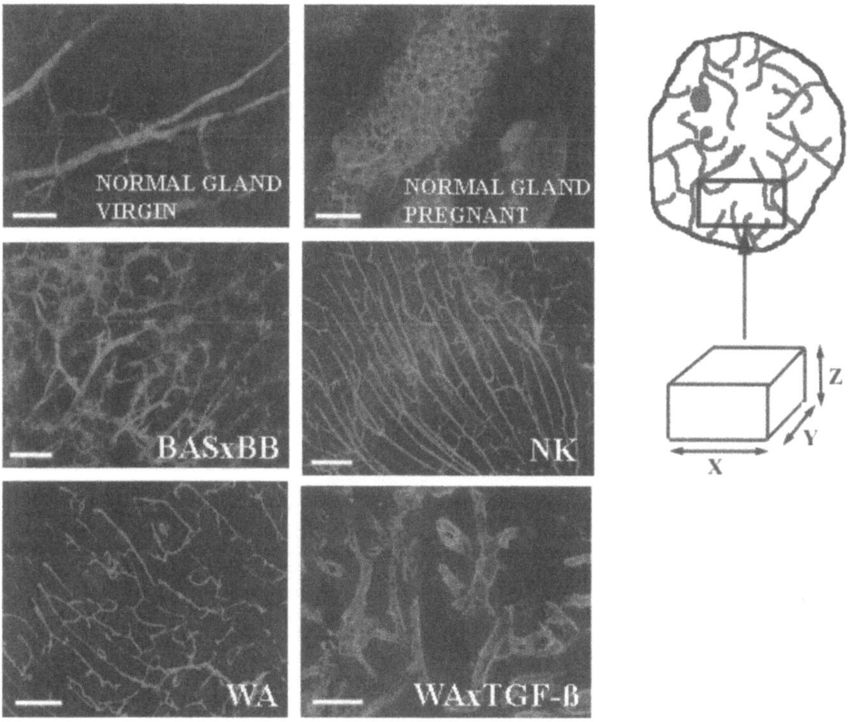

FIGURE 5.1. Genetic differences in initiating oncogenes leads to reproducible differences in vascular architecture. Syngeneic transgenic mice, carrying different initiating oncogenic mutations, were followed until mammary tumors developed to 1.0 cm^3. The mice were perfused with fluorescein isothiocyanate (FITC)-lectin conjugates and then perfusion fixed with buffered paraformaldehyde (4%) prior to analysis of vascular pattern by confocal microscopy. Panels include vascular profiles from normal and pregnant mammary glands, as well as from BAS (*BRCA-1* deficient), BB (*p53* deficient), NK (activated *erbB2/neu*), WA (*wnt10b* activated), and transforming growth factor-β (TGF-β) overexpression. The images represent confocal reconstructions of 3D images collected as described in the cartoon, and in the text.

found that measures of capillary density alone provide incomplete information as to vascular volume (capacity). When similar morphometric measurements were performed on a variety of tumors arising in distinct transgenic lines, we observed consistent differences in the degree and pattern of vasculature, that was predictive of the initiating tumorgenic mutation (Table 5.4).

These results show that simple counts of capillary profiles underestimate true vessel counts and that differences in vascular pattern were only clearly discernible by 3D reconstruction. Although traditional sectioning and staining for CD-31 (or other comparable endothelial markers) gave similar overall results with respect to vascular density (not shown), three limitations of this type of analysis became evident from our experiments: (1) many tumor capillaries do not uniformly express CD-31 or CD-34; (2) quantification of vascular density and volume from standard sections is more time-consuming, less reproducible, and prone to fixation and embedding artifacts; and (3) information concerning branching, network structure, and hierarchy is laborious or impossible to obtain from sections.

TABLE 5.4. Summary of tumor associated vascular patterns emerged from the analysis.

Tumor genotype	Observed pattern of tumor vasculature
BAS* × BB	Hierarchy in vessel diameter was preserved; vascular fragility was noted, areas of vessel rupture were evident, frequent branching and capillary networks were observed
NK	Long, thin vessels were observed with frequent branching; the hierarchy of vessel sizes was partially affected
WA	Wandering vessels were observed with frequent branching; the hierarchy of vessel sizes was partially affected
WA × TGF-β	Hierarchy largely lost, thick vascular channels less frequent than in WA alone.

* Indicated strains are described in Table 5.3 and in the text.

In summary, depending on the nature of the genetic transformation, we observed consistent changes in the vascular profile. These results support the idea of complex interaction between tumor cells and the tumor vasculature. Unlike previous analysis, 3D reconstruction demonstrated a clear link between vessel phenotype and the type of oncogenic transformation. Some of these changes related to changes in extracellular matrix proteins, particularly those associated with basement membrane structure (data not shown).

MODELS AND REPORTER SYSTEMS FOR ASSESSING ANGIOGENIC ACTIVITY IN MICE

Difficulties in identification of capillaries and small vessels provide a consistent limitation to quantitative analysis of angiogenesis in tissues. In particular, identification of vascular cells in tissue biopsies has been limited by availability of endothelial markers that are heterogeneously expressed in capillaries. As we have shown, such analysis is extremely limited in appreciating higher order vascular structures. In addition to 3D reconstruction of tumor explants, methodologic advances may soon provide access to such analysis noninvasively in the whole animal.

One approach is through the development of mice carrying "pre-labeled" endothelial cells. A particularly elegant example of this is use of autofluorescent proteins as reporter genes linked to endothelial-specific promoters. Brian Seed's group has shown the power of this approach by linking promoter elements from the VEGF gene to green fluorescent protein (GFP). Transgenic mice expressing this construct have been shown to produce fluorescent cells in response to anoxia and other proangiogenic cues (Fukumura et al, 1998). Such strategies could be linked to more sensitive and potentially less invasive visualization methodology.

Very promising avenues of research lie in the development of reporters that can be imaged in live animals. One avenue being developed by Michael Phelps and colleagues is the use of positron emission tomography (PET) instrumentation that can visualize small tumor masses in cells expressing transgenes (Gambhir et al, 1999; MacLaren et al, 1999). Though in its infancy, mathematical algorithms are being devised that may permit the analysis of bioluminescence directly in tissues. Luciferase reporter genes can be visualized in whole mice (Benaron et al, 1997).

This methodology has also been used to monitor adenoviral vector mediated gene expression in the mouse brain (Contag et al, 1997), and tumor cell clearance from therapeutically manipulated mice (Sweeney et al, 1999). Continued work in this area could refine these approaches from the current rather macroscopic limits. In the case of the new generation of micro-PET instruments, improvements in imaging reporters could bring resolution into the microliter (μm^3) range (Gambhir et al, 2000).

DISCUSSION

Though great strides have been made in identifying key regulators of angiogenesis, there is much more work to be done. The ability to effectively target mutations in mice is unsurpassed in any other vertebrate system, and affords the only method for analyzing the role of loss of function alleles in embryonic vascular development. This assay is becoming an essential tool for the analysis of gene function, and is particularly powerful in defining genetic pathways that regulate complex developmental events within the whole animal. Studies of transgenic animals are also facilitating advances in identification of efficacious therapeutic targets and in understanding the role of genetic modifiers. It appears fair to say that the impact of specific genes on angiogenesis, as opposed to other systems, depends critically on the ability to study the effects of loss-of-function mutations in mammalian development. The critical roles of the Tie receptor tyrosine kinases (Sato et al, 1995) and the more recent loss-of-function phenotypes of broadly expressed transcription factors like SMAD5 (Yang, 1999) and myocyte enhancer factor 2C (MEF) (Bi et al, 1999) would have been difficult to show without modern approaches to manipulating the genome of mice.

Given our data on the specific vascular phenotypes within tumors driven by diverse transforming genes, it is likely that much more work is needed to understand the role of vascular hierarchical structures in disease progression and therapy. Future efforts should focus more attention on the vascular phenotypes of transgenic models in which genetic background can be controlled. Because of their size and ease of manipulation, mouse models will play a major role in the development of new methodologies for less invasive high-resolution imaging of vascular development. Ultimately, all these approaches hold promise for moving the field toward a more comprehensive understanding of the molecular and cellular interactions that regulate the complex tissue interactions in animals and humans, and the critical role that angiogenesis plays in this process.

ACKNOWLEDGMENTS

The authors would like to thank Dr. Luisa Iruela-Arispe (UCLA) for collaboration on the 3D reconstructions and Dr. Philip Leder (Howard Hoghes Medical Institute and Harvard Medical School) for collaboration in the development of specific transgenic mouse models. The authors are grateful for support from the Department of Defense (DAMD 17-96-1-6095 and DAMD 17-99-9503), the National Institutes of Health (R01-CA86256-01), and a seed grant from the Jonsson Comprehensive Cancer Center, the Margaret E. Early Foundation, and the Ovarian Cancer Research Fund.

REFERENCES

Adams, R. H., Wilkinson, G. A., Weiss, C., Diella, F., Gale, N. W., Deutsch, U., Risau, W., and Klein, R. 1999. Roles of ephrinB ligands and EphB receptors in cardiovascular development: demarcation of arterial/venous domains, vascular morphogenesis, and sprouting angiogenesis. *Genes Dev.* 13:295–306.

Aird, W. C., Jahroudi, N., Weiler-Guettler, H., Rayburn, H. B., and Rosenberg, R. D. 1995. Human von Willebrand factor gene sequences target expression to a subpopulation of endothelial cells in transgenic mice. *Proc. Natl. Acad. Sci. USA* 92:4567–4571.

Arthur, H. M., Ure, J., Smith, A. J., Renforth, G., Wilson, D. I., Torsney, E., Charlton, R., Parums, D. V., Jowett, T., Marchuk, D. A., Burn, J., and Diamond, A. G. 2000. Endoglin, an ancillary TGF beta receptor, is required for extraembryonic angiogenesis and plays a key role in heart development. *Dev. Biol.* 217:42–53.

Auvinen, M., Laine, A., Paasinen-Sohns, A., Kangas, A., Kangas, L., Saksela, O., Andersson, L. C., and Holtta, E. 1997. Human ornithine decarboxylase-overproducing NIH3T3 cells induce rapidly growing, highly vascularized tumors in nude mice. *Cancer Res.* 57:3016–3025.

Bader, B. L., Rayburn, H., Crowley, D., and Hynes, R. O. 1998. Extensive vasculogenesis, angiogenesis, and organogenesis precede lethality in mice lacking all alpha v integrins. *Cell* 95:507–519.

Benaron, D. A., Contag, P. R., and Contag, C. H. 1997. Imaging brain structure and function, infection and gene expression in the body using light. *Philos Trans R Soc Lond. B. Biol. Sci.* 352:755–761.

Bergers, G., Javaherian, K., Lo, K. M., Folkman, J., and Hanahan, D. 1999. Effects of angiogenesis inhibitors on multistage carcinogenesis in mice. *Science* 284:808–812.

Bernstein, A., and Breitman, M. 1989. Genetic ablation in transgenic mice. *Mol. Biol. Med.* 6:523–530.

Bi, W., Drake, C. J., and Schwarz, J. J. 1999. The transcription factor MEF2C-null mouse exhibits complex vascular malformations and reduced cardiac expression of angiopoietin 1 and VEGF. *Dev. Biol.* 211:255–267.

Block, K. L., and Poncz, M. 1995. Platelet glycoprotein IIb gene expression as a model of megakaryocyte-specific expression. *Stem Cells* 13:135–145.

Boggio, K., Di Carlo, E., Rovero, S., Cavallo, F., Quaglino, E., Lollini, P. L., Nanni, P., Nicoletti, G., Wolf, S., Musiani, P., and Forni, G. 2000. Ability of systemic interleukin-12 to hamper progressive stages of mammary carcinogenesis in HER2/neu transgenic mice. *Cancer Res.* 60:359–364.

Brocard, J., Feil, R., Chambon, P., and Metzger, D. 1998. A chimeric Cre recombinase inducible by synthetic, but not by natural ligands of the glucocorticoid receptor. *Nucleic Acids Res.* 26:4086–4090.

Brown, L. F., Berse, B., Jackman, R. W., Tognazzi, K., Guidi, A. J., Dvorak, H. F., Senger, D. R., Connolly, J. L., and Schnitt, S. J. 1995. Expression of vascular permeability factor (vascular endothelial growth factor) and its receptors in breast cancer. *Hum. Pathol.* 26:86–91.

Cardiff, R. D., Anver, M. R., Gusterson, B. A., Hennighausen, L., Jensen, R. A., Merino, M. J., Rehm, S., Russo, J., Tavassoli, F. A., Wakefield, L. M., Ward, J. M., and Green, J. E. 2000. The mammary pathology of genetically engineered mice: the consensus report and recommendations from the Annapolis meeting. *Oncogene* 19:968–988.

Cardiff, R. D., Sinn, E., Muller, W., and Leder, P. 1991. Transgenic oncogene mice. Tumor phenotype predicts genotype. *Am. J. Pathol.* 139:495–501.

Carmeliet, P., and Collen, D. 1999. Role of vascular endothelial growth factor and vascular endothelial growth factor receptors in vascular development. *Curr. Top. Microbiol. Immunol.* 237:133–158.

Carmeliet, P., Dor, Y., Herbert, J. M., Fukumura, D., Brusselmans, K., Dewerchin, M., Neeman, M., Bono, F., Abramovitch, R., Maxwell, P., Koch, C. J., Ratcliffe, P., Moons, L., Jain, R. K., Collen, D., Keshert, E., and Keshet, E. 1998. Role of HIF-1 alpha in hypoxia-mediated apoptosis, cell proliferation and tumour angiogenesis. *Nature* 394:485–490.

Carmeliet, P., Lampugnani, M. G., Moons, L., Breviario, F., Compernolle, V., Bono, F., Balconi, G., Spagnuolo, R., Oostuyse, B., Dewerchin, M., Zanetti, A., Angellilo, A., Mattot, V., Nuyens, D., Lutgens, E., Clotman, F., de Ruiter, M. C., Gittenberger-de Groot, A., Poelmann, R., Lupu, F., Herbert, J. M., Collen, D., and Dejana, E. 1999a. Targeted deficiency or cytosolic truncation of the VE-cadherin gene in mice impairs VEGF-mediated endothelial survival and angiogenesis. *Cell* 98:147–157.

Carmeliet, P., Mackman, N., Moons, L., Luther, T., Gressens, P., Van Vlaenderen, I., Demunck, H., Kasper, M., Breier, G., Evrard, P., Muller, M., Risau, W., Edgington, T., and Collen, D. 1996. Role of tissue factor in embryonic blood vessel development. *Nature* 383:73–75.

Carmeliet, P., Ng, Y. S., Nuyens, D., Theilmeier, G., Brusselmans, K., Cornelissen, I., Ehler, E., Kakkar, V. V., Stalmans, I., Mattot, V., Perriard, J. C., Dewerchin, M., Flameng, W., Nagy, A., Lupu, F., Moons, L., Collen, D., D'Amore, P. A., and Shima, D. T. 1999b. Impaired myocardial angiogenesis and ischemic cardiomyopathy in mice lacking the vascular endothelial growth factor isoforms VEGF164 and VEGF188. *Nat. Med.* 5:495–502.

Christofori, G., and Hanahan, D. 1994. Molecular dissection of multi-stage tumorigenesis in transgenic mice. *Semin. Cancer Biol.* 5:3–12.

Contag, C. H., Spilman, S. D., Contag, P. R., Oshiro, M., Eames, B., Dennery, P., Stevenson, D. K., and Benaron, D. A. 1997. Visualizing gene expression in living mammals using a bioluminescent reporter. *Photochem. Photobiol.* 66:523–531.

Cormier, R. T., Hong, K. H., Halberg, R. B., Hawkins, T. L., Richardson, P., Mulherkar, R., Dove, W. F., and Lander, E. S. 1997. Secretory phospholipase Pla2g2a confers resistance to intestinal tumorigenesis. *Nat. Genet.* 17:88–91.

Coussens, L. M., Hanahan, D., and Arbeit, J. M. 1996. Genetic predisposition and parameters of malignant progression in K14-HPV16 transgenic mice. *Am. J. Pathol.* 149:1899–1917.

Cowan, P. J., Shinkel, T. A., Witort, E. J., Barlow, H., Pearse, M. J., and d'Apice, A. J. 1996. Targeting gene expression to endothelial cells in transgenic mice using the human intercellular adhesion molecule 2 promoter. *Transplantation* 62:155–160.

D'Amore, P. A., and Shima, D. T. 1996. Tumor angiogenesis: a physiological process or genetically determined? *Cancer Metastasis Rev.* 15:205–212.

Detmar, M., Brown, L. F., Schon, M. P., Elicker, B. M., Velasco, P., Richard, L., Fukumura, D., Monsky, W., Claffey, K. P., and Jain, R. K. 1998. Increased microvascular density and enhanced leukocyte rolling and adhesion in the skin of VEGF transgenic mice. *J. Invest. Dermatol.* 111:1–6.

Di Carlo, E., Diodoro, M. G., Boggio, K., Modesti, A., Modesti, M., Nanni, P., Forni, G., and Musiani, P. 1999. Analysis of mammary carcinoma onset and progression in HER-2/neu oncogene transgenic mice reveals a lobular origin. *Lab Invest.* 79:1261–1269.

Donehower, L. A., Harvey, M., Slagle, B. L., McArthur, M. J., Montgomery, C. A., Jr., Butel, J. S., and Bradley, A. 1992. Mice deficient for p53 are developmentally normal but susceptible to spontaneous tumours. *Nature* 356:215–221.

Dove, W. F., Cormier, R. T., Gould, K. A., Halberg, R. B., Merritt, A. J., Newton, M. A., and Shoemaker, A. R. 1998. The intestinal epithelium and its neoplasms: genetic, cellular and tissue interactions. *Philos. Trans. R. Soc. Lond. B. Biol. Sci.* 353:915–923.

Dumont, D. J., Gradwohl, G., Fong, G. H., Puri, M. C., Gertsenstein, M., Auerbach, A., and Breitman, M. L. 1994. Dominant-negative and targeted null mutations in the endothelial receptor tyrosine kinase, tek, reveal a critical role in vasculogenesis of the embryo. *Genes Dev.* 8:1897–1909.

Dumont, D. J., Jussila, L., Taipale, J., Lymboussaki, A., Mustonen, T., Pajusola, K., Breitman, M., and Alitalo, K. 1998. Cardiovascular failure in mouse embryos deficient in VEGF receptor-3. *Science* 282:946–949.

Eastin, W. C., Haseman, J. K., Mahler, J. F., and Bucher, J. R. 1998. The National Toxicology Program evaluation of genetically altered mice as predictive models for identifying carcinogens. *Toxicol. Pathol.* 26:461–473.

Fagan, K. A., Fouty, B. W., Tyler, R. C., Morris, K. G., Jr., Hepler, L. K., Sato, K., LeCras, T. D., Abman, S. H., Weinberger, H. D., Huang, P. L., McMurtry, I. F., and Rodman, D. M. 1999a. The pulmonary circulation of homozygous or heterozygous eNOS-null mice is hyperresponsive to mild hypoxia. *J. Clin. Invest.* 103:291–299.

Fagan, K. A., Tyler, R. C., Sato, K., Fouty, B. W., Morris, K. G., Jr., Huang, P. L., McMurtry, I. F., and Rodman, D. M. 1999b. Relative contributions of endothelial, inducible, and neuronal NOS to tone in the murine pulmonary circulation. *Am. J. Physiol.* 277:L472-L478.

Folkman, J., and Hanahan, D. 1991. Switch to the angiogenic phenotype during tumorigenesis. *Princess Takamatsu Symp.* 22:339–347.

Folkman, J., and Klagsbrun, M. 1987. Angiogenic factors. *Science* 235:442–427.

Fong, G. H., Rossant, J., Gertsenstein, M., and Breitman, M. L. 1995. Role of the Flt-1 receptor tyrosine kinase in regulating the assembly of vascular endothelium. *Nature* 376:66–70.

Fukumura, D., Xavier, R., Sugiura, T., Chen, Y., Park, E. C., Lu, N., Selig, M., Nielsen, G., Taksir, T., Jain, R. K., and Seed, B. 1998. Tumor induction of VEGF promoter activity in stromal cells. *Cell* 94:715–725.

Fulgham, D. L., Widhalm, S. R., Martin, S., and Coffin, J. D. 1999. FGF-2 dependent angiogenesis is a latent phenotype in basic fibroblast growth factor transgenic mice. *Endothelium* 6:185–195.

Gambhir, S. S., Barrio, J. R., Herschman, H. R., and Phelps, M. E. 1999. Assays for noninvasive imaging of reporter gene expression. *Nucl. Med. Biol.* 26:481–490.

Gambhir, S. S., Bauer, E., Black, M. E., Liang, Q., Kokoris, M. S., Barrio, J. R., Iyer, M., Namavari, M., Phelps, M. E., and Herschman, H. R. 2000. A mutant herpes simplex virus type 1 thymidine kinase reporter gene shows improved sensitivity for imaging reporter gene expression with positron emission tomography. *Proc. Natl. Acad. Sci. USA* 97:2785–2790.

Ganss, R., and Hanahan, D. 1998. Tumor microenvironment can restrict the effectiveness of activated antitumor lymphocytes. *Cancer Res.* 58:4673–4681.

George, E. L., Baldwin, H. S., and Hynes, R. O. 1997. Fibronectins are essential for heart and blood vessel morphogenesis but are dispensable for initial specification of precursor cells. *Blood* 90:3073–3081.

Giroux, S., Tremblay, M., Bernard, D., Cardin-Girard, J. F., Aubry, S., Larouche, L., Rousseau, S., Huot, J., Landry, J., Jeannotte, L., and Charron, J. 1999. Embryonic death of Mek1-deficient mice reveals a role for this kinase in angiogenesis in the labyrinthine region of the placenta. *Curr. Biol.* 9:369–372.

Go, C., Li, P., and Wang, X. 1999. Blocking transforming growth factor b signaling in transgenic epidermis accelerates chemical carcinogenesis: a mechanism associated with increased angiogenesis. *Cancer Res.* 59:2861–2868.

Good, D. J., Polverini, P. J., Rastinejad, F., Le Beau, M. M., Lemons, R. S., Frazier, W. A., and Bouck, N. P. 1990. A tumor suppressor-dependent inhibitor of angiogenesis is immunologically and functionally indistinguishable from a fragment of thrombospondin. *Proc. Natl. Acad. Sci. USA* 87:6624–6628.

Gory, S., Vernet, M., Laurent, M., Dejana, E., Dalmon, J., and Huber, P. 1999. The vascular endothelial-cadherin promoter directs endothelial-specific expression in transgenic mice. *Blood* 93:184–192.

Green, J. E., Shibata, M. A., Yoshidome, K., Liu, M. L., Jorcyk, C., Anver, M. R., Wigginton, J., Wiltrout, R., Shibata, E., Kaczmarczyk, S., Wang, W., Liu, Z. Y., Calvo, A., and Couldrey, C. 2000. The C3(1)/SV40 T-antigen transgenic mouse model of mammary cancer: ductal epithelial cell targeting with multistage progression to carcinoma. *Oncogene* 19:1020–1027.

Griep, A. E., Krawcek, J., Lee, D., Liem, A., Albert, D. M., Carabeo, R., Drinkwater, N., McCall, M., Sattler, C., Lasudry, J. G., and Lambert, P. F. 1998. Multiple genetic loci modify risk for retinoblastoma in transgenic mice. *Invest. Ophthalmol. Vis. Sci.* 39:2723–2732.

Grunstein, J., Roberts, W. G., Mathieu-Costello, O., Hanahan, D., and Johnson, R. S. 1999. Tumor-derived expression of vascular endothelial growth factor is a critical factor in tumor expansion and vascular function. *Cancer Res.* 59:1592–1598.

Hanahan, D., and Folkman, J. 1996. Patterns and emerging mechanisms of the angiogenic switch during tumorigenesis. *Cell* 86:353–364.

Harats, D., Kurihara, H., Belloni, P., Oakley, H., Ziober, A., Ackley, D., Cain, G., Kurihara, Y., Lawn, R., and Sigal, E. 1995. Targeting gene expression to the vascular wall in transgenic mice using the murine preproendothelin-1 promoter. *J. Clin. Invest.* 95:1335–1344.

Herbert, J. M., and Carmeliet, P. 1997. Involvement of u-PA in the anti-apoptotic activity of TGFbeta for vascular smooth muscle cells. *FEBS Lett.* 413:401–404.

Holash, J., Maisonpierre, P. C., Compton, D., Boland, P., Alexander, C. R., Zagzag, D., Yancopoulos, G. D., and Wiegand, S. J. 1999a. Vessel cooption, regression, and growth in tumors mediated by angiopoietins and VEGF. *Science* 284:1994–1998.

Holash, J., Wiegand, S. J., and Yancopoulos, G. D. 1999b. New model of tumor angiogenesis: dynamic balance between vessel regression and growth mediated by angiopoietins and VEGF. *Oncogene* 18:5356–5362.

Huang, Z. F., Higuchi, D., Lasky, N., and Broze, G. J., Jr. 1997. Tissue factor pathway inhibitor gene disruption produces intrauterine lethality in mice. *Blood* 90:944–951.

Iruela-Arispe, M. L., and Dvorak, H. F. 1997. Angiogenesis: a dynamic balance of stimulators and inhibitors. *Thromb. Haemost.* 78:672–677.

Iruela-Arispe, M. L., Vazquez, F., and Ortega, M. A. 1999. Antiangiogenic domains shared by thrombospondins and metallospondins, a new family of angiogenic inhibitors. *Ann. N. Y. Acad. Sci.* 886:58–66.

Kappel, A., Ronicke, V., Damert, A., Flamme, I., Risau, W., and Breier, G. 1999. Identification of vascular endothelial growth factor (VEGF) receptor-2 (Flk-1) promoter/enhancer sequences sufficient for angioblast and endothelial cell-specific transcription in transgenic mice. *Blood* 93:4284–4292.

Kawasaki, T., Kitsukawa, T., Bekku, Y., Matsuda, Y., Sanbo, M., Yagi, T., and Fujisawa, H. 1999. A requirement for neuropilin-1 in embryonic vessel formation. *Development* 126:4895–4902.

Kieser, A., Weich, H. A., Brandner, G., Marme, D., and Kolch, W. 1994. Mutant p53 potentiates protein kinase C induction of vascular endothelial growth factor expression. *Oncogene* 9:963–969.

Kim, H., and Muller, W. J. 1999. The role of the epidermal growth factor receptor family in mammary tumorigenesis and metastasis. *Exp. Cell. Res.* 253:78–87.

Kim, K. J., Li, B., Winer, J. Armanini, M., Gillet, N., Phillips, H. S., and Ferrara, N. 1993 Inhihition of vascular endothelial growth factor-induced angiogenesis suppresses tumor growth in vivo. *Nature* 362:841–844.

Kitsukawa, T., Shimono, A., Kawakami, A., Kondoh, H., and Fujisawa, H. 1995. Overexpression of a membrane protein, neuropilin, in chimeric mice causes anomalies in the cardiovascular system, nervous system and limbs. *Development* 121:4309–4318.

Korhonen, J., Lahtinen, I., Halmekyto, M., Alhonen, L., Janne, J., Dumont, D., and Alitalo, K. 1995. Endothelial-specific gene expression directed by the tie gene promoter in vivo. *Blood* 86:1828–1835.

Kozak, K. R., Abbott, B., and Hankinson, O. 1997. ARNT-deficient mice and placental differentiation. *Dev. Biol.* 191:297–305.

Krause, D. S., Mucenski, M. L., Lawler, A. M., and May, W. S. 1998. CD34 expression by embryonic hematopoietic and endothelial cells does not require c-Myb. *Exp. Hematol.* 26:1086–1092.

Kvanta, A., Steen, B., and Seregard, S. 1996. Expression of vascular endothelial growth factor (VEGF) in retinoblastoma but not in posterior uveal melanoma. *Exp. Eye Res.* 63:511–518.

Kwee, L., Burns, D. K., Rumberger, J. M., Norton, C., Wolitzky, B., Terry, R., Lombard-Gillooly, K. M., Shuster, D. J., Kontgen, F., and Stewart, C. 1995. Creation and characterization of E-selectin- and VCAM-1-deficient mice. *Ciba Found. Symp.* 189:17–28.

Kyriakides, T. R., Zhu, Y. H., Smith, L. T., Bain, S. D., Yang, Z., Lin, M. T., Danielson, K. G., Iozzo, R. V., LaMarca, M., McKinney, C. E., Ginns, E. I., and Bornstein, P. 1998. Mice that lack thrombospondin 2 display connective tissue abnormalities that are associated with disordered collagen fibrillogenesis, an increased vascular density, and a bleeding diathesis. *J. Cell Biol.* 140:419–430.

Lane, T. F., and Leder, P. 1997. Wnt10b transgenes direct hypermorphic development and transformation in mammary glands of male and female mice. *Oncogene* 15:2133–2144.

Lane, T. F., and Leder, P. (in preparation). Reduced expression of Brca1 leads to genomic instability and mammary tumor progression.

Larcher, F., Murillas, R., Bolontrade, M., Conti, C. J., and Jorcano, J. L. 1998. VEGF/VPF overexpression in skin of transgenic mice induces angiogenesis, vascular hyperpermeability and accelerated tumor development. *Oncogene* 17:303–311.

Lawler, J., Sunday, M., Thibert, V., Duquette, M., George, E. L., Rayburn, H., and Hynes, R. O. 1998. Thrombospondin-1 is required for normal murine pulmonary homeostasis and its absence causes pneumonia. *J. Clin. Invest.* 101:982–992.

Lawson, K. A., Dunn, N. R., Roelen, B. A., Zeinstra, L. M., Davis, A. M., Wright, C. V., Korving, J. P., and Hogan, B. L. 1999. Bmp4 is required for the generation of primordial germ cells in the mouse embryo. *Genes Dev.* 13:424–436.

Li, D. Y., Sorensen, L. K., Brooke, B. S., Urness, L. D., Davis, E. C., Taylor, D. G., Boak, B. B., and Wendel, D. P. 1999. Defective angiogenesis in mice lacking endoglin. *Science* 284:1534–1537.

Lifsted, T., Le Voyer, T., Williams, M., Muller, W., Klein-Szanto, A., Buetow, K. H., and Hunter, K. W. 1998. Identification of inbred mouse strains harboring genetic modifiers of mammary tumor age of onset and metastatic progression. *Int. J. Cancer* 77:640–644.

Lindahl, P., Johansson, B. R., Leveen, P., and Betsholtz, C. 1997. Pericyte loss and microaneurysm formation in PDGF-B-deficient mice. *Science* 277:242–245.

Lyden, D., Young, A. Z., Zagzag, D., Yan, W., Gerald, W., O'Reilly, R., Bader, B. L., Hynes, R. O., Zhuang, Y., Manova, K., and Benezra, R. 1999. Id1 and Id3 are required for neurogenesis, angiogenesis and vascularization of tumour xenografts. *Nature* 401:670–677.

MacLaren, D. C., Gambhir, S. S., Satyamurthy, N., Barrio, J. R., Sharfstein, S., Toyokuni, T., Wu, L., Berk, A. J., Cherry, S. R., Phelps, M. E., and Herschman, H. R. 1999. Repetitive, non-invasive imaging of the dopamine D2 receptor as a reporter gene in living animals. *Gene Ther.* 6:785–791.

Macleod, K. F., and Jacks, T. 1999. Insights into cancer from transgenic mouse models. *J. Pathol.* 187:43–60.

Maisonpierre, P. C., Suri, C., Jones, P. F., Bartunkova, S., Wiegand, S. J., Radziejewski, C., Compton, D., McClain, J., Aldrich, T. H., Papadopoulos, N., Daly, T. J., Davis, S., Sato, T. N., and Yancopoulos, G. D. 1997. Angiopoietin-2, a natural antagonist for Tie2 that disrupts in vivo angiogenesis. *Science* 277:55–60.

Maltepe, E., Schmidt, J. V., Baunoch, D., Bradfield, C. A., and Simon, M. C. 1997. Abnormal angiogenesis and responses to glucose and oxygen deprivation in mice lacking the protein ARNT. *Nature* 386:403–407.

Manenti, G., Gariboldi, M., Fiorino, A., Zanesi, N., Pierotti, M. A., and Dragani, T. A. 1997. Genetic mapping of lung cancer modifier loci specifically affecting tumor initiation and progression. *Cancer Res.* 57:4164–4166.

Martin, D. C., Sanchez-Sweatman, O. H., Ho, A. T., Inderdeo, D. S., Tsao, M. S., and Khokha, R. 1999. Transgenic TIMP-1 inhibits simian virus 40 T antigen-induced hepatocarcinogenesis by impairment of hepatocellular proliferation and tumor angiogenesis. *Lab. Invest.* 79:225–234.

Martin, J. S., Dickson, M. C., Cousins, F. M., Kulkarni, A. B., Karlsson, S., and Akhurst, R. J. 1995. Analysis of homozygous TGF beta 1 null mouse embryos demonstrates defects in yolk sac vasculogenesis and hematopoiesis. *Ann. N. Y. Acad. Sci.* 752:300–308.

Miller, D. L., Ortega, S., Bashayan, O., Basch, R., and Basilico, C. 2000. Compensation by fibroblast growth factor 1 (FGF1) does not account for the mild phenotypic defects observed in FGF2 null mice. *Mol. Cell Biol.* 20:2260–2268.

Mukhopadhyay, D., Tsiokas, L., and Sukhatme, V. P. 1995. Wild-type p53 and v-Src exert opposing influences on human vascular endothelial growth factor gene expression. *Cancer Res.* 55:6161–6165.

Muller, W. J., Sinn, E., Pattengale, P. K., Wallace, R., and Leder, P. 1988. Single-step induction of mammary adenocarcinoma in transgenic mice bearing the activated c-neu oncogene. *Cell* 54:105–115.

Norgaard, P., Law, B., Joseph, H., Page, D. L., Shyr, Y., Mays, D., Pietenpol, J. A., Kohl, N. E., Oliff, A., Coffey, R. J. J., Poulsen, H. S., and Moses, H. L. 1999. Treatment with farnesyl-protein transferase inhibitor induces regression of mammary tumors in transforming growth factor (TGF) alpha and TGF alpha/neu transgenic mice by inhibition of mitogenic activity and induction of apoptosis. *Clin. Cancer Res.* 5:35–42.

Oike, Y., Takakura, N., Hata, A., Kaname, T., Akizuki, M., Yamaguchi, Y., Yasue, H., Araki, K., Yamamura, K., and Suda, T. 1999. Mice homozygous for a truncated form of CREB-binding protein exhibit defects in hematopoiesis and vasculo-angiogenesis. *Blood* 93:2771–2779.

Ortega, M. A., and Iruela-Arispe, M. L. (submitted). Genetic modulation of thrombospondin-1 expression correlates with tumor susceptibility.

Parangi, S., Dietrich, W., Christofori, G., Lander, E. S., and Hanahan, D. 1995. Tumor suppressor loci on mouse chromosomes 9 and 16 are lost at distinct stages of tumorigenesis in a transgenic model of islet cell carcinoma. *Cancer Res.* 55:6071–6076.

Parangi, S., O'Reilly, M., Christofori, G., Holmgren, L., Grosfeld, J., Folkman, J., and Hanahan, D. 1996. Antiangiogenic therapy of transgenic mice impairs de novo tumor growth. *Proc. Natl. Acad. Sci. USA* 93:2002–2007.

Pierce, D. F. J., Gorska, A. E., Chytil, A., Meise, K. S., Page, D. L., Coffey, R. J. J., and Moses, H. L. 1995. Mammary tumor suppression by transforming growth factor beta 1 transgene expression. *Proc. Natl. Acad. Sci. USA* 9:4254–4258.

Rak, J., Mitsuhashi, Y., Bayko, L., Filmus, J., Shirasawa, S., Sasazuki, T., and Kerbel, R. S. 1995. Mutant ras oncogenes upregulate VEGF/VPF expression: implications for induction and inhibition of tumor angiogenesis. *Cancer Res.* 55:4575–4580.

Ravi, R., Mookerjee, B., Bhujwalla, Z. M., Sutter, C. H., Artemov, D., Zeng, Q., Dillehay, L. E., Madan, A., Semenza, G. L., and Bedi, A. 2000. Regulation of tumor angiogenesis by p53-induced degradation of hypoxia-inducible factor 1 alpha. *Genes Dev.* 14: 34–44.

Rohan, R. M., Fernandez, A., Udagawa, T., Yuan, J., and D'Amato, R. J. 2000. Genetic heterogeneity of angiogenesis in mice. *FASEB J.* 14:871–876.

Rosen, E. D., Chan, J. C., Idusogie, E., Clotman, F., Vlasuk, G., Luther, T., Jalbert, L. R., Albrecht, S., Zhong, L., Lissens, A., Schoonjans, L., Moons, L., Collen, D., Castellino,

F. J., and Carmeliet, P. 1997. Mice lacking factor VII develop normally but suffer fatal perinatal bleeding. *Nature* 390:290–294.
Saga, Y., Miyagawa-Tomita, S., Takagi, A., Kitajima, S., Miyazaki, J., and Inoue, T. 1999. MesP1 is expressed in the heart precursor cells and required for the formation of a single heart tube. *Development* 126:3437–3447.
Sapin, V., Dolle, P., Hindelang, C., Kastner, P., and Chambon, P. 1997. Defects of the chorioallantoic placenta in mouse RXRalpha null fetuses. *Dev. Biol.* 191:29–41.
Sato, T., Tozawa, Y., Deutsch, U., Wolburg-Buchholz, K., Fujiwara, Y., Gendron-Maguire, M., Gridley, T., Wolburg, H., Risau, W., and Qin, Y. 1995. Distinct roles of the receptor tyrosine kinases Tie-1 and Tie-2 in blood vessel formation. *Nature* 376:70–74.
Sauer, B. 1998. Inducible gene targeting in mice using the Cre/lox system. *Methods* 14:381–392.
Sawicki, J. A., Monks, B., and Morris, R. J. 1998. Cell-specific ecdysone-inducible expression of FLP recombinase in mammalian cells. *Biotechniques* 25:868–870, 872–875.
Schlaeger, T. M., Bartunkova, S., Lawitts, J. A., Teichmann, G., Risau, W., Deutsch, U., and Sato, T. N. 1997. Uniform vascular-endothelial-cell-specific gene expression in both embryonic and adult transgenic mice. *Proc. Natl. Acad. Sci. USA* 94:3058–3063.
Shalaby, F., Rossant, J., Yamaguchi, T. P., Gertsenstein, M., Wu, X. F., Breitman, M. L., and Schuh, A. C. 1995. Failure of blood-island formation and vasculogenesis in Flk-1-deficient mice. *Nature* 376:62–66.
Shi, Y. P., and Ferrara, N. 1999. Oncogenic ras fails to restore an in vivo tumorigenic phenotype in embryonic stem cells lacking vascular endothelial growth factor (VEGF). *Biochem. Biophys. Res. Commun.* 254:480–483.
Shweiki, D., Itin, A., Soffer, D., and Keshet, E. 1992. Vascular endothelial growth factor induced by hypoxia may mediate hypoxia-initiated angiogenesis. *Nature* 359:843–845.
Sternlicht, M. D., Bissell, M. J., and Werb, Z. 2000. The matrix metalloproteinase stromelysin-1 acts as a natural mammary tumor promoter. *Oncogene* 19:1102–1113.
Streit, M., Velasco, P., Brown, L. F., Skobe, M., Richard, L., Riccardi, L., Lawler, J., and Detmar, M. 1999. Overexpression of thrombospondin-1 decreases angiogenesis and inhibits the growth of human cutaneous squamous cell carcinomas. *Am. J. Pathol.* 155:441–452.
Suri, C., Jones, P. F., Patan, S., Bartunkova, S., Maisonpierre, P., Davis, S., Sato, T., and Yancopoulos, G. 1996. Requisite role of angiopoietin-1, a ligand for the TIE2 receptor, during embryonic angiogenesis. *Cell* 87:1171–1180.
Suri, C., McClain, J., Thurston, G., McDonald, D. M., Zhou, H., Oldmixon, E. H., and Sato, T. N. 1998. Increased vascularization in mice overexpressing angiopoietin-1. *Science* 282:468–471.
Sweeney, T. J., Mailander, V., Tucker, A. A., Olomu, A. B., Zhang, W., Cao, Y., Negrin, R. S., and Contag, C. H. 1999. Visualizing the kinetics of tumor-cell clearance in living animals. *Proc. Natl. Acad. Sci. USA* 96:12044–12049.
Sympson, C. J., Bissell, M. J., and Werb, Z. 1995. Mammary gland tumor formation in transgenic mice overexpressing stromelysin-1. *Semin. Cancer Biol.* 6:159–163.
Takahashi, Y., Bucana, C. D., Liu, W., Yoneda, J., Kitadai, Y., Cleary, K. R., and Ellis, L. M. 1996. Platelet-derived endothelial cell growth factor in human colon cancer angiogenesis: role of infiltrating cells. *J. Natl. Cancer Inst.* 88:1146–1151.
Taketo, M., Schroeder, A., Mobraaten, L., Gunning, K., Hanten, G., Fox, R., Roderick, T., Stewart, C., Lilly, F., Hansen, C. T., and Overbeek, P. A. 1991. FVB/N: an inbred mouse strain preferable for transgenic analyses. *Proc. Natl. Acad. Sci. USA* 88:2065–2069.
Tallquist, M. D., Soriano, P., and Klinghoffer, R. A. 1999. Growth factor signaling pathways in vascular development. *Oncogene* 18:7917–7932.
Talts, J. F., Wirl, G., Dictor, M., Muller, W. J., and Fässler, R. 1999. Tenascin-C modulates tumor stroma and monocyte/macrophage recruitment but not tumor growth or metastasis in a mouse strain with spontaneous mammary cancer. *J. Cell Sci.* 112:1855–1864.

Terry, R. W., Kwee, L., Baldwin, H. S., and Labow, M. A. 1997. Cre-mediated generation of a VCAM-1 null allele in transgenic mice. *Transgenic Res.* 6:349–356.

Theurillat, J. P., Hainfellner, J., Maddalena, A., Weissenberger, J., and Aguzzi, A. 1999. Early induction of angiogenetic signals in gliomas of GFAP-v-src transgenic mice. *Am. J. Pathol.* 154:581–590.

Thomasset, N., Lochter, A., Sympson, C. J., Lund, L. R., Williams, D. R., Behrendtsen, O., Werb, Z., and Bissell, M. J. 1998. Expression of autoactivated stromelysin-1 in mammary glands of transgenic mice leads to a reactive stroma during early development. *Am. J. Pathol.* 153:457–467.

Thompson, W. D., Li, W. W., and Maragoudakis, M. 1999. The clinical manipulation of angiogenesis: pathology, side-effects, surprises, and opportunities with novel human therapies. *J. Pathol.* 187:503–510.

Tyner, S. D., Choi, J., Laucirica, R., Ford, R. J., and Donehower, L. A. 1999. Increased tumor cell proliferation in murine tumors with decreasing dosage of wild-type p53. *Mol. Carcinog.* 24:197–208.

Visvader, J. E., Fujiwara, Y., and Orkin, S. H. 1998. Unsuspected role for the T-cell leukemia protein SCL/tal-1 in vascular development. *Genes Dev.* 12:473–479.

Vooijs, M., van der Valk, M., te Riele, H., and Berns, A. 1998. Flp-mediated tissue-specific inactivation of the retinoblastoma tumor suppressor gene in the mouse. *Oncogene* 17:1–12.

Wang, H. U., Chen, Z. F., and Anderson, D. J. 1998a. Molecular distinction and angiogenic interaction between embryonic arteries and veins revealed by ephrin-B2 and its receptor Eph-B4. *Cell* 93:741–753.

Wang, X. J., Greenhalgh, D. A., Jiang, A., He, D., Zhong, L., Brinkley, B. R., and Roop, D. R. 1998b. Analysis of centrosome abnormalities and angiogenesis in epidermal-targeted p53172H mutant and p53-knockout mice after chemical carcinogenesis: evidence for a gain of function. *Mol. Carcinog.* 23:185–192.

White, F. C., Benehacene, A., Scheele, J. S., and Kamps, M. 1997. VEGF mRNA is stabilized by ras and tyrosine kinase oncogenes, as well as by UV radiation—evidence for divergent stabilization pathways. *Growth Factors* 14:199–212.

Wigle, J. T., and Oliver, G. 1999. Prox1 function is required for the development of the murine lymphatic system. *Cell* 98:769–778.

Witty, J. P., Wright, J. H., and Matrisian, L. M. 1995. Matrix metalloproteinases are expressed during ductal and alveolar mammary morphogenesis, and misregulation of stromelysin-1 in transgenic mice induces unscheduled alveolar development. *Mol. Biol. Cell* 6:1287–1303.

Wu, H., Lee, S. H., Gao, J., Liu, X., and Iruela-Arispe, M. L. 1999. Inactivation of erythropoietin leads to defects in cardiac morphogenesis. *Development* 126:3597–3605.

Yamada, Y., Pannell, R., Forster, A., and Rabbitts, T. H. 2000. The oncogenic LIM-only transcription factor Lmo2 regulates angiogenesis but not vasculogenesis in mice. *Proc. Natl. Acad. Sci. USA* 97:320–324.

Yamagishi, H., Olson, E. N., and Srivastava, D. 2000. The basic helix-loop-helix transcription factor, dHAND, is required for vascular development. *J. Clin. Invest.* 105:261–270.

Yang, J. T., Rayburn, H., and Hynes, R. O. 1995. Cell adhesion events mediated by alpha 4 integrins are essential in placental and cardiac development. *Development* 121:549–560.

Yang, X. 1999. Angiogenesis defects and mesenchymal apoptosis in mice lacking SMAD5. *Development* 126:1571–1580.

Yuan, F., Chen, Y., Dellian, M., Safabakhsh, N., Ferrara, N., and Jain, R. K. 1996. Time-dependent vascular regression and permeability changes in established human tumor xenografts induced by an anti-vascular endothelial growth factor/vascular permeability factor antibody. *Proc. Natl. Acad. Sci. USA* 93:14765–14770.

Yuspa, S. H., Dlugosz, A. A., Cheng, C. K., Denning, M. F., Tennenbaum, T., Glick, A. B., and Weinberg, W. C. 1994. Role of oncogenes and tumor suppressor genes in multistage carcinogenesis. *J. Invest. Dermatol.* 103:90S-95S.

Zeng, X., Wert, S. E., Federici, R., Peters, K. G., and Whitsett, J. A. 1998. VEGF enhances pulmonary vasculogenesis and disrupts lung morphogenesis in vivo. *Dev. Dyn.* 21:215–227.

PART II

Vascular Morphogenesis in the Ovary

CHAPTER 6

Vascular Morphogenesis in the Ovary: Introduction and Overview

Hellmut G. Augustin

Vascular morphogenetic processes in the ovary occur cyclically as well as in association with pregnancy. Folliculogenesis and particularly the growth of the corpus luteum are intimately linked to the growth of blood vessels. This chapter gives an overview of the cyclic changes of the vasculature in the ovary. Focusing on the corpus luteum (CL), the temporal changes of the ovarian CL vasculature are discussed. The histomorphologic changes of the vasculature during the cyclic processes in the ovary are reviewed based on our current knowledge of the underlying molecular mechanisms of angiogenesis. Some comparative aspects between ovarian and tumor angiogenesis are incorporated to highlight similarities as well as differences between physiologic and pathologic angiogenesis. This introductory chapter is complemented by more specialized chapters on angiogenic cytokines (particularly heparin-binding growth factors) that regulate ovarian angiogenesis (Chapter 7); on the most critical regulator of ovarian angiogenesis, vascular endothelial growth factor (VEGF) (Chapter 8); on the hormonal regulation of angiogenesis and the regulation of the vascular system through the endocrine system (Chapter 9); and on the analysis of angiogenic processes in ovarian tumors (Chapter 10). Together, these chapters provide a comprehensive update of the current knowledge in the field of ovarian angiogenesis and demonstrate the functional relevance of ovarian angiogenesis for proper reproductive function and for a critical assessment of physiologic reproductive angiogenic processes in comparison to pathologic angiogenesis as it is associated with the growth of tumors.

The growth of new blood vessels through mechanisms of vasculogenesis (in situ differentiation of a primary capillary plexus originating from angioblastic stem cells) and angiogenesis (sprouting of capillaries from preexisting blood vessels) occurs primarily during embryonic and fetal development (for review see Carmeliet, 2000). The requirement for the growth of blood vessels in essentially all organs primarily reflects the fact that tissues can only be supplied to a very limited extent by diffusion (e.g., cartilage). Consequently, each increase in organ mass must be accompanied by an adequate growth of blood vessels. This notion also implies that angiogenesis in the adult is a primarily downregulated process simply due to the fact that there is very little physiologic growth and tissue remodeling in the healthy adult. The organs of the female reproductive system are among few tissues in the adult with significant turnover and remodeling, and consequently with significant physiologic angiogenesis. Of the different organs of the female reproductive system, the ovary has longest been recognized for its angiogenic processes (Bassett, 1943). More recently, cyclic angiogenic processes in the ovary were analyzed systematically to demonstrate distinct phases of blood vessel growth, vessel maturation, and eventually vessel regression, which together define the life cycle of the transient ovarian neovasculature (Reynolds et al, 1992; Abulafia and Sherer, 2000; Augustin, 2000).

Research into the understanding of the molecular mechanisms that govern the growth of new blood vessels has primarily been stimulated by the now widely accepted fact that the growth of tumors depends on angiogenesis. Work pursued over the last few years has shown that the inhibition of angiogenesis through pharmacologic or endogenous inhibitors of angiogenesis is emerging as a promising novel strategy in the war against cancer (Folkman, 1996; Harris, 1997; Augustin, 1998). A number of substances have entered clinical trials and several angiogenesis inhibitors are already in phase III clinical trials.

With most of the focus being on tumor angiogenic processes, the physiologic angiogenic processes in the female reproductive system have long been ignored. This is in contrast to the fact that the first angiogenic factor, later identified as the heparin binding growth factor fibroblast growth factor-2 (FGF-2) was identified in the ovary (Gospodarowicz et al, 1985). The ovary was among the first organs in which the most important angiogenic molecule VEGF was first characterized (Ravindranath et al, 1992; Shweiki et al, 1993). More recently, the first endogenous regulator of blood vessel regression, angiopoietin-2 (Ang-2) was first demonstrated to be abundantly expressed in the ovarian CL during luteolysis (Maisonpierre et al, 1997). These observations have contributed to shaping the field of reproductive angiogenesis as an active field of research with important implications for normal reproductive function as well as a paradigm for tumor angiogenic processes.

ANGIOGENESIS AROUND THE GROWING FOLLICLE

Small primordial follicles do not have a vascular network of their own. As follicles develop in the postpubertal female, they develop an antrum. It is at this time that a first capillary plexus ensheathing the growing follicle becomes detectable. This vascular sheath forms two concentric networks of vessels in the theca externa and interna. The vessels do not penetrate the membrana granulosa of the unruptured follicle. Particularly the growth of the inner network of capillaries in the theca interna coincides with a period of rapid growth and differentiation of the follicle (Richards, 1980). This observation has led to speculation that establishment of the vascular network around developing follicles may be rate limiting for the selection of the dominant follicle(s), i.e., the follicle(s) that is/are destined to ovulate. In turn, insufficient vascular supply could act as the trigger that leads to follicular atresia. Consistent with these hypotheses, preovulatory follicles of monkeys were found to have similar concentrations of gonadotropin-binding sites; however, only the follicle that was destined to ovulate became heavily labeled after intravenous injection of labeled gonadotropin (Zeleznik et al, 1981). These findings are consistent with an increased vascularity of the dominant follicle.

Extracts of the follicular wall were found to contain endothelial cell migratory and proliferative activity (Makris et al, 1984). Similarly, conditioned media from granulosa cells obtained from pregnant mare serum gonadotropin (PMSG)-treated rats stimulated endothelial cell proliferation (Koos, 1986). Preliminary characterization of the follicle-derived endotheliotropic activity described a chemoattractant associated with the molecular weight fraction of greater than 30 kd. Surprisingly, however, the presence of this activity was not correlated with follicular size or steroid concentrations in the media (Rone et al, 1993).

Taken together, the findings strongly suggest that the vascular supply plays a critical role in the selection of the dominant follicle that is destined to mature to ovulation. However, neither the endocrinologic nor the vascular changes leading to ovulation are sufficiently analyzed to provide conclusive evidence of the molecular mechanisms that lead to the selection of the dominant follicle.

HISTOMORPHOLOGIC ANALYSIS OF THE LIFE CYCLE OF THE VASCULATURE IN THE CORPUS LUTEUM

Following ovulation, the progesterone-secreting CL grows within the ovulatory cavity. The cyclic morphogenetic changes of the ovarian CL are associated with discrete phases of vessel growth, vessel maturation, and vessel regression. Together, these phases establish a physiologic life cycle of a transient vasculature (Figure 6.1). Functional experiments to manipulate reproductive angiogenesis have been performed in laboratory animals including rats and mice (Klauber et al, 1997a; Ferrara et al, 1998). The 4-day reproductive cycle of these species is associated with an intense angiogenic burst following ovulation. The short duration of the cycle, however, makes it difficult to discern discrete phases of vessel maturation and vessel regression that follow the angiogenic phase of the ovarian cycle. The following description of the morphologic changes of the vasculature throughout the ovarian cycle is primarily based on the analysis of large monovulatory species such as bovine (21-day cycle) and human (28-day cycle) (Augustin et al, 1995). The temporal difference between the bovine and the human ovarian cycle primarily reflects the longer duration of the follicle phase in humans, while the duration of the luteal phase is similar in both species.

Angiogenesis

The angiogenic phase of the CL cycle lasts for approximately one third of the ovarian cycle (human and bovine: 6 to 8 days; murine: 1 to 2 days). The CL forms during this time to grow from approximately 200 mg of ovulatory tissue toward a solid tissue mass of several grams. Considering that any growth of tissues beyond a critical diameter of approximately half a millimeter requires the growth of blood vessels, it can easily be appreciated that the growth of the CL must be accompanied by intense angiogenesis.

Macroscopically, the growing CL of larger species is characterized by its intense red coloration reflecting both hemorrhage and intense perfusion of the growing immature vascular network. The very early CL following ovulation is called corpus hemorrhagicum, whereas the growing CL is called corpus rubrum. Microscopically, classical sprouting angiogenesis can be found in the growing CL.

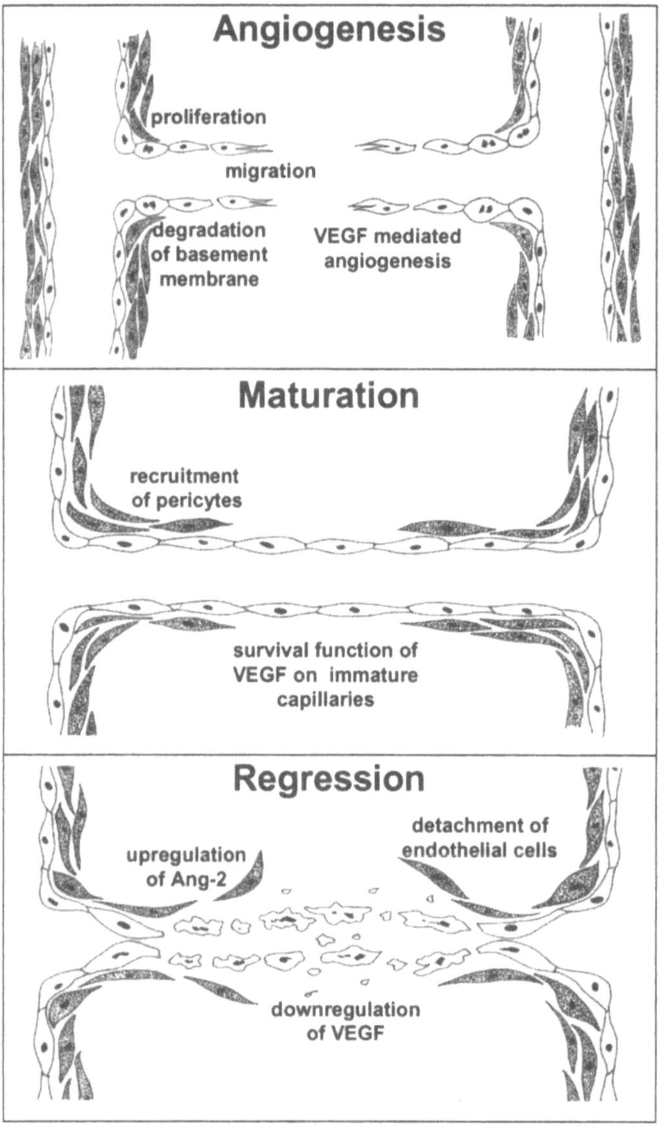

FIGURE 6.1. Schematic representation of the molecular mechanisms of angiogenesis throughout the ovarian cycle. *Angiogenesis:* The angiogenic growth phase of the ovarian cycle is primarily characterized by vascular endothelial growth factor (VEGF)-mediated sprouting angiogenesis, which involves local degradation of the surrounding extracellular matrix, endothelial cell migration and proliferation, as well as capillary network formation. Other cytokines such as fibroblast growth factor-2 (FGF-2) also play a role during ovarian angiogenesis, but VEGF appears to be the primary regulator of corpus luteum (CL) angiogenesis. *Maturation:* VEGF acts as a survival factor for the immature, not pericyte covered, neovasculature in the nonangiogenic mature CL. Approximately 60% of the microvascular network in the midstage CL is covered by pericytes reflecting a partially mature network of blood vessels. Pericyte-covered microvessels are independent of the survival factor function of VEGF. *Regression:* Luteolysis leads to the rapid disintegration of the vasculature of the CL. Downregulation of VEGF leads to the disintegration of immature microvessels that are not covered by pericytes. Concomitant upregulation of angiopoietin-2 (Ang-2) causes a destabilization of pericyte-covered microvessels that ultimately leads to the regression of these vessels as well. An increasing arteriolization of small arteries leads to a contractive occlusion of the vascular bed and a functional shutoff of the circulation. An influx of macrophages is responsible for the dissolution of the disintegrating tissue.

Sprouting capillaries invade the fibrin-rich ovulatory cavity. In rats, sprouting capillaries have been observed as early as 16 hours after ovulation (Matsushima et al, 1996). There is massive endothelial cell proliferation within hours after ovulation with as many as 36% of them incorporating ^{3}H-thymidine within a 1-hour pulse period (Gaede et al, 1985; Meyer and McGeachie, 1988). Similarly, most intense endothelial cell proliferation was also observed in the early luteal phase of human corpora lutea (Rodger et al, 1997) and in the cyclic ovary of marmosets (Young et al, 2000).

A systematic analysis of endothelial cell proliferation in the growing CL and in different malignant human tumors revealed that reproductive angiogenesis can be 4- to 20-fold more intense as the angiogenesis in some of the most malignant human tumors including glioblastomas (Eberhard et al, 2000). As many as 40% of CL microvessels were found to contain proliferating endothelial cells. In contrast, glioblastoma multiforme have on average a compartment of proliferating endothelial cells of close to 10%. Prostate and lung carcinomas have on average a proliferating endothelial cell compartment of only around 2% (Eberhard et al, 2000). This comparative analysis illustrates the intensity of angiogenesis during the normal ovarian cycle.

Endothelial cells of immature blood vessels have large, oval nuclei that line cavernous capillaries. Angiogenic endothelial cells are the most abundant cell population in the growing corpus rubrum accounting for as many as 50% of all cells. Despite the fact that follicle rupture is considered an inflammatory reaction involving cytokines such as interleukin-1 (IL-1) and tumor necrosis factor-α (TNF-α) (Vinatier et al, 1995), there is relatively little recruitment of inflammatory cells in the growing CL, suggesting that CL angiogenesis is primarily a noninflammatory event (particularly in comparison to tumor angiogenic processes). There is, however, a relatively strong influx of eosinophiles into the corpus hemorrhagicum immediately following ovulation (Höflinger, 1948). Interestingly, induction of ovarian eosinopenia by systemic administration of prednisolone was found to result in reduced progesterone production during the luteal phase of the ovarian cycle, presumably as a consequence of reduced CL angiogenesis (Murdoch and Steadman, 1991).

Blood Vessel Maturation

The mature CL is characterized by an intense yellow coloration reflecting its endocrine function as well as the maturation of the transient CL vasculature. The mature CL contains a dense network of capillaries. Each luteal cell has contact to at least two capillaries. Ultrastructurally, endothelial cells in the midstage CL can be identified as a discontinuous layer of cells with 1- to 2-μm gaps (Peters and McNatty, 1980). Maturation of the ovarian neovasculature is functionally characterized by the recruitment of mural cells. Microvessels recruit pericytes, which ensheath the capillaries and control the quiescent phenotype of the endothelial cells. Larger vessels recruit smooth muscle cells, which similarly stabilize a blood vessel and control its vasotonus. Approximately 60% of the microvessels in the mature midstage CL are covered by pericytes (as detected by α-smooth muscle actin staining) (Tsukada et al, 1996; Goede et al, 1998). This indicates that the CL vasculature acquires only partially a mature phenotype. Surprisingly, even the earliest neovasculature in the growing corpus rubrum has a pericyte coverage index of close to 60% (Goede et al, 1998; Eberhard et al, 2000). This illustrates that physiologic angiogenesis follows a well-orchestrated program during which endothelial cell proliferation and subsequent pericyte recruitment are intimately linked. This is very much in contrast to the disorganized angiogenic processes in human tumors. Most tumor types have a lower degree of microvessel maturation with pericyte recruitment indices around 15% (glioblastomas, renal cell carcinomas) and 40% (colon carcinomas, prostate carcinomas) (Eberhard et al, 2000).

Only little α-smooth muscle actin staining of larger vessels can be detected in the midstage CL, indicating that full arteriolization of capillaries, and thus full maturation, probably does not occur in the cyclic CL (Tsukada et al, 1996; Goede et al, 1998). This conclusion is also supported by the functional analysis of blood flow regulation in the midstage CL; microsphere distribution experiments have demonstrated that blood flow in the CL is primarily regulated by the systemic blood pressure. This indicates that an arteriolization of the capillary bed by ingrowth of smooth muscle cells takes places only to a very limited extent in the midstage CL (Wiltbank et al, 1990; Wiltbank, 1994).

Blood Vessel Regression

At the end of the ovarian cycle, regression of the cyclic CL is characterized by a rapid phase of tissue disintegration and involution over a few days (regressing CL) and a slow phase of tissue remodeling and connective tissue growth over several weeks (residual CL). Eventually, the CL will completely disappear, leaving a small hyaline scar (corpus albicans) (Hossain and O'Shea, 1981; Modlich et al, 1996).

Physiologic regression of blood vessels reflects a remarkable biologic phenomenon that occurs as rarely in the adult as physiologic angiogenic processes. The notion that blood vessel regression can occur as a physiologic process is also of interest in the context of tumor angiogenic processes. Tumor angiogenesis is a unidirectional process. Spontaneous blood vessel regression is usually not observed in tumors. Consequently, early experiments in the field of tumor antiangiogenesis were conceptualized as a strategy to inhibit further tumor growth by halting expansion of the tumor vascular network. Regression of an established vascular network was not necessarily considered an achievable option. The realization of vessel regression processes in the ovary as a physiologically existing entity as well as the identification of the first molecular regulator of vessel regression (Maisonpierre et al, 1997) has changed this perspective on the therapeutic potential of antiangiogenic tumor therapies. A number of experimental studies have now demonstrated that antiangiogenesis can successfully be aimed at inducing the regression of an existing tumor vascular bed, eventually leaving a small avascular dormant microtumor (Boehm et al, 1997).

Luteolysis leads to rapid functional inactivation of the CL and a reduction of its size to less than one third of its original size within 2 to 3 days. This process is particularly pronounced in the bovine species and occurs somewhat slower in humans. Vacuolation, hyalination, and cellular condensation can be seen in all cell types of the regressing CL. Dissolution of the blood vessels of the CL is associated with gradual rounding and condensation of endothelial cells of small vessels. At later stages, all luteal cells have disappeared, leaving arteriole-like remnants of blood vessels with an onion-shaped arrangement of surrounding myofibroblastic cells and heavily condensed endothelial cells with dense connective tissue in the residual CL. The residual CL is completely removed and replaced by connective tissue, leaving a small dark scar.

Ultrastructurally, one of the first signs of blood vessel regression after the induction of luteolysis is the dissociation of capillary endothelial cells from each other and from their underlying basement membrane, leading to plasma extravasation and edema formation (Modlich et al, 1996). As a consequence, capillary endothelial cells completely detach from the basement membrane as single cells or sheets of several cells. Loss of adhesion in anchorage-dependent cells, including endothelial cells, will ultimately lead to apoptosis (Re et al, 1994; Ruoslahti and Reed, 1994). Nevertheless, only very few apoptotic endothelial cells are identified by ultrastructural analysis of regressing CL or by the in situ detection of nucleosomal fragmentation products (Modlich et al, 1996; Fraser et al, 1999), suggesting that detaching endothelial cells are rapidly removed either by macrophages or simply by floating off into the circulation. Thus, the massive apoptosis observed during CL regression primarily reflects luteal cell and not endothelial cell apoptosis (Zheng et al, 1994; Bacci et al, 1996; Shikone et al, 1996).

Luteolysis is associated not just with tissue disintegration but also with substantial tissue remodeling. Morphologically evident within a few days after the induction of luteolysis, intensely α-smooth muscle actin–positive myofibroblastic cells accumulate around regressing small blood vessels, giving them a characteristic onion-shaped structure of concentric rings of smooth muscle cells. Proliferation of smooth muscle cells during this period appears to contribute to an occlusive contraction of small blood vessels that will lead to a complete shutdown of vascular blood flow. During a slow phase of tissue organization, all occluded blood vessels disappear as the residual CL organizes into the hyaline corpus albicans.

Macrophages play a major role in the organization of the disintegrating corpus luteum. Massive recruitment of macrophages occurs rapidly after the onset of luteolysis (Sawyer, 1995; Petrovska et al, 1996; Takaya et al, 1997). Macrophage recruitment in the regressing CL is associated with estrogen withdrawal (Naftalin et al, 1997) and prominent expression of monocyte chemoattractant protein-1 (MCP-1) (Townson et al, 1996; Goede et al, 1999).

MOLECULAR MECHANISMS OF CYCLIC ANGIOGENESIS IN THE OVARY

The analysis of the endocrine regulatory circuits in the ovary has been the focus of intense research in the field of endocrinology over the last 50 years. In contrast, the identification and characterization of angiogenic cytokines is a much younger field of research (Hanahan, 1997; Risau, 1997; Carmeliet, 2000). Despite the largely ignored importance of angiogenesis for normal ovarian function, angiogenesis researchers have extensively studied reproductive angiogenic processes (Reynolds et al, 1992; Abulafia and Sherer, 2000; Augustin, 2000). The ovarian CL was the first organ in which an angiogenic cytokine was identified (Jakob et al, 1977; Gospodarowicz and Thakral, 1978). This factor was later identified as the heparin-binding growth factor basic fibroblast growth factor (bFGF or FGF-2) (Gospodarowicz et al, 1985). Likewise, the most important angiogenic cytokine VEGF was characterized in the ovary (Ravindranath et al, 1992; Shweiki et al, 1993; Kamat et al, 1995). More recently, the first factor involved in angioregressive processes, angiopoietin-2 (Ang-2), was identified. Again, this factor was functionally characterized in the ovary (Maisonpierre et al, 1997).

Endocrine Regulation of Angiogenesis

The cyclic processes in the ovary are under endocrine control. Thus, it appears obvious to imply a functional relationship between the endocrine and the vascular system. A detailed account on the molecular and functional interplay between the endocrine and the vascular is summarized in Chapter 9.

Of the endocrine regulators, estrogens have been most extensively studied for their angiogenesis-inducing capacity. Estrogens are capable of inducing angiogenesis in vivo (Morales et al, 1995; Schnaper et al, 1996). Correspondingly, antiestrogens have been shown to exert angioinhibitory activity (Gagliardi and Collins, 1993). Causal evidence for the role of estrogens for angiogenesis could be generated in estrogen receptor–deficient mice in which angiogenesis cannot be induced by estrogens (Johns et al, 1996).

The angiostimulatory activities of estrogens may be mediated by direct endotheliotropic functions of estrogens or indirectly by modulating the expression of angiogenic cytokines in other cell types. Both mechanisms appear to be operative. Endothelial cells express estrogen receptors (Kim-Schulze et al, 1996). 17β-estradiol stimulates endothelial cell migration and proliferation and promotes the formation of capillary-like tubes in vitro (Morales et al, 1995). Antiestrogens inhibit FGF-2 as well as VEGF-induced endothelial cell proliferation (Gagliardi and Collins, 1993). The indirect angiogenesis-inducing capacity of estrogens manifests itself in their ability to control the expression of angiogenic cytokines in other cell types. Estrogens have been shown to regulate the expression of VEGF in a number of cell types (endometrial cells, macrophages) (McLaren et al, 1996; Shifren et al, 1996).

A number of other steroid hormones have also been tested for their angiomodulatory activity. The endogenous estrogen metabolite 2-methoxy estradiol (2-ME) has been shown to exert angioinhibitory activity (Fotsis et al, 1994; Klauber et al, 1997b). Likewise, medroxyprogesterone acetate (MPA) has antiangiogenic activity as was demonstrated in the rabbit cornea assay as well as in tumor experiments (Yamamoto et al, 1994).

Angiogenic Cytokines in the Cyclic Corpus Luteum

It has long been recognized that the small and large luteal cells are synthesizing and secreting mitogenic activity for endothelial cells (Koos, 1986; Rone and Goodman, 1990). Likewise, whole organ extracts of corpora lutea from different species have been shown to contain endothelial cell mitogenic and angiogenic activity (Gospodarowicz and Thakral, 1978; Sato et al, 1982; Rone et al, 1993).

A large number of angiogenic cytokines and more recently endogenous inhibitors of angiogenesis have been identified in the last 20 years (Table 6.1). These factors control physiologic angiogenic processes as they occur during developmental growth of tissues as well as pathologic angiogenic processes as they are associated with the growth of tumors. Most of these cytokines are not endothelial cell specific growth factors, but rather exert pleiotrophic effector functions on a number of cells and tissues. Cytokines with almost exclusive activity on endothe-

TABLE 6.1. Positive and negative endogenous regulators of angiogenesis.

Stimulators	Inhibitors
Peptide growth factors	*Peptide growth factors and proteolytic peptides*
Vascular endothelial growth factor-A,*† -B,† -C,† -D,† [-E], -F (VEGF-A,*† -B,† -C,† -D,† -E, -F)	Angiopoietin-2 (Ang-2)*† (in the absence of angiogenic activity)
Placenta growth factor (PlGF)†	Angiostatin†
Angiopoietin-1 (Ang-1)*†	Endostatin†
Angiopoietin-2 (Ang-2)*† (modulator in the presence of angiogenic activity)	16-kd prolactin fragment
Acidic fibroblast growth factor (FGF-1)*	Laminin peptides
Basic fibroblast growth factor (FGF-2)*	Fibronectin peptides
Platelet-derived growth factor (PDGF)	*Inhibitors of enzymatic activity*
Transforming growth factor-α (TGF-α)	Tissue metalloproteinase inhibitors (TIMP-1, -2, -3, -4)
Transforming growth factor-β (TGF-β)	Plasminogen activator inhibitors (PAI-1, -2)
Hepatocyte growth factor (HGF)	
Insulin-like growth factor-1 (IGF-1)	
Multifunctional cytokines / immune mediators	*Multifunctional cytokines / immune mediators*
Tumor necrosis factor-α (low dose)	Tumor necrosis factor-α (high dose)
Monocyte chemoattractant protein-1 (MCP-1)	Interferons
	Interleukin-12
CXC-chemokines	*CXC-chemokines*
	Platelet factor-4 (PF-4)
Interleukin-8 (IL-8)	Interferon-γ-inducible protein-10 (IP-10)
	Gro-β
Enzymes	*Extracellular matrix molecules*
Platelet-derived endothelial cell growth factor (PD-ECGF; thymidine phosphorylase)	Thrombospondin
Angiogenin (ribonuclease A homologue)	
Hormones	*Hormones/metabolites*
Estrogens*	2-Methoxyestradiol (2-ME)
Prostaglandin-E_1, -E_2*	Proliferin-related protein
Follistatin	
Proliferin	
Oligosaccharides	*Oligosaccharides*
Hyaluronan oligosaccharides	Hyaluronan, high molecular weight species
Gangliosides	

* Analyzed in the context of ovarian angiogenesis.
† Predominantly endothelial cell specific.

lial cells are the members of the VEGF family (see Chapter 8 for a detailed account on the role of VEGF for ovarian angiogenesis; also see Ferrara, 1999; Gale and Yancopoulos, 1999; Neufeld et al, 1999) and the angiopoietins (Davis and Yancopoulos, 1999; Gale and Yancopoulos, 1999). In addition to these factors, bFGF or FGF-2 has extensively been studied in the context of ovarian angiogenesis (for a detailed analysis of the role of FGFs and other heparin-binding growth factors see Chapter 7; also see Reynolds and Redmer, 1998, 1999).

Basic Fibroblast Growth Factor (FGF-2)

The heparin-binding growth factor FGF-2 was identified as the first angiogenic cytokine in the ovary (Jakob et al, 1977; Gospodarowicz and Thakral, 1978; Gospodarowicz et al, 1985). It was considered as one of the key regulators of angiogenesis in the early 1990s (prior to the identification of the specific angiogenic cytokines VEGFs and angiopoietins). FGF-2 is capable of stimulating endothelial cell proliferation and migration and potently induces angiogenesis in a number of in vivo angiogenesis assays. Surprisingly, mice with a targeted deletion of the FGF-2 have no apparent developmental phenotype that would indicate a critical role in physiologic angiogenesis (Dono et al, 1998; Ortega et al, 1998). These findings would suggest that FGF-2 may play a role during pathologic tumor angiogenesis but that it may be dispensable for physiologic developmental and reproductive angiogenesis. Alternatively, the lack of FGF-2 in FGF-2–deficient mice could functionally be compensated for by other members of the FGF family of growth factors.

FGF-2 protein is expressed in the CL albeit with small variations over the ovarian cycle (Neufeld et al, 1987; Stirling et al, 1991; Zheng et al, 1993; Reynolds and Redmer, 1998). Intriguingly, the endothelial cell mitogenic activity isolated from bovine corpora lutea can be neutralized by more than 80% with antibodies against FGF-2 (see Chapter 7 and Grazul-Bilska et al, 1992; Reynolds et al, 1994). Antibodies against acidic FGF (aFGF or FGF-1) inhibit the extracted endothelial cell mitogenic activity by 16% (Grazul-Bilska et al, 1992; Reynolds et al, 1994). Surprisingly, neutralizing antibodies against VEGF, which has been shown to be rate limiting for ovarian angiogenesis (see below and Chapter 8), inhibit the extracted endothelial cell mitogenic activity only by 30% (Doraiswamy et al, 1995). Future work will establish the functional relevance of the FGFs for reproductive angiogenesis as well as the molecular interplay between the FGFs and the VEGFs.

Vascular Endothelial Growth Factor/Vascular Permeability Factor (VEGF/VPF)

The most important angiogenic cytokine identified to date is the heparin-binding growth factor VEGF (see Chapter 8; also see Ferrara, 1999; Gale and Yancopoulos, 1999; Neufeld et al, 1999). The best studied member of the VEGF family is now called VEGF-A since it is a member of a growing family that now contains six members: VEGF-B is another endotheliotropic cytokine whose function is not well understood (Olofsson et al, 1999). VEGF-B–deficient mice are viable and fertile and have no apparent vascular phenotype (Olofsson et al, 1999). VEGF-C and VEGF-D are regulators of lymphangiogenesis, but are also capable of controlling hemangiogenic processes (Olofsson et al, 1999). VEGF-E is not a mammalian gene but has been identified as a viral homologue of VEGF that is expressed by Orf viruses and controls the hypervascular phenotype of cutaneous Orf virus lesions (Meyer et al, 1999). Lastly, VEGF-F has been identified as part of the human genome sequencing initiative but is functionally not yet characterized.

The classical VEGF-A is expressed in five different isoforms in the human. Of these $VEGF_{145}$, $VEGF_{165}$, $VEGF_{189}$, and $VEGF_{206}$ are heparin-binding growth factors, whereas $VEGF_{121}$ is not. VEGF was identified in 1983 as vascular permeability factor (VPF), a much more potent regulator of vascular permeability as classical permeability regulating molecules such as histamine and serotonin (Dvorak et al, 1999). It was not until the early 1990s that the angiogenesis-regulating function of VEGF/VPF was uncovered (Dvorak et al, 1999).

VEGF-A has a number of different endotheliotropic functions. In addition to its permeability regulating function in small vessels, it is mitogenic for endothelial cells in vitro as well as in vivo (Ferrara, 1999; Gale and Yancopoulos, 1999; Neufeld et al, 1999). Furthermore, it acts as an apoptosis-inhibiting survival factor for immature endothelial cells (Alon et al, 1995). VEGF exerts its endotheliotropic functions through almost exclusively endothelial cell expressed receptor tyrosine kinases, the VEGF receptors-1 and -2 (VEGFR-1 [older name: Flt-1]; VEGFR-2 [older name: KDR or Flk-1]. The functional interplay of VEGF with its receptors is not fully understood. VEGFR-2 appears to be the primary signal transducing receptor, whereas VEGFR-1 possibly acts as a modulatory receptor in endothelial cells that may regulate the bioavailability of VEGF.

The critical rate-limiting importance of VEGF for angiogenesis was impressively demonstrated by gene targeting experiments in mice. Mice with a deletion of a single VEGF allele (VEGF$^{+/-}$ mice) are not capable of developing, and they die early during development as a consequence of disturbed embryonic angiogenesis (Carmeliet et al, 1996; Ferrara et al, 1996). This observation was interpreted to reflect the critical importance of an exact dosage of VEGF. Loss of 50% genomic information is not compatible with life. Experiments with an overexpression of VEGF have yielded corresponding results leading to hypervascularized tissues of increased size. It is possibly for the strict requirement of controlling the dosage of VEGF that VEGF interactions with its receptor VEGFR-2 are controlled by a network of functional regulators that include VEGFR-1 as well as accessory molecules such as the neuropilins (Neufeld et al, 1999) and a splice variant of VEGFR-1 that exists as a potent dominant negative soluble receptor (sFlt-1).

VEGF-A is strongly expressed during the angiogenic period of the ovarian cycle (Ravindranath et al, 1992; Shweiki et al, 1993; Goede et al, 1998). The most conclusive evidence for the rate-limiting importance of VEGF for ovarian angiogenesis was recently demonstrated through a series of elegant functional in vivo experiments. The repeated intravenous administration of the functional antagonist of VEGF signaling sFlt-1 (soluble variant of VEGFR-1) in rats is capable of completely shutting down the ovarian cycle (see Chapter 8 and Ferrara et al, 1998). Soluble Flt-1 acts as a dominant negative receptor that binds VEGF with a much higher affinity than the signal transducing transmembrane receptor VEGFR-2 and thus prevents positive endotheliotropic VEGF signaling. Similar results have been obtained with neutralizing antibodies against VEGF (Fraser et al, 2000).

The regulation of ovarian VEGF-A expression has not yet been conclusively elucidated. Hypoxia acts as a strong inducer of VEGF-A transcriptional regulation in a number of cell populations (Shweiki et al, 1992; Claffey and Robinson, 1996) as well as stabilization of VEGF-A messenger RNA (mRNA) (Ikeda et al, 1995). Conceptually, hypoxia is believed to act as the primary inducer of VEGF-A during tumor angiogenesis. It is, however, questionable if hypoxia is the primary trigger of VEGF-A expression in the ovary. This conclusion is supported by several lines of evidence. First, the time course of VEGF-A expression in the ovary with prominent induction of VEGF-A transcription within hours after ovulation (Shweiki et al, 1993) suggests an expression of VEGF-A concomitant to luteal cell activation and proliferation rather than a delayed response following luteal proliferation and development of hypoxia. Second, the density of the vascular network in the mature CL with each luteal cell being in contact with several capillaries strongly suggests that the density of the vascular network in the CL primarily reflects the mature luteal cells' secretory functions rather than the possible hypoxia-driven nutritional needs of the organ. Third, the mature CL in multiovulatory laboratory animals is still relatively small (less than 1 mm in diameter), so it would hardly require a dense vascular network solely for nutritional purposes. However, even in these small laboratory animals, VEGF-A expression is detectable by in situ hybridization in a punctate pattern with similar intensity throughout the CL that does not correspond to possible oxygen gradients (Phillips et al, 1990; Shweiki et al, 1993).

Luteinized human granulosa cells have been shown to express high baseline levels of VEGF-A mRNA in culture (Yan et al, 1993). In these experiments, hypoxia was not analyzed for its ability to modulate granulosa cell VEGF-A expression. Treatment of nonhuman primates in the midluteal phase with gonadotropin-releasing hormone (GnRH) antagonists resulted in a significant downregulation of VEGF-A mRNA expression within 72 hours after treatment. Corresponding to these results, in vitro and in vivo experiments demonstrated that the gonadotropins follicle-stimulating hormone (FSH) and luteinizing hormone (LH) directly stimulate granulosa cell VEGF-A production (Christenson and Stouffer, 1997).

Several cell types have been shown to synthesize and secrete VEGF-A in response to ovarian steroids. Isolated human endometrial cells rapidly (1 hour) induce VEGF-A expression upon stimulation with estradiol (E_2), MPA, or a combination of E_2 and MPA (Shifren et al, 1996). Likewise, VEGF-A expression is induced in peritoneal fluid macrophages upon stimulation with ovarian steroids (McLaren et al, 1996). An in vivo analysis of the regulation of VEGF-A expression in rat mammary tumors identified upregulated levels of VEGF-A mRNA and protein upon estradiol treatment as well as downregulated levels of VEGF-A mRNA and protein upon injection of the aromatase inhibitor 4-hydroxyandrostenedione (4-OHA), which leads to reduced endogenous estrogen concentrations (Nakamura et al, 1996).

Other Angiogenesis Regulating Cytokines in the Ovary

Several other angiogenic cytokines have been identified in the ovary (Table 6.1) (for review see Beck and D'Amore, 1997; Goldberg and Rosen, 1997). These factors, however, are pleiotrophic cytokines whose primary biologic functions differ from their angiogenesis-inducing capacity.

Transforming growth factor-α (TGF-α) is a potent endothelial cell mitogen that has steroidogenic capacity. TGF-α is expressed in the cyclic CL where it was most prominently identified in midcycle large luteal cells by histochemical staining (Lobb and Dorrington, 1993).

Transforming growth factor-β (TGF-β_1 and TGF-β_2) has also been identified in the ovary by immunohistochemistry (Teerds and Dorrington, 1992). Intense TGF-β_1 staining and weak TGF-β_2 staining was detected around the growing follicle without any obvious regulation throughout the ovarian cycle. In the CL, moderate levels of TGF-β_1 and TGF-β_2 immunoreactivity were also detected that exhibited a heterogeneous expression pattern.

The activin-binding and -neutralizing glycoprotein follistatin has been shown to induce endothelial cell proliferation in vitro and angiogenesis in vivo (Kozian et al, 1997). Follistatin is expressed by ovarian luteal cells and its expression is restricted to the early and midluteal phase corresponding to the angiogenic phases of the CL cycle (Tisdall et al, 1994).

Molecular Mechanisms of Vessel Maturation

VEGF-A acts not just as an angiogenesis-inducing growth factor but also as a survival factor of endothelial cells in immature microvessels (Alon et al, 1995; Carmeliet, 2000). This was functionally demonstrated in conditionally VEGF overexpressing C6 gliomas. These tumors expressed VEGF under the control of the tetracycline promoter. Thus, removal of tetracycline from the drinking water of tet-VEGF C6 glioma–bearing mice leads to the sudden downregulation of VEGF production. Concomitantly with this downregulation of VEGF production, the immature vascular network in these tumors collapsed. Endothelial cells from microvessels that were not covered by perivascular mural pericytes detached and became apoptotic (Benjamin and Keshet, 1997). Additional experiments demonstrated that contact with pericytes renders endothelial cells independent of the survival factor activity of VEGF (Benjamin et al, 1998). The spatiotemporal changes of VEGF expression in the CL as well as the morphologic changes of the transient vasculature in the cyclic CL largely correspond to this experimental analysis of VEGF as a maintenance factor. A comparative analysis of VEGF expression in the different phases of the ovarian cycle revealed that VEGF is expressed at similar intensity in the angiogenic corpus rubrum as well as in the nonangiogenic mature CL. Both VEGF receptors can be detected throughout the ovarian cycle albeit with some quantitative changes (Goede et al, 1998). These findings correspond to the observation that the vasculature in the mature CL matures only partially through the recruitment of pericytes. Approximately 60% of the microvessels are covered by α-smooth muscle actin expressing pericytes, whereas 40% are not covered. These findings suggest that VEGF acts as a survival factor of the immature vasculature of the nonangiogenic mature CL.

Mechanisms of Vessel Regression

The analysis of the molecular mechanisms of luteolysis has been the focus of intense research for many years (for review see Messinis, 1997). Regression of the CL vasculature is in close functional relationship to the induction of luteolysis. Yet it is not understood if blood vessel regression in the CL is a cause or a consequence of luteolysis. Irrespective of this question, physiologic blood vessel regression in the ovary represents a unique biologic system whose detailed molecular understanding will be of utmost importance for the understanding and eventual therapeutic implementation of vessel regression as a strategy to interfere with the tumor-associated vasculature.

Luteolysis leads to a dramatic downregulation of VEGF-A expression in the ovary (Figure 6.1) (Goede et al, 1998). As outlined above, the sudden downregulation of VEGF-A is functionally interpreted as the removal of its survival factor function leading to the rapid disintegration of immature microvessels that are not covered by pericytes. Ultrastructurally, this microvessel dissociation process can be traced within the first few days following the onset of luteolysis. Endothelial cells detach from their basement membrane to eventually float off into the circulation. At the same time, a gradual arteriolization of small arteries leads to a contractive occlusion of the vasculature and consequently a shutoff of the circulation. Conceptually, the induction of vessel regression through the removal of a maintenance factor is quite surprising as one might have considered vessel regression as an active process that is induced by an on switch rather than an off switch.

In addition to the downregulation of VEGF as an inducing factor for vessel regression, the recently identified angiopoietins play a central role in controlling vessel maintenance, stabilization, and eventually regression. Ang-1 and Ang-2 have been identified as antagonistic ligands of the endothelial cell specifically expressed receptor tyrosine kinase Tie-2 (Davis and Yancopoulos, 1999; Gale and Yancopoulos, 1999). Ang-1 transduces a proendotheliotropic signal that leads to vessel maturation through the recruitment of pericytes. Endothelial cell–derived platelet-derived growth factor-B (PDGF-B) has been shown to act as a powerful pericyte recruiting agent (Lindahl et al, 1997; Benjamin et al, 1998). Yet the functional interplay between Ang-1 and PDGF-B has not been uncovered. Ang-1 has also been shown to be weakly angiogenic, but its primary role appears to be its function as a stabilizing factor. This stabilizing effect also antagonizes the permeability-inducing effect of VEGF (Thurston et al, 1999).

In contrast to the proendotheliotropic functions of Ang-1, Ang-2 appears to act as an antagonistic molecule that binds Tie-2 without transducing an activating signal. Thus, it acts as a competitive antagonist of Ang-1. These findings are supported by genetic manipulation experiments in mice. Ang-1–deficient mice die in utero (somewhat later than VEGF-deficient mice) as a consequence of an inability to form a mature functional network of blood vessels (Suri et al, 1996). The phenotype of Ang-1–deficient mice corresponds to the phenotype of transgenic mice that constitutively overexpress Ang-2, supporting the concept of a functional antagonism between Ang-1 and Ang-2 (Maisonpierre et al, 1997).

Ang-1 and Ang-2 mRNA is expressed in the angiogenic CL as well as in the mature nonangiogenic CL at approximately equimolar concentrations (Goede et al, 1998). Luteolysis is associated with a rapid and dramatic shift of the ratio of Ang-2 vs. Ang-1 toward Ang-2 (approximately 10-fold stronger expression of Ang-2 compared to Ang-1 [Goede et al, 1998]. Functionally, this overexpression of Ang-2 over Ang-1 can be interpreted as a destabilization of microvessels that are covered by pericytes. Consequently, given the simultaneous downregulation of VEGF, the Ang-2–mediated destabilization of pericyte-covered microvessels ultimately also leads to the regression of these vessels (Figure 6.1).

LYMPHANGIOGENESIS IN THE CYCLIC CORPUS LUTEUM

The ovary not only is a highly vascularized organ but also is endowed with a dense network of lymphatic vessels. Ovarian lymph formation is highest during the luteal phase, reflecting the high permeability of blood vessels in the CL with increased extravasation of plasma proteins.

Lymphangiogenesis in the ovary has received little attention thus far. In the rabbit, lymphatics are observed around the theca lutein cells but not around the granulosa cells 3 days after the injection of human chorionic gonadotropin (hCG) (Otsuki et al, 1987). A detailed analysis of rat CL morphogenesis revealed that lymph vessels are not immediately detectable in the CL after the first postbubertal ovulation. Lymphatics are, however, present in and around the CL growing after the second and subsequent ovulations (Ichikawa et al, 1987). These findings suggest that lymphangiogenesis in the CL does occur, but that it seems to follow rather than parallel the growth of blood vessels.

A detailed analysis of lymphatic angiogenesis in the CL in other species has not been performed. Lymphangiogenesis in the ovary may be an interesting subject for future study, particularly since the key regulatory molecules of lymphangiogenesis, lymphatic endothelial cell expression of Flt-4 (Kaipainen et al, 1995) and induction of lymphangiogenesis through the Flt-4 ligand VEGF-C (Jeltsch et al, 1997), have recently been identified.

PERSPECTIVE

A plethora of open questions related to the elucidation of the molecular mechanisms of blood vessel growth, vessel maturation, and vessel regression still remain to be answered. In contrast to unidirectional tumor models, physiologic cyclic processes as they occur in the female reproductive system offer a unique system to study all of these aspects. This will become particularly important as the analysis of molecular systems of blood vessel maturation and regression is just at its beginning. Furthermore, in addition to serving as an experimental model, the female reproductive system may itself be a target for novel therapeutic strategies through the manipulation of its vascular system.

The Corpus Luteum as an Angiogenesis Model

The VEGFs and the angiopoietins have been identified as key regulatory molecules of ovarian angiogenesis. Understanding of the molecular mechanisms of ovarian angiogenesis and the analysis of the morphologic changes of the vasculature throughout the ovarian cycle has laid the foundation for a wider appreciation of reproductive angiogenesis as a prerequisite of normal reproductive function as well as a paradigmatic model for tumor angiogenic processes. An increasingly longer list of novel molecular regulators of the angiogenic cascade including the ephrins, the neuropilins, transcription factors of angiogenesis, several adhesion molecules, and regulators of the proteolytic balance await a detailed analysis in the context of reproductive angiogenesis. Analytical experiments will have to be complemented by functional interventive studies as have been done for VEGF. Likewise, the molecular link between the endocrine and the vascular system is far from being satisfactorily understood. Lastly, a number of endogenous inhibitors of angiogenesis have been identified in recent years (Table 6.1). None of these molecules has yet been studied in the context of reproductive angiogenesis. The physiologic angiogenic processes in the female reproductive system offer a unique experimental system to answer questions related to angiogenesis, vessel maturation, and vessel regression that at the same time may help to shed further light on the mechanisms of tumor angiogenesis.

Ovarian Angiogenesis as a Target for Vascular Interventional Therapies

Interference with the vascular system of reproductive organs may become a therapeutic modality in the future. The use of assisted reproductive techniques has generated novel clinical problems that are caused by a disturbance of the ovarian vascular system. The ovarian hyperstimulation syndrome (OHSS) is the most serious complication of ovarian stimulation that may lead to cystic enlargement of the ovaries as well as ascites and pleural effusions. Patients with OHSS have elevated plasma levels of VEGF-A, and it is now established that an overproduction of VEGF-A is the pathogenetic cause of OHSS leading to severe hyperpermeability (McClure et al, 1994; Neulen et al, 1995; Rizk et al, 1997). This illustrates the importance of the vascular system for normal reproductive function and the possibility to therapeutically manipulate reproductive function by interfering with the ovarian vascular system.

REFERENCES

Abulafia, O., and Sherer, D. M. 2000. Angiogenesis of the ovary. *Am. J. Obstet. Gynecol.* 182:240–246.

Alon, T., Hemo, I., Itin, A., Pe'er, J., Stone, J., and Keshet, E. 1995. Vascular endothelial growth factor acts as a survival factor for newly formed retinal vessels and has implications for retinopathy of prematurity. *Nat. Med.* 1:1024–1028.

Augustin, H. G. 1998. Antiangiogenic tumour therapy: will it work? *Trends Pharmacol. Sci.* 19:216–222.

Augustin, H. G. 2000. Development of the vascular system in the corpus luteum. In: Risau W., Rubanyi G., eds. *Morphogenesis of Endothelium.* Harwood Academic Publishers, Amsterdam, pp. 237–255.

Augustin, H. G., Braun, K., Telemenakis, I., Modlich, U., and Kuhn, W. 1995. Ovarian angiogenesis: phenotypic characterization of endothelial cells in a physiological model of blood vessel growth and regression. *Am. J. Pathol.* 147:1–13.

Bacci, M. L., Barazzoni, A. M., Forni, M., and Costerbosa, G. L. 1996. In situ detection of apoptosis in regressing corpus luteum of pregnant sow: evidence of an early presence of DNA fragmentation. *Domest. Anim. Endocrinol.* 13:361–372.

Bassett, D. L. 1943. The changes in the vascular pattern of the ovary of the albino rat during the estrous cycle. *Am. J. Anat.* 73:251–291.

Beck, L., and D'Amore, P. A. 1997. Vascular development: cellular and molecular regulation. *FASEB J.* 11:365–373.

Benjamin, L. E., Hemo, I., and Keshet, E. 1998. A plasticity window for blood vessel remodelling is defined by pericyte coverage of the preformed endothelial network and is regulated by PDGF-B and VEGF. *Development* 125:1591–1598.

Benjamin, L. E., and Keshet, E. 1997. Conditional switching of vascular endothelial growth factor (VEGF) expression in tumors: induction of endothelial cell shedding and regression of hemangioblastoma-like vessels by VEGF withdrawal. *Proc. Natl. Acad. Sci. USA* 94:8761–8766.

Boehm, T., Folkman, J., Browder, T., and O'Reilly, M. S. 1997. Antiangiogenic therapy of experimental cancer does not induce acquired drug resistance. *Nature* 390:404–407.

Carmeliet, P. 2000. Mechanisms of angiogenesis and arteriogenesis. *Nat. Med.* 6:389–395.

Carmeliet, P., Ferreira, V., Breier, G., Pollefeyt, S., Kieckens, L., Gertsenstein, M., Fahrig, M., Vandenhoeck, A., Harpal, K., Eberhardt, C., Declercq, C., Pawling, J., Moons, L., Collen, D., Risau, W., and Nagy, A. 1996. Abnormal blood vessel development and lethality in embryos lacking a single VEGF allele. *Nature* 380:435–439.

Christenson, L. K., and Stouffer, R. L. 1997. Follicle-stimulating hormone and luteinizing hormone/chorionic gonadotropin stimulation of vascular endothelial growth factor production by macaque granulosa cells from pre- and periovulatory follicles. *J. Clin. Endocrinol. Metab.* 82:2135–2142.

Claffey, K. P., and Robinson, G. S. 1996. Regulation of VEGF/VPF expression in tumor cells: consequences for tumor growth and metastasis. *Cancer Metastasis Rev.* 15:165–176.

Davis, S., and Yancopoulos, G. D. 1999. The angiopoietins: Yin and Yang in angiogenesis. *Curr. Top. Microbiol. Immunol.* 237:173–185.

Dono, R., Texido, G., Dussel, R., Ehmke, H., and Zeller, R. 1998. Impaired cerebral cortex development and blood pressure regulation in FGF-2-deficient mice. *EMBO J.* 17:4213–4225.

Doraiswamy, V., Grazul-Bilska, A. T., Ricke, W. A., Redmer, D. A., and Reynolds, L. P. 1995. Immunoneutralization of angiogenic activity from ovine corpora lutea (CL) with antibodies against fibroblast growth factor (FGF)-2 and vascular endothelial growth factor (VEGF). *Biol. Reprod.* 52(suppl 1):112.

Dvorak, H. F., Nagy, J. A., Feng, D., Brown, L. F., and Dvorak, A. M. 1999. Vascular permeability factor/vascular endothelial growth factor and the significance of microvascular hyperpermeability in angiogenesis. *Curr. Top. Microbiol Immunol.* 237:97–132.

Eberhard, A., Kahlert, S., Goede, V., Hemmerlein, B., Plate, K. H., and Augustin, H. G. 2000. Heterogeneity of angiogenesis and blood vessel maturation in human tumors: implications for antiangiogenic tumor therapies. *Cancer Res.* 60:1388–1393.

Ferrara, N. 1999. Molecular and biological properties of vascular endothelial growth factor. *J. Mol. Med.* 77:527–543.

Ferrara, N., Carver-Moore, K., Chen, H., Dowd, M., Lu, L., O'Shea, K. S., Powell-Braxton, L., Hillan, K. J., and Moore, M. W. 1996. Heterozygous embryonic lethality induced by targeted inactivation of the VEGF gene. *Nature* 380:439–442.

Ferrara, N., Chen, H., Davis-Smyth, T., Gerber, H. P., Nguyen, T. N., Peers, D., Chisholm, V., Hillan, K. J., and Schwall, R. H. 1998. Vascular endothelial growth factor is essential for corpus luteum angiogenesis. *Nat. Med.* 4:336–340.

Folkman, J. 1996. Fighting cancer by attacking its blood supply. *Sci. Am.* 275:150–154.

Fotsis, T., Zhang, Y., Pepper, M. S., Adlercreutz, H., Montesano, R., Nawroth, P. P., and Schweigerer, L. 1994. The endogenous oestrogen metabolite 2-methoxyoestradiol inhibits angiogenesis and suppresses tumour growth. *Nature* 368:237–239.

Fraser, H. M., Dickson, S. E., Lunn, S. F., Wulff, C., Morris, K. D., Carroll, V. A., and Bicknell, R. 2000. Suppression of luteal angiogenesis in the primate after neutralization of vascular endothelial growth factor. *Endocrinology* 141:995–1000.

Fraser, H. M., Lunn, S. F., Harrison, D. J., and Kerr, J. B. 1999. Luteal regression in the primate: different forms of cell death during natural and gonadotropin-releasing hormone antagonist or prostaglandin analogue-induced luteolysis. *Biol. Reprod.* 61:1468–1479.

Gaede, S. D., Sholley, M. M., and Quattropani, S. L. 1985. Endothelial mitosis during the initial stages of corpus luteum neovascularization in the cycling adult rat. *Am. J. Anat.* 172:173–180.

Gagliardi, A. R., and Collins, D. C. 1993. Inhibition of angiogenesis by antiestrogens. *Cancer Res.* 153:533–535.

Gale, N. W., and Yancopoulos, G. D. 1999. Growth factors acting via endothelial cell-specific receptor tyrosine kinases: VEGFs, angiopoietins, and ephrins in vascular development. *Genes Dev.* 13:1055–1066.

Goede, V., Borgelli, L., Ziche, M., and Augustin, H. G. 1999. Induction of inflammatory angiogenesis by monocyte chemoattractant protein-1 (MCP-1). *Int. J. Cancer* 82:765–770.

Goede, V., Schmidt, T., Kimmina, S., Kozian, D. H., and Augustin, H. G. 1998. Analysis of blood vessel maturation processes during cyclic ovarian angiogenesis. *Lab. Invest.* 78:1385–1394.

Goldberg, I. D., and Rosen, E. M. 1997. *Regulation of Angiogenesis*. Basel: Birkhäuser.
Gospodarowicz, D., Cheng, J., Lui, G. M., Baird, A., Esch, F., and Böhlen, P. 1985. Corpus luteum angiogenic factor is related to fibroblast growth factor. *Endocrinology* 117:2383–2391.
Gospodarowicz, D., and Thakral, K. K. 1978. Production of a corpus luteum angiogenic factor responsible for proliferation of capillaries and neovascularization of the corpus luteum. *Proc. Natl. Acad. Sci. USA* 75:847–851.
Grazul-Bilska, A. T., Redmer, D. A., Killilea, S. D., Kraft, K. C., Reynolds L. P. 1992. Production of mitogenic factor(s) by ovine corpora lutea throughout the cycle. *Endocrinology* 130:3625–3632.
Hanahan, D. 1997. Signaling vascular morphogenesis and maintenance. *Science* 277:48–50.
Harris, A. L. 1997. Antiangiogenesis for cancer therapy. *Lancet* 349(suppl 2):13–15.
Höflinger, H. 1948. Das Ovar des Rindes in den verschiedenen Lebensperioden unter besonderer Berücksichtigung seiner funktionellen Feinstruktur. *Acta Anat* suppl 5. pp. 1–164.
Hossain, M. I., and O'Shea, J. D. 1981. Vascular changes during regression of the corpus luteum of the guinea pig. *Aust. J. Biol. Sci.* 34:649–660.
Ichikawa, S., Uchino, S., and Hirata, Y. 1987. Lymphatic and blood vasculature of the forming corpus luteum. *Lymphology* 20:73–83.
Ikeda, E., Achen, M. G., Breier, G., and Risau, W. 1995. Hypoxia-induced transcriptional activation and increased mRNA stability of vascular endothelial growth factor in C6 glioma cells. *J. Biol. Chem.* 270:19761–19766.
Jakob, W., Jentzsch, K. D., Mauersberger, B., and Oehme, P. 1977. Demonstration of angiogenesis-activity in the corpus luteum of cattle. *Exp. Pathol.* 13:231–236.
Jeltsch, M., Kaipainen, A., Joukov, V., Meng, X., Lakso, M., Rauvala, H., Swartz, M., Fukumura, D., Jain, R. K., and Alitalo, K. 1997. Hyperplasia of lymphatic vessels in VEGF-C transgenic mice. *Science* 276:1423–1425.
Johns, A., Freay, A. D., Fraser, W., Korach, K. S., and Rubanyi, G. M. 1996. Disruption of estrogen receptor gene prevents 17 beta estradiol-induced angiogenesis in transgenic mice. *Endocrinology* 137:4511–4513.
Kaipainen, A., Korhonen, J., Mustonen, T., van Hinsbergh, V. W., Fang, G. H., Dumont, D., Breitman, M., and Alitalo, K. 1995. Expression of the fms-like tyrosine kinase 4 gene becomes restricted to lymphatic endothelium during development. *Proc. Natl. Acad. Sci. USA* 92:3566–3570.
Kamat, B. R., Brown, L. F., Manseau, E. J., Senger, D. R., and Dvorak, H. F. 1995. Expression of vascular permeability factor/vascular endothelial growth factor by human granulosa and theca lutein cells: role in corpus luteum development. *Am. J. Pathol.* 146:157–165.
Kim-Schulze, S., McGowan, K. A., Hubchak, S. C., Cid, M. C., Martin, M. B., Kleinman, H. K., Greene, G. L., and Schnaper, H. W. 1996. Expression of an estrogen receptor by human coronary artery and umbilical vein endothelial cells. *Circulation* 94:1402–1407.
Klauber, N., Parangi, S., Flynn, E., Hamel, E., and D'Amato, R. J. 1997b. Inhibition of angiogenesis and breast cancer in mice by the microtubule inhibitors 2-methoxyestradiol and taxol. *Cancer Res.* 57:81–86.
Klauber, N., Rohan, R. M., Flynn, E., and D'Amato, R. J. 1997a. Critical components of the female reproductive pathway are suppressed by the angiogenesis inhibitor AGM-1470. *Nat. Med.* 4:443–446.
Koos, R. D. 1986. Stimulation of endothelial cell proliferation by rat granulosa cell-conditioned medium. *Endocrinology* 119:481–489.
Kozian, D. H., Ziche, M., and Augustin, H. G. 1997. The activin-binding protein follistatin regulates autocrine endothelial cell activity and induces angiogenesis. *Lab. Invest.* 76:267–276.

Lindahl, P., Johansson, B. R., Leveen, P., and Betsholtz, C. 1997. Pericyte loss and microaneurysm formation in PDGF-B-deficient mice. *Science* 277:242–245.

Lobb, D. K., and Dorrington, J. H. 1993. Transforming growth factor-alpha: identification in bovine corpus luteum by immunohistochemistry and Northern blot analysis. *Reprod. Fertil. Dev.* 5:523–529.

Maisonpierre, P. C., Suri, C., Jones, P. F., Bartunkova, S., Wiegand, S. J., Radziejewski, C., Compton, D., McClain, J., Aldrich, T. H., Papadopoulos, N., Daly, T. J., Davis, S., Sato, T. N., and Yancopoulos, G. D. 1997. Angiopoietin-2, a natural antagonist for Tie2 that disrupts in vivo angiogenesis. *Science* 277:55–60.

Makris, A., Ryan, K. J., Yasumizu, T., Hill, C. L., and Zetter, B. R. 1984. The nonluteal porcine ovary as a source of angiogenic activity. *Endocrinology* 115:1672–1677.

Matsushima, T., Fukuda, Y., Tsukada, K., and Yamanaka, N. 1996. The extracellular matrices and vascularization of the developing corpus luteum in rats. *J. Submicrosc. Cytol. Pathol.* 28:441–455.

McClure, N., Healy, D. L., Rogers, P. A., Sullivan, J., Beaton, L., Haning, R. V., Connolly, D. T., and Robertson, D. M. 1994. Vascular endothelial growth factor as capillary permeability agent in ovarian hyperstimulation syndrome. *Lancet* 344:235–236.

McLaren, J., Prentice, A., Charnock-Jones, D. S., Millican, S. A., Muller, K. H., Sharkey, A. M., and Smith, S. K. 1996. Vascular endothelial growth factor is produced by peritoneal fluid macrophages in endometriosis and is regulated by ovarian steroids. *J. Clin. Invest.* 98:482–489.

Messinis, I. E. 1997. Luteal function—luteolysis. *Ann N. Y. Acad. Sci.* 816:151–157.

Meyer, G. T., and McGeachie, J. K. 1988. Angiogenesis in the developing corpus luteum of pregnant rats: a stereologic and autoradiographic study. *Anat. Rec.* 222:18–25.

Meyer, M., Clauss, M., Lepple-Wienhues, A., Waltenberger, J., Augustin, H. G., Ziche, M., Lanz, C., Büttner, M., Rziha, H. J., and Dehio, C. 1999. A novel vascular endothelial growth factor encoded by Orf virus, VEGF-E, mediates angiogenesis via signalling through VEGFR-2 (KDR) but not VEGFR-1 (Flt-1) receptor tyrosine kinases. *EMBO J.* 18:363–374.

Modlich, U., Kaup, F. J., and Augustin, H. G. 1996. Cyclic angiogenesis and blood vessel regression in the ovary: blood vessel regression during luteolysis involves endothelial cell detachment and vessel occlusion. *Lab. Invest.* 74:771–780.

Morales, D. E., McGowan, K. A., Grant, D. S., Maheshwari, S., Bhartiya, D., Cid, M. C., Kleinman, H. K., and Schnaper, H. W. 1995. Estrogen promotes angiogenic activity in human umbilical vein endothelial cells in vitro and in a murine model. *Circulation* 91:755–763.

Murdoch, W. J., and Steadman, L. E. 1991. Investigations concerning the relationship of ovarian eosinophilia to ovulation and luteal function in the sheep. *Am. J. Reprod. Immunol.* 25:81–87.

Naftalin, D. M., Bove, S. E., Keyes, P. L., and Townson, D. H. 1997. Estrogen withdrawal induces macrophage invasion in the rabbit corpus luteum. *Biol. Reprod.* 56:1175–1180.

Nakamura, J., Savinov, A., Lu, Q., and Brodie, A. 1996. Estrogen regulates vascular endothelial growth/permeability factor expression in 7,12-dimethylbenz(a)anthracene-induced rat mammary tumors. *Endocrinology* 137:5589–5596.

Neufeld, G., Cohen, T., Gengrinovitch, S., and Poltorak, Z. 1999. Vascular endothelial growth factor (VEGF) and its receptors. *FASEB J.* 13:9–22.

Neufeld, G., Ferrara, N., Schweigerer, L., Mitchell, R., and Gospodarowicz, D. 1987. Bovine granulosa cells produce basic fibroblast growth factor. *Endocrinology* 121:597–603.

Neulen, J., Yan, Z., Raczek, S., Weindel, K., Keck, C., Weich, H. A., Marme, D., and Breckwoldt, M. 1995. Human chorionic gonadotropin–dependent expression of vascular endothelial growth factor/vascular permeability factor in human granulosa cells: importance in ovarian hyperstimulation syndrome. *J. Clin. Endocrinol. Metab.* 80:1967–1971.

Olofsson, B., Jeltsch, M., Eriksson, U., and Alitalo, K. 1999. Current biology of VEGF-B and VEGF-C. *Curr. Opin. Biotech.* 10:528–535.
Ortega, S., Ittmann, M., Tsang, S. H., Ehrlich, M., and Basilico, C. 1998. Neuronal defects and delayed wound healing in mice lacking fibroblast growth factor 2. *Proc. Natl. Acad. Sci. USA* 95:5672–5677.
Otsuki, Y., Magari, S., and Sugimoto, O. 1987. Fine structure and morphometric analysis of lymphatic capillaries in the developing corpus luteum of the rabbit. *Lymphology* 20:64–72.
Peters, H., and McNatty, K. P. 1980. Morphology of the ovary. In: Peters, H., and McNatty, K. P., eds. *The Ovary: a Correlation of Structure and Function in Mammals.* Granada Publishing, London, pp. 12–35.
Petrovska, M., Dimitrov, D. G., and Michael, S. D. 1996. Quantitative changes in macrophage distribution in normal mouse ovary over the course of the estrous cycle examined with an image analysis system. *Am. J. Reprod. Immunol.* 36:175–183.
Phillips, H. S., Hains, J., Leung, D. W., and Ferrara, N. 1990. Vascular endothelial growth factor is expressed in rat corpus luteum. *Endocrinology* 127:965–967.
Ravindranath, N., Little-Ihrig, L., Phillips, H. S., Ferrara, N., and Zeleznik, A. J. 1992. Vascular endothelial growth factor messenger ribonucleic acid expression in the primate ovary. *Endocrinology* 131:254–260.
Re, F., Zanetti, A., Sironi, M., Polentarutti, N., Lanfrancone, L., Dejana, E., and Colotta, A. 1994. Inhibition of anchorage-dependent cell spreading triggers apoptosis in cultured human endothelial cells. *J. Cell Biol.* 127:537–546.
Reynolds, L. P., and Redmer, D. A. 1998. Expression of the angiogenic factors, basic fibroblast growth factor and vascular endothelial growth factor, in the ovary. *J. Anim. Sci.* 76:1671–1681.
Reynolds, L. P., and Redmer, D. A. 1999. Growth and development of the corpus luteum. *J. Reprod. Fertil* 54(suppl):181–191.
Reynolds, L. P., Grazul-Bilska, A. T., Killilea, S. D., and Redmer, D. A. 1994. Mitogenic factors of corpora lutea. *Prog. Growth Factor Res.* 5:159–175.
Reynolds, L. P., Killilea, S. D., and Redmer, D. A. 1992. Angiogenesis in the female reproductive system. *FASEB J.* 6:886–892.
Richards, J. S. 1980. Maturation of ovarian follicles: actions and interactions of pituitary and ovarian hormones on follicular cell differentiation. *Physiol. Rev.* 60:51–89.
Risau, W. 1997. Mechanisms of angiogenesis. *Nature* 386:671–674.
Rizk, B., Aboulghar, M., Smitz, J., and Ron-El, R. 1997. The role of vascular endothelial growth factor and interleukins in the pathogenesis of severe ovarian hyperstimulation syndrome. *Hum. Reprod. Update* 3:255–266.
Rodger, F. E., Young, F. M., Fraser, H. M., and Illingworth, P. J. 1997. Endothelial cell proliferation follows the mid-cycle luteinizing hormone surge, but not human chorionic gonadotrophin rescue, in the human corpus luteum. *Hum. Reprod.* 12:1723–1729.
Rone, J. D., and Goodman, A. L. 1990. Preliminary characterization of angiogenic activity in media conditioned by cells from luteinized rat ovaries. *Endocrinology* 127:2821–2828.
Rone, J. D., Halvorson, L. M., and Goodman, A. L. 1993. Ovarian angiogenesis in rabbits: endotheliotrophic chemoattractant activity from isolated follicles and dispersed granulosa cells. *J. Reprod. Fertil.* 97:359–365.
Ruoslahti, E., and Reed, J. C. 1994. Anchorage dependence, integrins, and apoptosis. *Cell* 77:477–478.
Sato, E., Ishibashi, T., and Koide, S. S. 1982. Inducement of blood vessel formation by ovarian extracts from mice injected with gonadotropins. *Experientia* 15:1248–1249.
Sawyer, H. R. 1995. Structural and functional properties of the corpus luteum of pregnancy. *J. Reprod. Fertil. Suppl.* 49:97–110.
Schnaper, H. W., McGowan, K. A., Kim-Schulze, S., and Cid, M. C. 1996. Oestrogen and endothelial cell angiogenic activity. *Clin. Exp. Pharmacol. Physiol.* 23:247–250.

Shifren, J. L., Tseng, J. F., Zaloudek, C. J., Ryan, I. P., Meng, Y. G., Ferrara, N., Jaffe, R. B., and Taylor, R. N. 1996. Ovarian steroid regulation of vascular endothelial growth factor in the human endometrium: implications for angiogenesis during the menstrual cycle and in the pathogenesis of endometriosis. *J. Clin. Endocrinol. Metab.* 81:3112–3118.

Shikone, T., Yamoto, M., Kokawa, K., Yamashita, K., Nishimori, K., and Nakano, R. 1996. Apoptosis of human corpora lutea during cyclic luteal regression and early pregnancy. *J. Clin. Endocrinol. Metab.* 81:2376–2380.

Shweiki, D., Itin, A., Neufeld, G., Gitay-Goren, H., and Keshet E. 1993. Patterns of expression of vascular endothelial growth factor (VEGF) and VEGF receptors in mice suggest a role in hormonally regulated angiogenesis. *J. Clin. Invest.* 91:2235–2243.

Shweiki, D., Itin, A., Soffer, D., and Keshet, E. 1992. Vascular endothelial growth factor induced by hypoxia may mediate hypoxia-initiated angiogenesis. *Nature* 359:843–845.

Stirling, D., Waterman, M. R., and Simpson, E. R. 1991. Expression of mRNA encoding basic fibroblast growth factor (bFGF) in bovine corpora lutea and cultured luteal cells. *J. Reprod. Fertil.* 91:1–8.

Suri, C., Jones, P. F., Patan, S., Bartunkova, S., Maisonpierre, P. C., Davis, S., Sato, T. N., and Yancopoulos, G. D. 1996. Requisite role of angiopoietin-1, a ligand for the TIE2 receptor, during embryonic angiogenesis. *Cell* 27:1171–1180.

Takaya, R., Fukaya, T., Sasano, H., Suzuki, T., Tamura, M., and Yajima, A. 1997. Macrophages in normal cycling human ovaries; immunohistochemical localization and characterization. *Hum. Reprod.* 12:1508–1512.

Teerds, K. J., and Dorrington, J. H. 1992. Immunohistochemical localization of transforming growth factor-beta 1 and -beta 2 during follicular development in the adult rat ovary. *Mol. Cell. Endocrinol.* 84:R7–R13.

Thurston, G., Suri, C., Smith, K., McClain, J., Sato, T. N., Yancopoulos, G. D., and McDonald, D. M. 1999. Leakage-resistant blood vessels in mice transgenically overexpressing angiopoietin-1. *Science* 286:2511–2514.

Tisdall, D. J., Hudson, N., Smith, P., and McNatty, K. P. 1994. Localization of ovine follistatin and alpha and beta A inhibin mRNA in the sheep ovary during the oestrous cycle. *J. Mol. Endocrinol.* 12:181–193.

Townson, D. H., Warren, J. S., Flory, C. M., Naftalin, D. M., and Keyes, P. L. 1996. Expression of monocyte chemoattractant protein-1 in the corpus luteum of the rat. *Biol. Reprod.* 54:513–520.

Tsukada, K., Matsushima, T., and Yamanaka, N. 1996. Neovascularization of the corpus luteum of rats during the estrus cycle. *Pathol. Int.* 46:408–416.

Vinatier, D., Dufour, P., Tordjeman-Rizzi, N., Prolongeau, J. F., Depret-Moser, S., and Monnier, J. C. 1995. Immunological aspects of ovarian function: role of the cytokines. *Eur. J. Obstet. Gynecol. Reprod. Biol.* 63:155–168.

Wiltbank, M. C. 1994. Cell types and hormonal mechanisms associated with mid-cycle corpus luteum function. *J. Anim. Sci.* 72:1873–1883.

Wiltbank, M. C., Gallagher, K. P., Christensen, A. K., Brabec, R. K., and Keyes, P. L. 1990. Physiological and immunocytochemical evidence for a new concept of blood flow regulation in the corpus luteum. *Biol. Reprod.* 42:139–149.

Yamamoto, T., Terada, N., Nishizawa, Y., and Petrow, V. 1994. Angiostatic activities of medroxyprogesterone acetate and its analogues. *Int. J. Cancer* 56:393–399.

Yan, Z., Weich, H. A., Bernart, W., Breckwoldt, M., and Neulen, J. 1993. Vascular endothelial growth factor (VEGF) messenger ribonucleic acid (mRNA) expression in luteinized human granulosa cells in vitro. *J. Clin. Endocrinol. Metab.* 77:1723–1725.

Young, F. M., Rodger, F. E., Illingworth, P. J., and Fraser, H. M. 2000. Cell proliferation and vascular morphology in the marmoset corpus luteum. *Hum. Reprod.* 15:557–566.

Zeleznik, A. J., Schuler, H. M., and Reichert, L. E. 1981. Gonadotropin-binding sites in the rhesus monkey ovary: role of the vasculature in the selective distribution of human chorionic gonadotropin to the preovulatory follicle. *Endocrinology* 109:356–362.

Zheng, J., Fricke, P. M., Reynolds, L. P., and Redmer, D. A. 1994. Evaluation of growth, cell proliferation, and cell death in bovine corpora lutea throughout the estrous cycle. *Biol. Reprod.* 51:623–632.
Zheng, J., Redmer, D. A., and Reynolds, L. P. 1993. Vascular development and heparin-binding growth factors in the bovine corpus luteum at several stages of the estrous cycle. *Biol. Reprod.* 49:1177–1189.

CHAPTER 7

Growth Factors During Ovarian Angiogenesis

Anna T. Grazul-Bilska, Dale A. Redmer, and Lawrence P. Reynolds

One of the few organs in adult organisms where capillaries are known to proliferate at regular intervals is the ovary.

O. Hudlická (1984)

As recognized by Hudlická, capillary growth in normal adult tissues is indeed a rare event, and the endothelium represents an extremely stable population of cells with a low mitotic activity (Denekamp, 1984; Hudlická, 1984). This observation is not surprising, because physiologic angiogenesis is usually associated with tissue growth or repair, and most normal adult tissues are relatively stable (Hudlická, 1984; Klagsbrun and D'Amore, 1991).

In contrast, the tissues of the adult female reproductive system, including the uterus, placenta, and ovary, grow extremely rapidly and at regular intervals (Reynolds et al, 1992, 1994, 2000; Redmer and Reynolds, 1996; Reynolds and Redmer, 1998; Ricke et al, 1999). For example, during its growth phase, which lasts about 8 to 10 days in large mammals, the ovarian corpus luteum (CL) doubles in size and cell numbers every 60 to 70 hours (Reynolds et al, 1994, 2000). This phenomenal growth rate is associated with an extremely high rate of cell proliferation, and is equaled only by the fastest growing tumors (Baserga, 1985; Reynolds et al, 1994, 2000). Unlike that of tumors, however, growth of the female reproductive tissues is normally a self-limiting and highly ordered process (Reynolds et al, 2000).

To support the phenomenal rate of tissue growth, microvascular growth and development also are extremely rapid in the female reproductive tissues, and they are highly vascular when mature (Reynolds et al, 1992, 1994, 2000; Reynolds and Redmer, 1995, 1998; Redmer and Reynolds, 1996). For example, most (≈50–85%) of luteal cell proliferation occurs in the microvascular compartment (Jablonka-Shariff et al, 1993; Reynolds et al, 1994; Nicosia et al, 1995; Christenson and Stouffer, 1996; Reynolds and Redmer, 1998; Ricke et al, 1999). As a result, in the mature CL microvascular pericytes and endothelial cells compose ≈50% to 70% of the total cell population (Farin et al, 1986; Lei et al, 1991; Reynolds et al, 2000). In association with their high vascularity, the female reproductive tissues receive some of the greatest rates of blood flow, per unit of tissue, of any adult organ and also exhibit a high metabolic rate (Reynolds, 1986; Swann and Bruce, 1987; Adair et al, 1990; Reynolds and Redmer, 1995, 1998; Redmer and Reynolds, 1996).

OVARIAN ANGIOGENESIS AND ANGIOGENIC FACTORS

During the first half of the 20th century, numerous investigators noticed this high vascularity of the ovarian tissues, and concluded that angiogenesis must be a critical component of follicular and luteal function (for a review, see Redmer and Reynolds, 1996, and Reynolds et al, 2000). In evaluating vascular development in the human ovary, Clark (1900) concluded, "The vital impulse to growth in the theca interna depends not on a maintenance of its primitive blood supply, but upon a decided increase of that supply." Similarly, in their classic study of the vascular system of the chicken ovary, Nalbandov and James (1949) observed, "The spatial relationship of the growing follicles to the surrounding blood vessels is such that it is logical to conclude that an adequate vascular system is essential for the growth of a follicle."

More recent work has supported the concept that maintenance of the follicular vasculature is important for maintaining follicular health. For example, dominant follicles of sheep and monkeys have not only a more vascular theca but also an increased uptake of serum gonadotropins compared with other antral follicles, whereas vascularity of atretic follicles is reduced (Hay et al, 1976; diZerega and Hodgen, 1980; McNatty et al, 1981; Zeleznik et al, 1981). Moor and Seamark (1986) observed that early atretic follicles of sheep will regenerate when placed in vitro, and concluded that, in vivo, atresia was caused by decreased vascularity, which resulted in limited access of the follicles to nutrients, substrates, and tropic hormones in vivo resulting from decreased vascularity. Greenwald (1989) found that one of the earliest signs of atresia in hamsters was reduced DNA synthesis of follicular endothelial cells, which was associated with reduced follicular vascularity. Similarly, we have observed a cessation in proliferation of thecal endothelial cells, associated with a decrease in thecal vascularity, soon after the onset of atresia in bovine, ovine, and porcine follicles (Jablonka-Shariff et al, 1994; P.M. Fricke, J.J. Ford, D.A. Redmer, and L.P. Reynolds, unpublished observations).

Thus, increased thecal vascularity may be a primary determinant of follicular dominance, and, conversely, reduced thecal vascularity appears to be an important component of follicular atresia. In support of these concepts, in sheep and cows only the granulosa cells of the estrogen-active, presumably healthy, preovulatory follicles produce angiogenic activity in vitro, whereas granulosa cells of the estrogen-inactive, presumably atretic, follicles do not (Figure 7.1) (Taraska et al, 1989; Redmer et al, 1991). Theca, on the other hand, produced angiogenic activity regardless of whether the follicle was healthy or atretic (Figure 7.1) (Taraska et al, 1989; Redmer et al, 1991). These authors concluded that while the theca of all the antral follicles produces angiogenic activity, production of angiogenic factors by the granulosa cells may contribute to follicular dominance or to maintaining the health of the preovulatory follicles (Taraska et al, 1989; Redmer et al, 1991).

Microvascular development of the follicular wall becomes even more extensive after ovulation, in association with vascularization of the CL (Redmer and

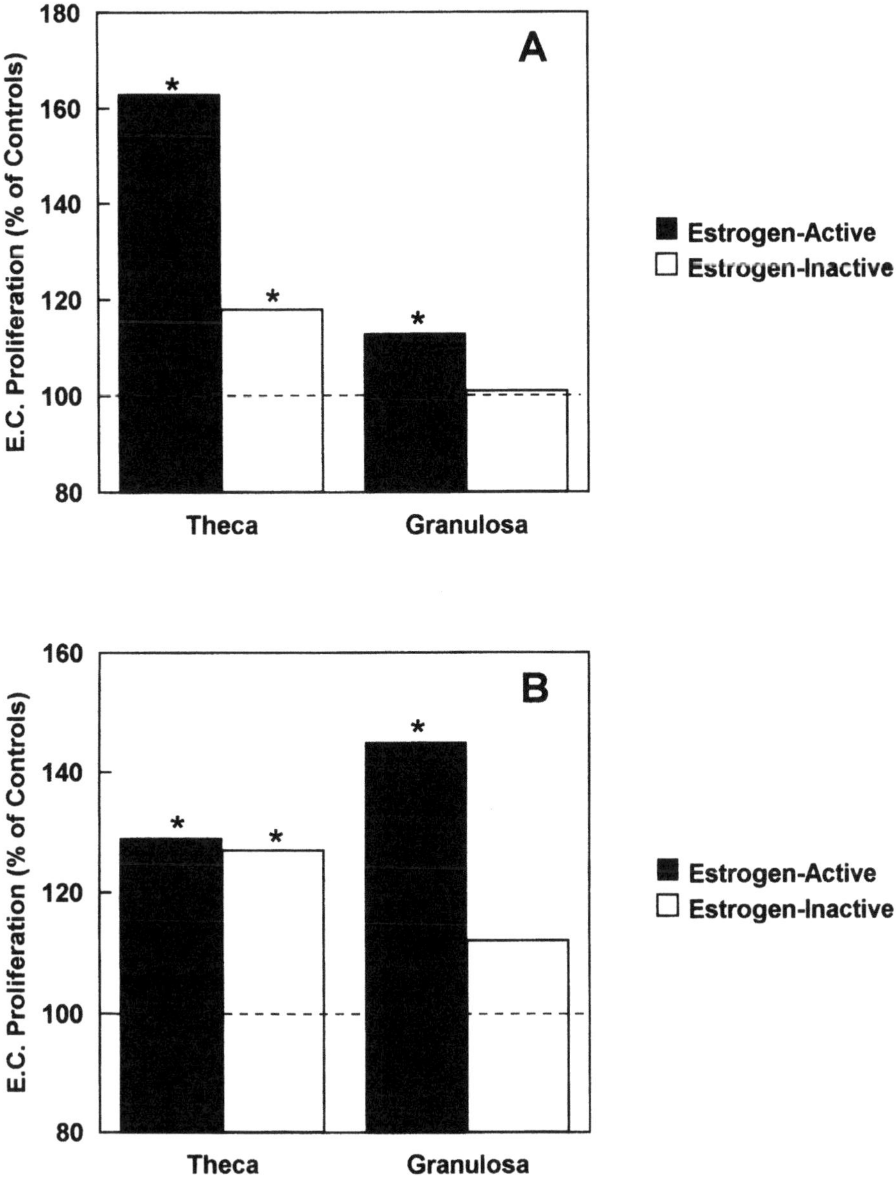

FIGURE 7.1. Effects of theca (T)- and granulosa (G)-conditioned media from (A) ewes and (B) cows on proliferation of endothelial cells (E.C.). For these studies, theca and granulosa from preovulatory-sized follicles ($n = 9$ ewes for panel A and $n = 17$ cows for panel B) were cultured for 24 hours in serum-free media, and the conditioned media from these cultures were subsequently evaluated for their ability to stimulate proliferation of a bovine aortic endothelial cell line. Follicles were classified as estrogen-active if the ratio of estrogen to progesterone (ng/mL : ng/mL) in follicular fluid was ≥1. Asterisks above the bars indicate that they differed ($p < .01$) from unconditioned media controls. (A: data from Taraska et al, 1989 (with permission); B: data from Redmer et al, 1991 (with permission).)

Reynolds, 1996; Reynolds and Redmer, 1998). The CL becomes so vascular, in fact, that the majority of the parenchymal (steroidogenic) cells of the mature CL are in contact with one or more capillaries (Dharmarajan et al, 1985; Reynolds et al, 1992; Redmer and Reynolds, 1996; Reynolds and Redmer, 1998). The mature CL also receives most of the ovarian blood supply, and ovarian blood flow is highly correlated with the rate of progesterone secretion (Niswender and Nett, 1988; Reynolds, 1986; Reynolds et al, 1994). Conversely, inadequate luteal function has been associated with decreased luteal vascularization in primates and cows (Jones et al, 1970; Smith et al, 1996).

Ovarian follicles and CL have been known for more than 20 years to contain angiogenic activity (for review, see Redmer and Reynolds, 1996, and Reynolds et al, 2000). In addition, tissues or cultured cells from ovarian follicles and CL contain numerous growth factors and cytokines, including epidermal growth factor, fibroblast growth factors, insulin-like growth factors, nerve growth factor, transforming growth factors, tumor necrosis factors, and vascular endothelial growth factors (Koos, 1993; Reynolds et al, 1994, 2000; Redmer and Reynolds, 1996). Thus, a bewildering array of growth factors, all of which may influence angiogenesis, is present in ovarian tissues.

To determine which are the major angiogenic factors of the CL, we evaluated in vitro production of angiogenic activity by growing, early cycle CL as well as mature CL during the estrous cycle and pregnancy (reviewed by Reynolds et al, 1994; Redmer and Reynolds, 1996). Across a number of these studies in cows, pigs, and sheep, a major finding was that all of the activity present in media conditioned by luteal tissues binds quite strongly to heparin-affinity columns (Reynolds et al, 1992; Redmer and Reynolds, 1996). Based on this observation, we hypothesized that the major ovarian angiogenic factors belong to one of the families of heparin-binding angiogenic factors, namely the fibroblast growth factors (FGFs) or the vascular endothelial growth factors (VEGFs). This hypothesis is consistent with the suggestion that the FGF and VEGF are probably key mediators of the angiogenic process in a variety of tissues (Klagsbrun and D'Amore, 1991; Ferrara and Davis-Smyth, 1997; Neufeld et al, 1999).

The angiogenic activity produced by bovine and ovine CL not only binds to heparin-affinity columns but elutes at about the same NaCl concentrations as acidic FGF (aFGF, or FGF-1) and basic FGF (bFGF, or FGF-2), and we have shown in several studies that aFGF and bFGF proteins are present in bovine and ovine CL (Grazul-Bilska et al, 1993, 1995b; Zheng et al, 1993; Jablonka-Shariff et al, 1997). We also have shown that aFGF and bFGF stimulate proliferation of ovine luteal cells from several stages of the estrous cycle, but especially those from early in the cycle (Grazul-Bilska et al, 1995a). In addition, Stirling et al (1991) have shown that the pattern of expression of bFGF messenger RNA (mRNA) by bovine CL closely follows that of angiogenic activity production (Redmer et al, 1988), and also is stimulated by luteinizing hormone (LH) as is secretion of angiogenic activity (Redmer et al, 1987).

When we attempted to immunoneutralize the endothelial mitogenic activity produced by the cow, pig, and sheep CL, antibodies against aFGF neutralized a relatively small amount (≈16%) of the angiogenic activity, whereas antibodies against bFGF neutralized a large amount (≈82%) of this activity (Grazul-Bilska et al, 1992a,b, 1993, 1995b; Ricke et al, 1995). We concluded, therefore, that bFGF (but not aFGF) was probably a major FGF involved in luteal angiogenesis

(Grazul-Bilska et al, 1993, 1995b; Reynolds et al, 1994; Redmer and Reynolds, 1996).

If bFGF is an important luteal angiogenic factor, however, it was not clear why similar amounts of its protein or mRNA were present across all stages of estrous cycle (Stirling et al, 1991; Grazul-Bilska et al, 1992b, 1993; Zheng et al, 1993; Ricke et al, 1995), since luteal angiogenesis occurs primarily early in the estrous cycle (Zheng et al, 1993; Ricke et al, 1995; Redmer and Reynolds, 1996). We therefore evaluated luteal expression of two major FGF receptors (FGFR), FGFR-1, also known as flg (fms-like gene), and FGFR-2, also known as bek (bacterially-expressed kinase), both of which contain intracellular tyrosine kinase domains. FGFR-1 was expressed at relatively high levels, and localization to the luteal parenchymal cells was greatest during mid-cycle and was dramatically reduced during luteal regression (Doraiswamy et al, 1998). In contrast, strong staining for FGFR-1 was present in luteal endothelial cells at all stages, and was especially abundant in those of the larger luteal microvessels late in the estrous cycle. Although luteal FGFR-2 was present at relatively low levels, it was localized to parenchymal cells at all stages but was present in the vasculature, primarily the larger microvessels, only late in the estrous cycle (Doraiswamy et al, 1998). Interestingly, many of the larger luteal microvessels are maintained during luteal regression (Redmer and Reynolds, 1996; Reynolds and Redmer, 1998). If FGF inhibits luteal cell death, as has been shown for other cell types (Tilly et al, 1992; Yasuda et al, 1995), a selective increase in FGFR in the endothelial cells of the larger microvessels would explain how they are maintained while the other luteal tissues regress. We have therefore hypothesized that FGF may affect not only luteal cell proliferation but also luteal cell death, thereby regulating luteal cell turnover and function (Redmer and Reynolds, 1996; Reynolds and Redmer, 1998).

A portion of the angiogenic activity produced by bovine and ovine corpora lutea also elutes from heparin-affinity columns at about the same NaCl concentrations as would be expected for VEGF (Grazul-Bilska et al, 1993, 1995b). To evaluate the expression of VEGF in the ovine corpora lutea, we cloned and sequenced ovine VEGF complementary DNA (cDNA) and developed a sensitive ribonuclease protection assay to quantify VEGF mRNA (Redmer et al, 1996). Although VEGF mRNA was present in ovine CL throughout the estrous cycle, its levels were greatest early in the cycle, when luteal vascularization also is occurring (Redmer and Reynolds, 1996; Reynolds and Redmer, 1998; Reynolds et al, 2000).

When we attempted to immunoneutralize the endothelial mitogenic activity produced by ovine CL, only about 30% of the activity was neutralized with anti-VEGF, whereas 70% to 100% was neutralized with anti-bFGF (Figure 7.2A) (Doraiswamy et al, 1995). In contrast, the same VEGF antibody neutralized 65% of the endothelial chemotactic activity produced by the ovine CL early in the estrous cycle, while the bFGF antibody neutralized only 25% of this activity (Figure 7.2B) (Doraiswamy et al, 1995). Only a small amount (10–25%) of the endothelial chemotactic activity produced by midcycle CL was neutralized by treatment with anti-VEGF or anti-bFGF (Figure 7.2B). These observations are consistent with the greatest levels of VEGF mRNA being present in the early cycle CL (Redmer et al, 1996) and also with the suggestion that VEGF plays a major role in luteal vascular development (Redmer and Reynolds, 1996; Reynolds and Redmer, 1998; Reynolds et al, 2000).

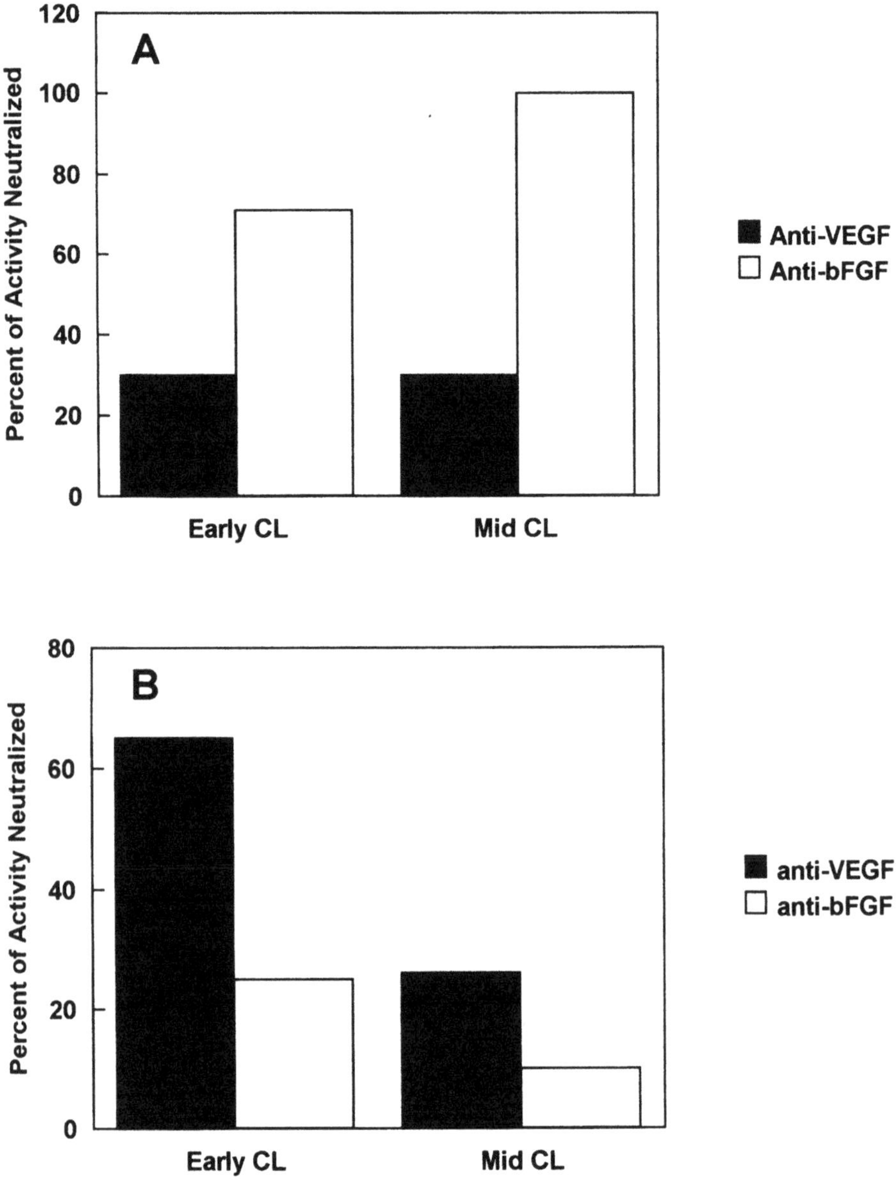

FIGURE 7.2. Effects of immunoneutralizing vascular endothelial growth factor (VEGF) and basic fibroblast growth factor (bFGF) antibodies on (A) endothelial mitogenic activity and (B) endothelial chemotactic activity produced by sheep CL from early in the estrous cycle (day 5; $n = 16$ ewes) or midcycle (day 10; $n = 9$ ewes). Endothelial mitogenic activity (A) was diminished 30% to 31% by anti-VEGF and 71% to 100% by anti-bFGF. For early CL, endothelial cell chemotactic activity (B) was diminished 65% by anti-VEGF and 25% by anti-bFGF 85%; for midcycle CL, the antibodies had only moderate effects (10% to 25% reduction) on endothelial chemotactic activity. For details of the immunoneutralization procedures, see Grazul-Bilska et al (1992a,b, 1993). (Data from Doraiswamy et al, 1995 (Abstract), and Doraiswamy, 1998.)

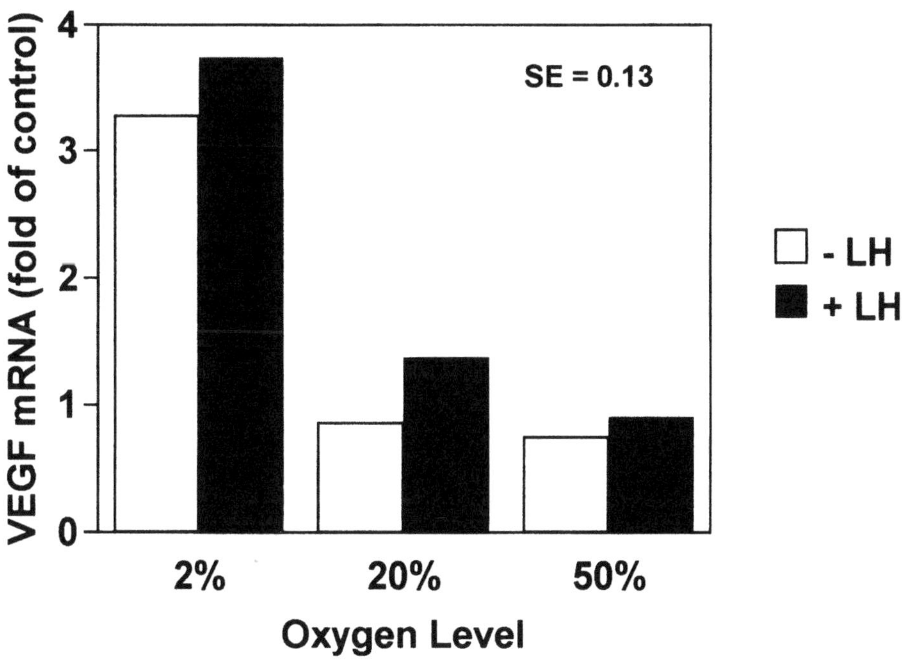

FIGURE 7.3. Effects of luteinizing hormone (LH) and O_2 on VEGF messenger RNA (mRNA) expression by ovine luteal cells in culture. Luteal cells were obtained by dispersion of corpora lutea from ewes ($n = 13$) on day 5 or 10 after estrus. Across all O_2 levels, LH stimulated ($p < .01$) expression of VEGF mRNA. Similarly, regardless of whether or not LH was present, low O_2 stimulated a threefold increase ($p < .01$), whereas high O_2 caused a 30% decrease ($p < .05$) in VEGF mRNA. The data are expressed as fold of control (e.g., 2-fold greater than control), which was a pooled sample of RNA from ovine liver. Concentration of LH in the +LH incubations was 100 ng/mL. (Data from Toutges et al, 1999 (Abstract).)

Additional observations provide further support for the hypothesis that VEGF is an important regulator of luteal vascular development. For example, in monkey CL VEGF mRNA also is present in greatest amounts in the early luteal compared with the late luteal phase and, in addition, is reduced by treatment with a gonadotropin-releasing hormone (GnRH) antagonist (Ravindranath et al, 1992). Others have found VEGF mRNA or protein in luteal tissues or luteinized granulosa cells (Phillips et al, 1990; Yan et al, 1993; Kamat et al, 1995; Goede et al, 1998). These observations provide further support for the hypothesis that VEGF is an important regulator of luteal vascular development.

Recently, we have observed that luteinizing hormone (LH) and oxygen (O_2) regulate VEGF mRNA expression in ovine luteal cells (Figure 7.3) (Toutges et al, 1999). These studies were based on the observation that human chorionic gonadotropin or LH can induce VEGF mRNA expression in preovulatory rat follicles and in cultured bovine granulosa cells that are luteinizing (Garrido et al, 1993; Koos, 1995). If VEGF is the major luteal angiogenic factor, its regulation by LH would make sense because LH is an important luteotropic factor and is critical for normal luteal development and function (Niswender and Nett, 1988; Quintal-Franco et al, 1999). Additionally, O_2 is a potent stimulator of VEGF expression across a number of cell and tissue types, and this is consistent with the concept that metabolic demand is the primary factor regulating vascular development of tissues (Adair et al, 1990; Neufeld et al, 1999). Because the developing CL resembles a healing wound, and thus would be expected to be hypoxic until the luteal parenchymal lobules become well vascularized (Silvester and Luck, 1999), it also is logical to conclude that O_2 level would be a major regulator of luteal VEGF expression, and our data support this conclusion (Figure 7.3).

To determine the cell types expressing VEGF during luteal vascularization, we utilized an affinity purified polyclonal antibody against a 15-amino acid peptide from the N-terminal region of ovine VEGF (Redmer et al, 1996). Consistent with the mRNA and immunoneutralization studies, localization of VEGF protein appeared to be greatest during early luteal development and least late in the estrous cycle, during luteal regression (Figure 7.4) (Reynolds and Redmer, 1998; Doraiswamy, 1998). In addition, VEGF was expressed by connective tissue cells of the luteal capsule and connective tissue tracts, by cells within the luteal parenchymal lobules, and by cells of the luteal arterioles (Figure 7.4). The cells within the parenchymal lobules were closely associated with the luteal capillaries and, based on their anatomic location, appeared to be capillary pericytes (Figure 7.4). Similarly, the luteal arteriolar cells containing VEGF appeared to be vascular smooth muscle (Figure 7.4). We have confirmed that these VEGF-expressing cells are indeed capillary pericytes and vascular smooth muscle by colocalization of VEGF and smooth muscle α-actin, which is a relatively specific marker of cells of the vascular smooth muscle lineage, including pericytes (Hirschi and D'Amore, 1996; Doraiswamy, 1998).

Thus, at least in the ovine CL, VEGF protein is localized to thecal-derived connective tissue tracts, arterioles, and capillary pericytes. Neeman et al (1997) found that follicular development is accompanied by elevated levels of VEGF mRNA. Consistent with these observations, we recently have shown that VEGF is expressed exclusively in the theca and not the granulosa of preovulatory bovine and ovine follicles, and also that the VEGF-expressing thecal pericytes invade the granulosa layer within hours after ovulation, even before the thecal endothelial cells (Doraiswamy, 1998). In addition, we recently have shown that a substantial proportion of the proliferating microvascular cells of the early CL are indeed pericytes (K. Fischer, D.A. Redmer, and L.P. Reynolds, unpublished observations). Interestingly, under hypoxic conditions, VEGF itself is mitogenic for pericytes (Yamagishi et al, 1999).

The thecal-derived microvessels are responsible for vascularization of the developing CL (Zheng et al, 1993; Redmer and Reynolds, 1996; Reynolds and Redmer, 1998). Within the last decade it has been suggested that the primary role of perivascular cells, and especially capillary pericytes, is to regulate endothelial cell function and angiogenesis during tissue growth and development (Antonelli-Orlidge et al, 1989; Hirschi and D'Amore, 1996). In addition, vascular smooth muscle and pericytes have been shown to contain or produce VEGF (Shifren et al, 1994; Aiello et al, 1995; Nomura et al, 1995). These observations, taken together with studies showing that VEGF is critical for normal vascular development, are consistent with the hypothesis that thecal-derived perivascular cells may direct vascularization of the ovine CL via production of VEGF.

We also have recently shown that granulosa cells of preovulatory ovine follicles produce a factor that stimulates migration of pericytes (Redmer et al, 1999). Because bFGF is a potent stimulator of pericyte migration (H.E. MacDonald, D.A.

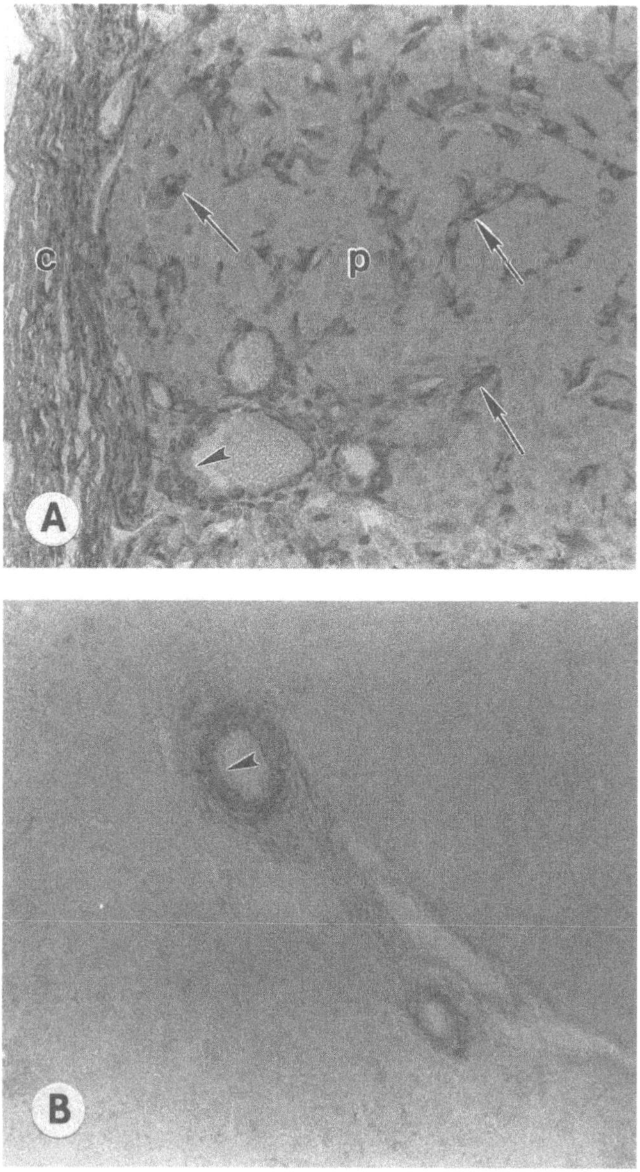

FIGURE 7.4. Immunohistochemical localization of VEGF (gray-black staining) in (A) early cycle (day 2 after estrus) and (B) late cycle (day 15 after estrus) sheep CL. In the developing CL (A), note that the cells expressing VEGF protein (arrows) are located in the thecal-derived luteal capsule (c) as well as within the granulosa-derived parenchymal lobule (p). In the regressing CL (B), VEGF protein was confined primarily to the luteal arterioles. In both sections, note that the arteriolar cells containing VEGF protein are distinct from the arteriolar endothelial cells (arrowheads). By using dual immunofluorescence, we recently have confirmed that the VEGF-containing cells within the luteal parenchymal lobules are microvascular pericytes (Doraiswamy, 1998). (From Reynolds and Redmer, 1998 (with permission).)

Redmer, and L.P. Reynolds, unpublished observations), and may be produced by granulosa cells (Neufeld et al, 1987; Wordinger et al, 1993), we have hypothesized that, during the periovulatory period, bFGF secreted by the granulosa cells may stimulate invasion of thecal pericytes. These observations have led us to propose a new model for the role of angiogenic growth factors in vascularization of the ovine CL (Figure 7.5).

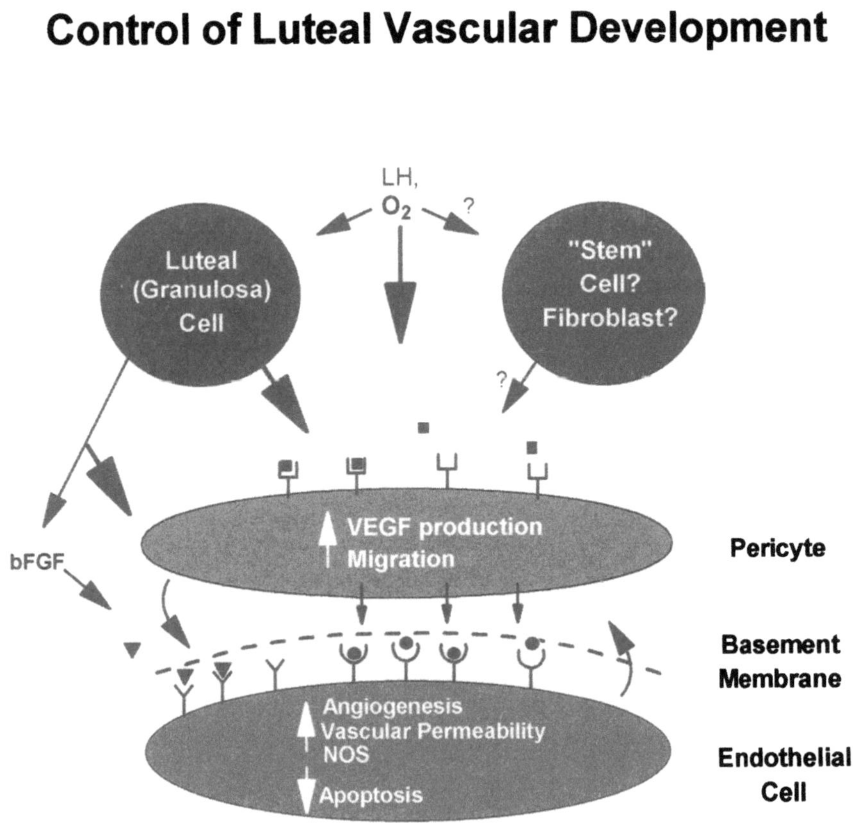

FIGURE 7.5. Current model for the role of angiogenic growth factors in vascularization of the ovine CL. Basic fibroblast growth factor (bFGF) and vascular endothelial growth factor (VEGF) seem to be the major luteal angiogenic growth factors. Granulosa cells and the luteal cells derived from them are capable of producing bFGF, which may regulate not only migration of pericytes but also endothelial cell survival (via inhibition of cell death). Similarly, VEGF, derived from the perivascular cells, may stimulate not only luteal angiogenesis but also luteal vascular permeability and blood flow (via increased nitric oxide synthase [NOS] production). Although LH stimulates production of angiogenic factors by luteal tissues, whether this effect is direct or mediated by luteal parenchymal or other cells is unknown. Oxygen tension seems to be a major regulator of luteal VEGF production; again, whether this effect is direct or indirect is not known. (Adapted from Reynolds and Redmer, 1998.)

IMPLICATIONS FOR LUTEAL FUNCTION AND FERTILITY

Inadequate or abnormal luteal function occurs in domestic animals during various physiologic conditions, including puberty, resumption of ovarian cyclicity postpartum, the beginning of the breeding season, and after ram-induced ovulation (Garverick and Smith, 1986; Hunter, 1991; Garverick et al, 1992; Foley, 1996). In addition, abnormal CL can be formed following the induction of ovulation by GnRH-treatment of anestrus ewes or postpartum cows (Keisler and Keisler, 1989; Hunter, 1991). In humans, inadequate luteal function occurs relatively frequently, and is recognized as a luteal-phase defect or deficiency, or a short luteal phase (Soules et al, 1977; Andrews, 1979; Smith et al, 1985).

Luteal dysfunctions include inadequate progesterone production, premature luteolysis, or persistent CL (Garverick and Smith, 1986; Hunter, 1991; Garverick et al, 1992; Foley, 1996). Insufficient progesterone secretion is unable to maintain the appropriate length of the luteal phase during the estrous cycle or pregnancy (Garverick and Smith, 1986; Hunter, 1991; Garverick et al, 1992; Foley, 1996). In cows, inadequate progesterone production has been associated with infertility and early embryonic death especially early in gestation (Lamming et al, 1989; Short et al, 1990). In humans, luteal-phase defect is a factor in as high as 20% of infertility cases and 60% of repeated first-trimester abortions (Andrews, 1979; Darland, 1985; Karamardian and Grimes, 1992).

As mentioned previously, there is some evidence to suggest a relationship between luteal dysfunction and reduced luteal vascularity (Jones et al, 1970; Smith et al, 1996). However, the question remains: Is reduced luteal vascularity secondary to luteal insufficiency or, conversely, is inadequate luteal vascularization a cause of luteal dysfunction? Although cause and effect are difficult to establish, two recent studies have shown that inhibition of VEGF action in the developing CL leads to both reduced luteal vascularity and luteal progesterone production (Ferrara et al, 1998; Fraser et al, 2000). These authors concluded that VEGF expression plays a critical role in luteal function. Further, it is tempting to speculate that lack of inappropriate follicular or luteal VEGF expression may contribute to ovarian dysfunction and thereby be an important cause of infertility. If this proves to be the case, regulation of ovarian VEGF expression could be a novel and powerful method for regulating fertility.

CONCLUSION

Although it has long been apparent that development of the microvasculature is critical for normal ovarian function, in the last two decades we have gained new insights into the role of the microvasculature in follicular and luteal development and function. For example, we now know that the extent of angiogenesis and microvascular perfusion may be primary factors in maintaining follicular health and dominance. We also know that inhibition of luteal angiogenesis can severely compromise luteal function and thus no doubt would have a dramatic effect on fertility. Moreover, we have identified at least some of the major factors that regulate ovarian angiogenesis, for example the FGF and VEGF families of proteins (Reynolds and Redmer, 1998). No doubt other angiogenic factors, such as the angiopoietins and their receptors, also will be important in ovarian angiogenesis, as they already have been shown to be important for angiogenesis in other tissues (Hanahan, 1997; Davis and Yancopoulos, 1999).

It remains to be seen if we can use this knowledge or, for example, that gained from studies of angiogenic and antiangiogenic factors in tumors (Barinaga, 1997; Hanahan, 1997), to regulate fertility. Regulation of fertility by manipulating either follicular or luteal angiogenesis seems reasonable, however, because, as mentioned above, ovarian tissues are some of the few adult tissues that exhibit angiogenesis and thus should provide a relatively unique target for angiogenic regulators. We would also suggest that it is critical to understand the potential negative effects of various angiogenic regulators, many of which will be used to treat various pathologic conditions such as tumors, on function of the female reproductive system, including not only the ovary but the uterus and placenta, which also exhibit dramatic angiogenesis and are highly vascular. In addition, we are convinced that future studies of ovarian angiogenesis will make a substantial contribution to our understanding of not only ovarian function but also growth and development of tissues in general.

ACKNOWLEDGMENTS

We wish to acknowledge our collaborators S. Derek Killilea and Robert M. Moor, our technicians James D. Kirsch and Kim C. Kraft, and our former graduate students Vinayak Doraiswamy, Paul M. Fricke, William A. Ricke, Albina Jablonka-Shariff, and Jing Zheng, without whom this work could not have been accomplished. Our studies have been supported in part by competitive grants from the National Research Initiative of the U.S. Department of Agriculture, by the U.S. National Science Foundation, and also by projects 1780, 1782, and 1795 of the North Dakota Agricultural Experiment Station.

REFERENCES

Adair, T. H., Gay, W. J., and Montani, J.-P. 1990. Growth regulation of the vascular system: evidence for a metabolic hypothesis. *Am. J. Physiol.* 259:R393-R404.

Aiello, L. P., Northrup, J. M., Keyt, B. A., Takagi, H., and Iwamoto, M. A. 1995. Hypoxic regulation of vascular endothelial growth factor in retinal cells. *Arch. Ophthalmol.* 113:1538–1534.

Andrews, W. C. 1979. Luteal phase defects. *Fertil. Steril.* 32:501–509.

Antonelli-Orlidge, A., Smith, S. R., and D'Amore, P. A. 1989. Influence of pericytes on capillary endothelial cell growth. *Am. Rev. Respir. Dis.* 140:1129–1131.

Barinaga, M. 1997. Cancer research: designing therapies that target tumor blood vessels. *Science* 275:482–484.

Baserga, R. 1985. *The Biology of Cell Reproduction*. Cambridge, MA: Harvard University Press.

Christenson, L. K., and Stouffer, R. L. 1996. Proliferation of microvascular endothelial cells in the primate corpus luteum during the menstrual cycle and simulated early pregnancy. *Endocrinology* 137:367–374.

Clark, J. G. 1900. The origin, development and degeneration of the blood-vessels of the human ovary. *Johns Hopkins Hosp. Rep.* 9:593–676.

Darland, N. W. 1985. Infertility associated with luteal phase defect. *J. Obstet. Gynecol. Neonatal Nurs.* 14:212–217.

Davis, S., and Yancopoulos, G. D. 1999. The angiopoietins: Yin and Yang in angiogenesis. *Curr. Topics Microbiol. Immunol.* 237:173–185.

Denekamp, J. 1984. Vasculature as a target for tumour therapy. In: Hammersen, F., and Hudlicka, O., eds. *Progress in Applied Microcirculation*, vol. 4. Karger, Basel, pp. 28–38.

Dharmarajan, A. M., Bruce, N. W., and Meyer, G. T. 1985. Quantitative ultrastructural characteristics relating to transport between luteal cell cytoplasm and blood in the corpus luteum of the pregnant rat. *Am. J. Anat.* 172:87–99.

diZerega, G. S., and Hodgen, G. D. 1980. Fluorescence localization of luteinizing hormone/human chorionic gonadotropin uptake in the primate ovary. II. Changing distribution during selection of the dominant follicle. *J. Clin. Endocrinol. Metab.* 51:903–907.

Doraiswamy, V. D. 1998. Angiogenesis in the ovine ovary: expression of vascular endothelial growth factor (VEGF) and fibroblast growth factor (FGF). *Dissertation.* North Dakota State University, Fargo.

Doraiswamy, V., Grazul-Bilska, A. T., Ricke, W. A., Redmer, D. A., and Reynolds, L. P. 1995. Immunoneutralization of angiogenic activity from ovine corpora lutea (CL) with antibodies against fibroblast growth factor (FGF)-2 and vascular endothelial growth factor (VEGF). *Biol. Reprod.* 52(suppl 1):112.

Doraiswamy, V., Knutson, D. L., Grazul-Bilska, A. T., Redmer, D. A., and Reynolds, L. P. 1998. Fibroblast growth factor receptor (FGFR)-1 and -2 in the ovine corpus luteum throughout the estrous cycle. *Growth Factors* 16:125–135.

Farin, C. E., Moeller, C. L., Sawyer, H. R., Gamboni, F., and Niswender, G. D. 1986. Morphometric analysis of cell types in the ovine corpus luteum throughout the estrous cycle. *Biol. Reprod.* 35:1299–1308.

Ferrara, N., Chen, H., Davis-Smyth, T., Gerber, H. P., Nguyen, T. N., Peers, D., Chisolm, V., Hillan, K. J., and Schwall, R. H. 1998. Vascular endothelial growth factor is essential for corpus luteum angiogenesis. *Nat. Med.* 4:336–340.

Ferrara, N., and Davis-Smyth, T. 1997. The biology of vascular endothelial growth factor. *Endoc. Rev.* 18:4–25.

Foley, G. L. 1996. Pathology of the corpus luteum of cows. *Theriogenology* 45:1413–1428.

Fraser, H. M., Dickson, S. E., Lunn, S. F., Wulff, C., Morris, K. D., Carroll, V. A., and Bicknell, R. 2000. Suppression of luteal angiogenesis in the primate after neutralization of vascular endothelial growth factor. *Endocrinology* 141:995–1000.

Garrido, C., Saule, S., and Gospodarowicz, D. 1993. Transcriptional regulation of vascular endothelial growth factor gene expression in ovarian bovine granulosa cells. *Growth Factors* 8:109–117.

Garverick, H. A., and Smith, M. F. 1986. Mechanisms associated with subnormal luteal function. *J. Anim. Sci.* 62(suppl 2):92–105.

Garverick, H. A., Zollers, W. G., Jr., and Smith, M. F. 1992. Mechanisms associated with corpus luteum lifespan in animals having normal or subnormal luteal function. *Anim. Reprod. Sci.* 28:111–124.

Goede, V., Schmidt, T., Kimmina, S., Kozian, D., and Augustin, H. G. 1998. Analysis of blood vessel maturation processes during cyclic ovarian angiogenesis. *Lab. Invest.* 78:1385–1394.

Grazul-Bilska, A. T., Redmer, D. A., Jablonka-Shariff, A., Biondini, M. E., and Reynolds, L. P. 1995a. Proliferation and progesterone production of ovine luteal cells from several stages of the estrous cycle: effects of fibroblast growth factors (FGF) and luteinizing hormone (LH). *Can. J. Physiol. Pharmacol.* 73:491–500.

Grazul-Bilska, A. T., Redmer, D. A., Killilea, S. D., Kraft, K. C., and Reynolds, L. P. 1992b. Production of mitogenic factor(s) by ovine corpora lutea throughout the estrous cycle. *Endocrinology* 130:3625–3632.

Grazul-Bilska, A. T., Redmer, D. A., Killilea, S. D., Zheng, J., and Reynolds, L. P. 1993. Initial characterization of endothelial mitogens produced by bovine corpora lutea from the estrous cycle. *Biochem. Cell Biol.* 71:270–277.

Grazul-Bilska, A. T., Redmer, D. A., Zheng, J., Killilea, S. D., and Reynolds, L. P. 1995b. Initial characterization of mitogenic activity of ovine corpora lutea from early pregnancy. *Growth Factors* 12:131–144.

Grazul-Bilska, A. T., Reynolds, L. P., Slanger, W. D., and Redmer, D. A. 1992a. Production of heparin-binding angiogenic factor(s) by bovine corpora lutea during pregnancy. *J. Anim. Sci.* 70:254–262.

Greenwald, G. S. 1989. Temporal and topographic changes in DNA synthesis after induced follicular atresia. *Biol. Reprod.* 41:175–181.

Hanahan, D. 1997. Signaling vascular morphogenesis and maintenance. *Science* 277:48–50.

Hay, M. F., Cran, D. G., and Moor, R. M. 1976. Structural changes occurring during atresia in sheep ovarian follicles. *Cell Tissue Res.* 169:515–529.

Hirschi, K. K., and D'Amore, P. A. 1996. Pericytes in the microvasculature. *Cardiovasc. Res.* 32:687–696.

Hudlická, O. 1984. Development of microcirculation: capillary growth and adaptation. In: Renkin, E. M., and Michel, C. C., eds. *Handbook of Physiology*, sec. 2, vol. IV, part 1. Waverly Press, Baltimore, pp. 165–216.

Hunter, M. G. 1991. Characteristics and causes of the inadequate corpus luteum. *J. Reprod. Fertil. Suppl.* 43:91–99.

Jablonka-Shariff, A., Fricke, P. M., Grazul-Bilska, A. T., Reynolds, L. P., and Redmer, D. A. 1994. Size, number, cellular proliferation, and atresia of gonadotropin-induced follicles in ewes. *Biol. Reprod.* 51:531–540.

Jablonka-Shariff, A., Grazul-Bilska, A. T., Redmer, D. A., and Reynolds, L. P. 1993. Growth and cellular proliferation of ovine corpora lutea throughout the estrous cycle. *Endocrinology* 133:1871–1879.

Jablonka-Shariff, A., Grazul-Bilska, A. T., Redmer, D. A., and Reynolds, L. P. 1997. Cellular proliferation and fibroblast growth factors in the corpus luteum during early pregnancy in ewes. *Growth Factors* 14:15–23.

Jones, G. S., Maffezzoli, R. D., Strott, C. A., Ross, G. T., and Kaplan, G. 1970. Pathophysiology of reproductive failure after clomiphene-induced ovulation. *Am. J. Obstetric. Gynecol.* 108:847–867.

Kamat, B. R., Brown, L. F., Manseau, E. J., Senger, D. R., and Dvorak, H. F. 1995. Expression of vascular permeability factor/vascular endothelial growth factor by human granulosa and theca lutein cells: role in corpus luteum development. *Am. J. Pathol.* 146:157–165.

Karamardian, L. M., and Grimes, D. A. 1992. Luteal phase deficiency: effect of treatment on pregnancy rates. *Am. J. Obstet. Gynecol.* 167:1391–1398.

Keisler, D. H., and Keisler, L. W. 1989. Formation and function of GnRH-induced subnormal corpora lutea in cyclic ewes. *J. Reprod. Fertil.* 87:265–273.

Klagsbrun, M., and D'Amore, P. A. 1991. Regulators of angiogenesis. *Annu. Rev. Physiol.* 53:217–239.

Koos, R. D. 1993. Ovarian angiogenesis. In: Adashi, E. Y., and Leung, P. C. K., eds. *The Ovary.* Raven Press, New York, pp. 433–453.

Koos, R. D. 1995. Increased expression of vascular endothelial growth/permeability factor in the rat ovary following an ovulatory gonadotropin stimulus: potential roles in follicle rupture. *Biol. Reprod.* 52:1426–1435.

Lamming, G. E., Darwash, A. O., and Back, H. L. 1989. Corpus luteum function in dairy cows and embryo mortality. *J. Reprod. Fertil. Suppl.* 37:245–252.

Lei, Z. M., Chegini, N., and Rao, C. V. 1991. Quantitative cell composition of human and bovine corpora lutea from various reproductive states. *Biol. Reprod.* 44:1148–1156.

McNatty, K. P., Dobson, C., Gibb, M., Kieboom, L. E., and Thurley, D. C. 1981. Accumulation of luteinizing hormone, oestradiol and androstenedione by sheep ovarian follicles in vivo. *J. Endocrinol.* 91:99–109.

Moor, R. M., and Seamark, R. F. 1986. Cell signaling, permeability, and microvasculatory changes during antral follicle development in mammals. *J. Dairy Sci.* 69:927–943.

Nalbandov, A. V., and James, M. F. 1949. The blood-vascular system of the chicken ovary. *Am. J. Anat.* 85:347–377.

Neeman, M., Abramovitch, R., Schiffenbauer, Y. S., and Tempel, C. 1997. Regulation of angiogenesis by hypoxic stress: from solid tumours to the ovarian follicle. *Int. J. Exp. Pathol.* 78:57–70.

Neufeld, G., Cohen, T., Gengrinovitch, S., and Poltorak, Z. 1999. Vascular endothelial growth factor (VEGF) and its receptors. *FASEB J.* 13:9–22.

Neufeld, G., Ferrara, N., Schweigerer, L., Mitchell, R., and Gospodarowicz, D. 1987. Bovine granulosa cells produce basic fibroblast growth factor. *Endocrinology* 121:597–603.

Nicosia, S. V., Diaz, J., Nicosia, R. F., Saunders, B. O., and Muro-Cacho, C. 1995. Cell proliferation and apoptosis during development and aging of the rabbit corpus luteum. *Ann. Clin. Lab. Sci.* 25:143–157.

Niswender, G. D., and Nett, T. M. 1988. The corpus luteum and its control. In: Knobil, E., and Neill, J. D., et al, eds. *The Physiology of Reproduction.* Raven Press, New York, pp. 489–525.

Nomura, M., Yamagishi, S., Harada, S., Hayashi, Y., Yamashima, T., Yamashita, J., and Yamamoto, H. 1995. Possible participation of autocrine and paracine vascular endothelial growth factors in hypoxia-induced proliferation of endothelial cells and pericytes. *J. Biol. Chem.* 270:28316–28324.

Phillips, H. S., Hains, J., Leung, D. W., and Ferrara, N. 1990. Vascular endothelial growth factor is expressed in rat corpus luteum. *Endocrinology* 127:965–967.

Quintal-Franco, J. A., Kojima, F. N., Melvin, E. J., Lindsey, B. R., Zanella, E., Fike, K. E., Wehrman, M. E., Clopton, D. T., and Kinder, J. E. 1999. Corpus luteum development and function in cattle with episodic release of luteinizing hormone pulses inhibited in the follicular and early luteal phases of the estrous cycle. *Biol. Reprod.* 61:921–926.

Ravindranath, N., Little-Ihrig, L., Phillips, H. S., Ferrara, N., and Zeleznik, A. J. 1992. Vascular endothelial growth factor messenger ribonucleic acid expression in the primate ovary. *Endocrinology* 131:254–260.

Redmer, D. A., Dai, Y., Li, J., Charnock-Jones, D. S., Smith, S. K., Reynolds, L. P., and Moor, R. M. 1996. Characterization and expression of vascular endothelial growth factor (VEGF) in the ovine corpus luteum. *J. Reprod. Fertil.* 108:157–165.

Redmer, D. A., Grazul, A. T., Kirsch, J. D., and Reynolds, L. P. 1988. Angiogenic activity of bovine corpus luteum at several stages of luteal development. *J. Reprod. Fertil.* 82:627–634.

Redmer, D. A., Kirsch, J. D., and Grazul, A. T. 1987. In vitro production of angiotropic factor by bovine corpus luteum partial characterization of activities that are chemotactic and mitogenic foe endothelial cells. In: Mahesh, V. B., Dhindsa, D. S., Anderson, E., and Kalra, S. P., eds. *Regulation of Ovarian and Testicular Function.* Advances in Experimental Medical Biology, vol. 219. Plenum Press, New York, pp. 683–688.

Redmer, D. A., Kirsch, J. D., and Reynolds, L. P. 1991. Production of mitogenic factors by cell types of estrogen-active and estrogen-inactive preovulatory bovine follicles. *J. Anim. Sci.* 69:237–245.

Redmer, D. A., McDonald, H. E., Klemin, P. L., Kirsch, J. D., Kraft, K. C., Johnson, M. L., Reynolds, L. P., and Grazul-Bilska, A. T. 1999. Media conditioned by granulosa cells from bovine follicles stimulates pericyte migration. *Biol. Reprod.* 60(suppl 1):172.

Redmer, D. A., and Reynolds, L. P. 1996. Angiogenesis in the ovary. *Rev. Reprod.* 1:182–192.

Reynolds, L. P. 1986. Utero-ovarian interactions during early pregnancy: role of conceptus-induced vasodilation. *J. Anim. Sci.* 62(suppl 2):47–61.

Reynolds, L. P., Grazul-Bilska A. T., and Redmer D. A. 2000. Angiogenesis in the corpus luteum. *Endocrine* 12:1–10.

Reynolds, L. P., Killilea, S. D., and Redmer, D. A. 1992. Angiogenesis in the female reproductive system. *FASEB J.* 6:886–892.

Reynolds, L. P., Killilea, S. D., Grazul-Bilska, A. T., and Redmer, D. A. 1994. Mitogenic factors of corpora lutea. *Prog. Growth Factor Res.* 5:159–175.

Reynolds, L. P., and Redmer, D. A. 1995. Utero-placental vascular development and placental function. *J. Anim. Sci.* 73:1839–1851.

Reynolds, L. P., and Redmer D. A. 1998. Expression of the angiogenic factors, basic fibroblast growth factor (bFGF) and vascular endothelial growth factor (VEGF), in the ovary. *J. Anim. Sci.* 76:1671–1681.

Ricke, W. A., Redmer, D. A., and Reynolds, L. P. 1995. Initial characterization of mitogenic factors produced by porcine corpora lutea throughout the estrous cycle. *Biol. Reprod.* 52(suppl 1):112.

Ricke, W. A., Redmer, D. A., and Reynolds, L. P. 1999. Growth and cellular proliferation of porcine corpora lutea throughout the oestrous cycle. *J. Reprod. Fertil.* 117:369–377.

Shifren, J. L., Doldi, N., Ferrara, N., Mesiano, S., and Jaffe, R. B. 1994. In the human fetus, vascular endothelial growth factor is expressed in epithelial cells and myocytes, but not vascular endothelium: Implications for mode of action. *J. Clin. Endocrinol. Metab.* 79:316–322.

Short, R. E., Bellows, R. A., Staigmiller, R. B., Berardinelli, J. G., and Custer, E. E. 1990. Physiological mechanisms controlling anestrus and infertility in postpartum beef cattle. *J. Anim. Sci.* 68:799–816.

Silvester, L. M., and Luck, M. R. 1999. Distribution of extracellular matrix components in the developing ruminant corpus luteum: a wound repair hypothesis for luteinization. *J. Reprod. Fertil.* 116:187–198.

Smith, G. D., Sawyer, H. R., Mirando, M. A., Griswold, M. D., Sadhu, A., and Reeves, J. J. 1996. Steady-state luteinizing hormone receptor messenger ribonucleic acid levels and endothelial cell composition in bovine normal- and short-lived corpora lutea. *Biol. Reprod.* 55:902–909.

Smith, S. K., Lenton, E. A., Landren, B. M., and Cooke, I. D. 1985. Is the short luteal phase a defective luteal phase? *Ann. NY Acad. Sci.* 442:387–390.

Soules, M. R., Wiebe, R. H., Askel, S., and Hammond, C. B. 1977. The diagnosis and therapy of luteal phase deficiency. *Fertil. Steril.* 29:1033–1037.

Stirling, D. T., Waterman, M. F., and Simpson, E. R. 1991. Expression of mRNA encoding basic fibroblast growth factor (bFGF) in bovine corpora lutea and cultured cells. *J. Reprod. Fertil.* 91:1–8.

Swann, R. T., and Bruce, N. W. 1987. Oxygen consumption, carbon dioxide production and progestagen secretion in the intact ovary of the day-16 pregnant rat. *J. Reprod. Fertil.* 80:599–605.

Taraska, T., Reynolds, L. P., and Redmer, D. A. 1989. In vitro secretion of angiogenic activity by ovine follicles. In: Hirshfield, A. N., ed. *Growth Factors and the Ovary.* Plenum Press, New York, pp. 267–272.

Tilly, J. L., Billig, H., Kowalski, K. I., and Hsueh, A. J. W. 1992. Epidermal growth factor and basic fibroblast growth factor suppress the spontaneous onset of apoptosis in cultured rat ovarian granulosa cells and follicles by a tyrosine kinase-dependent mechanism. *Mol. Endocrinol.* 6:1942–1950.

Toutges, M. J., Grazul-Bilska, A. T., Kirsch, J. D., Reynolds, L. P., and Redmer, D. A. 1999. Effects of luteinizing hormone (LH) and oxygen (O_2) levels on vascular endothelial growth factor (VEGF) mRNA expression by ovine luteal cells in culture. *Biol. Reprod.* 60(suppl 1):280.

Wordinger, R. J., Brun-Zinkernagel, A.-M., and Chang, I.-F. C. 1993. Immunohistochemical localization of basic fibroblast growth factor (bFGF) within growing and atretic mouse ovarian follicles. *Growth Factors* 9:279–289.

Yamagishi, S.-I., Yonekura, H., Yamamoto, Y., Fujimori, H., Sakurai, S., Tanake, N., and Yamamoto, H. 1999. Vascular endothelial growth factor acts as a pericyte mitogen under hypoxic conditions. *Lab. Invest.* 79:501–509.

Yan, Z., Weich, H. A., Bernart, W., Breckwoldt, M., and Neulen, J. 1993. Vascular endothelial growth factor (VEGF) messenger ribonucleic acid (mRNA) expression in luteinized human granulosa cells in vitro. *J. Clin. Endocrinol. Metab.* 77:1723–1725.

Yasuda, T., Grinspan, J., Stern, J., Franceschini, B., Bannerman, P., and Pleasure, D. 1995. Apoptosis occurs in the oligodendroglial lineage, and is prevented by basic fibroblast growth factor. *J. Neurosci. Res.* 40:306–317.

Zeleznik, A. J., Schuler, H. M., and Reichert, L. E. 1981. Gonadotropin-binding sites in the rhesus monkey ovary: role of the vasculature in the selective distribution of human chorionic gonadotropin to the preovulatory follicle. *Endocrinology* 109:356–361.

Zheng, J., Redmer, D. A., and Reynolds, L. P. 1993. Vascular development and heparin-binding growth factors in the bovine corpus luteum at several stages of the estrous cycle. *Biol. Reprod.* 49:1177–1189.

CHAPTER **8**

Vascular Endothelial Growth Factor: A Key Regulator of Physiologic Angiogenesis

Napoleone Ferrara

The development of a vascular supply is essential not only for organ development and differentiation during embryogenesis but also for wound healing and reproductive functions in the adult (Folkman, 1995). Angiogenesis is also implicated in the pathogenesis of a variety of disorders, including proliferative retinopathies, age-related macular degeneration, tumors, rheumatoid arthritis, and psoriasis (Garner, 1994; Folkman, 1995).

Many potential regulators of angiogenesis have been identified, including acidic fibroblast growth factor (aFGF), basic FGF (bFGF), transforming growth factor-α (TGF-α), TGF-β, hepatocyte growth factor/scatter factor (HGF/SF), turmor necrosis factor-α (TNF-α), angiogenin, interleukin-8 (IL-8), etc. (Folkman and Shing, 1992; Risau, 1997) and the angiopoietins, the ligands of the Tie-2 receptor (Suri et al, 1996; Maisonpierre et al, 1997). There is compelling evidence that vascular endothelial growth factor (VEGF), referred to also as VEGF-A, is an essential regulator of normal and abnormal angiogenesis, including that associated with tumors and intraocular neovascular syndromes (Kim et al, 1993; Aiello et al, 1994, 1995). Various VEGF inhibitors are being currently tested in numerous clinical trials (Ferrara and Alitalo, 1999). VEGF appears to be the most critical factor for the development of the vascular system, as inactivation of even a single VEGF allele results in embryonic lethality. For a comprehensive review of the molecular and biologic properties of VEGF, the reader is referred to several recent review articles on this topic (Ferrara, 1999; Neufeld et al, 1999; Ortega et al, 1999). Other chapters in this book address the role of other members of the VEGF gene family, such as placenta growth factor (PlGF), VEGF-B, VEGF-C and VEGF-D. This chapter mainly focuses on the role of VEGF-A in physiologic angiogenesis, with an emphasis on the female reproductive tract.

BIOLOGIC EFFECTS OF VEGF

VEGF was characterized as a mitogen for vascular endothelial cells derived from arteries, veins, and lymphatics (Ferrara and Henzel, 1989; Plouet et al, 1989; Conn et al, 1990; Ferrara and Davis-Smyth, 1997). However, studies have reported mitogenic effects of VEGF also on a few nonendothelial cell types, such as retinal pigment epithelial cells (Guerrin et al, 1995), pancreatic duct cells (Oberg-Welsh et al, 1997), and Schwann cells (Sondell et al, 1999). VEGF also elicits a pronounced angiogenic response in a variety of in vivo models including the chick chorioallantoic membrane (Leung et al, 1989; Plouet et al, 1989).

Also, VEGF functions as a survival factor for cultured endothelial cells in serum-depleted conditions (Gerber et al, 1998a,b). Consistent with such pro-survival activity, VEGF induces expression of the antiapoptotic proteins Bcl-2 and A1 in human endothelial cells (Gerber et al, 1998a). Alon et al (1995) have provided evidence that VEGF is a survival factor for immature retinal vessels, and suggested that pericyte coverage is the critical event that determines whether or not enolotheoliol cells are dependent on VEGF for survivol.

VEGF is known also as vascular permeability factor (VPF) based on its ability to induce vascular leakage in the guinea pig skin (Dvorak et al, 1995). Dvorak (1986) proposed that an increase in microvascular permeability to proteins is a crucial step in angiogenesis. A recent study has indicated that members of the Src family are differentially involved in mediating VEGF-dependent permeability versus angiogenesis (Eliceiri et al, 1999). Although mice lacking individual Src family kinases (SFKs) showed normal angiogenesis, mice deficient in pp60c-src or pp62c-yes showed no VEGF-induced vascular permeability, yet fyn$^{-/-}$ mice displayed normal permeability. Therefore, VEGF-mediated angiogenesis requires SFK activity in general, whereas the vascular permeability-enhancing activity of VEGF specifically depends on the SFKs src and yes (Eliceiri et al, 1999).

Bates et al have shown that VEGF also induces an increase in hydraulic conductivity of isolated microvessels (Bates and Curry, 1996) and that such effect is mediated by increased calcium influx (Bates and Curry, 1997). Other studies have also suggested that VEGF also induces fenestrations in endothelial cells, both in vivo (Roberts and Palade, 1995, 1997) and in vitro (Esser et al, 1998).

Characteristics of the VEGF Isoforms

By alternative exon splicing, several VEGF isoforms can be generated (Leung et al, 1989; Houck et al, 1991; Tischer et al, 1991; Shima et al, 1996). Native VEGF is a basic, heparin-binding, homodimeric glycoprotein of 45,000 daltons (Ferrara and Henzel, 1989). These properties correspond to those of $VEGF_{165}$, the major isoform (Houck et al, 1992). $VEGF_{121}$ is a weakly acidic polypeptide that fails to bind to heparin (Houck et al, 1992). $VEGF_{189}$ and $VEGF_{206}$ are more basic and bind to heparin with greater affinity than $VEGF_{165}$ (Houck et al, 1992). $VEGF_{121}$ is a freely diffusible protein; $VEGF_{165}$ is also secreted, although a significant fraction remains bound to the cell surface and the extracellular matrix (ECM). In contrast, $VEGF_{189}$ and $VEGF_{206}$ are almost completely sequestered in the ECM (Park et al, 1993). However, these isoforms may be released in a soluble form by heparin or heparinase, suggesting that their binding site is represented by proteoglycans containing heparin-like moieties. The long forms may be released also by plasmin following cleavage at the COOH terminus. This action generates a bioactive proteolytic fragment having molecular weight of ~34,000 daltons (Houck et al, 1992). Keyt et al (1996a) have shown that the bioactive product of plasmin action is composed of the first 110 NH_2-terminal amino acids of VEGF. These findings suggest that the VEGF proteins may become available to endothelial cells by at least two different mechanisms: as freely diffusible proteins ($VEGF_{121}$, $VEGF_{165}$), or following protease activation and cleavage of the longer isoforms. However, loss of heparin binding, whether it is due to alternative splicing of RNA or plasmin cleavage, results in a substantial loss of mitogenic activity for vascular endothelial cells (Keyt et al, 1996b).

Interestingly, a recent study has shown that only $VEGF_{164}$ is able to fully rescue a tumorigenic phenotype in $VEGF^{-/-}$ cells (Grunstein et al, 2000). $VEGF_{120}$ partially rescued and $VEGF_{188}$ failed completely to rescue tumor expansion, indicating that $VEGF_{164}$ has optimal characteristics of bioavailability, combined with high biologic potency. Recent studies have demonstrated that $VEGF_{121}$ fails to bind neuropilin-1, an isoform-specific VEGF receptor that presents $VEGF_{165}$ to its signaling receptors in a manner that enhances the effectiveness of the signal transduction cascade (Soker et al, 1998). Poltorak et al (1997) have provided evidence for the existence of an additional alternatively spliced isoform containing exons 1 to 6 and 8 of the VEGF gene, $VEGF_{145}$.

THE VEGF RECEPTORS

The Flt-1 (VEGFR-1) and Flk-1/KDR (VEGFR-2) Tyrosine Kinases

Binding Characteristics and Structural Properties

Two VEGF receptor tyrosine kinases (RTKs) have been identified. The Flt-1 (fms-like tyrosine kinase) (de Vries et al, 1992) and KDR (kinase domain region) (Terman et al, 1992) receptors bind VEGF with high affinity. The murine homologue of KDR is known as Flk-1 and shares 85% sequence identity with human KDR (Matthews et al, 1991). Both Flt-1 and KDR/Flk-1 have seven immunoglobulin (Ig)-like domains in the extracellular domain (ECD), a single transmembrane region and a tyrosine kinase domain, which is interrupted by a kinase-insert domain (Shibuya et al, 1990; Matthews et al, 1991; Terman et al, 1991). Flt-1 has the highest affinity for VEGF, with a dissociation constant (K_d) of approximately 10 to 20pM (de Vries et al, 1992). KDR has a lower affinity for VEGF, with a (K_d) of approximately 75 to 125pM (Terman et al, 1992). A cDNA coding an alternatively spliced soluble form of Flt-1 (sFlt-1), lacking the seventh Ig-like domain, transmembrane sequence, and the cytoplasmic domain, has been identified in human umbilical vein endothelial cells (Kendall and Thomas, 1993; Kendall et al, 1996). An additional member of the family of RTKs with seven Ig-like domains in the ECD is Flt-4 (VEGFR-3) (Galland et al, 1992; Pajusola et al, 1992; Finnerty et al, 1993), which, however, is not a receptor for VEGF but rather binds a newly identified ligand called VEGF-C or VEGF-related peptide (VRP) (Joukov et al, 1996; Lee et al, 1996). VEGF-C has been shown to be involved in the regulation of lymphatic angiogenesis (Jeltsch et al, 1997; Enholm et al, 1998) and is described in Chapter 2.

Previous studies have mapped the binding site for VEGF to the second immunoglobulin-like domain of Flt-1 and KDR (Davis-Smyth et al, 1996). Deletion of the second domain of Flt-1 abolished the binding of VEGF. Introduction of the second domain of KDR into an Flt-1 mutant lacking the homologous domain restored VEGF binding (Davis-Smyth et al, 1996, 1998). However, the ligand specificity was characteristic of the KDR receptor, since the mutant failed to bind PlGF. Wiesmann et al (1997) have solved the crystal structure of a VEGF-

Flt-1 domain 2 complex. These studies have shown domain 2 in a predominantly hydrophobic interaction with the poles of the VEGF dimer.

Signal Transduction

VEGF has been shown to induce the phosphorylation of at least 11 proteins in bovine aortic endothelial cells (Guo et al, 1995). Phospholipase C (PLC)-γ, and two proteins that associate with PLC-γ were phosphorylated in response to VEGF. Furthermore, immunoblot analysis for mediators of signal transduction that contain SH-2 domains demonstrated that VEGF induces phosphorylation of phosphatidylinositol 3-kinase, ras guanosine triphosphatase (GTPase) activating protein (GAP), and several others. These studies, however, did not identify which VEGF receptors are involved in these events. It has been suggested that nitric oxide (NO) mediates, at least in part, the mitogenic effect of VEGF on cultured microvascular endothelium isolated from coronary venules (Morbidelli et al, 1996).

Several studies have indicated that Flt-1 and KDR have different signal transduction properties (Waltenberger et al, 1994; Seetharam et al, 1995). Porcine aortic endothelial cells lacking endogenous VEGF receptors display chemotaxis and mitogenesis in response to VEGF when transfected with a plasmid coding for KDR (Waltenberger et al, 1994). In contrast, transfected cells expressing Flt-1 lack such responses (Waltenberger et al, 1994; Seetharam et al, 1995). Flk-1/KDR undergoes strong ligand-dependent tyrosine phosphorylation in intact cells, while Flt-1 reveals a weak or undetectable response (Waltenberger et al, 1994; Seetharam et al, 1995). Also, VEGF stimulation results in weak tyrosine phosphorylation that does not generate any mitogenic signal in transfected NIH 3T3 cells expressing Flt-1 (Seetharam et al, 1995).

These findings agree with other studies showing that PlGF, which binds with high affinity to Flt-1 but not to Flk-1/KDR, lacks direct mitogenic or permeability-enhancing properties or the ability to effectively stimulate tyrosine phosphorylation in endothelial cells (Park et al, 1994). Interestingly, however, high concentrations of PlGF expected to saturate the Flt-1 sites are able to potentiate the activity of VEGF, both in vivo and in vitro (Park et al, 1994). These findings led to the suggestion that Flt-1 may be not primarily a signaling receptor but rather a "decoy" receptor, able to regulate in a negative fashion the activity of VEGF on the vascular endothelium by sequestering and rendering this ligand less available to Flk-1/KDR (Park et al, 1994). Subsequent studies indicated that Flt-1 is able to interact with various signal transducing proteins, including the p85 subunit of the PI3 kinase and the mitogen-activated protein (MAP) kinase, and in some circumstances generate a mitogenic signal, at least in transfected cell lines (Waltenberger et al, 1994; Cunningham et al, 1995; Landgren et al, 1998).

However, recent studies have provided evidence in support of the earlier hypothesis. Hiratsuka et al (1998) have shown that a targeted mutation resulting in a Flt-1 receptor lacking the tyrosine kinase (TK) domain but able to bind VEGF does not result in lethality or any defect in embryonic development and angiogenesis in mice, while complete inactivation of the receptor results in early embryonic lethality (see below). Furthermore, endothelial cells isolated from Flt-1($TK^{-/-}$) animals display a normal mitogenic response to VEGF. These results suggest that Flt-1 plays its main role in embryonic angiogenesis as a ligand-binding molecule, rather than as a signal transducing receptor, at least as defined by conventional criteria. Interestingly, recent studies have shown that the juxtamembrane region of Flt-1 contains a repressor sequence that constitutively inhibits VEGF-dependent signal transduction (Gille et al, 2000b). Deletion of such a repressor sequence confers on Flt-1 the ability to mediate VEGF-dependent endothelial cell migration and PI3′ kinase activation (Gille et al, 2000b).

Taken together, these studies support the hypothesis that, interaction with Flk-1/KDR is essential to induce the full spectrum of VEGF biologic responses. In contrast, the significance of Flk-1 in the adult vasculature is still unclear. In further support of this conclusion, VEGF mutants that bind selectively to Flk-1/KDR, are fully active endothelial cell mitogens (Keyt et al, 1996b; Gille et al, 2001) and antiidiotypic antibodies that activate Flk-1/KDR promote tumor angiogenesis (Ortega, 1997). Furthermore, Flk-1/KDR activation has been shown to be required for the antiapoptotic effects of VEGF for human umbilical vein endothelial cells (HUVE) cells in serum-free conditions (Gerber et al, 1998b). Such a prosurvival effect of VEGF is mediated by the PI3 kinase/Akt pathway (Gerber et al, 1998b). Recently, Ogawa et al (1998) have isolated a polypeptide with about 25% amino acid identity to mammalian VEGF, encoded by a gene previously identified (Lyttle et al, 1994) in the genome of the Orf virus, a parapoxvirus that affects sheep and goats and occasionally humans. This polypeptide, named VEGF-E, is a potent mitogen and permeability-enhancing factor. Remarkably, VEGF-E binds KDR/Flk-1 and induces its autophosphorylation to almost the same extent as $VEGF_{165}$, but fails to bind Flt-1 (Ogawa et al, 1998; Meyer et al, 1999).

Soker et al (1998) have demonstrated the existence of an additional VEGF receptor that binds $VEGF_{165}$ but not $VEGF_{121}$. This isoform-specific VEGF-binding site is identical to human neuropilin-1, a receptor for the collapsin/semaphorin family that mediates neuronal cell guidance (Fujisawa and Kitsukawa, 1998). When coexpressed in cells with KDR, neuropilin-1 enhanced the binding of $VEGF_{165}$ to KDR and $VEGF_{165}$-mediated chemotaxis. Conversely, inhibition of $VEGF_{165}$ binding to neuropilin-1 inhibits its binding to KDR and its mitogenic activity for endothelial cells. These findings suggest that neuropilin-1 may present $VEGF_{165}$ to the KDR receptor in a manner that enhances the effectiveness of KDR-mediated signal transduction. So far there is no evidence that neuropilin-1 signals follow VEGF binding. As previously mentioned, these findings may contribute to explain the greater mitogenic potency of $VEGF_{165}$ compared to $VEGF_{121}$. Migdal et al (1998) have shown that PlGF-2, a heparin-binding isoform of PlGF (Maglione et al, 1993), also binds to neuropilin-1.

Role of VEGF and Its Receptors in Physiologic Angiogenesis

Embryonic and Early Postnatal Development

Gene targeting studies have demonstrated that both Flt-1 and Flk-1/KDR are essential for development of the embryonic vasculature in mice (Fong et al, 1995; Shalaby et al, 1995). However, their respective roles in endothelial cell proliferation and differentiation appear to be distinct. Mouse embryos homozygous for a targeted mutation in the Flt-1 locus died in utero between day 8.5 and 9.5 (Fong et al, 1995). Endothelial cells developed in both embryonic and extraembryonic sites but failed to organize in normal vascular channels. As previously noted, more recent studies revealed that normal blood vessel development and survival may occur in the absence of the tyrosine kinase domain of Flt-1 (Hiratsuka et al, 1998). Flk-$1^{-/-}$ mice lacked vasculogenesis and also failed to develop blood islands. Hematopoietic precursors were severely disrupted and organized blood vessels failed to develop throughout the embryo or the yolk sac, resulting in death in utero between day 8.5 and 9.5 (Shalaby et al, 1995).

Two independent studies (Carmeliet et al, 1996; Ferrara et al, 1996) have generated direct evidence for the role played by VEGF in embryonic vasculogenesis and angiogenesis. Inactivation of a single VEGF allele in mice resulted in embryonic lethality between day 11 and 12. The $VEGF^{+/-}$ embryos were growth retarded and also exhibited a number of developmental anomalies. The forebrain region appeared significantly underdeveloped. In the heart region, the outflow region was grossly malformed; the dorsal aortae were rudimentary, and the thickness of the ventricular wall was markedly decreased. The yolk sac revealed a markedly reduced number of nucleated red blood cells within the blood islands. Significant defects in the vasculature of other tissues and organs including placenta and nervous system were observed. In situ hybridization demonstrated the expression of VEGF mRNA in heterozygous embryos. Thus, the $VEGF^{+/-}$ phenotype is due to gene dosage and not to maternal imprinting.

More recently, Carmeliet et al (1999) reported an isoform-specific knockout of the VEGF gene. By Cre-*loxP* technology, they excised exons 6 and 7 of the VEGF gene in embryonic stem cells and thus generated mice that express exclusively $VEGF_{120}$. Interestingly, 50% of the $VEGF_{120/120}$ mice died shortly after delivery, while the remainder died within 2 weeks. The survivors demonstrated impaired myocardial contractility, heart enlargement, and defective angiogenesis leading to ischemic cardiomyopathy. These findings suggest that the action of the heparin-binding isoforms of VEGF cannot be replaced by $VEGF_{120}$.

To determine the role of VEGF in postnatal development, two independent approaches were recently employed. Inducible, Cre-*loxP*-mediated, gene targeting or administration of a soluble VEGF receptor chimeric protein [mFlt (1–3)-IgG] were used to inactivate VEGF in early postnatal life (Gerber et al, 1999a). Mice containing the "floxed" VEGF allele were bred to a strain transgenic for Cre recombinase controlled by an interferon-inducible promoter (MX-1) (Kuhn et al, 1995). Partial inhibition of VEGF achieved by this inducible gene targeting system resulted in increased mortality, stunted body growth, and impaired organ development, most notably of the liver. Administration of mFlt (1–3)-IgG, which achieves a considerably higher degree of VEGF inhibition, resulted in nearly complete growth arrest when the treatment was initiated at day 1 or day 8 postnatally. Also, such treatment was also accompanied by rapid lethality. Decreased levels of proliferation of various cell types were detected in all organs studied. Ultrastructural analysis documented alterations in endothelial and other cell types. Histologic and biochemical changes consistent with liver and renal failure were observed. Endothelial cells isolated from the liver of mFlt (1–3)-IgG–treated neonates demonstrated increased apoptotic index, indicating that VEGF is required not only for proliferation but also for survival of endothelial cells. Furthermore, these findings are consistent with a critical dependence of glomerular development on VEGF (Simon et al, 1995; Kitamoto et al, 1997). However, VEGF inhibition resulted in less significant alterations as the animal matured and the dependence on VEGF was eventually lost sometime after the fourth postnatal week. Administration of mFlt (1–3)-IgG to juvenile mice failed to induce apoptosis in liver endothelial cells (Gerber et al, 1999a).

Endochondral Bone Formation

Endochondral bone formation is a fundamental mechanism for longitudinal bone growth during vertebrate development (Poole, 1991). Cartilage, an avascular tissue, is replaced by bone in a process named endochondral ossification. During this process, the epiphyseal growth plate undergoes morphogenesis. A region of resting chondrocytes differentiates into a zone of proliferating chondrocytes that then hypertrophies and finally undergoes apoptotic cell death, while being replaced by bone. Such a sequence of events relies on the precise coupling of chondrogenesis (cartilage production) with osteogenesis (bone formation). During this process blood vessel invasion from the metaphysis coincides with mineralization of the extracellular matrix (ECM), apoptosis of hypertrophic chondrocytes, ECM degradation, and bone formation.

Recently, the role of VEGF in endochondral bone formation has been examined (Gerber et al, 1999b). These studies have shown that the VEGF mRNA is expressed by hypertrophic chondrocytes in the epiphyseal growth plate. To determine the functional significance of VEGF, a soluble VEGF receptor was administered systemically to 24-day-old mice. Following such treatment, blood vessel invasion was almost completely suppressed, concomitant with impaired trabecular bone formation. Recruitment and/or differentiation of chondroclasts, which express gelatinase B/matrix metalloproteinase-9 (Vu et al, 1998), and resorption of terminal chondrocytes was substantially decreased. Although proliferation, differentiation, and maturation of chondrocytes were apparently normal, resorption of hypertrophic chondrocytes was inhibited, resulting in a three- to sixfold expansion of the hypertrophic chondrocyte zone (Gerber et al, 1999b).

Interestingly, a similar, phenotype was obtained, when VEGF was deleted in the cartilage of developing mice by means of Cre-*loxP*-mediated, tissue-specific gene ablation (Haigh et al, 2000). In this system, the cartilage-specific collagen 2a promoter drives the expression of Cre-recombinase, which in turn induces the VEGF knockout. These *loxP* sites were introduced into the genome of mice flanking exon 3 of the mouse VEGF gene (Gerber et al, 1999a). In this model, a two- to threefold increase in the number of hypertrophic chondroyctes at embryonic day 17.5 was observed (Haigh et al, 2000).

These findings indicate that VEGF-dependent blood vessels is essential for coupling cartilage resorption with bone formation. Therefore, the vasculature carries the essential cellular and humoral signals required for correct growth plate morphogenesis. However, cessation of the anti-VEGF treatment was followed by capillary invasion, restoration of bone growth, resorption of the hypertrophic cartilage, and normalization of the growth plate architecture. Interestingly, VEGF receptor expression was localized not only to vascular endothelial cells. Osteoblasts also strongly expressed Flt-1 but not Flk-1/KDR. In this context, it is interesting to point that an earlier study found VEGF to have chemotactic effects on bovine osteoblasts (Midy and Plouet, 1994). It is tempting to speculate that these effects are mediated by Flt-1. These finding indicate that VEGF-mediated capillary invasion is a critical signal that regulates growth plate morphogenesis and triggers cartilage remodeling, indicating that VEGF is an essential coordinator of chondrocyte death, chondroclast function, extracellular matrix remodeling, angiogenesis, and bone formation in the growth plate.

Female Reproductive Tract Angiogenesis

Follicular growth and the development and endocrine function of the ovarian corpus luteum (CL) are dependent on the proliferation of new capillary vessels (Bassett, 1943). Subsequently, the blood vessels regress, suggesting the coordinated action of inducers and inhibitors of angiogenesis in the course of the ovarian cycle (Maisonpierre et al, 1997; Goede et al, 1998). Earlier studies suggested the release of angiogenic factors from developing CL (Gospodarowicz and Thakral, 1978). Therefore, the identification of the regulators of CL angiogenesis has been objects of intense investigation. Previous studies have shown the VEGF messenger RNA (mRNA) is temporally and spatially related to the proliferation of blood vessels in the rat, mouse, and primate ovary and in the rat uterus, suggesting that VEGF may be a mediator of the cyclical growth of blood vessels that occurs in the female reproductive tract (Phillips et al, 1990; Ravindranath et al, 1992; Cullinan-Bove and Koos, 1993; Shweiki et al, 1993). The availability of an effective inhibitor of rodent VEGF, such as a truncated soluble Flt-1 receptor [mFlt (1–3)-IgG] has made it possible to test directly this hypothesis (Davis-Smyth et al, 1996; Ferrara et al, 1998). Treatment with mFlt (1–3)-IgG resulted in virtually complete suppression of CL angiogenesis in a rat model of hormonally induced ovulation (Ferrara et al, 1998). This effect was associated with inhibition of CL development and progesterone release. Also, failure of maturation of the endometrium was observed, probably reflecting suppression of ovarian steroid production plus a direct inhibition of locally produced VEGF. Areas of ischemic necrosis were demonstrated in the CL of treated animals. However, no effect on the preexisting ovarian vasculature was observed. These findings indicate that VEGF-mediated angiogenesis is essential for CL development and endocrine function. A similar inhibition of CL and uterine angiogenesis has been observed in a toxicologic study in cynomolgus monkeys treated with a humanized anti-VEGF monoclonal antibody (rhuMAb VEGF) (Ryan et al, 1999). Furthermore, Fraser et al (2000) have recently shown that administration of an anti-VEGF neutralizing antibody also inhibits CL development in the marmoset. Treatment, initiated at the time of ovulation and continued for 3 days or 10 days, resulted in blockade of development of the normally extensive capillary bed. Luteal function was also markedly compromised by the treatment. These findings support the notion that VEGF-mediated angiogenesis is an essential component of luteal function.

POTENTIAL ROLE OF VEGF IN OVARIAN DISORDERS

There is compelling evidence that VEGF is a key mediator of both angiogenesis and enhanced vascular permeability characteristic of ovarian malignancies (Senger et al, 1983; Dvorak et al, 1995; Sowter et al, 1997; Mesiano et al, 1998). The possible role of VEGF in the pathogenesis of a variety of nonmalignant ovarian disorders has also been the object of intense investigation over the last several years. A series of studies has examined the potential role of VEGF in the pathogenesis of ovarian hyperstimulation syndrome (OHSS). This potentially fatal condition may follow medical induction of ovulation with gonadotropins. Although the mechanisms that lead to the development of the syndrome are not clearly understood, two major aspects are ovarian enlargement, with multiple follicular cysts and corpora lutea, and increased vascular permeability, leading to extravasation of protein-rich fluid out of the intravascular space. The massive third space created by a shift of fluid in the peritoneal cavity is responsible for most of the circulatory and excretory problems, which include hypovolemia, leading to anuria or hypotension in the most severe form (Engel et al, 1972; Haning et al, 1985).

Goldsman et al (1995) postulated the existence of a vasoactive substance present in the follicular fluid that is responsible for the symptoms. An earlier study examined the possibility that VEGF may be implicated in the pathogenesis of this syndrome by virtue of its activity as a direct capillary permeability-enhancing agent (McClure et al, 1994). Two similar peaks of permeability-enhancing activity were seen in OHSS ascites and in liver ascites spiked with VEGF. However, no activity was seen in control liver ascites. Incubation with VEGF antiserum decreased activity in the two OHSS peaks by 79% and 65% and the two spiked liver peaks by 49% and 50%. Control serum produced 24% and 27%, and 17% and 0% reductions, respectively. These findings led the authors to conclude that the major capillary permeability agent in OHSS ascites fluid is VEGF.

Similarly, Levin et al (1998) have shown that follicular fluid from OHSS patients contains permeability-enhancing activity that can be blocked by anti-VEGF neutralizing antibodies. However, other studies have cast doubt on the hypothesis that VEGF may be a causative factor in the vascular permeability associated with OHSS. Kobayashi et al (1998) proposed the kallikrein-kinin system, rather than VEGF, is primarily responsible for such increased permeability. According to these studies, the VEGF level in the peritoneal fluid of OHSS patients was lower than in those with ovarian carcinoma. Furthermore, Trasylol blocked the permeability-enhancing properties of the OHSS fluid, pointing to the involvement of the kallikrein-kinin system.

Clearly, a puzzling issue is the fact that the VEGF levels in the serum or peritoneal fluid of OHSS patients, albeit increased relative to control, are still relatively low, typically below 1 ng/mL (Agrawal et al, 1998; D'Ambrogio et al, 1999). While these VEGF concentrations are capable of exerting significant biologic effects, it is unlikely that they may be responsible for the massive vascular leakage characteristic of the syndrome. Similar or even considerably higher VEGF levels have been detected in patients affected by a variety of malignancies, not associated with ascites (Salven et al, 1997, 1998; Takigawa et al, 1998). Similarly, conflicting findings have been associated with the potential value of VEGF measurements as a prognostic factor for OHSS. Agrawal et al (1998) have suggested that serum VEGF levels in patients undergoing ovulation induction are a prognostic factor for the development of OHSS. However, other investigators have failed to find any significant difference between the serum VEGF levels of patients who developed moderate forms of OHSS and women without any symptoms of the syndrome (D'Ambrogio et al, 1999). The authors concluded that serum VEGF levels, before starting gonadotropin treatment, are not predictive of the subsequent development of moderate forms of ovarian hyperstimulation syndrome (D'Ambrogio et al, 1999). It is possible that, while VEGF alone is insufficient to cause the symptoms of OHSS, it may synergize with a number of inflammatory cytokines, including IL-2 and IL-6, that are known to be elevated as well (Aboulghar et al, 1999).

It is tempting to speculate that the investigation of the role of VEGF will further our understanding of the pathogenesis of several disorders of ovarian function that profoundly impair fertility. The hyperplasia and hypervascularity of polycystic ovary syndrome (Goldziher and Green, 1962) could be due at least in part to increased production of VEGF by theca-lutein cells. In this context, upregulation of VEGF mRNA has been demonstrated in the ovaries of patients affected by this syndrome (Kamat et al, 1995). Furthermore, several studies have documented the increase in the levels of VEGF in the peritoneal fluid of patients with endometriosis, a condition associated with infertility and pain and characterized by endometrial implants in the peritonal cavity (Shifren et al, 1996). Activated macrophages heve been shown to be a major source of VEGF and this production is regulated by sex steroids (McLaren et al, 1996). Peritoneal fluid concentrations of VEGF are significantly higher in women with moderate to severe endometriosis than in women with minimal to mild endometriosis or no disease. Thus, VEGF may be important in both physiologic and pathologic angiogenesis of human endometrium (McLaren et al, 1996; Shifren et al, 1996). More recent studies have confirmed the high VEGF levels in the peritoneal cavity of endometriosis patients, together with an increase in inflammatory cytokines, such as IL-6 (Mahnke et al, 2000) and IL-8 (Fasciani et al, 2000).

PERSPECTIVES

There is compelling evidence that the actions of VEGF on the vascular endothelium are complex and by no means limited to the induction of growth. Recent studies have emphasized the role of VEGF in preventing apoptosis of endothelial cells and regression of blood vessels. Such a maintenance function appears to be developmentally regulated, since it is critically dependent on the age of the animal. Therefore, a process of maturation occurs in endothelial cells such that VEGF eventually is no longer essential for survival. This "switch" seems to take place in the mouse around the fourth postnatal week, and in the fully developed animal VEGF may be required primarily for active angiogenic processes such as corpus luteum development or wound healing. The molecular nature of this switch remains largely to be determined.

There is strong evidence that VEGF is a key regulator of angiogenesis/permeability in the female reproductive tract. Increased VEGF levels are likely to play an important role in the pathogenesis of several major malignant and nonmalignant conditions, including endometriosis, polycystic ovary syndrome, and perhaps OHSS. It is likely that VEGF inhibitors will be tested in gynecologic clinical trials in the future.

REFERENCES

Aboulghar, M. A., Mansour, R. T., Serour, G. I., El Helw, B. A., and Shaarawy, M. 1999. Elevated levels of interleukin-2, soluble interleukin-2 receptor alpha, interleukin-6, soluble interleukin-6 receptor and vascular endothelial growth factor in serum and ascitic fluid of patients with severe ovarian hyperstimulation syndrome. *Eur. J. Obstet. Gynecol. Reprod. Biol.* 87:81–85.

Agrawal, R., Conway, G., Sladkevicius, P., Tan, S. L., Engmann, L., Payne, N., Bekir, J., Campbell, S., and Jacobs, H. 1998. Serum vascular endothelial growth factor and Doppler blood flow velocities in in vitro fertilization: relevance to ovarian hyperstimulation syndrome and polycystic ovaries. *Fertil. Steril.* 70:651–658.

Aiello, L. P., Avery, R. L., Arrigg, P. G., Keyt, B. A., Jampel, H. D., Shah, S. T., Pasquale, L. R., Thieme, H., Iwamoto, M. A., Park, J. E., Nguyen, H., Aiello, L. M., Ferrara, N., and King, G. L. 1994. Vascular endothelial growth factor in ocular fluid of patients with diabetic retinopathy and other retinal disorders. *N. Engl. J. Med.* 331:1480–1487.

Aiello, L. P., Pierce, E. A., Foley, E. D., Takagi, H., Chen, H., Riddle, L., Ferrara, N., King, G. L., and Smith, L. E. 1995. Suppression of retinal neovascularization in vivo by inhibition of vascular endothelial growth factor (VEGF) using soluble VEGF-receptor chimeric proteins. *Proc. Natl. Acad. Sci. USA* 92:10457–10461.

Alon, T., Hemo, I., Itin, A., Pe'er, J., Stone, J., and Keshet, E. 1995. Vascular endothelial growth factor acts as a survival factor for newly formed retinal vessels and has implications for retinopathy of prematurity. *Nat. Med.* 1:1024–1028.

Bassett, D. L. 1943. The changes in the vascular pattern of the ovary of the albino rat during the estrous cycle. *Am. J. Anat.* 73:251–259.

Bates, D. O., and Curry, F. E. 1996. Vascular endothelial growth factor increases hydraulic conductivity of isolated perfused microvessels. *Am. J. Physiol.* 271:H2520–H2528.

Bates, D. O., and Curry, F. E. 1997. Vascular endothelial growth factor increases microvascular permeability via a Ca(2+)-dependent pathway. *Am. J. Physiol.* 273:H687–H694.

Carmeliet, P., Ferreira, V., Breier, G., Pollefeyt, S., Kieckens, L., Gertsenstein, M., Fahrig, M., Vandenhoeck, A., Harpal, K., Eberhardt, C., Declercq, C., Pawling, J., Moons, L., Collen, D., Risau, W., and Nagy, A. 1996. Abnormal blood vessel development and lethality in embryos lacking a single VEGF allele. *Nature* 380:435–439.

Carmeliet, P., Ng, Y.-S., Nuyens, D., Theilmeier, G., Brusselmans, K., Cornelissen, I., Ehler, E., Kakkar, V. V., Stalmans, I., Mattot, V., Perriard, J.-C., Dewerchin, M., Flameng, W., Nagy, A., Lupu, F., Moons, L., Collen, D., D'Amore, P. A., and Shima, D. T. 1999. Impaired myocardial angiogenesis and ischemic cardiomyopathy in mice lacking the vascular endothelila growth factor isoforms VEGF164 and VEGF188. *Nat. Med.* 5:495–502.

Conn, G., Bayne, M. L., Soderman, D. D., Kwok, P. W., Sullivan, K. A., Palisi, T. M., Hope, D. A., and Thomas, K. A. 1990. Amino acid and cDNA sequences of a vascular endothelial cell mitogen that is homologous to platelet-derived growth factor. *Proc. Natl. Acad. Sci. USA* 87:2628–2632.

Cullinan-Bove, K., and Koos, R. D. 1993. Vascular endothelial growth factor/vascular permeability factor expression in the rat uterus: rapid stimulation by estrogen correlates with estrogen-induced increases in uterine capillary permeability and growth. *Endocrinology* 133:829–837.

Cunningham, S. A., Waxham, M. N., Arrate, P. M., and Brock, T. A. 1995. Interaction of the Flt-1 tyrosine kinase receptor with the p85 subunit of phosphatidylinositol 3-kinase. Mapping of a novel site involved in binding. *J. Biol. Chem.* 270:20254–20257.

D'Ambrogio, G., Fasciani, A., Monti, M., Cattani, R., Genazzani, A. R., and Artini, P. G. 1999. Serum vascular endothelial growth factor levels before starting gonadotropin treatment in women who have developed moderate forms of ovarian hyperstimulation syndrome. *Gynecol. Endocrinol.* 13:311–315.

Davis-Smyth, T., Chen, H., Park, J., Presta, L. G., and Ferrara, N. 1996. The second immunoglobulin-like domain of the VEGF tyrosine kinase receptor Flt-1 determines ligand binding and may initiate a signal transduction cascade. *EMBO J.* 15:4919–4927.

Davis-Smyth, T., Presta, L. G., and Ferrara, N. 1998. Mapping the charged residues in the second immunoglobulin-like domain of the vascular endothelial growth factor/placenta growth factor receptor Flt-1 required for binding and structural stability. *J. Biol. Chem.* 273:3216–3222.

de Vries, C., Escobedo, J. A., Ueno, H., Houck, K., Ferrara, N., and Williams, L. T. 1992. The fms-like tyrosine kinase, a receptor for vascular endothelial growth factor. *Science* 255:989–991.

Dvorak, H. F. 1986. Tumors: wounds that do not heal. Similarities between tumor stroma generation and wound healing. *N. Engl. J. Med.* 315:1650–1659.

Dvorak, H. F., Brown, L. F., Detmar, M., and Dvorak, A. M. 1995. Vascular permeability factor/vascular endothelial growth factor, microvascular hyperpermeability, and angiogenesis. *Am. J. Pathol.* 146:1029–1039.

Eliceiri, B. P., Paul, R., Schwartzberg, P. L., Hood, J. D., Leng, J., and Cheresh, D. A. 1999. Selective requirement for Src kinases during VEGF-induced angiogenesis and vascular permeability. *Mol. Cell.* 4:915–924.

Engel, T., Jewelewicz, R., Dyrenfurth, I., Speroff, L., and Vande Wiele, R. L. 1972. Ovarian hyperstimulation syndrome: report of one case with notes on pathogenesis and treatment. *Am. J. Obstet. Gynecol.* 112:1052–1057.

Enholm, B., Jussila, L., Karkkainen, M., and Alitalo, K. 1998. Vascular endothelial growth factor-C: a growth factor for lymphatic and blood vessel endothelial cells. *Trends Cardiovasc. Med.* 8:292–296.

Esser, S., Wolburg, K., Wolburg, H., Breier, G., Kurzchalia, T., and Risau, W. 1998. Vascular endothelial growth factor induces endothelial fenestrations in vitro. *J. Cell Biol.* 140:947–959.

Fasciani, A., D'Ambrogio, G., Bocci, G., Monti, M., Genazzani, A. R., and Artini, P. G. 2000. High concentrations of the vascular endothelial growth factor and interleukin-8 in ovarian endometriomata. *Mol. Hum. Reprod.* 6:50–54.

Ferrara, N. 1999. Role of vascular endothelial growth factor in the regulation of angiogenesis. *Kidney Int.* 56:794–814.

Ferrara, N., and Alitalo, K. 1999. Clinical applications of angiogenic growth factors and their inhibitors. *Nat. Med.* 5:1359–1364.

Ferrara, N., Carver Moore, K., Chen, H., Dowd, M., Lu, L., O'Shea, K. S., Powell Braxton, L., Hillan, K. J., and Moore, M. W. 1996. Heterozygous embryonic lethality induced by targeted inactivation of the VEGF gene. *Nature* 380:439–442.

Ferrara, N., Chen, H., Davis-Smyth, T., Gerber, H.-P., Nguyen, T.-N., Peers, D., Chisholm, V., Hillan, K. J., and Schwall, R. H. 1998. Vascular endothelial growth factor is essential for corpus luteum angiogenesis. *Nat. Med.* 4:336–340.

Ferrara, N., and Davis-Smyth, T. 1997. The biology of vascular endothelial growth factor. *Endocr. Rev.* 18:4–25.

Ferrara, N., and Henzel, W. J. 1989. Pituitary follicular cells secrete a novel heparin-binding growth factor specific for vascular endothelial cells. *Biochem. Biophys. Res. Commun.* 161:851–858.

Finnerty, H., Kelleher, K., Morris, G. E., Bean, K., Merberg, D. M., Kriz, R., Morris, J. C., Sookdeo, H., Turner, K. J., and Wood, C. R. 1993. Molecular cloning of murine FLT and FLT4. *Oncogene* 8:2293–2298.

Folkman, J. 1995. Angiogenesis in cancer, vascular, rheumatoid and other disease. *Nat. Med.* 1:27–31.

Folkman, J., and Shing, Y. 1992. Angiogenesis. *J. Biol. Chem.* 267:10931–10934.

Fong, G. H., Rossant, J., Gertsenstein, M., and Breitman, M. L. 1995. Role of the Flt-1 receptor tyrosine kinase in regulating the assembly of vascular endothelium. *Nature* 376:66–70.

Fraser, H. M., Dickson, S. E., Lunn, S. F., Wulff, C., Morris, K. D., Carroll, V. A., and Bicknell, R. 2000. Suppression of luteal angiogenesis in the primate after neutralization of vascular endothelial growth factor. *Endocrinology* 141:995–1000.

Fujisawa, H., and Kitsukawa, T. 1998. Receptors for collapsin/semaphorins. *Curr. Opin. Neurobiol.* 8:587–592.

Galland, F., Karamysheva, A., Mattei, M. G., Rosnet, O., Marchetto, S., and Birnbaum, D. 1992. Chromosomal localization of FLT4, a novel receptor-type tyrosine kinase gene. *Genomics* 13:475–478.

Garner, A. 1994. Vascular diseases. In: Garner, A. and Klintworth, J. K., eds. *Pathobiology of Ocular Diseases*, 2nd ed. Marcel Dekker, New York, pp. 1625–1710.

Gerber, H. P., Dixit, V., and Ferrara, N. 1998a. Vascular endothelial growth factor induces expression of the antiapoptotic proteins Bcl-2 and A1 in vascular endothelial cells. *J. Biol. Chem.* 273:13313–13316.

Gerber, H. P., Hillan, K. J., Ryan, A. M., Kowalski, J., Keller, G.-A., Rangell, L., Wright, B. D., Radtke, F., Aguet, M., and Ferrara, N. 1999a. VEGF is required for growth and survival in neonatal mice. *Development* 126:1149–1159.

Gerber, H. P., McMurtrey, A., Kowalski, J., Yan, M., Keyt, B. A., Dixit, V., and Ferrara, N. 1998b. VEGF regulates endothelial cell survival by the PI3-kinase/Akt signal transduction pathway. Requirement for Flk-1/KDR activation. *J. Biol. Chem.* 273:30366–30343.

Gerber, H. P., Vu, T. H., Ryan, A. M., Kowalski, J., Werb, Z., and Ferrara, N. 1999b. VEGF couples hypertrophic cartilage remodeling, ossification and angiogenesis during endochondral bone formation. *Nat. Med.* 5:623–628.

Gille, H., Kowalski, J., Li, B., Moffat, B., Zioncheck, T., Peleltier, N., and Ferrara, N. 2001. Signaling and biological effects mediated by VEGFR-1 and VEGFR-2. A reassessment using novel receptor-specific VEGF mutants. *J. Biol. Chem.* 276:3222–3230.

Gille, H., Kowalski, J., Yu, L., Chen, H., Pisabarro, M. T., Davis-Smyth, T., and Ferrara, N. 2000b. A repressor sequence in the juxtamembrane domain of Flt-1 (VEGFR-1) constitutively inhibits VEGF-dependent PI 3 kinase activation and endothelial cell migration. *EMBO J.* 19:4064–4073.

Goede, V., Schmidt, T., Kimmina, S., Kozian, D., and Augustin, H. G. 1998. Analysis of blood vessel maturation processes during cyclic ovarian angiogenesis. *Lab. Invest.* 78:1385–1394.

Goldsman, M., Pedram, A., Dominguez, C., Ciuffardi, I., Levin, E., and Asch, R. H. 1995. Increased capillary permeability induced by human follicular fluid: a hypothesis for an ovarian origin of the hyperstimulation syndrome. *Fertil. Steril.* 63:268–272.

Goldziher, J. W., and Green, J. A. 1962. The polycistic ovary. I. Clinical and histologic features. *J. Clin. Endocrinol. Metab.* 22:325–332.

Gospodarowicz, D., and Thakral, K. K. 1978. Production of a corpus luteum angiogenic factor responsible for proliferation of capillaries and neovascularization of the corpus luteum. *Proc. Natl. Acad. Sci. USA* 75:847–851.

Grunstein, J., Masbad, J. J., Hickley, R., Giordano, F., and Johnson, R. S. 2000. Isoforms of vascular endothelial growth factor act in a coordinated fashion to recruit and expand tumor vasculature. *Mol. Cell. Biol.* 20:7282–7291.

Guerrin, M., Moukadiri, H., Chollet, P., Moro, F., Dutt, K., Malecaze, F., and Plouet, J. 1995. Vasculotropin/vascular endothelial growth factor is an autocrine growth factor for human retinal pigment epithelial cells cultured in vitro. *J. Cell. Physiol.* 164:385–394.

Guo, D., Jia, Q., Song, H. Y., Warren, R. S., and Donner, D. B. 1995. Vascular endothelial cell growth factor promotes tyrosine phosphorylation of mediators of signal transduction that contain SH2 domains. Association with endothelial cell proliferation. *J. Biol. Chem.* 270:6729–6733.

Haigh, J. J., Gerber, H. P., Ferrara, N., and Wagner, E. F. 2000. Conditional inactivation of VEGF-A in areas of collagen2a1 expression results in embryonic lethality in the heterozygous state. *Development* 127:1445–1453.

Haning, R. J., Strawn, E., and Nolten, W. 1985. Pathophysiology of the ovarian hyperstimulation syndrome. *Obstet. Gynecol.* 166:220–224.

Hiratsuka, S., Minowa, O., Kuno, J., Noda, T., and Shibuya, M. 1998. Flt-1 lacking the tyrosine kinase domain is sufficient for normal development and angiogenesis in mice. *Proc. Natl. Acad. Sci. USA* 95:9349–9354.

Houck, K. A., Ferrara, N., Winer, J., Cachianes, G., Li, B., and Leung, D. W. 1991. The vascular endothelial growth factor family: identification of a fourth molecular species and characterization of alternative splicing of RNA. *Mol. Endocrinol.* 5:1806–1814.

Houck, K. A., Leung, D. W., Rowland, A. M., Winer, J., and Ferrara, N. 1992. Dual regulation of vascular endothelial growth factor bioavailability by genetic and proteolytic mechanisms. *J. Biol. Chem.* 267:26031–26037.

Jeltsch, M., Kaipainen, A., Joukov, V., Meng, X., Lakso, M., Rauvala, H., Awartz, M., Fukumura, D., Jain, R. K., and Alitalo, K. 1997. Hyperplasia of lymphatic vessels in VEGF-C transgenic mice. *Science* 276:1423–1425.

Joukov, V., Pajusola, K., Kaipainen, A., Chilov, D., Lahtinen, I., Kukk, E., Saksela, O., Kalkkinen, N., and Alitalo, K. 1996. A novel vascular endothelial growth factor, VEGF-C, is a ligand for the Flt4 (VEGFR-3) and KDR (VEGFR-2) receptor tyrosine kinases. *EMBO J.* 15:290–298.

Kamat, B. R., Brown, L. F., Manseau, E. J., Senger, D. R., and Dvorak, H. F. 1995. Expression of vascular permeability factor/vascular endothelial growth factor by human granulosa and theca lutein cells. Role in corpus luteum development. *Am. J. Pathol.* 146:157–165.

Kendall, R. L., and Thomas, K. A. 1993. Inhibition of vascular endothelial cell growth factor activity by an endogenously encoded soluble receptor. *Proc. Natl. Acad. Sci. USA* 90:10705–10709.

Kendall, R. L., Wang, G., and Thomas, K. A. 1996. Identification of a natural soluble form of the vascular endothelial growth factor receptor, FLT-1, and its heterodimerization with KDR. *Biochem. Biophys. Res. Commun.* 226:324–328.

Keyt, B. A., Berleau, L. T., Nguyen, H. V., Chen, H., Heinsohn, H., Vandlen, R., and Ferrara, N. 1996a. The carboxyl-terminal domain (111–165) of vascular endothelial growth factor is critical for its mitogenic potency. *J. Biol. Chem.* 271:7788–7795.

Keyt, B. A., Nguyen, H. V., Berleau, L. T., Duarte, C. M., Park, J., Chen, H., and Ferrara, N. 1996b. Identification of vascular endothelial growth factor determinants for binding KDR and FLT-1 receptors. Generation of receptor-selective VEGF variants by site-directed mutagenesis. *J. Biol. Chem.* 271:5638–5646.

Kim, K. J., Li, B., Winer, J., Armanini, M., Gillett, N., Phillips, H. S., and Ferrara, N. 1993. Inhibition of vascular endothelial growth factor-induced angiogenesis suppresses tumor growth in vivo. *Nature* 362:841–844.

Kitamoto, Y., Tokunaga, H., and Tomita, K. 1997. Vascular endothelial growth factor is an essential molecule for mouse kidney development: glomerulogenesis and nephrogenesis. *J. Clin. Invest.* 99:2351–2357.

Kobayashi, H., Okada, Y., Asahina, T., Gotoh, J., and Terao, T. 1998. The kallikrein-kinin system, but not vascular endothelial growth factor, plays a role in the increased vascular permeability associated with ovarian hyperstimulation syndrome. *J. Mol. Endocrinol.* 20:363–374.

Kuhn, R., Schwenk, F., Aguet, M., and Rajewsky, K. 1995. Inducible gene targeting in mice. *Science* 269:1427–1429.

Landgren, E., Schiller, P., Cao, Y., and Claesson-Welsh, L. 1998. Placenta growth factor stimulates MAP kinase and mitogenicity but not phospholipase C-gamma and migration of endothelial cells expressing Flt 1. *Oncogene* 16:359–367.

Lee, J., Gray, A., Yuan, J., Luoh, S. M., Avraham, H., and Wood, W. I. 1996. Vascular endothelial growth factor-related protein: a ligand and specific activator of the tyrosine kinase receptor Flt4. *Proc. Natl. Acad. Sci. USA* 93:1988–1992.

Leung, D. W., Cachianes, G., Kuang, W. J., Goeddel, D. V., and Ferrara, N. 1989. Vascular endothelial growth factor is a secreted angiogenic mitogen. *Science* 246:1306–1309.

Levin, E. R., Rosen, G. F., Cassidenti, D. L., Yee, B., Meldrum, D., Wisot, A., and Pedram, A. 1998. Role of vascular endothelial cell growth factor in ovarian hyperstimulation syndrome. *J. Clin. Invest.* 102:1978–1985.

Lyttle, D. J., Fraser, K. M., Flemings, S. B., Mercer, A. A., and Robinson, A. J. 1994. Homologs of vascular endothelial growth factor are encoded by the poxvirus orf virus. *J. Virol.* 68:84–92.

Maglione, D., Guerriero, V., Viglietto, G., Ferraro, M. G., Aprelikova, O., Alitalo, K., Del Vecchio, S., Lei, K. J., Chou, J. Y., and Persico, M. G. 1993. Two alternative mRNAs coding for the angiogenic factor, placenta growth factor (PlGF), are transcribed from a single gene of chromosome 14. *Oncogene* 8:925–931.

Mahnke, J. L., Dawood, M. Y., and Huang, J. C. 2000. Vascular endothelial growth factor and interleukin-6 in peritoneal fluid of women with endometriosis. *Fertil. Steril.* 73:166–170.

Maisonpierre, P. C., Suri, C., Jones, P. F., Bartunkova, S., Wiegend, S. J., Radziejewski, C., Compton, D., McClain, J., Aldrich, T. H., Papadopulos, N., Daly, T. J., Davis, S., Sato, T. N., and Yancopoulos, G. D. 1997. Angiopoietin-2, a natural antagonist for Tie-2 that disrupts in vivo angiogenesis. *Science* 277:55–60.

Matthews, W., Jordan, C. T., Gavin, M., Jenkins, N. A., Copeland, N. G., and Lemischka, I. R. 1991. A receptor tyrosine kinase cDNA isolated from a population of enriched primitive hematopoietic cells and exhibiting close genetic linkage to c-kit. *Proc. Natl. Acad. Sci. USA* 88:9026–9030.

McClure, N., Healy, D. L., Rogers, P. A., Sullivan, J., Beaton, L., Haning, R. V., Connolly, D. T., and Robertson, D. M. 1994. Vascular endothelial growth factor as a capillary permeability agent in ovarian hyperstimulation syndrome. *Lancet* 344:235–236.

McLaren, J., Prentice, A., Charnock-Jones, D. S., Millican, S. A., Muller, K. H., Sharkey, A. M., and Smith, S. K. 1996. Vascular endothelial growth factor is produced by peritoneal fluid macrophages in endometriosis and is regulated by ovarian steroids. *J. Clin. Invest.* 98:482–489.

Mesiano, S., Ferrara, N., and Jaffe, R. B. 1998. Role of vascular endothelial growth factor in ovarian cancer: inhibition of ascites formation by immunoneutralization. *Am. J. Pathol.* 153:1249–1256.

Meyer, M., Clauss, M., Lepple-Wienhues, A., Waltenberger, J., Augustin, H. G., Ziche, M., Lanz, C., Buttner, M., Rziha, H. J., and Dehio, C. 1999. A novel vascular endothelial growth factor encoded by Orf virus, VEGF-E, mediates angiogenesis via signalling through VEGFR-2 (KDR) but not VEGFR-1 (Flt-1) receptor tyrosine kinases. *EMBO J.* 18:363–374.

Midy, V., and Plouet, J. 1994. Vasculotropin/vascular endothelial growth factor induces differentiation in cultured osteoblasts. *Biochem. Biophys. Res. Commun.* 199:380–386.

Migdal, M., Huppertz, B., Tessler, S., Comforti, A., Shibuya, M., Reich, R., Baumann, H., and Neufeld, G. 1998. Neuropilin-1 is a placenta growth factor-2 receptor. *J. Biol. Chem.* 273:22272–22278.

Morbidelli, L., Chang, C. H., Douglas, J. G., Granger, H. J., Ledda, F., and Ziche, M. 1996. Nitric oxide mediates mitogenic effect of VEGF on coronary venular endothelium. *Am. J. Physiol.* 270:H411–H415.

Neufeld, G., Cohen, T., Gengrinovitch, S., and Poltorak, Z. 1999. Vascular endothelial growth factor (VEGF) and its receptors. *FASEB J.* 13:9–22.

Oberg-Welsh, C., Sandler, S., Andersson, A., and Welsh, M. 1997. Effects of vascular endothelial growth factor on pancreatic duct cell replication and the insulin production of fetal islet-like cell clusters in vitro. *Mol. Cell. Endocrinol.* 126:125–132.

Ogawa, S., Oku, A., Sawano, A., Yamaguchi, S., Yazaki, Y., and Shibuya, M. 1998. A novel type of vascular endothelial growth factor, VEGF-E (NZ-7 VEGF), preferentially utilizes KDR/Flk-1 receptor and carries a potent mitotic activity without heparin-binding domain. *J. Biol. Chem.* 273:31273–31282.

Ortega, N., Hutchings, H., and Plouet, J. 1999. Signal relays in the VEGF system. *Front. Biosci.* 4:D141–D152.

Ortega, N., Jonca, F., Vincent, S., Favard, C., Ruchoux, M. M., and Plouet, J. 1997. Systemic activation of the vascular endothelial growth factor receptor KDR/flk-1 selectively triggers endothelial cells with an angiogenic phenotype. *Am. J. Pathol.* 151:1215–1224.

Pajusola, K., Aprelikova, O., Korhonen, J., Kaipainen, A., Pertovaara, L., Alitalo, R., and Alitalo, K. 1992. FLT4 receptor tyrosine kinase contains seven immunoglobulin-like loops and is expressed in multiple human tissues and cell lines. *Cancer. Res.* 52:5738–5743.

Park, J. E., Chen, H. H., Winer, J., Houck, K. A., and Ferrara, N. 1994. Placenta growth factor. Potentiation of vascular endothelial growth factor bioactivity, in vitro and in vivo, and high affinity binding to Flt-1 but not to Flk-1/KDR. *J. Biol. Chem.* 269:25646–25654.

Park, J. E., Keller, H.-A., and Ferrara, N. 1993. The vascular endothelial growth factor isoforms (VEGF): differential deposition into the subepithelial extracellular matrix and bioactivity of extracellular matrix-bound VEGF. *Mol. Biol. Cell* 4:1317–1326.

Phillips, H. S., Hains, J., Leung, D. W., and Ferrara, N. 1990. Vascular endothelial growth factor is expressed in rat corpus luteum. *Endocrinology* 127:965–967.

Plouet, J., Schilling, J., and Gospodarowicz, D. 1989. Isolation and characterization of a newly identified endothelial cell mitogen produced by AtT20 cells. *EMBO J.* 8:3801–3808.

Poltorak, Z., Cohen, T., Sivan, R., Kandelis, Y., Spira, G., Vlodavsky, I., Keshet, E., and Neufeld, G. 1997. VEGF145, a secreted vascular endothelial growth factor isoform that binds to extracellular matrix. *J. Biol. Chem.* 272:7151–7158.

Poole, A. R. 1991. *Cartilage: Molecular Aspects.* CRC Press, Boca Raton, FL, pp. 179–211.

Ravindranath, N., Little-Ihrig, L., Phillips, H. S., Ferrara, N., and Zeleznik, A. J. 1992. Vascular endothelial growth factor messenger ribonucleic acid expression in the primate ovary. *Endocrinology* 131:254–260.

Risau, W. 1997. Mechanisms of angiogenesis. *Nature* 386:671–674.

Roberts, W. G., and Palade, G. E. 1995. Increased microvascular permeability and endothelial fenestration induced by vascular endothelial growth factor. *J. Cell Sci.* 108:2369–2379.

Roberts, W. G., and Palade, G. E. 1997. Neovasculature induced by vascular endothelial growth factor is fenestrated. *Cancer Res.* 57:765–772.

Ryan, A. M., Eppler, D. B., Hagler, K. E., Bruner, R. H., Thomford, P. J., Hall, R. L., Shopp, G. M., and O'Neill, C. A. 1999. Preclinical safety evaluation of rhuMAbVEGF, an antiangiogenic humanized monoclonal antibody. *Toxicol. Pathol.* 27:78–86.

Salven, P., Ruotsalainen, T., Mattson, K., and Joensuu, H. 1998. High pre-treatment serum level of vascular endothelial growth factor (VEGF) is associated with poor outcome in small-cell lung cancer. *Int. J. Cancer* 79:144–146.

Salven, P., Teerenhovi, L., and Joensuu, H. 1997. A high pretreatment serum vascular endothelial growth factor concentration is associated with poor outcome in non-Hodgkin's lymphoma. *Blood* 90:3167–3172.

Seetharam, L., Gotoh, N., Maru, Y., Neufeld, G., Yamaguchi, S., and Shibuya, M. A. 1995. Unique signal transduction from FLT tyrosine kinase, a receptor for vascular endothelial growth factor VEGF. *Oncogene* 10:135–147.

Senger, D. R., Galli, S. J., Dvorak, A. M., Perruzzi, C. A., Harvey, V. S., and Dvorak, H. F. 1983. Tumor cells secrete a vascular permeability factor that promotes accumulation of ascites fluid. *Science* 219:983–985.

Shalaby, F., Rossant, J., Yamaguchi, T. P., Gertsenstein, M., Wu, X. F., Breitman, M. L., and Schuh, A. C. 1995. Failure of blood-island formation and vasculogenesis in Flk-1-deficient mice. *Nature* 376:62–66.

Shibuya, M., Yamaguchi, S., Yamane, A., Ikeda, T., Tojo, A., Matsushime, H., and Sato, M. 1990. Nucleotide sequence and expression of a novel human receptor-type tyrosine kinase (flt) closely related to the fms family. *Oncogene* 8:519–527.

Shifren, J. L., Tseng, J. F., Zaloudek, C. J., Ryan, I. P., Meng, Y. G., Ferrara, N., Jaffe, R. B., and Taylor, R. N. 1996. Ovarian steroid regulation of vascular endothelial growth factor in the human endometrium: implications for angiogenesis during the menstrual cycle and in the pathogenesis of endometriosis. *J. Clin. Endocrinol. Metabol.* 81:3112–3118.

Shima, D. T., Kuroki, M., Deutsch, U., Ng, Y. S., Adamis, A. P., and D'Amore, P. A. 1996. The mouse gene for vascular endothelial growth factor. Genomic structure, definition of the transcriptional unit, and characterization of transcriptional and post-transcriptional regulatory sequences. *J. Biol. Chem.* 271:3877–3883.

Shweiki, D., Itin, A., Neufeld, G., Gitay-Goren, H., and Keshet, E. 1993. Patterns of expression of vascular endothelial growth factor (VEGF) and VEGF receptors in mice suggest a role in hormonally-mediated angiogenesis. *J. Clin. Invest.* 91:2235–2243.

Simon, M., Grone, H. J., Johren, O., Kullmer, J., Plate, K. H., Risau, W., and Fuchs, E. 1995. Expression of vascular endothelial growth factor and its receptors in human renal ontogenesis and in adult kidney. *Am. J. Physiol.* 268:F240–F250.

Soker, S., Takashima, S., Miao, H. Q., Neufeld, G., and Klagsbrun, M. 1998. Neuropilin-1 is expressed by endothelial and tumor cells as an isoform-specific receptor for vascular endothelial growth factor. *Cell* 92:735–745.

Sondell, M., Lundborg, G., and Kanje, M. 1999. Vascular endothelial growth factor has neurotrophic activity and stimulates axonal outgrowth, enhancing cell survival and Schwann cell proliferation in the peripheral nervous system. *J. Neurosci.* 19:5731–5740.

Sowter, H. M., Corps, A. N., Evans, A. L., Clark, D. E., Charnock-Jones, D. S., and Smith, S. K. 1997. Expression and localization of the vascular endothelial growth factor family in ovarian epithelial tumors. *Lab. Invest.* 77:607–614.

Suri, C., Jones, P. F., Patan, S., Bartunkova, S., Maisonpierre, P. C., Davis, S., Sato, T. N., and Yancopoulos, G. D. 1996. Requisite role of angiopoietin-1, a ligand for the TIE2 receptor, during embryonic angiogenesis. *Cell* 87:1171–1180.

Takigawa, N., Segawa, Y., Fujimoto, N., Hotta, K., and Eguchi, K. 1998. Elevated vascular endothelial growth factor levels in sera of patients with lung cancer. *Anticancer Res.* 18:1251–1254.

Terman, B. I., Carrion, M. E., Kovacs, E., Rasmussen, B. A., Eddy, R. L., and Shows, T. B. 1991. Identification of a new endothelial cell growth factor receptor tyrosine kinase. *Oncogene* 6:1677–1683.

Terman, B. I., Dougher Vermazen, M., Carrion, M. E., Dimitrov, D., Armellino, D. C., Gospodarowicz, D., and Böhlen, P. 1992. Identification of the KDR tyrosine kinase as a receptor for vascular endothelial cell growth factor. *Biochem. Biophys. Res. Commun.* 187:1579–1586.

Tischer, E., Mitchell, R., Hartman, T., Silva, M., Gospodarowicz, D., Fiddes, J. C., and Abraham, J. A. 1991. The human gene for vascular endothelial growth factor. Multiple protein forms are encoded through alternative exon splicing. *J. Biol. Chem.* 266:11947–11954.

Vu, T. H., Shipley, J. M., Bergers, G., Berger, J. E., Helms, J. A., Hanahan, D., Shapiro, S. D., Senior, R. M., and Werb, Z. 1998. MMP-9/gelatinase B is a key regulator of growth plate angiogenesis and apoptosis of hypertrophic chondrocytes. *Cell* 93:411–422.

Waltenberger, J., Claesson Welsh, L., Siegbahn, A., Shibuya, M., and Heldin, C. H. 1994. Different signal transduction properties of KDR and Flt1, two receptors for vascular endothelial growth factor. *J. Biol. Chem.* 269:26988–26995.

Wiesmann, C., Fuh, G., Christinger, H. W., Eigenbrot, C., Wells, J. A., and de Vos, A. M. 1997. Crystal structure at 1.7 A resolution of VEGF in complex with domain 2 of the Flt-1 receptor. *Cell* 91:695–704.

CHAPTER 9

Hormonal Regulation of Gonadal Angiogenesis

H. William Schnaper and Constance E. Runyan

In the reproductive system, the rapid growth and regression of decidual tissue depends on concomitant increases and decreases in the extent of the vascular network supporting that tissue. This suggests that, like reproductive organ function, angiogenesis and vessel regression are regulated by gonadal steroids. This chapter considers the evidence supporting this hypothesis and discusses potential mechanisms by which these hormones may modulate vascular cell function.

GONADAL HORMONES AND VESSEL DENSITY

A preponderance of evidence supports the notion that sex steroids regulate the number of blood vessels present in reproductive tissue. There is increased flow to the ovary or corpus luteum (CL) at midcycle (Tempel and Neeman, 1999), suggesting that the vascularity of these structures is increased during times at which high concentrations of estrogen are present. However, examination of estrogen effects on larger vessels of the systemic circulation have indicated that vessel caliber may vary with circulating estradiol during the menstrual cycle (Hashimoto et al, 1995; Hegele-Hartung et al, 1997). Thus, rather than indicating angiogenesis, increased flow could reflect the relaxing effect of estrogen on vascular smooth muscle tone, mediated by endothelial nitric oxide production (Darkow et al, 1997). Further, studies using traditional techniques to assess vessel density have yielded conflicting results (Machiarelli et al, 1991; Rogers et al, 1998). However, direct evidence to support increased vascularization has been provided by Augustin and colleagues (1995), who studied vascularity during three phases of the menstrual cycle. Early in the cycle, vascularity increases, accompanied by increased indices of angiogenic activity. The dense vascular network of the second, plateau, phase gives way to loss of vascularity during the regression phase. Subsequent data from others (Suzuki et al, 1998) have confirmed this finding. Within hours of breakdown of the follicular basement membrane after ovulation, capillaries invade the developing CL (Tsukuda et al, 1996). Additional data indicate that cyclic angiogenesis occurs in the expanding endometrium as well (Jones, 1991; Kaiserman-Abramof and Padykula, 1989).

Angiogenesis after fertilization also plays a major role in sustaining the gravid state. Initial follicular vascularization (Collins et al, 1991) is important in this process. Capillary sprouts rapidly invade, doing so within 12 hours of ovulation in the rabbit. In the absence of fertilization the number of sprouts decreases, but sprouting increases by the third day of gestation (Meyer and McGeachie, 1988). The relationship between the monthly cycling of estrogen and the timing of vessel formation (Rodger et al, 1997), and the requirement for luteinization to maintain pregnancy, support a role for gonadal steroids in these events.

Finally, a parallel regulatory system may be present in male reproductive tissue, since testosterone stimulates prostatic angiogenesis in castrated rats (Franck-Lissbrant et al, 1998).

REGULATION OF ANGIOGENESIS IN REPRODUCTIVE TISSUES

Angiogenesis is a tightly regulated process that is governed by a balance of supporting and inhibiting forces. A series of events occurs to regulate this process. After degradation of the basement membrane that surrounds an existing vessel, the endothelial cells migrate into the stromal space, attaching to the interstitial matrix. The cells then proliferate and organize into a new capillary loop. With formation and remodeling of a new basement membrane, the vessel is mature (Folkman and Haudenschild, 1980). Some authors have proposed an alternative mechanism of vessel growth in decidual tissue, a process of elongation termed intussusception (Rogers et al, 1998). It is important to emphasize that while the cell activities of protease production, attachment to surfaces, migration, proliferation, and organization can be modeled separately in vitro, they do not necessarily occur as temporally discrete processes in vivo. However, each function contributes to angiogenesis, and each can be regulated to modulate vessel formation.

This modulation is controlled by production and activity of a number of cytokines and growth factors. For example, follicular cells secrete a substance that is chemotactic for endothelial cells (Rone et al, 1993). Production of this substance varies during the menstrual cycle. Such regulation provides an indirect mechanism by which gonadal steroids may regulate angiogenesis. Consistent with this hypothesis, various potentially angiogenic factors have been identified in cycling tissue. Perhaps the two best studied of these are vascular endothelial growth factor (VEGF) and basic fibroblast growth factor (bFGF). Either of these mediators may stimulate endothelial cell attachment, proliferation, migration, or organization. VEGF is expressed in decidual tissue throughout at least the first half of the cycle (Goede et al, 1998), although the timing of its disappearance is uncertain (Shifren et al, 1996; Yamamoto et al, 1997). Mice that are treated with soluble Flt-1 (VEGF receptor type 1), an inhibitor of VEGF stimulation, are infertile (Ferrara et al, 1998). In contrast, mice that are homozygous for deletion of bFGF are fertile (Zhou et al, 1998). While these results suggest that VEGF is more important than bFGF for reproductive angiogenesis, such a conclusion is not certain. Gonadal steroids and bFGF clearly interact in modulating angiogenesis, as tested in a corneal micropocket assay (Yamamoto et al, 1994). Further, VEGF is expressed in the ovary for much of the menstrual cycle, yet angiogenesis occurs only during the first half of the cycle. It is therefore likely that VEGF provides only a part of the stimulus for the complex phenomenon of tissue vascularization.

Additional angiogenic factors or their receptors have been found to be cyclically regulated. Angiogenin, an activator of endothelial cells, increases in the ovary during the menstrual cycle and is strongly expressed in the CL (Lee et al, 1999). Expression of Tie-2, the angiopoietin receptor, increases with follicular maturation and uterine development. The ratio of angiopoietin-2 to angiopoietin-1 increases during regression; this change in ratio is commonly associated with blood vessel regression (Goede et al, 1998). Thus, cyclical regulation of angiogenesis may be attributed in part to cyclical production of proangiogenic factors.

This regulation is further effected by changes in proteins other than growth factors. Secreted protein, acidic and rich in cysteine (SPARC), and thrombospondin (TSP), are expressed in the CL. SPARC, which at least in fragment form is proangiogenic, is present in both follicular cells and the developing CL. TSP, which is antiangiogenic, is present in the avascular follicle but is confined to relatively avascular areas of the CL (Bagavandoss et al, 1998). TSP production is cell cycle dependent, with expression by stromal cells varying inversely with angiogenic activity. In vitro, progesterone stimulates stromal cell synthesis of TSP (Iruela-Arispe et al, 1996). Mifepristone (RU-486), a competitive antagonist of progesterone that also has antiestrogen effects, appears to suppress bFGF production by the endometrium (Grow et al, 1998). Such an action could represent an event that, when not opposed by other, proangiogenic stimuli, would contribute to termination of vascularization and the onset of the regression phase.

The correlation between these events and production of gonadal steroids is not coincidental; both angiogenesis and vessel regression could result from the effects of these hormones on growth factor production by decidual tissue. For example, endometrial cells produce VEGF in response to treatment with estradiol (Hyder et al, 1996). Again, in vivo correlation is complex; in one study estrogen *decreases* endometrial VEGF expression and progestins *increase* VEGF expression (Greb et al, 1997). Other angiogenic factors are produced by steroid-stimulated uterine tissue as well. Administration of estradiol to ovariectomized mice rapidly induces expression of messenger RNA (mRNA) for erythropoietin, a protein that induces both erythrocyte production and angiogenesis, in the uterus (Yasuda et al, 1998). Another potentially angiogenic factor is released by the endometrium after stimulation with Norplant (Subakir et al, 1996). Platelet-derived endothelial cell growth factor (PDECGF) is expressed in the endometrium and decreases with ovulation, then increases again immediately after ovulation (Fujimoto et al, 1999). This suggests that PDECGF expression correlates inversely with estrogen expression or may be induced by progesterone. With all of these responses, temporal correlation between angiogenic factor production and various cyclical events or changes in hormone levels does not imply specific causality. Multiple events could be responsible for observed changes; events during the luteal phase could result from a prolonged increase in estradiol levels or from production of progesterone. It is apparent, however, that changes in sex hormones may regulate angiogenesis indirectly by stimulating production of modulating factors by reproductive tissue.

DIRECT EFFECTS OF STEROIDS ON ANGIOGENESIS

All of the findings discussed thus far represent indirect effects in which gonadal steroids stimulate reproductive tissue to modulate angiogenic activity. Experimental data demonstrate that these hormones also have direct effects on endothelial cell angiogenic activity. Although the hormonal requirements for decidual tissue growth make it difficult to examine vessel growth in these tissues in the absence of gonadal steroids, other systems have been used to evaluate possible direct effects. Estrogen is essential to support vascularization of a basement membrane extract injected subcutaneously in ovariectomized mice (Morales et al, 1995). As described above, angiogenesis can be broken down into component endothelial cell activities. Studies of endothelial cell behavior in vitro have shown that physiologic concentrations of estradiol enhance endothelial cell proliferation, migration, attachment to various substrata, and organization into capillary-like structures (Morales et al, 1995). These are dose-dependent effects (Morales et al, 1995; Banerjee et al, 1997) that are apparent at estrogen concentrations near those achieved in the systemic circulation during ovulation and the luteal phase (Naftolin and Tolis, 1978) but are maximal in vitro at slightly higher concentrations. Interestingly, at even higher concentrations, estradiol inhibits angiogenic activity of cultured endothelial cells (Jaggers et al, 1996; Banerjee et al, 1997). Although the tissue levels that are achieved under physiologic conditions are uncertain, it is likely that in some circumstances they are higher than systemic levels. Thus, variations in local estrogen production may establish gradients that regulate not only the occurrence of angiogenesis but also its geometry.

Other steroid hormones beside estradiol affect angiogenesis. Retinoids have variable effects, including inhibition through modulation of extracellular matrix protease production (Lansink et al, 1998). In the corneal pocket angiogenesis assay, several steroids were found to be antiangiogenic. The corticosteroid dexamethasone and the progestins medroxyprogesterone acetate and melengestrol are more inhibitory than dehydronorgestrol acetate, which in turn is more inhibitory than megestrol acetate (Yamamoto et al, 1994). In general, progestins decrease angiogenesis, whereas estrogens are enhancing.

The mechanisms of these regulatory effects on angiogenic function are complex. *Proliferation* may be affected by means similar to those described for estrogens and progestins in regulating various aspects of progression through the cell cycle by other types of cells. Preliminary results suggest that endothelial cells placed in culture medium supplemented with charcoal-stripped serum accumulate in G0/G1 (McGuire and Schnaper, 1999). Although it is not clear whether this cell cycle arrest reflects a loss of direct estrogen stimulation or decreased growth factor response, studies in neoplastic tissue support the notion that estrogen may be a cell cycle progression factor (Leung and Potter, 1987; Musgrove and Sutherland, 1994). Endothelial cell *attachment* may be enhanced through increased integrin expression or avidity. *Migration* is modulated in neural and cancer cells by estrogen (Fujimoto et al, 1996; Williams et al, 1999) and a similar mechanism could be present in endothelial cells. *Protease activity* also is important in angiogenesis. Estradiol modulates extracellular matrix proteases in tumor cells (Xing and Rabbani, 1999). In cultured endothelial cells, estrogen effects appear to vary depending on culture conditions. When the cells are confluent and therefore quiescent, treatment with 1 ng/mL estradiol decreases the ratio of signal intensity comparing mRNA for the urokinase-type and tissue-type plasminogen activators (uPA and tPA) to that for the plasminogen activator inhibitor, PAI-1, putatively favoring protease inhibition. In cells that are subconfluent or where the endothelial cell monolayer has been disrupted by scraping, estradiol increases the relative ratio to favor the proteases over the inhibitors (Schnaper et al, 1999a,b).

An important aspect of cell function is maintenance of a dynamic state that prevents activation of the cell death machinery. Incomplete or insufficient signals, including loss of anchorage (Re et al, 1994), insufficient external stimulation (Raff et al, 1993), or direct induction of apoptotic signals (Cryns and Yuan, 1998), may activate pathways that induce programmed cell death. VEGF has an antiapoptotic function in endothelial cells (Spyridopoulos et al, 1997a), and expression of $\alpha_v\beta_3$ integrin prevents apoptosis in endothelial cells participating in angiogenesis (Brooks et al, 1994). These are potential mechanisms by which estradiol may help "rescue" endothelial cells from apoptosis (Spyridopoulos et al, 1997b).

Progesterone may either oppose estrogen effects on the vasculature or have direct inhibitory effects on endothelial cell function. Reports describe disparate effects on endothelium-dependent vessel relaxation (Miller and Vanhoutte, 1991; Hegele-Hartung et al, 1997; Minshall et al, 1998). Angiostatic effects of medroxyprogesterone acetate have been described (Yamamoto et al, 1994). These may be mediated, at least in part, by decreased endothelial cell proliferation; progesterone, like estrogen withdrawal, causes cell cycle arrest in G1 (Vazquez et al, 1999).

GONADAL STEROIDS AND THEIR RECEPTORS

Estrogen (Figure 9.1) is produced primarily from testosterone by the action of aromatase (also called estrogen synthase), an enzyme that is present in gonadal tissue, but also is present in adipose and smooth muscle tissue (Brodie et al, 1997) as well as in the vessel wall (Diano et al, 1999; Nakamura et al, 1999). Thus, men have more substrate for estrogen synthesis and, despite lacking ovaries, have a small but significant amount of estrogen present systemically at all times. In contrast, women have less substrate; after menopause, they have lower systemic estrogen levels than do men. There are multiple estrogen metabolites; the activity of these variants is determined by the status of various side groups. The most common systemic estrogen in humans, 17β-estradiol, also is the most active natural estrogen in assays evaluating effects on gonadal tissue (e.g., endometrial cell proliferation). More recently, it has become clear that estrogen effects are not limited to the female reproductive tract (Schnaper, 1999). Estrogens may have varied effects on cellular responses, depending on the nature of the target tissue.

By far the dominant, naturally occurring progestin is progesterone, synthesized from pregnenolone by the action of 3β-hydroxysteroid dehydrogenase. It is not produced in high quantities during the ovulatory estrogen spike, but is expressed strongly during the luteal phase and during gestation. In many functions, progesterone counters the effects of estrogens. Thus, the two hormones strike a fine balance that can be rapidly modulated by altering the relative concentrations of each hormone and its metabolites.

The receptors for estrogens, progestins, and androgens are classically defined as nuclear receptors that serve as ligand-inducible transcription factors. These receptors bind small, cell membrane–permeable hormones and are then translocated to the nucleus where they bind to DNA sequences in gene promoters, modulating transcriptional activity. The effects of different hormone metabolites on specific transcriptional activities vary because of subtle differences in the structure of the ligand-receptor complex (McDonnell et al, 1995). These differences provide a further degree of subtlety and complexity in regulating cell function.

Nuclear receptors are classified into two subfamilies based on the DNA sequences (hormone response elements) to which they bind (Lowe et al, 1998). The progesterone receptor (PR) is a typical steroid receptor, binding to the

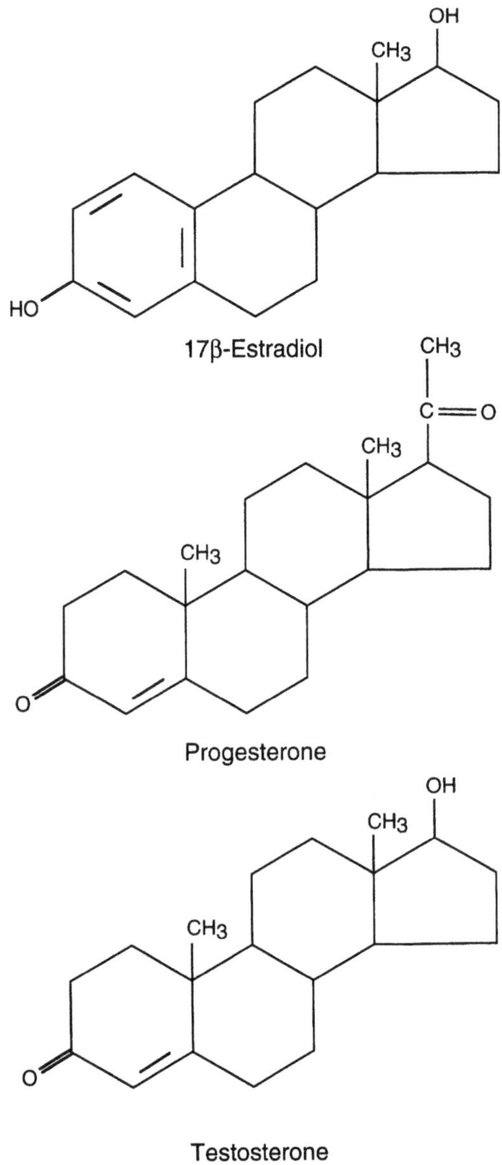

FIGURE 9.1. Chemical structure of the most common gonadal steroid hormones.

sequence TGAACATGTTCT. The palindromic nature of this sequence recognizes the mirror-image homodimer formed by the activated PR (Mangelsdorf et al, 1995). The estrogen receptor (ER) belongs to a second group of receptors that may form heterodimers with other members of its subfamily. These other members include the thyroid hormone, vitamin D, and retinoic acid receptors, as well as members of the steroid hormone orphan receptor subgroup. The estrogen response element (ERE) has a different sequence and structure from that recognized by the PR: AGGTCAnnnTGACCT. In contrast to some members of this

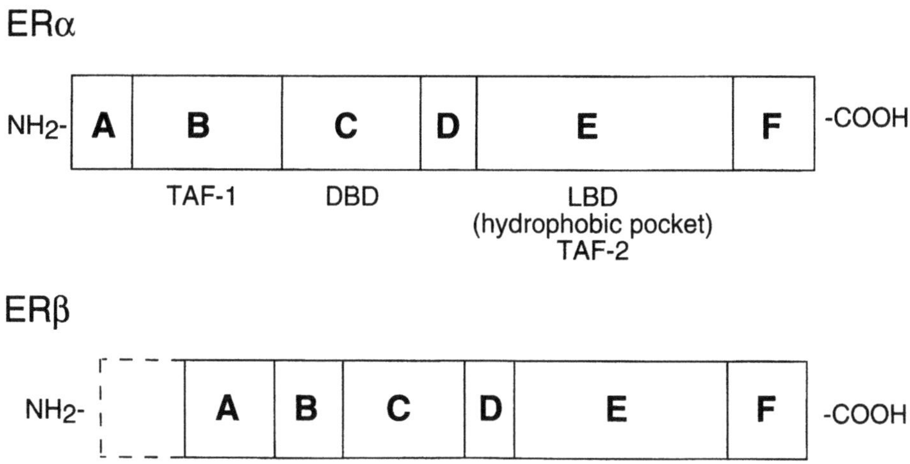

FIGURE 9.2. Domain structure of the estrogen receptors. The broken line in ERβ indicates the longer isoform derived from alternative start site for transcription (see text). TAF, transcriptional activating function; DBD, DNA-binding domain; LBD, ligand-binding domain.

second receptor subfamily, the ER binding sequence is palindromic, consistent with functional ER homodimerization (Kumar and Chambon, 1988).

The structure of these receptors defines their function. A number of segments have been defined based on differential conservation and designated A through F, as illustrated by the diagram of ER sequence structure in Figure 9.2. The amino-terminal domain of these receptors has the least conservation and therefore significantly influences specificity. The DNA-binding domain (C) is the most highly conserved domain among the nuclear receptors. It includes two zinc fingers, and a less-conserved carboxy-terminal sequence that determines relative specificity for different gene promoter *cis*-acting elements. The α-type estrogen receptor (ERα) has been well characterized (Green et al, 1986; Greene et al, 1986). Two domains (B and E) possess *t*ranscriptional *a*ctivating *f*unction and are therefore called TAF-1 and TAF-2, respectively. The E domain also is the hormone-binding domain and may mediate dimerization. A second ER, ERβ, has a roughly similar structure (Figure 9.2), including a DNA-binding domain with 96% homology to the ERα C domain and a hormone-binding domain that is 53% homologous to the ERα E domain. After translocation to the nucleus, the ER supports assembly of a transcription-regulatory complex through interaction of other proteins with the TAF sites. Transcription regulatory proteins that contribute to ER function include ER-associated protein (ERAP)-160 (Halachmi et al, 1994), steroid receptor coactivator (SRC)-1 (Shiau et al, 1998), and the CREB-binding protein p300 (Kraus and Kadonaga, 1998).

While this paradigm clearly represents a common mechanism of ER action, alternative pathways also may be observed. It has been suggested that neither estrogen binding nor receptor dimerization is essential for ER activation (Furlow et al, 1993), which might instead occur via growth factor–stimulated signal transduction (Kato et al, 1995). In addition, recent data suggest that ER may modulate cell function without being translocated to the nucleus (Razandi et al, 1999). The

ERβ may interact indirectly with gene promoter AP-1 sites (Paech et al, 1997). Possible roles for these alternative mechanisms will be considered below.

The progesterone receptor is structurally similar to ER (Loosfelt et al, 1986). Two distinct isoforms are produced from the same gene. Isoform A is somewhat shorter than B, and results from utilization of an intragenic promoter that yields a shorter transcript, lacking a portion of the 5′ sequence (Savouret et al, 1994). Both isoform promoters are estrogen-inducible (Kastner et al, 1990). These isoforms have different effects on the structure of complexes formed with the target DNA (Prendergast et al, 1996). They may be differentially expressed in target tissues, causing different target organs to respond differently to the same hormone. For example, the ratio of expression of the two isoforms has been found to vary in the primate CL through different phases of the menstrual cycle (Duffy et al, 1997).

Testosterone is synthesized from androstenediol by the action of 3β-hydroxysteroid dehydrogenase. The DNA-binding domain of the human androgen receptor shows high homology to other members of the steroid receptor family and the receptor is structurally similar to both the glucocorticoid receptor and the PR (Brinkmann et al, 1989).

ENDOTHELIAL CELL EXPRESSION OF SEX STEROID HORMONE RECEPTORS

Classically, gonadal hormones were felt to be targeted to reproductive tissues. In the past 10 years, however, it has become increasingly apparent that many other tissues express receptors for, and respond to, these hormones. These tissues include those of the cardiovascular system. Cells in the vessel wall express such receptors. Initially identified in vascular smooth muscle cells, ER eventually were described for human umbilical vein endothelial cells (HUVECs) (Kim-Schulze et al, 1996; Venkov et al, 1996), for human coronary artery endothelial cells (HCAECs) (Kim-Schulze et al, 1996), and subsequently in several other populations of cultured endothelial cells isolated from different species (Bayard et al, 1995). The HUVEC and HCAEC receptors showed specific, saturable binding with a similar dissociation constant (K_d) of approximately 5 nM. The receptor number was determined to be approximately 5×10^4 per cell. In comparison with the intensity of ER staining in cultured breast cancer cells, the estimates of receptor number appear to be somewhat larger than anticipated. The relatively low abundance of ER in endothelial cells means that radioligand-binding studies are performed at the lower limits of assay sensitivity, confounding efforts to draw fine distinctions among cells cultured from men or women, or from different tissues (Kim-Schulze et al, 1996).

ER in endothelial cells are functional by two tests: they bind to the vitellogenin ERE in electrophoretic mobility shift assays (Kim-Schulze et al, 1996; Venkov et al, 1996) and induce expression of transiently transfected ERE-luciferase constructs in estrogen-treated endothelial cells (Chen et al, 1999). A critical regulator of estrogen receptor expression is estrogen itself. Estradiol stimulates expression of mRNA for ERα (Kim-Schulze et al, 1996). This holds true not only for endothelial cells in culture, but also for vascular tissues; vessels from premenopausal women show more ERα protein than do vessels from postmenopausal women or men (Losordo et al, 1994). However, it should be noted that estrogen effects on ER expression are tissue-specific; ER is downregulated by estradiol in breast cancer cell lines (Saceda et al, 1988).

A role for the ERα in endothelial cell function is supported by Johns and colleagues (1996), who measured the local plasma volume after bFGF-stimulated vascularization of subcutaneously injected basement membrane extract. An increase in vascularization over control conditions (that is, in comparison with animals injected with extract that did not contain bFGF) did not occur in ERα-deficient mice, even with estrogen supplementation. This result suggests that ERα plays an important role in bFGF-induced angiogenesis. In contrast to this study of angiogenesis, another study examined the role of estrogen in facilitating intimal healing. Previously, it had been shown that estradiol decreases intimal thickening after carotid injury (Chen et al, 1996). When similar studies were performed in estrogen receptor–deficient mice, the improvement with estrogen treatment was similar (Iafrati et al, 1997).

One possible explanation for the difference in these responses in the absence of ERα is that ER is not needed for intimal thickening. Alternatively, it is possible that enhanced endothelial cell angiogenic activity is more dependent on the ERα, whereas decreased intimal thickening (inhibition of vascular smooth muscle cell proliferation) is more a function of ERβ action. Efforts to definitively show the presence of ERβ in endothelial cells, as opposed to VSMC, have been mixed. Several groups have suggested that endothelial cells express ERβ, based on detection of ERβ mRNA by reverse transcription–polymerase chain reaction (RT-PCR) (Kuiper et al, 1997). However, detection of ERβ protein is expressed in endothelial cells by specific binding of anti-ERβ antibody has not yet been reported. In gel shift assays, supershift using anti-ERα antibody is complete (Figure 9.3), suggesting that all of the active ER-ERE complexes contain at least one ERα molecule.

More recently, an endothelial progesterone receptor has been described in cultured human endothelial cells derived from several different tissues, as well as in situ. A K_d of 6.1 nM, and as many as 10^5 receptors per cell, were estimated by Scatchard analysis. The functional nature of the endothelial PR was confirmed by activation of a progesterone-response element-luciferase reporter construct transfected into the cells (Vazquez et al, 1999). Splice variants have been described. In some cases, these variants may express as dominant-negative proteins, interfering with function of the classical receptors (Hodges et al, 1999).

MECHANISM OF HORMONE ACTION ON ENDOTHELIAL CELL FUNCTION

Steroid hormones classically have been felt to regulate cell function by a "genomic" mechanism of action, binding to a response element in the gene promoter region. The presence of this mechanism in endothelial cells is supported by the observation that an ERE-luciferase construct transfected into endothelial cells is activated by estradiol treatment. This pathway could account for increases in endothelial cell expression of nitric oxide synthase protein and mRNA (Kleinert et al, 1998), cyclooxygenase-1 (Jun et al, 1998) or interleukin-6 (Keck et al, 1998), and for decreased endothelin-1 production (Akashita et al, 1998).

One genomic effect may serve as an amplification loop for growth factor signaling. Estradiol stimulates delayed activation of the ERK mitogen-activated

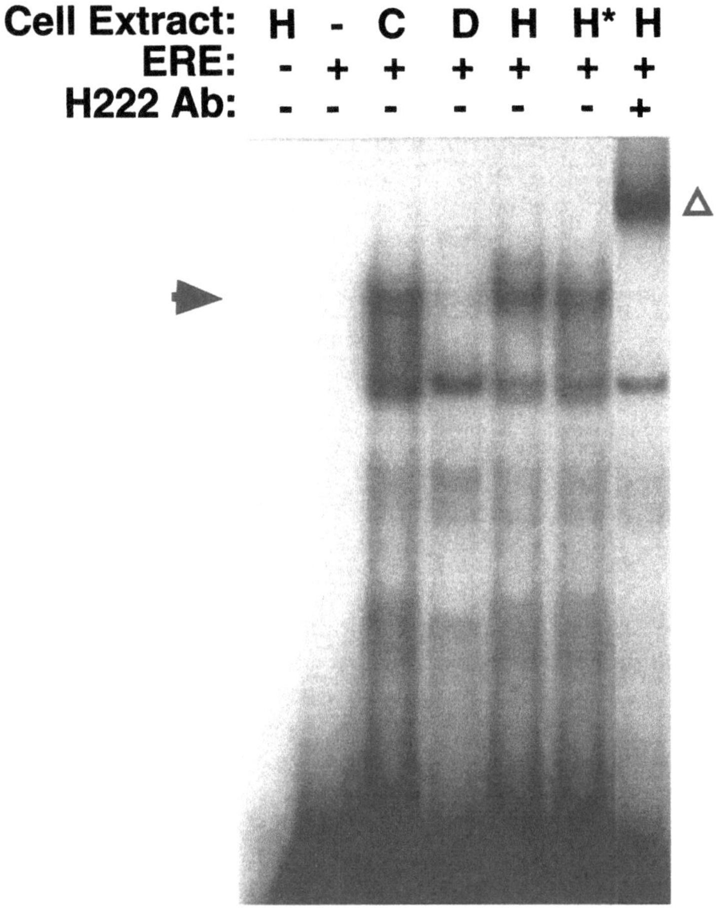

FIGURE 9.3. Electrophoretic mobility shift assay (EMSA) for estrogen receptor–estrogen response element (ER-ERE) interaction. Nuclear extracts were harvested from estrogen-treated human umbilical vein endothelial cells (H), MCF-7 breast cancer cells or MDA-MB.231, ER-negative breast cancer cells (D); or from human umbilical vein endothelial cells (HUVEC) that were not treated with estrogen (H*). Binding was assessed by EMSA using a labeled, vitellogenin ERE sequence. The solid arrowhead identifies a band that results from a shifted complex in all lanes containing ERE plus nuclear extract, except for the extract from the ER-negative MDA cells. In other experiments, specificity of binding was confirmed by competition with unlabeled ERE. The specific shifted complex is supershifted by incubation with a specific anti-ERα antibody (H222), as indicated by the open triangle. (Figure adapted with permission from Kim-Schulze et al, 1996.)

protein (MAP) kinase pathway; activity peaks at 3 hours, coinciding with increased autophosphorylation of the bFGF receptor. Activation of ERK occurs within 5 minutes when naive endothelial cells are treated with media conditioned by endothelial cells that have been exposed to estradiol for 3 hours, a response that is reduced by pretreatment of the conditioned media with anti-bFGF antibody. Expression of bFGF mRNA begins to increase within 30 to 50 minutes of estradiol treatment. These findings describe an autocrine loop for bFGF production that is stimulated by estradiol (Kim-Schulze et al, 1998). It is noteworthy that several investigators have reported that ERK MAP kinases are capable of activat-

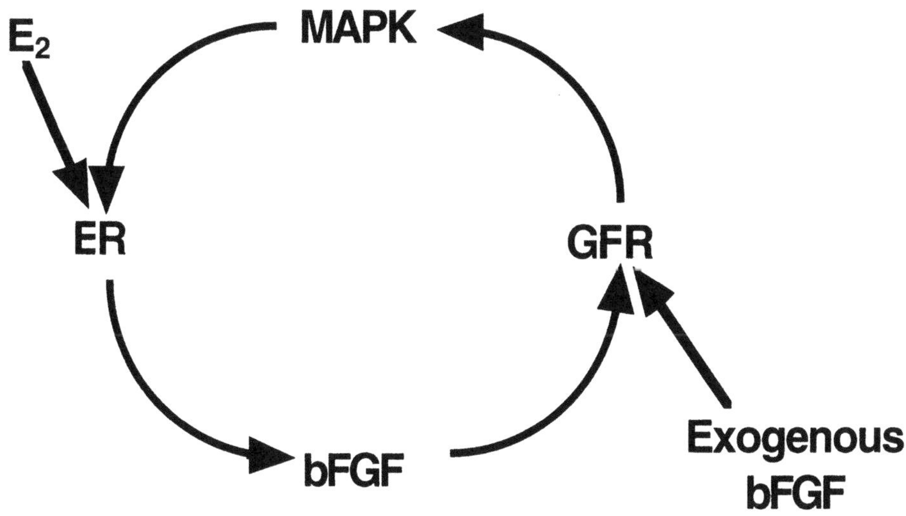

FIGURE 9.4. Simplified scheme showing autocrine loop by which estrogen receptor binding could amplify growth factor responses. E_2, 17β-estradiol; ER, estrogen receptor; bFGF, basic fibroblast growth factor; MAPK, ERK mitogen-activated protein kinase.

ing ER in the absence of estradiol (Kato et al, 1995), although the implications of this finding remain controversial (Joel et al, 1998). This potential autocrine loop (Figure 9.4) could explain, at least in part, the apparent amplification of growth factor responsiveness that occurs with estrogen treatment (Schnaper et al, 1999a,b).

However, these classical steroid hormone effects only partly describe estrogen signaling events in endothelial cells. It has been known for some time that estrogen rapidly increases reflex vessel relaxation (White et al, 1995) through the activation of nitric oxide synthase (NOS) (Darkow et al, 1997). This action is mediated by estrogen effects on ERK MAP kinase that require only minutes of stimulation (Chen et al, 1999). Superphysiologic concentrations of estrogen have been shown to activate cyclic adenosine monophosphate (cAMP) in nonendothelial cells (Aronica et al, 1994), and to increase intracellular [Ca^{2+}] and subsequently to mediate translocation of endothelial NOS in endothelium (Goetz et al, 1999). Neither the timing nor the nature of these events are explained by gene activation. Similarly, estradiol induces endothelial cells to release bFGF when the cells are cultured on basement membrane proteins. The release occurs in 2 hours, more rapidly than the amount of time required for bFGF synthesis in the autocrine loop described above, and it is not blocked by cycloheximide, indicating that no new protein synthesis is required (Albuquerque et al, 1998). Although estradiol has been shown to rapidly induce Ca^{2+} mobilization (Goetz et al, 1999), an intracellular calcium chelator was evaluated but did not block release. The protein kinse C (PKC) inhibitor calphostin C blocks release, as does the specific ER antagonist ICI182,780 (Albuquerque et al, 1998). Taken together, these data suggest signaling interaction among the ER, PKC, and integrins that bind to specific basement membrane proteins. Further support of a role for integrins is found in a report suggesting that estradiol enhances the activation of focal adhesion kinase, $p125^{FAK}$, in endothelial cells (Alvarez et al, 1997).

Potential mechanisms of nongenomic signal transduction by ER have been described in other systems. Estradiol rapidly stimulates ERK MAP kinase in a cultured breast cancer cell line (Migliaccio et al, 1996). Overexpression studies indicate that small fractions of both ERα and ERβ may be associated with the cell membrane by an unknown mechanism. These receptors can transduce a variety of signals more commonly associated with growth factor–stimulated signaling (Razandi et al, 1999). The applicability of these findings to endothelial cells is presented under intense investigation (Russell et al, 2000; Simoncini et al, 2000).

BALANCE OF FORCES GOVERNING ANGIOGENESIS

As with most biologic systems, hormonal regulation of angiogenesis involves a finely tuned balance among various forces (Table 9.1). We already have considered the largely opposing effects of estrogens and progestins on a variety of endothelial cell activities. Unopposed estrogen might be strongly angiogenic. In contrast, estrogen and progesterone together might yield a spectrum of signals whose net effect is determined in part by the local environment, permitting maintenance of the vascularized state. Estrogen withdrawal might favor apoptosis (Spyridopoulos et al, 1997b) as part of decidual tissue regression. Interruption of this balance, as is the case with Norplant or mifepristone, has complicated effects depending on how each tissue system responds to therapeutic manipulation.

In addition, there are a variety of natural and synthetic inhibitors of estrogen activity. Synthetic antiestrogens vary in their effect on tissue. For example, tamoxifen is antiestrogenic with regard to breast tissue, but has more typically proestrogenic effects on vascular function (Rutqvist and Mattson, 1993). A natural "antiestrogen" is the estrogen metabolite, 2-methoxyestradiol (2-ME), which

Table 9.1. Actions of sex steroids on angiogenesis.

Agent	Action	Reference
Proangiogenic		
Estrogen	Angiogenesis (bFGF/Matrigel)	Johns et al, 1996; Morales et al, 1995
	Attachment	Morales et al, 1995
	Proliferation	Alvarez et al, 1997; Morales et al, 1995
	Migration	Morales et al, 1995
	Organization	Morales et al, 1995
	Antiapoptosis	Spyridopoulos et al, 1997b
	Integrin expression	Cid et al, 1994
Testosterone	Angiogenesis	Franck-Lissbrant et al, 1998
Antiangiogenic		
Progesterone	Antiproliferative	Vazquez et al, 1999
	Thrombospondin-1 (TSP-1) production	Iruela-Arispe et al, 1996
Medroxyprogesterone acetate	Corneal angiogenesis	Yamamoto et al, 1994
Mifepristone	Decreased bFGF synthesis	Grow et al, 1998
2-methoxyestradiol	Antiproliferative	Klauber et al, 1997
	Proapoptotic	Tsukamoto et al, 1998
	Antiangiogenic	Yue et al, 1997
Dexamethasone	Corneal angiogenesis	Yamamoto et al, 1994

opposes many of the effects of 17β-estradiol on endothelial cells. 2-ME induces apoptosis (Tsukamoto et al, 1998) and inhibits angiogenesis by increasing expression of Fas ligand and activating the c-Jun N-terminal kinase (JNK) MAP kinase pathway (Yue et al, 1997). 2-ME inhibits bovine capillary endothelial cell proliferation by attaching to the colchicine-binding site of tubulin (Klauber et al, 1997). This leads to cell-cycle arrest in mitosis (Attalla et al, 1996), in contrast to the arrest in G0/G1 observed with hormone depletion (McGuire and Schnaper, 1999) or with progesterone treatment.

Studies of cultured endothelial cells have found similar effects of gonadal hormones regardless of differences in the origin of the cells with regard to species, vessel caliber, or whether the vessel was arterial, venous, or capillary. In contrast, in many cases results of estrogen treatment are opposite in endothelial and vascular smooth muscle cells. In a given cell type, the culture conditions appear to be an important determinant of estrogen receptor expression and function. These observations suggest that the steroids have a general effect on the amplitude of responsiveness rather than a specific inducing action on one pathway. Consistent with this hypothesis, some estrogen effects (e.g., expression of uPA and PAI-1 mRNA) depend on the status of the cell at the time it is treated. The failure of bFGF to induce angiogenesis in ERα-deficient mice suggests that estrogen-ER interaction provides a cofactor-like activity that supports responsiveness. In view of the fact that many estrogen effects on angiogenesis are similar to growth factor effects, the data suggest a novel role for estrogen/ER in signal transduction. Rather than activating genes or specific signaling pathways, estrogens and progestins could be involved in "crosstalk" with other signaling pathways to determine the outcome of endothelial cell stimulation.

PERSPECTIVE

As described above, aromatase is present in the vascular wall (Diano et al, 1999) and its expression may be enhanced by estrogen (Nakamura et al, 1999). Local estrogen production may thus be an important paracrine event in vascular homeostasis in two respects: first, it may be a critical regulator of decidual angiogenesis and vascular healing; and second, local tissue concentrations could vary significantly from systemic levels. It is possible that estrogen concentrations in the ovary could at times be so high that they are antiangiogenic. This provides an additional layer of complexity for understanding regulation of decidual tissue angiogenesis.

Finally, estrogen is widely viewed as having protective benefits against the development of atherosclerosis (Mendelsohn and Karas, 1999). It is unlikely that this benefit came about because of direct genetic selection pressure, since the most significant manifestations of atherosclerotic cardiovascular disease occur well after child-bearing age and have been a clinical problem largely with increased longevity and rich diets over the past several hundred years. However, many of the endothelial cell functions that are important in angiogenesis—endothelial cell attachment, migration, proliferation, organization, etc.—also are important in promoting intimal healing of large vessels. Therefore, the relative protection that premenopausal women receive from their hormonal status most likely reflects an unanticipated benefit of a system that developed as a means of regulating reproductive angiogenesis.

REFERENCES

Akashita, M., Kozaki, K., Eto, M., Yoshizumi, M., Ishkawa, M., and Toba, K. 1998. Estrogen attenuates endothelin-1 production by bovine endothelial cells via estrogen receptor. *Biochem. Biophys. Res. Commun.* 251:17–21.

Albuquerque, M. L. C., Akiyama, S. K., and Schnaper, H. W. 1998. Human coronary artery endothelial cell secretion of basic fibroblast growth factor is enhanced by matrix proteins, 17β-estradiol, and a PKC signaling pathway. *Exp. Cell Res.* 245:163–169.

Alvarez, R. J., Gips, S. J., Moldovan, N., Wilhide, C. C., Milliken, E. E., and Hoang, A. T. 1997. 17β-estradiol inhibits apoptosis of endothelial cells. *Biochem. Biophys. Res. Commun.* 237:372–381.

Aronica, S. M., Kraus, W. L., and Katzenellenbogen, B. S. 1994. Estrogen action via the cAMP signaling pathway: stimulation of adenylate cyclase, and cAMP-regulated gene transcription. *Proc. Natl. Acad. Sci. USA* 91:8517–8521.

Attalla, H., Mäkela, T. P., Adlercreutz, H., and Anderson, L. C. 1996. 2-Methoxyestradiol arrests cells in mitosis without depolymerizing tubulin. *Biochem. Biophys. Res. Commun.* 228:467–473.

Augustin, H. G., Braun, K., Telemenakis, I., Modlich, U., and Kuhn, W. 1995. Ovarian angiogenesis. Phenotypic characterization of endothelial cells in a physiological model of blood vessel growth and regression. *Am. J. Pathol.* 147:339–351.

Bagavandoss, P., Sage, E. H., and Vernon, R. B. 1998. Secreted protein, acidic and rich in cysteine (SPARC) and thrombospondin in the developing follicle and corpus luteum of the rat. *J. Histochem. Cytochem.* 46:1043–1049.

Banerjee, S. K., Campbell, D. R., Weston, A. P., and Banerjee, D. K. 1997. Biphasic estrogen response on bovine adrenal medulla capillary endothelial cell adhesion, proliferation and tube formation. *Mol. Cell. Biochem.* 177:97–105.

Bayard, F., Clamens, S., Delsol, G., Blaes, N., Maret, A., and Faye, J. C. 1995. Oestrogen synthesis, oestrogen metabolism and functional oestrogen receptors in bovine aortic endothelial cells. *Ciba Found. Symp.* 191:122–132.

Brinkmann, A. O., Faber, P. W., van Rooij, H. C. J., Kuiper, G. G. J. M., Ris, C., Klassen, P., van der Korput, J. A., Voorhorst, M. M., van Laar, J. H., Mulder, E., et al. 1989. The human androgen receptor: domain structure, genomic organization and regulation of expression. *J. Steroid Biochem.* 34:307–310.

Brodie, A., Lu, Q., and Nakamura, J. 1997. Aromatase in the normal breast and breast cancer. *J. Steroid Biochem. Mol. Biol.* 61:281–286.

Brooks, P. C., Montgomery, A. M., Rosenfeld, M., Reisfeld, R. A., Hu, T., Klier, G., and Cheresh, D. A. 1994. Integrin alpha v beta 3 antagonists promote tumor regression by inducing apoptosis of angiogenic blood vessels. *Cell* 79:1157–1164.

Chen, S.-J., Li, H., Durand, J., Oparil, S., and Chen, Y.-F. 1996. Estrogen reduces myointimal proliferation after balloon injury of rat carotid artery. *Circulation* 93:577–584.

Chen, Z., Yuhanna, I. S., Galcheva-Gargova, Z., Karas, R. H., Mendelsohn, M. E., and Shaul, P. W. 1999. Estrogen receptor α mediates the nongenomic activation of endothelial nitric oxide synthase by estrogen. *J. Clin. Invest.* 104:401–406.

Cid, M. C., Esparza, J., Schnaper, H. W., et al. 2000. Estradiol enhances endothelial cell interactions with extracellular matrix proteins via an increase in integrin expression and function. *Angiogenesis* 3:271–280.

Collins, W., Jurkovic, D., Bourne, T., Kurjak, A., and Campbell, S. 1991. Ovarian morphology, endocrine function and intra-follicular blood flow during the peri-ovulatory period. *Hum. Reprod.* 6:319–324.

Cryns, V., and Yuan, J. 1998. Proteases to die for. *Genes Dev.* 12:1551–1570.

Darkow, D. J., Lu, L., and White, R. E. 1997. Estrogen relaxation of coronary artery smooth muscle is mediated by nitric oxide and cGMP. *Am. J. Physiol.* 272:H2765–H2773.

Diano, S., Horvath, T. L., Mor, G., Register, T., Adams, M., Harada, N., and Naftoklin, F. 1999. Aromatase and estrogen receptor immunoreactivity in the coronary arteries of monjeys and human subjects. *Menopause* 6:21–28.

Duffy, D. M., Wells, T. R., Haluska, G. J., and Stouffer, R. L. 1997. The ratio of progesterone receptor isoforms changes in the monkey corpus lutuem during the luteal phase of the menstrual cycle. *Biol. Reprod.* 57:693–699.

Ferrara, N., Chen, H., Davis-Smyth, T., Gerber, H.-P., Nguyen, T.-N., Peers, D., Chisholm, V., Hillan, K. J., and Schwall, R. H. 1998. Vascular endothelial growth factor is essential for corpus luteum angiogenesis. *Nat. Med.* 4:336–340.

Folkman, J., and Haudenschild, C. 1980. Angiogenesis in vitro. *Nature* 288:551–556.

Franck-Lissbrant, I., Häggstrom, S., Damber, J.-E., and Bergh, A. 1998. Testosterone stimulates angiogenesis and vascular regrowth in the ventral prostate in castrated adult rats. *Endocrinology* 139:451–456.

Fujimoto, J., Hori, M., Ichigo, S., Morishita, S., and Tamaya, T. 1996. Estrogen activates migration potential of endometrial cancer cells through basement membrane. *Tumor Biol.* 17:48–57.

Fujimoto, J., Sakaguchi, H., Hirose, R., and Tamaya, T. 1999. Expression of platelet-derived endothelial cell growth factor (PD-ECGF) related to angiogenesis in ovarian endometriosis. *J. Clin. Endocrinol. Metab.* 84:359–362.

Furlow, J. D., Murdoch, F. E., and Gorski, J. 1993. High affinity binding of the estrogen receptor to a DNA response element does not require homodimer formation or estrogen. *J. Biol. Chem.* 268:12519–12525.

Goede, V., Schmidt, T., Kimmina, S., Kozian, D., and Augustin, H. G. 1998. Analysis of blood vessel maturation processes during cyclic ovarian angiogenesis. *Lab. Invest.* 78:1385–1394.

Goetz, R. M., Thatte, H. S., Prabhakar, P., Cho, M. R., Michel, T., and Golan, D. E. 1999. Estradiol induces the calcium-dependent translocation of endothelial nitric oxide synthase. *Proc. Natl. Acad. Sci. USA* 96:2788–2793.

Greb, R. R., Heikinheimo, O., Williams, R. F., Hodgen, G. D., and Goodman, A. L. 1997. Vascular endothelial growth factor in primate endometrium is regualted by oestrogen-receptor and progesterone-receptor ligands in vivo. *Hum. Reprod.* 12:1280–1292.

Green, S., Kumar, V., Krust, A., Walter, P., and Champon, P. 1986. Structural and functional domains of the estrogen receptor. *Cold Spring Harb. Symp. Quant. Biol.* 51:751–758.

Greene, G. L., Gilna, P., Waterfield, M. A. B., Hort, Y., and Shine, J. 1986. Sequence and expression of human estrogen receptor complementary DNA. *Science* 231:1150–1154.

Grow, D. R., Reece, M. T., Hsiu, J. G., Adams, L., Newcomb, P., Williams, R. F., and Hodgen, G. D. 1998. Chronic antiprogestin therapy produces a stable atrophic endometrium with decreased fibroblast growth factor: a 1-year primate study on contraception and amenorrhea. *Fertil. Steril.* 69:936–943.

Halachmi, S., Marden, E., Martin, G., MacKay, H., Abbondanza, C., and Brown, M. 1994. Estrogen receptor-associated proteins: possible mediatros of hormone-induced transcription. *Science* 264:1455–1458.

Hashimoto, M., Akishita, M., Eto, M., Ishikawa, M., Kozaki, K., Toba, K., Sagara, Y., Taketani, Y., Orimo, H., and Ouchi, Y. 1995. Modulation of endothelium-dependent flow-mediated dilation of the brachial artery by sex and menstrual cycle. *Circulation* 92:3431–3435.

Hegele-Hartung, C., Fritzmeier, K.-H., and Diel, P. 1997. Effects of a pure antiestrogen on estrogen-mediated alterations of blood flow and progesterone receptor expression in the aorta of ovariectomized rabbits. *J. Steroid Biochem.* 63:237–249.

Hodges, Y. K., Richer, J. K., Horwitz, K. B., and Horwitz, L. D. 1999. Variant estrogen and progesterone receptor messages in human vascular smooth muscle. *Circulation* 99:2688–2693.

Hyder, S. M., Stancel, G. M., Chiapetta, C., Murthy, L., Boetter-Tong, H. L., and Makela, S. 1996. Uterine expression of vascular endothelial growth factor is increased by estradiol and tamoxifen. *Cancer Res.* 56:3954–3960.

Iafrati, M. D., Karas, R. H., Aronovitz, M., King, M., Sullivan, T. R., Jr., Lubahn, D. B., O'Donnell, T. F., Korach, K. S., and Mendelsohn, M. E. 1997. Estrogen inhibits the vascular injury response in estrogen receptor α-deficient mice. *Nat. Med.* 3:545–548.

Iruela-Arispe, M., Porer, P., Bornstein, P., and Sage, E. H. 1996. Thrombospondin-1, an inhibitor of angiogenesis, is regulated by progesterone in the human endometrium. *J. Clin. Invest.* 97:403–412.

Jaggers, D. C., Collins, W. P., and Milligan, S. R. 1996. Potent inhibitory effects of steroids in an *in vitro* model of angiogenesis. *J. Endocrinol.* 150:457–464.

Joel, P. B., Traish, A. M., and Lannigan, D. A. 1998. Estradiol-induced phosphorylation of serine 118 in the estrogen receptor is independent of p42/p44 mitogen-activated protein kinase. *J. Biol. Chem.* 273:13317–13323.

Johns, A., Freay, A. D., Fraser, W., Korach, K. S., and Rubanyi, G. M. 1996. Disruption of estrogen receptor gene prevents 17 beta estradiol-induced angiogenesis in transgenic mice. *Endocrinology* 137:4511–4513.

Jones, G. S. 1991. Luteal phase defect. A review of pathophysiology. *Curr. Opin. Obstet. Gynecol.* 3:641–648.

Jun, S. S., Chen, Z., Pace, M. C., and Shaul, P. W. 1998. Estrogen upregulates cyclooxygenase-1 gene expression in ovine fetal pulmonary artery endothelium. *J. Clin. Invest.* 102:176–183.

Kaiserman-Abramof, I. R., and Padykula, H. A. 1989. Angiogenesis in the postovulatory primate endometrium: the coiled arteriolar system. *Anat. Rec.* 224:479–489.

Kastner, P., Krust, A., Turcotte, B., Stropp, U., Tora, L., Gronemeyer, H., and Chambon, P. 1990. Two distinct estrogen-regulated promoters generate transcripts encoding the two functionally different human progesterone receptor forms A and B. *EMBO J.* 9:1603–1614.

Kato, S., Endoh, H., Masuhiro, Y., Kitamoto, T., Uchiyama, S., Sasaki, H., Masushige, S., Gotoh, Y., Nishida, E., Kawashima, H., Metzger, D., and Chambon, P. 1995. Activation of the estrogen receptor through phosphorylation by mitogen-activated protein kinase. *Science* 270:1491–1494.

Keck, C., Herchenbach, D., Pfisterer, J., and Breckwoldt, M. 1998. Effects of 17β-estradiol and progesterone on interleukin-6 production and proliferation of human umbilical vein endothelial cells. *Exp. Clin. Endocrinol. Diabetes* 106:334–339.

Kim-Schulze, S., Lowe Jr., W. L., and Schnaper, H. W. 1998. Estrogen induces mitogen-activated protein kinase (ERK 1/2) activity in endothelial cells via an autocrine loop involving basic fibroblast growth factor. *Circulation* 98:413–421.

Kim-Schulze, S., McGowan, K. A., Hubchak, S. C., Cid, M. C., Martin, M. B., Kleinman, H. K., Greene, G. L., and Schnaper, H. W. 1996. Expression of an estrogen receptor by human coronary artery and umbilical vein endothelial cells. *Circulation* 94:1402–1407.

Klauber, N., Parangi, S., Flynn, E., Hamel, E., and D'Amato, R. J. 1997. Inhibition of angiogenesis and breast cancer in mice by the microtubule inhibitors 2-methoxyestradiol and taxol. *Cancer Res.* 57:81–86.

Kleinert, H., Wallerath, T., Euchenhofer, C., Ihrig-Biedert, I. Li H., and Förstermann, U. 1998. Estrogens incrase transcription of the human endothelial NO synthase gene. Analysis of the transcription factors involved. *Hypertension* 31:582–588.

Kraus, W. L., and Kadonaga, J. T. 1998. p300 and estrogen receptor cooperatively activate transcription via differential enhancement of initiation and reinitiation. *Genes Dev.* 12:331–342.

Kuiper, G. G., Carlsson, B., Grandien, K., Enmark, E., Haggblad, J., Nilsson, S., and Gustafsson, J. A. 1997. Comparison of the ligand binding specificity and transcript tissue distribution of estrogen receptors alpha and beta. *Endocrinology* 138:863–870.

Kumar, V., and Chambon, P. 1988. The estrogen receptor binds tightly to its responsive element as a ligand-induced homodimer. *Cell* 55:145–156.

Lansink, M., Koolwijk, P., van Hinsbergh, V., and Kooistra, T. 1998. Effect of steroid hormones and retinoids on the formation of capillary-like tubular structures of human microvascular endothelial cells in fibrin matrices is relatd to urokinse expression. *Blood* 92:927–938.

Lee, H. S., Lee, I. S., Kang, T. C., Jeong, G. B., and Chang, S. I. 1999. Angiogenin is involved in morphological changes and angiogenesis in the ovary. *Biochem. Biophys. Res. Commun.* 257:182–186.

Leung, B. S., and Potter, A. H. 1987. Mode of estrogen action on cell proliferation in CAMA-1 cells: II. Sensitivity of G1 phase population. *J. Cell. Biochem.* 34:213–225.

Loosfelt, H., Atger, M., Misrahi, M., Guichon-Mantel, A., Meriel, C. F. L., Logeat, F., Benarous, R., and Milgrom, E. 1986. Cloning and sequence analysis of rabbit progesterone-receptor complementary cDNA. *Proc. Natl. Acad. Sci. USA* 83:9045–9049.

Losordo, D. W., Kearney, M., Kim, E. A., Jekanowski, J., and Isner, J. M. 1994. Variable expression of the estrogen receptor in normal and atherosclerotic coronary arteries of premenopausal women. *Circulation* 89:1501–1510.

Lowe, W. L., Pestell, R. G., Madison, L. D., and Jameson, J. L. 1998. Mechanisms of hormone action. In: Jameson, J. L., ed. *Principles of Molecular Medicine*. Humana Press, Totowa, NJ, pp. 419–431.

Machiarelli, G., Nottola, S. A., Vizza, E., Kikuta, A., Murakami, T., and Motta, P. M. 1991. Ovarian microvasculature in normal and hCG stimulated rabbits. A study of vascular corrosion casts with particular regard to the interstitium. *J. Submicrosc. Cytol. Pathol.* 23:391–395.

Mangelsdorf, D. J., Thummel, C., Beato, M., Herrlich, P., Schutz, G., Umesono, K., Blumberg, B., Kastner, P., Mark, M., Chambon, P., et al. 1995. The nuclear receptor superfamily: the second decade. *Cell* 83:835–839.

McDonnell, D. P., Clemm, D. L., Hermann, T., Goldman, M. E., and Pike, J. W. 1995. Analysis of estrogen receptor function in vitro reveals three distinct classes of antiestrogens. *Mol. Endocrinol.* 9:659–669.

McGuire, J. K., and Schnaper, H. W. 1999. Estradiol induces cyclin-associated kinase activity in vascular endothelial cells. *Pediatr. Res.* 45:42A.

Mendelsohn, M. E., and Karas, R. H. 1999. The protective effects of estrogen on the cardiovascular system. *N. Engl. J. Med.* 340:1801–1811.

Meyer, G. T., and McGeachie, J. K. 1988. Angiogenesis in the developing corpus luteum of pregnant rats: a stereologic and autoradiographic study. *Anat. Rec.* 222:18–25.

Migliaccio, A., Di Domenico, M., Castoria, G., de Falco, A., Bontempo, P., Nola, E., and Auricchio , F. 1996. Tyrosine kinase/p21^{ras}/MAP-kinase pathway activation by estradiol-receptor complex in MCF-7 cells. *EMBO J.* 15:1292–1300.

Miller, V. M., and Vanhoutte, P. M. 1991. Progesterone and modulation of endothelium-dependent responses in canine coronary arteries. *Am. J. Physiol.* 261:R1022–R1027.

Minshall, R. D., Stanczyk, F. Z., Miyagawa, K., Uchida, B., Axthelm, M., Novy, M., and Hermsmeyer, K. 1998. Ovarian steroid protection against coronary artery hyperreactivity in rhesus monkeys. *J. Clin. Endocrinol. Metab.* 83:649–659.

Morales, D. E., McGowan, K. A., Grant, D. S., Maheshwari, S., Bhartiya, D., Cid, M. C., Kleinman, H. K., and Schnaper, H. W. 1995. Estrogen promotes angiogenic activity in human umbilical vein endothelial cells in vitro and in a murine model. *Circulation* 91:755–763.

Musgrove, E. A., and Sutherland, R. L. 1994. Cell cycle control by steroid hormones. *Semin. Cancer Biol.* 5:381–389.
Naftolin, F., and Tolis, G. 1978. Neuroendocrine regulation of the menstrual cycle. *Clin. Obstet. Gynecol.* 21:17–29.
Nakamura, J., Lu, Q., Aberdeen, G., Albrecht, E., and Brodie, A. 1999. The effect of estrogen on aromatase and vascular endothjelial growth factor messenger ribonucleic acid in the normal nonhuman primate mammary gland. *J. Clin. Endocrinol. Metab.* 84:1432–1437.
Paech, K., Webb, P., Kuiper, G. G., Nilsson, S., Gustafsson, J., Kushner, P., and Scanlan, T. S. 1997. Differential ligand activation of estrogen receptors ERα and ERβ at AP1 sites. *Science* 277:1508–1510.
Prendergast, P., Pan, Z., and Edwards, D. P. 1996. Progesterone receptor-induced bending of its target DNA: distinct effects of the A and B receptor forms. *Mol. Endocrinol.* 10:393–407.
Raff, M. C., Barres, B. A., Burne, J. F., Coles, H., Ishizaki, Y., and Jacobson, M. D. 1993. Programmed cell death and the control of cell survival: lessons from the nervous system. *Science* 262:695–700.
Razandi, M., Pedram, A., Greene, G. L., and Levin, E. R. 1999. Cell membrane and nuclear estrogen receptors (ERs) originate from a single transcript: studies of ERα and ERβ expressed in Chinese hamster ovary cells. *Mol. Endocrinol.* 13:307–319.
Re, F., Zanetti, A., Sironi, M., Polentarutti, N., Lanfrancone, L., Dejana, E., and Colotta, A. 1994. Inhibition of anchorage-dependent cell spreading triggers apoptosis in cultured human endothelial cells. *J. Cell Biol.* 127:537–546.
Rodger, F. R., Young, F. M., Fraser, H. M., and Illingworth, P. J. 1997. Endothelial cell proliferation follows the mid-cycle luteinizing hormone surge, but not human chorionic gonadotropin rescue, in the human corpus luteum. *Hum. Reprod.* 12:1723–1729.
Rogers, P. A. W., Lederman, F., and Taylor, N. 1998. Endometrial microvascular growth in normal and dysfunctional states. *Hum. Reprod. Update* 4:503–508.
Rone, J. D., Halvorson, L. M., and Goodman, A. L. 1993. Ovarian angiogenesis in rabbits: endotheliotrophic chemoattractant activity from isolated follicles and dispersed granulosa cells. *J. Reprod. Fertil.* 97:359–365.
Russell, K. S., Haynes, M. P., Caulin-Glaser, T., et al. 2000. Estrogen stimulates heat shock protein 90 binding to endothelial nitric oxide synthase in human vascular endothelial cells. *J. Biol. Chem.* 275:5026–5030.
Rutqvist, L. E., and Mattson, A. 1993. Cardiac and thrombolic morbidity among postmenopausal women with early-stage breast cancer in a randomized trial of adjuvant tamoxifen. *J. Natl. Cancer Inst.* 85:1398–1406.
Saceda, M., Lippman, M. E., Chambon, P., and Lindsey, R. L., Ponglikitmongkol, M., Puente, M., and Martin, M. B. 1988. Regulation of the estrogen receptor in MCF-7 cells by estradiol. *Mol. Endocrinol.* 2:1157–1162.
Savouret, J. F., Chauchereau, A., Misrahi, M., Lescop, P., Mantel, A., Bailly, A., and Milgrom, E. 1994. The progesterone receptor. Biological effects of progestins and antiprogestins. *Hum. Reprod.* 9(suppl):7–11.
Schnaper, H. W. 1999. Estrogen: it's not just for reproduction any more. *Kidney Int.* 55:1577–1579.
Schnaper, H. W., Albuquerque, M. L. C., McGuire, J., Kim-Schulze, S., Runyan, C., and Hubchak, S. C. 1999a. Modulation of endothelial cell function by estrogen. In: Rubanyi, G. M., ed. *Angiogenesis in Health and Disease*. Marcel Dekker, New York, pp. 175–190.
Schnaper, H. W., McGuire, J., Runyan, C., and Hubchak, S. C. 1999b. Sex steroids and the endothelium. *Curr. Med. Chem.* 7:519–531.
Shiau, A. K., Barstad, D., Loria, P. M., Cheng, L., Kushner, P. J., Agard, D. A., and Greene, G. L. 1998. The structural basis of estrogen receptor/coactivator recognition and the antagonism of this interaction by tamoxifen. *Cell* 95:927–937.

Shifren, J. L., Tseng, J. F., Zaloudek, C. J., Ryan, I. P., Meng, Y. G., Ferrara, N., Jaffe, R. B., and Taylor, R. N. 1996. Ovarian steroid regulation of vascular endothelial growth factor in the human endometrium: implications for angiogenesis during the menstrual cycle and in the pathogenesis of endometriosis. *J. Clin. Endocrinol. Metab.* 81:3112–3118.

Simoncini, T., Hafezi-Moghadam, A., Brazil, D. P., et al. 2000. Interaction of oestrogen receptor with the regulatory subunit of phosphatidylinositol-3-OH kinase. *Nature* 407:538–541.

Spyridopoulos, I., Brogi, E., Kearney, M., Sullivan, A. B., Cetrulo, C., Isner, J. M., and Losorodo, D. W. 1997b. Vascular endothelial growth factor inhibits endothelial cell apoptosis induced by tumor necrosis factor-alpha: balance between growth and death signals. *J. Mol. Cell. Cardiol.* 29:1321–1330.

Spyridopoulos, I., Sullivan, A. B., Kearney, M., Isner, J. M., and Losordo, D. W. 1997a. Estrogen receptor-mediated inhibition of human endothelial cell apoptosis. Estradiol as a survival factor. *Circulation* 95:1505–1514.

Subakir, S. B., Hadisaputra, W., Handoyo, A. E., and Affandi, B. 1996. Endometrial angiogenic response in norplant users. *Hum. Reprod.* 11(suppl):51–55.

Suzuki, T., Sasano, H., Takaya, R., Fukaya, T., Yajima, A., and Nagura, H. 1998. Cyclic changes of vasculature and vascular phenotypes in normal huan ovaries. *Hum. Reprod.* 13:953–959.

Tempel, C., and Neeman, M. 1999. Perfusion of the rat ovary: application of pulsed arterial spin labeling MRI. *Magn. Reson. Med.* 41:113–123.

Tsukamoto, A., Kaneko, Y., Yoshida, Y., Han, K., Ichinose, M., and Kimura, S. 1998. 2-Methoxyestradiol, an endogenous metabolite of estrogen, enhances apoptosis and β-galactosidase expression in vascular endothelial cells. *Biochem. Biophys. Res. Commun.* 248:9–12.

Tsukuda, K., Matsushima, T., and Yamanaka, N. 1996. Neovascularization of the corpus luteum of rats during the estrus cycle. *Pathol. Int.* 46:408–416.

Vazquez, F., Rodriguez-Manzaneque, J. C., Lydon, J. P., Edwards, D. P., O'Malley, B. W., and Iruela-Arispe, M. L. 1999. Progesterone regulates proliferation of endothelial cells. *J. Biol. Chem.* 274:2185–2192.

Venkov, C. D., Rankin, A. B., and Vaughan, D. E. 1996. Identification of authentic estrogen receptor in cultured endotehlial cells. A potential mechanism for steroid hormone regulation of endothelial function. *Circulation* 94:727–733.

White, R. E., Darkow, D. J., and Falvo Lang, J. L. 1995. Estrogen relaxes coronary arteries by opening BKCachannels through a cGMP-dependent mechanism. *Circ. Res.* 77:936–942.

Williams, S., Leventhal, C., Lemmon, V., Nedergaard, M., and Goldman, S. A. 1999. Estrogen promotes the initial migration and inception of NgCAM-dependent calcium-signaling by new neurons of the adult songbird brain. *Mol. Cell. Neurosci.* 13:51–55.

Xing, R. H., and Rabbani, S. A. 1999. Transcriptional regulation of urokinase (uPA) gene expression in breast cancer cells: role of DNA methylation. *Int. J. Cancer* 81:443–450.

Yamamoto, S., Konishi, I., Tsuruta, Y., Nanbu, K., Mandai, M., and Kuroda, H. 1997. Expression of vascular endothelual growth factor (VEGF) during follicular and corpus luteum formation in the human ovary. *Gynecol. Endocrinol.* 11:371–381.

Yamamoto, T., Terada, N., Nishizawa, Y., and Petrow, V. 1994. Angiostatic activities of medroxyprogesterone acetate and its analogues. *Int. J. Cancer* 56:393–399.

Yasuda, Y., Masuda, S., Chikuma, M., Inoue, K., Nagao, M., and Sasaki, R. 1998. Estrogen-dependent production of erythropoietin in uterus and its implication in uterine angiogenesis. *J. Biol. Chem.* 273:25381–25387.

Yue, T.-L., Wang, X., Louden, C. S., Gupta, S., Pillarisetti, K., Gu, J.-L., Hart, T. K., Lysko, P. G., and Feuerstein, G. Z. 1997. 2-methoxyestradiol, an endogenous estrogen metabolite, induces apoptosis in endothelial cells and inhibits angiogenesis: possible role for stress-activated preotein kinase signaling pathway and Fas expression. *Mol. Pharmacol.* 51:951–962.

Zhou, M., Sutliff, R. L., Paul, R. J., Lorenz, J. N., Hoying, J. B., Haudenschild, C. C., Yin, M., Coffin, J. D., Kong, L., Kranias, E. G., Luo, W., Boivin, G. P., Duffy, J. J., Pawlowski, S. A., and Doetschman, T. 1998. Fibroblast growth factor 2 control of vascular tone. *Nat. Med.* 4:201–207.

CHAPTER 10

Angiogenesis, Vascular Imaging, and Therapeutic Approaches in Ovarian Tumors

Elise C. Kohn and Steven K. Libutti

Angiogenesis has multiple critical purposes in the normal human ovary as described in the Chapters 7, 8, and 9. While the component processes underlying neovascularization may be similar between physiologic and tumor-associated angiogenesis (Kohn and Liotta, 1995), angiogenesis has a very different purpose in the context of malignant and nonmalignant ovarian neoplasms. Physiologic angiogenesis supplies nutrients necessary for normal cyclical ovarian function, production, and maintenance of the placenta during pregnancy, and development of the fetus. Those same vessels provide an outbound route for hormones and growth factors of ovarian origin and cellular and subcellular waste products. However, in malignancy an additional function is attributed to the vascular tree: provision of the conduit for tumor cell dissemination. This latter function allows completion of the key step that separates nonmalignant tumors from malignant tumors—the potential for invasion and development of tumor satellites. These metastatic deposits are the primary cause of death in advanced ovarian cancers of germ cell, stromal, and epithelial origin. The potential for tumor cell invasion is by itself inadequate for the process of tumor dissemination in the absence of vascular access. Thus, angiogenesis can be considered a rate-limiting process for normal ovarian function as well as in ovarian malignancies.

The importance of angiogenesis in cancer and malignant invasion was described and demonstrated by Folkman (Folkman et al, 1974, Folkman, 1976) and Liotta (Liotta et al, 1974, 1976). They recognized that both nutritional support and access to the systemic circulation were dependent on neovascularization (Folkman, 1971). The capacity for nutritional diffusion has been measured and found to be limited beyond a diffusional area of 0.15 mm (Folkman et al, 1974). Thus, a proliferative process, normal or malignant, will require induction of neovessels to support growth beyond that limit. The local milieu is regulated by the varied cells present, their individual needs, and their contributions, demonstrating the importance of the local microenvironment to the success of the physiologic process, cancer dissemination included. Understanding how these events occur and how they interact, especially in the process of ovarian malignancies, has been the point of intense research recently, mostly focused on the epithelial ovarian tumors, the most common and most deadly ovarian neoplasms.

That new blood vessels develop from existing vascular trunks has been long known, but advances in the understanding of the molecular mechanisms of angiogenesis have led to new insights. Tumor neovascularization initiates from host vessels either as an invasive sprout from an existing vessel or by de novo branching, after which vessels are formed and blood flow is established (Folkman et al, 1989; Jain et al, 1998). Vascular initiation in malignancy, as in physiologic angiogenesis, is in response to local elaboration of cytokines, growth factors, and soluble matrix molecules produced by immune, malignant, stromal, and normal cells (Rosen and Goldberg, 1995a,b; Hanahan and Folkman, 1996; Klein et al, 1997; Neufeld et al, 1999). This results in an imbalance between pro- and antiangiogenic factors. Basic fibroblast growth factor (bFGF) and vascular endothelial growth factor (VEGF) are among the most potent angiostimulatory factors for normal and tumor-associated neovascularization. The family of VEGF proteins now consists of five family members (Bussolino et al, 1997), but for the discussion below, unless otherwise indicated, VEGF will mean VEGF-A (VEGF/vascular permeability factor, VPF). It has been shown experimentally that endothelial activity is a tightly regulated process. If critical signals are absent or abnormal, the endothelial cell may receive a cell death message, and subsequently undergo apoptosis (Holmgren, et al, 1995). A critical action of VEGF is to induce production of bcl-2, an anti-apoptotic gene (Gerber et al, 1998). The maintenance of a balance between proliferation and apoptosis was shown experimentally to be a hallmark of quiescent vasculature. This balance is shifted by the de novo and increased elaboration of proangiogenic cytokines by the tumor cells and in response to the tumor by local stromal components.

Many studies have described the phenotype of neovessel formation as physiologic invasion. Indeed, the multiple steps outlined above involve most of the same cassette of proliferation and invasion genes used by invasive and metastatic tumor cells (Kohn and Liotta, 1995). Endothelial cells turn on a proliferative and invasive phenotype upon receipt of an angiostimulatory message. This is necessary for the production of the branch or sprout initiating the new vascular tree. Endothelial cells change their attachment and spreading on their local basement membrane, secrete proteolytic enzymes locally to degrade that scaffolding, and then migrate into the stroma in the process of new vessel germination. Proliferation is an associated and necessary event. This is demonstrated in the new in vivo model system in which a human skin graft is placed on a severe combined immunodeficiency disease (SCID) or nude mouse (Beck and D'Amore, 1997). Viability of the graft requires anastomosis of the human and mouse vessels, demonstrating the requirement for network formation. The final steps are the development and organization of a local basement membrane and recruitment of pericytes and smooth muscle cells for the larger vessels (Brooks et al, 1995; Chakravarthy and Gardiner, 1999). Pericytes are necessary for vessel stabilization and for final maturation and are not seen on the vast majority of tumor-associated microvessels (Benjamin et al, 1999; Chakravarthy and Gardiner, 1999). In tumor neovasculature, this results in a basement membrane coating that is not complete or organized and intraendothelial junctions that are loose, both of which add to the leaky and tortuous vasculature seen in tumors (Jain, 1990; Jain et al, 1998).

The endothelium is a dynamic organ in the ovary, relatively quiescent only in prepubescent and postmenopausal women. The cyclical shift in the balance between proliferative and degenerative signals has been shown to be critical in the ovary. A shift in this balance by endogenous or exogenous molecules is known to result in the vascular remodeling occurring during corpus luteum (CL) development and dissolution (Modlich et al, 1996; Neeman et al, 1997; Goede et al, 1998). VEGF has been shown to be an important endogenously produced molecule, driving the vessel permeability and density associated with follicle development. Local response by endothelial cells to the ovarian stromal production of VEGF results in their contribution to the ovulatory process. Continued local production of VEGF is necessary for the maturation of the very vascular CL. There is recapitulation of this process in endometriotic explants wherein there is prominent VEGF production and evidence for an autocrine loop (McLaren et al, 1996a,b; Smith 1997). Thus, neovascularization in the normal ovary constitutes a unique set of molecular and biologic events during normal physiologic events and in nonmalignant reproductive pathology.

OVARIAN CANCER IS AN ANGIOGENIC TUMOR

Several lines of evidence support the importance of angiogenesis in the process of ovarian cancer progression and suggest that it may be a critical factor in ovarian cancer development. This progress is based on both in vitro and in vivo studies. In vitro and xenograft studies have lagged behind our recognition of the importance of angiogenesis in clinical ovarian cancer, but new and enlightening data are emerging. Neeman's group (Schiffenbauer et al, 1997) hypothesized that there was a selective tumor promoting event in the postmenopausal ovary and demonstrated that event to be regulation of VEGF. They studied sham or ovariectomized mice implanted with ovarian cancer cell line spheroids. These tumor spheroids require neovascularization in order to form viable and proliferative tumors in the host animals. The investigators showed that the take of the ovarian cancer spheroids in nonovariectomized mice was diminished compared to that in the ovariectomized mice as shown by blocking antibody experiments. They demonstrated that this was due to reduced VEGF production in the ovariectomized mice with resultant blockade of neovascularization. Luteinizing hormone (LH) and follicle-stimulating hormone (FSH) were implicated as stimulants of VEGF production in the ovariectomized mice. The authors thus hypothesize that the high concentrations of LH and FSH in the postmenopausal ovary may be permissive for ovarian cancer through their associated release of the tight regulation of VEGF and other critical angiostimulatory cytokines, facilitating the progression of a transforming event. Progesterone has also been shown to have a proangiogenic activity (Vazquez et al, 1999) in vitro. Endothelial cells were shown to have functional progesterone receptors, the activation of which resulted in inhibited endothelial proliferation. This was confirmed using an aortic re-endothelialization assay in wild-type and progesterone receptor knockout mice.

Progesterone exerted its activity in part by downregulation of retinoblastoma protein phosphorylation and altered expression of cyclins E and A. The loss of the progesterone signal in postmenopausal women may also be permissive to the angiogenic switch of early cancer progression. Thus, the dynamics of angiogenesis during normal ovarian cycling and the loss of physiologic regulation of angiogenesis occurring with menopause come together to show how neovascularization is a critical partner of the normal and malignant ovary. A misregulation of this partnership may be a simple first step in the promotion of ovarian cancer.

TABLE 10.1. Putative markers of angiogenesis in ovarian cancer.

Marker	Location	Function/predictor
bFGF	Urine	Proangiogenic cytokine/increased concentration in ovarian cancer
VEGF	Tumor cells	Proangiogenic cytokine/increase associated with poorer DFS and OS
TP	Tumor cells	Proangiogenic enzyme/increase associated with poor prognosis
MMPs	Tumor and stromal cells	Proinvasive enzymes/increase associated with increased grade, aggressiveness, and poor outcomes
Microvessels	Tumor	Neovasculature/increase associated with poor DFS and OS

bFGF, basic fibroblast growth factor; DFS, disease-free survival; MMPs, matrix metalloproteinases; OS, overall survival; TP, thymidine phosphorylase.

Clinical correlative studies have provided the translational link of the requirement found in vitro for angiogenesis to its potential importance as a prognostic or even diagnostic marker for the presence of ovarian malignancy (Table 10.1). Several techniques have been developed for qualitative assessment of vascularization in patient specimens. While results of these experiments have strengthened our belief that angiogenesis is important in normal ovary, ovarian diseases, and cancers, they have not been validated as of yet as useful techniques to predict the development of malignancy or its response to therapeutic interventions.

Quantitation of high microvessel density and/or microvessel area initially was shown experimentally to correlate with the potential for development of pulmonary metastases by experimental metastasis induction, and more recently was shown to correlate with metastatic potential (Liotta et al, 1974; Weidner et al, 1991). Microvessel number or vessel density within the tumor has been correlated with poorer disease-free outcome in all cancers tested and with poor overall survival in some cancers, including epithelial ovarian cancer (Weidner et al, 1991; Hollingsworth et al, 1995; Weidner and Folkman, 1996). Hollingsworth and colleagues (1995) were the first to demonstrate a role for angiogenesis in epithelial ovarian cancer. Microvessel area and density was quantitated in a cohort of advanced-stage epithelial ovarian cancer patients for whom long-term follow-up was available and treatment regimens were similar cisplatin-containing regimens. Immunohistochemical staining with anti-CD34, a pleiotrophic marker of stem cells including vascular stem cells, correlated with a shorter time to disease relapse and a shorter overall survival for patients with high vascular density. This was confirmed by Schoell and coworkers (1977), who addressed the predictive role of microvascular density with a different approach. They demonstrated that microvessel density was the unique phenotypic difference between two small advanced ovarian cancer patient cohorts well matched for age, stage, grade, and treatment, but differing only in overall survival—less than 18 months vs. greater than 6 years.

Darai and colleagues (1998) studied microvessel density using anti-CD31, recognizing the intercellular adhesion marker platelet endothelial cell adhesion molecule (PECAM-1)/CD-31. They found a positive correlation between increasing CD31 expression and stage of disease, histologic subtype, degree of histologic differentiation, and survival of patients with invasive epithelial ovarian cancer, but not for those patients with low malignant potential disease. Similarly, Orre et al (1998) found increased microvessel density in mucinous tumors using CD31 and

CD34 expression. Microvessel density was greater in mucinous tumors at all stages compared with serous tumors. Abulafia and colleagues (Abulafia et al, 1997; Abulafia and Sherer, 2000) evaluated microvessel counts in primary and matched metastatic ovarian tumor samples using factor VIII staining. Microvessel counts in omental metastasis but not in primary ovarian tumors correlated with preoperative CA-125 levels and patient survival.

Many growth factors and cytokines have been identified that have growth stimulatory properties for ovarian cancer. Endothelial cells respond to many of the same growth stimulants as do the ovarian cancer cells, but in a paracrine fashion (Ferrara et al, 1992; Ferrara, 1995; Singh et al, 1995; Bikfalvi et al, 1997; Neufeld et al, 1999). Several growth factors, such as insulin-like growth factor-1, epidermal growth factor (EGF), and transforming growth factor-β (TGF-β) may form autocrine cancer growth loops, and also form paracrine angiogenesis loops. Among the most potent angiostimulatory agents are VEGF, acidic and basic FGFs, EGF, and TGF-β. The activity later defined as basic FGF (bFGF), for example, was first identified by Gospodarowicz and Thakral (1978) as an angiogenic factor isolated from the CL. Crickard and associates (1994) showed abundant message for bFGF in three ovarian cancer cell lines as well as expression of FGF receptors and binding sites for bFGF, suggesting an autocrine ovarian cancer loop. The powerful mitogenic stimulus of bFGF on local endothelial cells means that local bFGF production by the ovarian cancer cells can create a strong local angiogenic influence by a paracrine loop.

VEGF, also known as vascular permeability factor (VPF), was described initially by Dvorak's group (Senger et al, 1983) as the factor that stimulated the production of ascites in ovarian cancer xenografts as a paracrine effect. This observation has been supported by measurement of VEGF in ascitic fluids of ovarian cancer patients and in cyst fluids from benign and malignant cysts (Barton et al, 1997; Zebrowski et al, 1999). VEGF concentration was highest in malignant ascites, but was also elevated in cyst fluids. Quantitative plasma or urine measurements of secreted growth factors or other markers and both immunohistochemical and in situ assessment of factor expression have been correlated with clinical parameters of epithelial ovarian cancer (Nguyen et al, 1993, 1994; Singh et al, 1995; Dirix et al, 1997). Nguyen and colleagues (1994) measured bFGF in the urine of patients with varied cancers and found the ovarian cancer cohort to have one of the highest concentrations of urinary bFGF. This has not yet been found either to be predictive of disease or outcome in ovarian cancer or other cancers, but may be a useful surrogate marker to follow treatment with antiangiogenic therapies. On the contrary, VEGF quantity and expression has been shown consistently in multiple tumors, including ovarian cancer, to be correlated with outcome or increasing disease (Garzetti et al, 1999; Orre and Rogers, 1999). It has been proposed that circulating VEGF may be a useful marker for the differentiation of benign and malignant ovarian neoplasm and that it may be useful prognostically (Abu-Jawdeh et al, 1996; Barton et al, 1997; Tempfer et al, 1998; Ghen et al, 1999; Zebrowski et al, 1999). One study showed a disparity in the predictability of VEGF compared with CA-15 (Obermair et al, 1999). VEGF expression by in situ hybridization was shown to be statistically significantly and independently higher in stage I and II epithelial ovarian cancer patients with shorter relapse-free and overall survival (Paley et al, 1997). Similar results were found in a smaller study where there was a more than doubled overall survival

(8 vs. 60 mos, $p = .058$) in patients with advanced-stage serous ovarian cancer with low or absent VEGF (Hartenbach et al, 1997).

VEGF is reportedly selectively active on endothelial cells. However, in situ hybridization and immunohistochemical studies have identified epithelial ovarian cancer cells as capable of making and secreting VEGF to target local neovasculature (Boocock et al, 1995; Abu-Jawdeh et al, 1996). This further supports the critical interactive role of angiogenesis in ovarian cancer progression. Interestingly, Boocock and colleagues reported the expression of the VEGF receptors Flt and kinase derived receptor (KDR) on the membrane of ovarian cancer cell lines. This suggests that VEGF may create an autocrine loop in ovarian cancer.

The family of matrix metalloproteinases, now over 20 in number, are a group of enzymes that degrade the extracellular matrix of the basement membrane (Chambers and Matrisian, 1997). They have been shown to be important in malignant invasion as well as in angiogenesis. They are both therapeutic targets (discussed below) as well as putative biomarkers of cancer. While not validated as secreted and circulating biomarkers in ovarian cancer, there are data to indicate their production by ovarian cancer cells and endothelial cells (Garzetti et al, 1995; Kohn and Liotta, 1995; Kohn et al, 1995; Fishman et al, 1997). Garzetti and colleagues (1995) showed that ovarian cancer expressed gelatinase matrix metalloproteinase 2/MMP-2/gelatinase A (gelA/MMP-2)–, which was significantly correlated with risk of recurrence independently of grade, nodal status, or residual disease after debulking. Studies to date have not demonstrated measurement of circulating gelatinase A itself to be a prognostic marker. Another enzyme shown to have prognostic significance and to be proangiogenic is thymidine phosphorylase, also known as platelet-derived endothelial cell growth factor (Reynolds et al, 1994; Hawighorst et al, 1997; Hata et al, 1999). Expression of this protein correlated with areas of high blood velocity on Doppler flow analysis and was shown to be associated with a significantly worse outcome in high expressing patients. Therefore, neovascularization is important globally in normal premenopausal ovarian physiology and in ovarian cancer.

VASCULAR IMAGING

Traditional measures of responses to chemotherapeutic agents have consisted mainly of physical examination and x-ray–based methodology. Decreases in the measured bi-dimensional diameters of tumors as seen on computed tomography (CT) scans are used as a gauge of the efficacy of the treatment being studied. These methods do not give information regarding the changes in the tumor physiology. With respect to antiangiogenic therapy, measurement of the changes in the tumor's metabolism and blood flow may give important information regarding the efficacy of the treatment (Libutti et al, 1999).

New imaging methods are being developed that can provide information on the physiologic changes taking place in a tumor during therapy. The ability to monitor such changes could allow the course and duration of therapy to be adjusted while treatment is still being administered. It might also be possible to predict, after a single course of therapy, whether the patient's tumor is likely to respond to further therapy (Gupta et al, 1999). This could potentially save patients from debilitating and unproductive treatment, or permit a shift to more efficacious drugs. The concept of physiologic imaging becomes even more important when the target of therapy shifts from the tumor cell to the tumor-associated vasculature.

Tumor biopsies have been employed and vascular density can be measured histologically. However, this is an invasive procedure that requires easily accessible tissue. Therefore, there has been increasing interest in noninvasive imaging tests that can be used to measure tumor blood flow, tumor blood volume, changes in tumor metabolism, and changes in tumor vascular density (Eskey et al, 1994; Hattori et al, 1994; Hering et al, 1995). Such modalities could be used to better assess the effectiveness of new agents directed against the tumor vasculature.

If an agent is designed to inhibit new vessel formation and to decrease the overall vascularity of a tumor, it follows that the blood flow to the tumor as well as the total blood volume in the tumor should decrease. Positron emission tomography (PET) has been used in a variety of clinical settings to measure both blood flow and blood volume (Brudin et al, 1992; Schelbert, 1994), as well as rates of glucose metabolism. Much of this work has been done in the heart and in the brain, but recently attempts have been made to apply these techniques to the study of neoplastic tissue (Sagar et al, 1993; Leenders, 1994). Three radiolabeled pharmaceuticals are of particular interest to make such measurements—HO^{15} to measure blood flow, ^{11}CO to measure blood volume, and ^{18}F-fluorodeoxyglucose (^{18}FDG) to approximate glucose metabolism (Figure 10.1).

The radiopharmaceutical HO^{15} is used to make accurate measurements of blood flow to the heart, brain, and other tissues. This radioactive form of water is freely diffusable in tissue, and by injecting it as an intravenous bolus and monitoring its rapid delivery, diffusion, and washout from a tissue, one can determine flow in absolute units of milliliter per minute per gram of tissue. The short half-life (minutes) of HO^{15} and rapid delivery and clearance allow for short scanning times (<5 minutes), with only minimal radiation exposure to the patient. With regard to neoplasms, investigators have shown changes in the blood flow to breast tumors (Mankoff et al, 1999). The intensity of each pixel in the image of the tumor is proportional to local blood flow (mL/min/g).

Radiolabeled carbon monoxide (^{11}CO) is an excellent red blood cell label (binding tightly to the hemoglobin in red blood cells), permitting quantitative measurement of red blood cell volume in tissues. By taking one venous blood sample from the patient for calibration, it is possible to make a three-dimensional image whose brightness measures the absolute concentration of red blood cells throughout the body (including the tumor). Because the PET camera is so sensitive (capable of measuring picograms of radiopharmaceutical in vivo), only minute amounts of carbon monoxide (far too small to have any deleterious pharmacologic effect) need be administered, usually by inhalation. The 20-minute half-life of the radioisotope permits ample imaging time, but ensures relatively rapid removal from the body. In a typical ^{18}CO image the brightness of each pixel is proportional to the number of red blood cells within the pixel's volume.

Studies have shown that malignant lesions demonstrate elevated glycolysis when compared to normal tissues (DiChiro, 1986; Conti et al, 1996). By using the radiolabeled glucose analogue ^{18}FDG, investigators have documented increased uptake of FDG in cancer tissues (Ito et al, 1992). PET scanning with ^{18}FDG has been used to image a variety of malignant neoplasms including breast cancer, head and neck tumors, lung cancer, lymphoma, melanoma, ovarian tumors, bone cancers, and colorectal carcinomas.

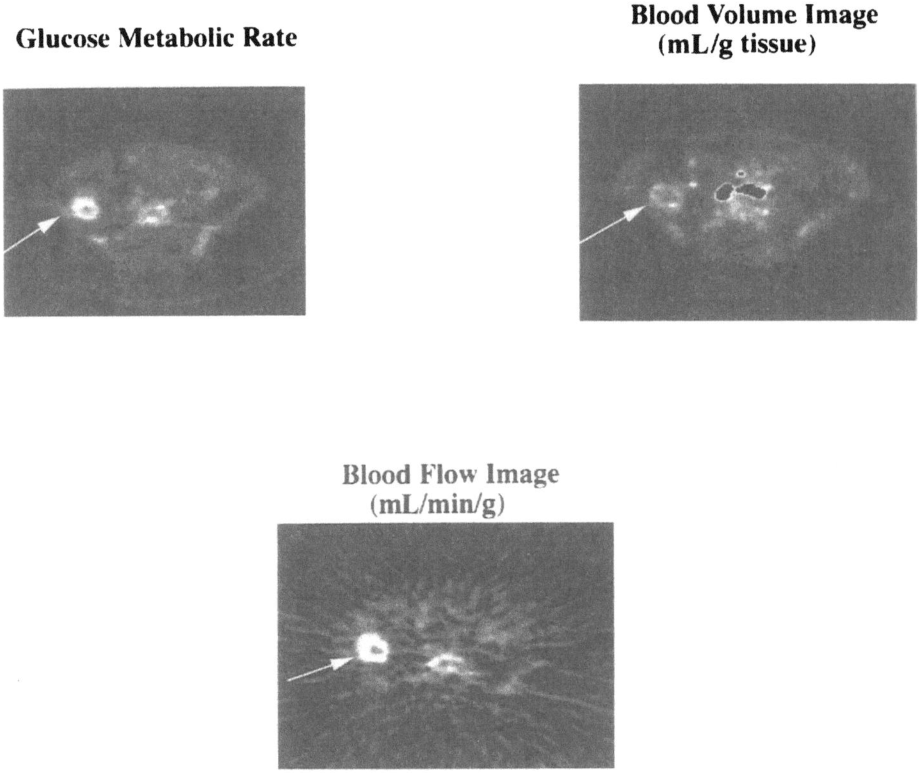

FIGURE 10.1. Clockwise from the upper left are positron emission tomography (PET) images of the same retroperitoneal tumor (arrow) using ^{18}FDG, ^{11}CO, and HO^{15}. These isotopes can measure metabolic rate, blood volume, and blood flow, respectively. The pixel intensity is directly proportional to the magnitude of the parameter being measured. (Courtesy of Dr. Steve Bacharach.)

The mechanism whereby ^{18}FDG is selectively accumulated in neoplastic tissue is based primarily on increased uptake of ^{18}FDG and conversion to ^{18}FDG-6-phosphate. The inability of neoplastic cells to further metabolize ^{18}FDG-6-phosphate causes the molecule to be trapped in the cells, facilitating accurate imaging. The intensity of an image is proportional to the amount of ^{18}FDG that has entered the cells, begun the first steps of the metabolic conversion of sugar to energy, and then been trapped. The reasons why so many tumors avidly accumulate glucose are only partly understood. The energy needs of rapidly dividing cells obviously plays a role. In addition, as the blood flow to a tumor is decreased, the supply of oxygen and nutrients should be decreased as well. This should result in a shift from aerobic to anaerobic metabolism and this also may impact on glucose metabolism of the tumor cells. For these reasons, it may be important to measure both tumor blood flow and metabolism (Libutti et al, 1999). In fact, the ability of PET to make such in vivo measurements of approximate glucose metabolism may yield further understanding of the energy requirements and the methods of energy production of tumors. It is for these reasons that functional imaging (i.e., imaging of tumor physiology and biochemical functions) may be such an important adjunct to conventional imaging methods such as CT, which portray primarily anatomy.

An additional imaging technique that complements PET is dynamic enhanced magnetic resonance imaging (DEMRI), which involves the rapid administration of a gadolinium-based contrast agent followed by rapid analysis of signal intensity (Libutti et al, 1999). Patterns of MRI contrast uptake within tumors correlate with microvessel density (van Dijke et al, 1996). Although some variability has been noted in the strength of this correlation, it should be possible in individual subjects to monitor changes in microvessel density using DEMRI, even if absolute densities cannot be determined. This is particularly true for the early phases of enhancement after injection when the agent is principally intravascular and has not yet leaked into the interstitial space.

Most investigators have concentrated on MRI mammography where differences in enhancement patterns have proven to be reliable predictors of the malignancy of a lesion (Buadu et al, 1996; Buckley et al, 1997). Others have investigated the role of dynamic contrast enhanced MRI in uterine cervical cancer, myocardium, and liver metastases (Pearlman et al, 1995; Brasch et al, 1997; Hawighorst et al, 1997). The development of an intravascular contrast agent appears to be useful in the assessment of microvessel density during dynamic studies.

Using image-processing algorithms established for the detection of malignancy by MRI, it might be possible to assess changes of perfusion and microvessel density over time in lesions being treated with agents that target the tumor vasculature. Although absolute measurements of flow are difficult, comparative quantification can be used to assess changes in tumor perfusion and microvascular density. Diffusion weighted MRI allows an assessment of net perfusion and diffusion and is widely used in the assessment of cerebral vascular accidents. This technique may also reflect tumor perfusion, as diffusion would be expected to decrease as flow decreases.

Noninvasive imaging tests may provide for physiologic and biochemical monitoring of responses to anticancer agents. The ability to detect changes in tumor blood flow and metabolism may be very useful in the evaluation of agents directed against the tumor vasculature. Currently, we are evaluating PET and dynamic MRI as tests to follow the activity of antiangiogenic agents in clinical trials at the Clinical Center of the National Institutes of Health (Libutti et al, 1999). If these techniques are validated, they may permit a more rapid evaluation of novel agents leading to modified trial designs. Our goal is to utilize these powerful techniques to screen agents more rapidly and efficiently. Continued research in this field will likely achieve this goal.

THERAPEUTIC APPROACHES

The concept of antiangiogenic therapy was initially proposed by Folkman (1997), but only came to clinical trial in the 1990s. The general class of antiangiogenesis inhibitors contains agents of multiple targets and mechanisms (Table 10.2). Many of the agents were originally defined for other mechanisms or activities. For example, the taxane family of agents are first-line chemotherapeutic agents against ovarian cancer and breast cancer (Kohn et al, 1994a; McGuire et al, 1996), but in the new low-dose format they may have antiangiogenic activity (Belotti et al, 1996; Klauber et al, 1997). CAI (carboxyamido-triazole) and the matrix metalloproteinase inhibitor class were identified in studies directed at regulation of malignant invasion (Davies et al, 1993; Kohn et al, 1994b,c, 1995; Eccles et al, 1996). The

TABLE 10.2. Anti-angiogenic agents by functional category.

Class/mechanism	Phase	Example agents (not comprehensive)
Matrix metalloproteinase inhibitors	I–III	BB-94/Marimastat, Col3, doxycycline, AG3340, AE941
Integrin inhibition	I/II	$\alpha_v\beta_3$ peptide antagonists
	I/II	anti-$\alpha_v\beta_3$ antibody
Ion channel inhibition	I	Squalamine
	I/II/III	CAI
Growth factor regulation	I/II	Su-5416 (to VEGFR)
	I/II/III	Su-101 (to PDGFR)
	I/II/III	Anti-EGFR
	I/II/III	Interferons α,β
Cytoskeleton inhibitors	I/II/III (?)	Low-dose paclitaxel and docetaxel
Cell cycle dysregulation	I/II	TNP-470
Unknown/under study	I/II/III	Thalidomide
	I	2-Methoxyestradiol
	I/II	Angiostatin and endostatin

EGFR, epidermal growth factor receptor; PDGFR, platelet-derived growth factor receptor; VEGFR, vascular endothelial growth factor receptor.

recognition that angiogenesis requires the same cassette of genes and activities for the production of the vascular sprout and neovessel as are used in the formation of invasive metastases suggests that agents that are antiinvasive should be anti-angiogenic (Kohn and Liotta, 1995). Progress has been made in the treatment of epithelial ovarian cancer, and survival improvement has been seen. Unlike other cancers where that improvement can be attributed at least initially to stage migration, there is no reliable early detection for ovarian cancer and the screening that is available has not been shown preferentially to identify early stage cancers, necessary for stage migration. Thus, the progress to date has been due to improved surgery, better chemotherapeutics, and better supportive care. Much room remains for new therapeutic directions such as antiangiogenesis therapies. Current studies argue that angiogenesis may be a potent prognostic factor for all stages of ovarian cancer and its dynamic biology suggests it is a good target for therapeutic intervention.

Many putative antiangiogenic agents are now under clinical investigation. Only three, the taxanes, thalidomide, and doxycycline, have received some form of approval by the U.S. Food and Drug Administration, but not for their antiangiogenic activity. Paclitaxel, approved for and widely used in the initial treatment of ovarian cancer, has been shown to be angioinhibitory by several investigators (Belotti et al, 1996; Klauber et al, 1997). Taxotere has been reported in abstract form to facilitate inhibition of angiogenesis. Thalidomide, approved in the United States for use in leprosy, was shown to block corneal neovascularization in rabbits (D'Amato et al, 1994). The bioactivity in leprosy or in cancer has not been definitively linked back to its ability to inhibit angiogenesis. However, its teratogenicity of blockade of bone growth and maturation may be due to loss or reduction of the very angiogenic region at the epiphyseal plate (Vu et al, 1998). The rest of the plethora of over 100 putative antiangiogenic agents remain in clinical or preclinical testing.

Several common themes emerge when reviewing these agents. The primary selection for antiangiogenesis agent efficacy is xenograft models, therefore selectivity against neovessels. As expected little disease specificity has been observed to date in these animal models. This is logical given that cancer requires some amount of angiogenesis, and for most cancers at least one marker of angiogenic activity has been found to correlate with patient outcome. While the family of antiangiogenic agents has been shown to have activity in vitro and in vivo in xenografts or angiogenesis assays, only rare partial or complete clinical responses have been demonstrated in clinical trials. This disparity may be due to drug activity against true neovessels, those that are recruited during the relatively short time course under which the animal models are run in comparison to patient tumor vessels that may be considerably older. It is unknown whether cancers have true neovessels with rapid turnover, or have immature vessels that may be phenotypically similar to neovessels. Many of the new antiangiogenesis agents have been demonstrated in the laboratory to be cytostatic to endothelial cells, although some also have cytostatic activity against tumor cells. This cytostatic activity might not be expected to yield measurable disease reductions clinically, but might translate into disease stabilization, flattening of the tumor growth rate curve, and improved progression-free and/or overall survival.

Several agents have reached the clinic for testing against ovarian cancer and many more are in broad accrual phase I studies. The first oral antiangiogenesis agent to enter clinical trial was CAI, initially discovered as an inhibitor of invasion and then shown to inhibit angiogenesis (Kohn and Liotta, 1990; Kohn et al, 1992, 1994b,c, 1995; Luzzi et al, 1998). It was initially demonstrated to be cytostatic against multiple tumor cell types including several ovarian cancer cell lines and to have anticancer, antiangiogenic, and antimetastatic activity in nude mouse ovarian cancer xenografts with oral administration. Oral administration of CAI results in plasma concentrations in the range targeted by the laboratory studies and was found to cause disease stabilization ranging from 2 to greater than 15 months in patients with advanced cancers and up to 14 months in women with advanced refractory ovarian cancer (Kohn et al, 1996, 1997). In most patients, progression of disease occurred at existing sites of disease rather than at new metastatic sites. CAI was shown to cause disease stabilization in several ovarian cancer patients during the phase I clinical trial and is now in phase II trial for patients with relapsed or refractory disease, as well as in phase III for lung cancer and in phase I combination with full-dose paclitaxel.

Other studies are under consideration. Thalidomide was tested in phase II trial at the Royal Marsden Hospital showing poor general results, and no patients with ovarian cancer attained a response; three patients attained disease stabilization longer than 3 months on this 100 mg/day dose (Eisen et al, 2000). Several matrix metalloproteinase inhibitors (MMPIs) have been developed and taken to clinical trial (Table 10.2) (Davies et al, 1993; Brown, 1994; Wang et al, 1994; Brown and Giavazzi, 1995; Taraboletti et al, 1995; Anderson et al, 1996; Eccles et al, 1996; Watson et al, 1996; Gatto et al, 1999; Rowinsky et al, 2000). These agents, like CAI, fall into the category of agents that are antiinvasive, against both the malignant invasion of cancer dissemination as well as the physiologic invasion associated with neovascularization in cancer and other pathologies. Disease stabilization has been demonstrated with several agents including stabilization of CA-125 for 1 month for marimastat and measurable disease stabilization by Bay 12-9566. This agent was initiated in a phase III trial in ovarian cancer that has now been closed without completion. Squalamine, an inhibitor of the endothelial cell specific sodium-hydrogen exchanger (NHE-3) has now entered clinical trial targeted to ovarian cancer (Sills et al, 1998). No active clinical trials in ovarian cancer using MMPIs are logged in the National Cancer Institute (NCI) Physician's Data Query (PDQ) as of this writing. Ovarian cancer remains a rich avenue for the use of these drugs, especially since there is abundant VEGF production and angiogenesis in the solid-phase form.

ANTIANGIOGENESIS THERAPY: HOW AND WHEN TO USE

The above example agents are but a few of an explosive number of new agents targeted against invasion and angiogenesis to reach clinical trial in the final decade of the 20th century. It is not yet clear how best to use these agents or how to best assess their benefit. Studies have shown angiogenesis to precede invasion in carcinoma-in-situ of the breast, and to be evident in hyperplastic breast and prostate specimens prior to a detectable malignant change (Kohn and Liotta, 1995). This would suggest that antiangiogenesis agents, with the proper caveats, may be useful chemoprevention agents as well as therapeutics. Issues still to be addressed in this growing clinical application include identification of the best time to use antiangiogenic agents, how to determine the balance between benefit and risk, in what combinations will this family of drugs best work, and what cancers or patient subsets will benefit most from these agents.

Validated and reliable, specific, and sensitive biomarkers and/or imaging modalities such as those described above will be required for investigators to address these issues and to properly focus the use of these agents. As of now, no such biomarkers have been validated; however, the progress in vascular imaging techniques appears to be a valuable new tool to use in these therapeutic clinical trials. Studies, such as those described above, have linked angiogenesis with outcome in many cancers including ovarian cancer. These approaches are now being incorporated prospectively into clinical trials to validate their discrimination capacity. Proof of principle for this conjecture would be observed, for example, if patients with high microvessel counts, high circulating VEGF, and/or high matrix metalloproteinase (MMP) serum concentrations are shown to respond to treatment with antiinvasive or antiangiogenic agents with longer stabilization times or disease reduction, or alteration of those putative marker levels. Alternatively, demonstrating a reduction in markers of vascular flow by novel imaging modalities during a course of antiangiogenic therapy, with a shift to angiogenic disease during disease progression could be used as validation. However, since angiogenesis is a microenvironment event, it is possible that VEGF or MMPs may only be elevated locally and there may be intratumoral heterogeneity that prevents discrimination in the circulation. In this case, the imaging modalities yield more important primary diagnostic and followable results.

New clinical paradigms are being tested wherein antiangiogenesis and antiinvasive agents are being used in an adjuvant mode. The hypothesis underlying adjuvant testing is the hypothesis that the disease stabilization of these agents may cause stabilization of microscopic, undetectable disease. This low-volume and small-quantity disease may have minimal angiogenic and invasive potential and may be kept in check with antiangiogenic and antiinvasive agents with a reduced toxicity compared with use of chemotherapeutic agents in the same setting. If cancer deposits can be maintained in the prevascularized or minimally vascularized and noninvasive state, patients may achieve a prolonged time to detectable relapse. Whether this translates into overall disease survival will need to be tested.

CONCLUSION

Ovarian cancer is the leading cause of death from gynecologic cancer, but progress in therapeutic intervention, imaging modalities, surgical technique, and translational research have together resulted in improved disease-free and overall survival. The recognition of the significant impact of neovascularization in the initiation and development of ovarian cancer has appropriately targeted it for the application of the new class of antiangiogenic agents and angiogenesis imaging modalities. It is hoped that the addition of these new directions to the ovarian cancer armamentarium will improve outcome further and will also be applied toward earlier diagnosis and prevention approaches.

REFERENCES

Abu-Jawdeh, G. M., Faix, J. D., Niloff, J., Tognazzi, K., Manseau, E., and Dvorak, H. F. 1996. Strong expression of vascular permeability factor (vascular endothelial growth factor) and its receptors in ovarian borderline and malignant neoplasms. *Lab. Invest.* 74:1105–1115.

Abulafia, O., and Sherer, D. M. 2000. Angiogenesis of the ovary. *Am. J. Obstet. Gynecol.* 18:240–246.

Abulafia, O., Triest, W. E., and Sherer, D. M. 1997. Angiogenesis in primary and metastatic epithelial ovarian carcinoma. *Am. J. Obstet. Gynecol.* 177:541–547.

Anderson, I. C., Shipp, M. A., Docherty, A. J., and Teicher, B. A. 1996. Combination therapy including a gelatinase inhibitor and cytotoxic agent reduces local invasion and metastasis of murine Lewis lung carcinoma. *Cancer Res.* 56:715–718.

Barton, D. P., Cai, A., Wendt, K., Young, M., Gamero, A., and De Cesare, S. 1997. Angiogenic protein expression in advanced epithelial ovarian cancer. *Clin. Cancer Res.* 3:1579–1586.

Beck, L. J., and D'Amore, P. A. 1997. Vascular development: cellular and molecular regulation. *FASEB J.* 11:365–373.

Belotti, D., Vergani, V., Drudis, T., Borsotti, P., Pitelli, M. R., Viale, G., Giavazzi, R., and Taraboletti, G. 1996. The microtubule-affecting drug paclitaxel has antiangiogenic activity. *Clin. Cancer Res.* 2:1843–1849.

Benjamin, L., Golijanin, D., Itin, A., Poide, D., and Keshet, E. 1999. Selective ablation of immature blood vessels in established human tumors follows vascular endothelial growth factor withdrawl. *J. Clin. Invest.* 103:159–165.

Bikfalvi, A., Klein, S., Pintucci, G., and Rifkin, D. B. 1997. Biological roles of fibroblast growth factor 2. *Endocr. Rev.* 18:26–45.

Boocock, C. A., Charnock-Jones, S., Sharkey, A. M., McLaren, J., Barker, P. J., and Wright, K. A. 1995. Expression of vascular endothelial growth factor and its receptors *flt* and KDR in ovarian carcinoma. *J. Natl. Cancer Inst.* 87:506–516.

Brasch, R., Pham, C., Shames, D., Roberts, T., van Dijke, K., van Bruggen, N., Mann, J., Ostrowitzki, S., and Melnyk, O. 1997. Assessing tumor angiogenesis using macromolecular MRI imaging contrast media. *J. Magn. Reson. Imaging* 7:68–74.

Brooks, P. C., Stromblad, S., Klemke, R., Visscher, D., Sarkar, F. H., and Cheresh, D. A. 1995. Anti-integrin alpha v beta 3 blocks human breast cancer growth and angiogenesis in human skin. *J. Clin. Invest.* 96:1815–1818.

Brown, P. D. 1994. Preclinical and clinical studies on the matrix metalloproteinase inhibitor, batimastat (BB94). *Clin. Exp. Metastasis* 12:23.

Brown, P. D., and Giavazzi, R. 1995. Matrix metalloproteinase inhibition: a review of anti-tumour activity. *Ann. Oncol.* 6:967–974.

Brudin, L., Rhodes, C., Valind, S., Buckingham, P., Jones, T., and Hughes, J. 1992. Regional structure-function correlations in chronic obstructive lung disease measured with positron emission tomography. *Thorax* 47:914–921.

Buadu, L., Murakami, J., Murayama, S., Hashiguchi, N., Sakai, S., Masuda, K., Toyoshima, S., Kuroki, S., and Ohno, S. 1996. Breast lesions: correlation of contrast medium enhancement patterns on MRI images with histopathologic findings and tumor angiogenesis. *Radiology* 200:639–649.

Buckley, D., Drew, P., Mussurakis, S., Monson, J., and Horsman, A. 1997. Microvessel density of invasive breast cancer assessed by dynamic Gd-DTPA enhanced MRI. *J. Magn. Reson. Imaging* 7:461–464.

Bussolino, F., Mantovani, A., and Persico, G. 1997. Molecular mechanisms of blood vessel formation. *Trends Biochem. Sci.* 22:251–256.

Chakravarthy, U., and Gardiner, T. A. 1999. Endothelium-derived agents in pericyte function/dysfunction. *Prog. Retin. Eye Res.* 18:511–527.

Chambers, A. F., and Matrisian, L. M. 1997. Changing views of the role of matrix metalloproteinases in metastasis. *J. Natl. Cancer Inst.* 89:160–170.

Conti, P., Lilien, D., Hawley, K., Keppler, J., Grafton, S., and Bading, J. 1996. PET and [18F]-FDG in oncology: a clinical update. *Nucl. Med. Biol.* 23:717–735.

Crickard, K., Gross, J. L., Crickard, U., Yoonessi, M., Lele, S., Herblin, W. F., and Eidsvoog, K. 1994. Basic fibroblast growth factor and receptor expression in human ovarian cancer. *Gynecol. Oncol.* 55:277–284.

D'Amato, R., Loughnan, M., Flynn, E., and Folkman, J. 1994. Thalidomide is an inhibitor of angiogenesis. *Proc. Natl. Acad. Sci. USA* 91:4082–4085.

Darai, E., Bringuier, A.-F., Walker-Combrouze, F., Fauconnier, A., Couvelard, A., Feldmann, G., Madelenat, P., and Scoazec, J.-Y. 1998. CD31 expression in benign, borderline, and malignant epithelial ovarian tumors: an immunohistochemical and serological analysis. *Gynecol. Oncol.* 71:122–127.

Davies, B., Brown, P. D., East, N., Crimmin, M. J., and Balkwill, F. R. 1993. A synthetic matrix metalloproteinase inhibitor decreases tumor burden and prolongs survival of mice bearing human ovarian carcinoma xenografts. *Cancer Res.* 53:2087–2091.

DiChiro, G. 1986. Positron emission tomography using [18F]flurodeoxyglucose in brain tumors: a powerful diagnostic and prognostic tool. *Invest. Radiol.* 22:360–371.

Dirix, L. Y., Vermeulen, P. B., Pawinski, A., Prove, A., Benoy, I., De Pooter, C., Martin, M., and Oosterom, A. T. 1997. Elevated levels of the angiogenic cytokines basic fibroblast growth factor and vascular endothelial growth factor in sera of cancer patients. *Br. J. Cancer* 76:238–243.

Eccles, S. A., Box, G. M., Court, W. J., Bone, E. A., Thomas, W., and Brown, P. D. 1996. Control of lymphatic and hematogenous metastasis of a rat mammary carcinoma by the matrix metalloproteinase inhibitor batimastat (BB-94). *Cancer Res.* 56:2815–2822.

Eisen, T., Boshoff, C., Mak, I., Sapunar, F., Vaughan, M., Pyle, L., Johnston, S., Ahern, R., Smith, I. E., and Gore, M. E. 2000. Continuous low dose thalidomide: a phase II study in advanced melanoma, renal cell, ovarian and breast cancer. *Br. J. Cancer* 82:812–817.

Eskey, C., Wolmark, N., McDowell, C., Domach, M., and Jain, R. 1994. Residence time distributions of various tracers in tumors: implications for drug delivery and blood flow measurement. *J. Natl. Cancer Inst.* 86:293–299.

Ferrara, N. 1995. The role of vascular endothelial growth factor in pathological angiogenesis. *Breast Cancer Res. Treat.* 36:127–137.

Ferrara, N., Houck, K., Jakeman L., and Leung, D. W. 1992. Molecular and biological properties of the vascular endothelial growth factor family of proteins. *Endocr. Rev.* 13:18–32.

Fishman, D. A., Bafetti L. M., Banionis, S., Kearns, A. S., Chilukuri, K., and Stack, M. S. 1997. Production of extracellular matrix-degrading proteinases by primary cultures of human epithelial ovarian carcinoma cells. *Cancer* 80:1457–1463.

Folkman, J. 1971. Tumor angiogenesis: therapeutic implications. *N. Engl. J. Med.* 285: 1182–1186.

Folkman, J. 1976. The vascularization of tumors. *Sci. Am.* 234:58–64.

Folkman, J. 1997. Angiogenesis and angiogenesis inhibition: an overview. *Exs.* 79:1–8.

Folkman, J., Hochberg, M., and Knighton, D. 1974. Self-regulation of growth in three dimensions: the role of surface area limitations. In: Clarkson, B., and Baserga, R., eds. *Control of Proliferation in Animal Cells.* Cold Spring Harbor Press, 833–842.

Folkman, J., Karol, W., Ingber, D., and Hanahan, D. 1989. Induction of angiogenesis during the transition from hyperplasia to neoplasia. *Nature* 339:58–61.

Garzetti, G. G., Ciavattini, A., Lucarini, G., Goteri, G., de Nictolis, M., Garbisa, S., Masiero, L., Romanini, C., and Graziella, B. 1995. Tissue and serum metalloproteinase (MMP-2) expression in advanced ovarian serous cystoadenocarcinomas: clinical and prognostic implications. *Anticancer Res.* 15:2799–2804.

Garzetti, G. G., Ciavattini, A., Lucarini, G., Pugnaloni, A., De Nictolis, M., Amati, S., Romanini, C., and Biagini, G. 1999. Vascular endothelial growth factor expression as a prognostic index in serous ovarian cystadenocarcinomas: relationship with MIB1 immunostaining. *Gynecol. Oncol.* 73:396–401.

Gatto, C., Rieppi, M., Borsotti, P., Innocenti, S., Ceruti, R., Drudis, T., Scanziani, E., Casazza, A. M., Taraboletti, G., and Giavazzi, R. 1999. BAY 1-9566, a novel inhibitor of matrix metalloproteinases with antiangiogenic activity. *Clin. Cancer Res.* 5:3603–3607.

Gerber, H. P., Dixit, V., and Ferrara, N. 1998. Vascular endothelial growth factor induces expression of the antiapoptotic proteins Bcl- and A1 in vascular endothelial cells. *J. Biol. Chem.* 273:13313–13316.

Ghen, C. A., Cheng W. F., Lee C. N., Chen T. M., Kung C. C., Hsieh F. J., and Hsieh, C. Y. 1999. Serum vascular endothelial growth factor in epithelial ovarian neoplasms: correlation with patient survival. *Gynecol. Oncol.* 74:234–240.

Goede, V., Schmidt, T., Kimmina, S., Kozian, D., and Augustin, H. G. 1998. Analysis of blood vessel maturation processes during cyclic ovarian angiogenesis. *Lab. Invest.* 78:1385–1394.

Gospodarowicz, D., and Thakral, K. K. 1978. Production of a corpus luteum angiogenic factor responsible for proliferation of capillaries and neovascularization of the corpus luteum. *Proc. Natl. Acad. Sci. USA* 75:847–851.

Gupta, N., Rogers, J., Rai, A., Bishop, H., Gunel, E., and Graber, G. 1999. Tumor FDG uptake as a prognostic marker of treatment response and patient outcome in lung cancer. *J. Nucl. Med.* Abstract 136P.

Hanahan, D., and Folkman, J. 1996. Patterns and emerging mechanisms of the angiogenic switch during tumorigenesis. *Cell* 86:353–364.

Hartenbach, E. M., Olson, T. A., Goswitz, D., Mohanraj, D., Twiggs, Carson, L. F., and Ramakrishnan, S. 1997. Vascular endothelial growth factor (VEGF) expression and survival in human epithelial ovarian carcinomas. *Cancer Lett.* 121:169–175.

Hata, K., Kamikawa, T., Arao, S., Tashiro, H., Katabuchi, H., Okamura, H., Fujiwaki, R., Miyazaki, K., and Fukumoto, M. 1999. Expression of the thymidine phosphorylase gene in epithelial ovarian cancer. *Br. J. Cancer* 79:1848–1854.

Hattori, H., Miyoshi, T., Okada, J., Yoshikawa, K., Arimizu, N., and Hattori, N. 1994. Tumor blood flow measured using dynamic computed tomography. *Invest. Radiol.* 29:873–876.

Hawighorst, H., Knapstein, P., Weikel, W., Knopp, M., Knof, A., Brix, G., Schaeffer, U., Wilkens, C., Schoenberg, S., Essig, M., Vaupel, P., and van Kaick, G. 1997. Angiogenesis of uterine cervical carcinoma characterization by pharmacokinetic magnetic resonance parameters and histological microvessel density with correlation to lymphatic involvement. *Cancer Res.* 57:4777–4786.

Hering, E., Blekkenhorst, G., and Jones, D. 1995. Tumor blood flow measurements using coincidence counting on patients treated with neutrons. *Int. J. Radiat. Oncol. Biol. Phys.* 32:129–135.

Hollingsworth, H. C., Kohn, E. C., Steinberg, S. M., Rothenberg, M. L., and Merino, M. J. 1995. Tumor angiogenesis in advanced stage ovarian carcinoma. *Am. J. Pathol* 147:9–19.

Holmgren, L., O'Reilly M. S., and Folkman, J. 1995. Dormancy of micrometastases: balanced proliferation and apoptosis in the presence of angiogenesis suppression. *Nat. Med.* 1:149–153.

Ito, K., Kato, T., and Tadokoro, M. 1992. Recurrent rectal cancer and scar: differentiation with PET and MR imaging. *Radiology* 182:549–552.

Jain, R. K. 1990. Vascular and interstitial barriers to delivery of therapeutic agents in tumors. *Cancer Metastasis. Rev.* 9:253–266.

Jain, R. K., Safabakhsh, N., Sckell, A., Chen, Y., Jiang, P., Benjamin, L., Yuan, F., and Keshet, E. 1998. Endothelial cell death, angiogenesis, and microvacular function after castration in an androgen-dependent tumor: role of vascular endothelial growth factor. *Proc. Natl. Acad. Sci. USA* 95:10820–10825.

Klauber, N., Parangi, S., Flynn, E., Hamel, E., and D'Amato, R. J. 1997. Inhibition of angiogenesis and breast cancer in mice by the microtubule inhibitors—methoxyestradiol and taxol. *Cancer Res.* 57:81–86.

Klein, S., Roghani, M., and Rifkin, D. B. 1997. Fibroblast growth factors as angiogenesis factors: new insights into their mechanism of action. *Exs.* 79:159–192.

Kohn, E. C., Alessandro, R., Spoonster, J., Wersto, R., and Liotta, L. A. 1995. Angiogenesis: role of calcium-mediated signal transduction. *Proc. Natl. Acad. Sci. USA* 92:1307–1311.

Kohn, E. C., Felder, C. C., Jacobs, W., Holmes, K. A., Day, A. F., Freer, R., and Liotta, L. A. 1994c. Structure function analysis of signal and growth inhibition by carboxyamido-triazole, CAI. *Cancer Res.* 54:935–942.

Kohn, E. C., Figg, W. D., Sarosy, G. A., Bauer, K. S., Davis, P. A., Soltis, M. J., Thompkins, A., Liotta, L. A., and Reed, E. 1997. Phase I trial of micronized formulation CAI in patients with refractory solid tumors: pharmacokinetics, clinical outcome and comparison of formulations. *J. Clin. Oncol* 15:1985–1993.

Kohn, E. C., Jacobs, W., Kim, Y. S., Alessandro, R., Stetler-Stevenson, W. G., and Liotta, L. A. 1994b. Calcium influx modulates expression of matrix metalloproteinase (7 kDa type IV collagenase, gelatinase A) expression. *J. Biol. Chem.* 269:21505–21511.

Kohn, E. C., and Liotta, L. A. 1990. L65158, a novel antiproliferative and antimetastasis agent. *J. Natl. Cancer Inst.* 8:54–60.

Kohn, E. C., and Liotta, L. A. 1995. Molecular insights into cancer invasion: strategies for prevention and intervention. *Cancer Res.* 55:1856–1862.

Kohn, E. C., Reed, E., Sarosy, G., Christian, M., Link, C. J., Cole, K., Figg, W. D., Davis, P. A., Jacob, J., Goldspiel, B., and Liotta, L. A. 1996. Clinical investigation of a cytostatic calcium influx inhibitor in patients with refractory cancers. *Cancer Res.* 56:569–573.

Kohn, E. C., Sandeen, M. A., and Liotta, L. A. 1992. In vivo efficacy of a novel inhibitor of selected signal transduction pathways including calcium, arachidonate, and inositol phosphates. *Cancer Res.* 52:3208–3212.

Kohn, E. C., Sarosy, G., Bicher, A., Link, C., Christian, M., Ognibene, F. P., Cunnion, R., Steinberg, S., Adamo, D. O., and Davis, P. 1994a. Dose intense taxol: high response rate in patients with platinum-resistant recurrent ovairan cancer. *J. Natl. Cancer Inst.* 86:18–24.

Leenders, K. 1994. PET—blood flow and oxygen consumption in brain tumors. *J. Neurooncol.* 22:269–273.

Libutti, S., Choyke, P., Carrasquillo, J., Bacharach, S., and Neumann, R. 1999. Monitoring responses to antiangiogenic agents using noninvasive imaging tests. *Cancer J. Sci. Am.* 5:252–256.

Liotta, L. A., Kleinerman, J., and Saidel, G. 1974. Quantitative relationships of intravascular tumor cells: tumor vessels and pulmonary metastases following tumor implantation. *Cancer Res.* 34:997–1003.

Liotta, L. A., Saidel, M. G., and Kleinerman, J. 1976. The significance of hematogenous tumor cell clumps in the metastatic process. *Cancer Res.* 36:889–894.

Luzzi, K. J., Varghese, H. J., MacDonald, I. C., Kohn, E. C., Schmidt, E. E., Morris, V. L., Chambers, F., and Groom, A. C. 1998. Inhibition of angiogenesis in liver metastases by carboxyamidotriazole (CAI): computer assisted analysis. *Angiogenesis* 2:373–379.

Mankoff, D., Dunnwald, L., Gralow, J., Ellis, G., Charlop, A., and Livingston, R. 1999. Changes in blood flow and metabolism in locally advance breast cancer in response to presurgical chemotherapy. *J. Nucl. Med.* 40(abstract):137P.

McGuire, W. P., Hoskins, W. J., Brady, M. F., Kucera, P. R., Partridge, E. E., and Look, K. Y. 1996. Cyclophosphamide and cisplatin compared with paclitaxel and cisplatin in patients wtih stage III and stage IV ovarian cancer. *N. Engl. J. Med.* 334:1–6.

McLaren, J., Prentice, A., Charnock-Jones, D. S., and Smith, S. K. 1996a. Vascular endothelial growth factor (VEGF) concentrations are elevated in peritoneal fluid of women with endometriosis. *Hum. Reprod.* 11:220–223.

McLaren, J., Prentice, A., Charnock-Jones, D. S., Millican, S. A., Muller, K. H., Sharkey, A. M., and Smith, S. K. 1996b. Vascular endothelial growth factor is produced by peritoneal fluid macrophages in endometriosis and is regulated by ovarian steroids. *J. Clin. Invest.* 98:482–489.

Modlich, U., Kaup, F.-J., and Augustin, H. G. 1996. Cyclic angiogenesis and blood vessel regression in the ovary: blood vessel regression during luteolysis involves endothelial cell detachment and vessel occlusion. *Lab. Invest.* 74:771–780.

Neeman, M., Abramovitch, R., Schiffenbauer, Y. S., and Tempel, C. 1997. Regulation of angiogenesis by hypoxic stress: from solid tumors to the ovarian follicle. *Int. J. Exp. Pathol.* 78:57–70.

Neufeld, G., Cophen, T., Gengrinovitch, S., and Poltorak, Z. 1999. Vascular endothelial growth factor (VEGF) and its receptors. *FASEB J.* 13:9–22.

Nguyen, M., Watanabe H., and Budson, A. E. 1994. Elevated levels of an angiogenic peptide, basic fibroblast growth factor, in the urine of patients with a wide spectrum of cancers. *J. Natl. Cancer Inst.* 86:356–361.

Nguyen, M., Watanabe, H., Budson, A. E., Richie, J. P., and Folkman, J. 1993. Elevated levels of the angiogenic peptide basic fibroblast growth factor in urine of bladder cancer patients. *J. Natl. Cancer Inst.* 85:241–242.

Obermair, A., Wasicky, R., Kaider, A., Preyer, O., Losch, A., Leodolter, S., and Kolbl, H. 1999. Prognostic significance of tumor angiogenesis in epithelial ovarian cancer. *Cancer Lett.* 138:175–182.

Orre, M., Lofti-Miri, M., Mamers, P., and Rogers, P. A. W. 1998. Increased microvessel density in mucinous compared with malignant serous and benign tumors of the ovary. *Br. J. Cancer* 77:2204–2209.

Orre, M., and Rogers, P. A. 1999. VEGF, VEGFR-1, VEGFR-2, microvessel density and endothelial cell proliferation in tumours of the ovary. *Int. J. Cancer* 84:101–108.

Paley, P. J., Staskus, K. A., Gebhard, K., Mohanraj, D., Twiggs, L. B., Carson, L. F., and Ramakrishnan, S. 1997. Vascular endothelial growth factor expression in early stage ovarian carcinoma. *Cancer* 80:98–106.

Pearlman, J., Hibberd, M., Chuang, M., Harada, K., Lopez, J., Gladstone, S., Friedman, M., Sellke, F., and Simons, M. 1995. Magnetic resonance mapping demonstrates benefits of VEGF-induced myocardial angiogenesis. *Nat. Med.* 1:1085–1089.

Reynolds, K., Farzaneh, F., Collins, W. P., Campbell, S., Bourne, T. H., Lawton, F., Moghaddam, A., Harris, A. L., and Bicknell, R. 1994. Association of ovarian malignancy with expression of platelet-derived endothelial cell growth factor. *J. Natl. Cancer Inst.* 86:1234–1238.

Rosen, E. M., and Goldberg, I. D. 1995a. Regulation of scatter factor (hepatocyte growth factor) production by tumor-stroma interaction. *Exs.* 74:17–31.

Rosen, E. M., and Goldberg, I. D. 1995b. Scatter factor and angiogenesis. *Adv. Cancer Res.* 67:257–279.

Rowinsky, E. K., Humphrey, R., Hammond, A., Aylesworth, C., Smetzer, L., Hidalgo, M., Morrow, M., Smith, L., Garner, A., Sorensen, J. M., Von Hoff, D. D., and Eckhardt, S. G. 2000. Phase I and pharmacologic study of the specific matrix metalloproteinase inhibitor BAY 1-9566 on a protracted oral daily dosing schedule in patients with solid malignancies. *J. Clin. Oncol* 18:178–186.

Sagar, S., Klassen, G., Barclay, K., and Aldrich, J. 1993. Tumour blood flow: measurement and manipulation for therapeutic gain. *Cancer Treat. Rev.* 19:299–349.

Schelbert, H. 1994. Blood flow and metabolism by PET. *Cardiol. Clin.* 12:303–315.

Schiffenbauer, Y. S., Abramovitch, R., Meir, G., Nevo, N., Holzinger, M., Itin, A., Keshet, E., and Neeman, M. 1997. Loss of ovarian function promotes angiogenesis in human ovarian carcinoma. *Proc. Natl. Acad. Sci. USA* 94:13203–13208.

Schoell, W. M., Pieber, D., Reich, O., Lahousen, M., Janicek, M., Guecer F., and Winter, R. 1977. Tumor angiogenesis as a prognostic factor in ovarian carcinoma: quantification of endothelial immunoreactivity by image analysis. *Cancer* 80:2257–2262.

Senger, D. R., Galli, S. J., Dvorak, A. M., Pernuzzi, C. A., Harvey, V. S., and Dvorak, H. F. 1983. Tumor cells secrete a vascular permeability factor that promotes accumulation of ascites fluid. *Science* 219:983–985.

Sills, A. K., Williams, J. I., Tyler, B. M., Epstein, D. S., Sipos, E. P., Davis, J. D., McLane, M. P., Pitchford, S., Cheshire, K., Gannon, F. H., Kinney, W. A., Chao, T. L., Donowitz, M., Laterra, J., Zasloff, M., and Brem, H. 1998. Squalamine inhibits angiogenesis and solid tumor growth in vivo and perturbs embryonic vasculature. *Cancer Res.* 58:2784–2792.

Singh, R. K., Gutman, M., Bucana, C. D., Sanchez, R., Llansa M., and Fidler, I. J. 1995. Interferons α and β down-regulate the expression of basic fibroblast growth factor in human carcinomas. *Proc. Natl. Acad. Sci. USA*. 92:4562–4566.

Smith, S. K. 1997. Angiogenesis. *Semin. Reprod. Endocrinol.* 158:221–227.

Taraboletti, G., Garofalo, A., Belotti, D., Drudis, T., Borsotti, P., Scanziani, E., Brown, P. D., and Giavazzi, R. 1995. Inhibition of angiogenesis and murine hemangioma growth by batimastat, a synthetic inhibitor of matrix metalloproteinases. *J. Natl. Cancer Inst.* 87:293–298.

Tempfer, C., Obermair, A., Hefler, L., Haeusler, G., Gitsch, G., and Kainz, C. 1998. Vascular endothelial growth factor serum concentrations in ovarian cancer. *Obstet. Gynecol.* 92:360–363.

van Dijke, C., Brasch, R., Roberts, T., Weidner, N., Mathur, A., Shames, D., Mann, J., Demsar, F., Lang, P., and Schwickert, H. 1996. Mammary carcinoma model: correlation of macromolecular contrast-enhanced MRI imaging characterizations of tumor microvasculature and histologic capillary density. *Radiology* 198:813–818.

Vazquez, F., Rodriguez-Manzaneque, J. C., Lydon, J. P., Edwards, D. P., O'Malley, B. W., and Iruela-Arispe, M. L. 1999. Progesterone regulates proliferation of endothelial cells. *J. Biol. Chem.* 274:2185–2192.

Vu, T. H., Shipley, J. M., Bergers, G., Berger, J. E., Helms, J. A., Hanahan, D., Shapiro, S. D., Senior, R. M., and Werb, Z. 1998. MMP-9/gelatinase B is a key regulator of growth plate angiogenesis and apoptosis of hypertrophic chondrocytes. *Cell* 93:411–422.

Wang, X., Fu, X., Brown, P. D., Crimmin, M. J., and Hoffman, R. M. 1994. Matrix metalloproteinase inhibitor BB-94 (Batimastat) inhibits human colon tumor growth and spread in a patient-like orthotopic model in nude mice. *Cancer Res.* 54:4726–4728.

Watson, S. A., Morris, T. M., Parsons, S. L., Steele, R. J., and Brown, P. D. 1996. Therapeutic effect of the matrix metalloproteinase inhibitor, batimastat, in a human colorectal cancer ascites model. *Br. J. Cancer* 74:1354–1358.

Weidner, N., and Folkman, J. 1996. Tumor vascularity as a prognostic factor in cancer. In: DeVita, V. T., Hellman, S., and Rosenberg, S. A., eds. *Important advances in Oncology* 1996. Lippincott-Raven, Philadelphia, pp. 167–190.

Weidner, N., Semple, J. P., Welch, W. R., and Folkman, J. 1991. Tumor angiogenesis and metastasis—correlation in invasive breast carcinoma. *N. Engl. J. Med.* 324:1–8.

Zebrowski, B. K., Liu, W., Ramirez, K., Akagi, Y., Mills, G. B., and Ellis, L. M. 1999. Markedly elevated levels of vascular endothelial growth factor in malignant ascites. *Ann. Surg. Oncol.* 6:373–378.

PART III

Vascular Morphogenesis in the Uterus

CHAPTER 11

The Vascular System in the Endometrium: Introduction and Overview

Peter A.W. Rogers and Caroline E. Gargett

The chapters in this section describe endometrial vascular structure and function, with particular emphasis on normal physiologic events such as growth, implantation, and menstruation, as well as pathologic situations such as endometriosis, menorrhagia, breakthrough bleeding, adenomyosis, and fibroids. The role that estrogen and progesterone play in regulating the endometrial vasculature during these processes is discussed, as are various other factors that regulate vascular function at a local level. The recent sudden increase in understanding of how blood vessel growth and regression is regulated, coupled with the major involvement of uterine vasculature in almost all clinically relevant gynecologic disorders, has opened exciting new possibilities for medical treatment. The challenge for scientists and clinicians working in this field will be to harness the promise that this knowledge brings to develop effective new therapies. Possible novel treatments that could be considered include use of angiogenesis inhibitors to prevent endometriosis, and vessel stabilization and maturation factors to reduce vessel fragility and breakthrough bleeding. The latest information in this rapidly developing field is detailed in the chapters in this section, with contributions from three of the leading groups currently working on different aspects of the regulation of endometrial vascular structure and function.

The vascular system is a complex organ that plays a central role in a number of functions critical to the survival of the organism. These include oxygen transport, nutrition, excretion, fluid balance, homeostasis, and immune defense. In most adult vascular beds, stability and maintenance of homeostasis are primary functions of the vasculature. In these organs, the vasculature changes very little during adult life. In contrast to the vascular system, the mammalian reproductive tract is a relatively recent evolutionary addition. Reproductive function in mammals is primarily orchestrated by the sex steroids estrogen and progesterone. This is particularly the case at the level of the uterus, where estrogen and progesterone directly or indirectly regulate most biologic functions, and in particular growth and differentiation. As a consequence of this, the endometrial vascular bed is constantly responding to the cyclical fluctuations in sex steroids, resulting in repeated alterations in angiogenesis, regression, hemostasis, blood flow, oxygenation, nutrient supply, and tissue fluid balance.

ENDOMETRIAL VASCULAR STRUCTURE

Human endometrial vascular structure has been described in a number of studies (Markee, 1940; Ramsey, 1982; Ferenczy, 1987). The arcuate arteries within the myometrium give rise to the radial arteries that supply the endometrium. After passing through the myometrial-endometrial junction, the radial arteries split to form the smaller basal arteries that supply the basalis (the basal layer of the endometrium) and the spiral arterioles, which continue up toward the endometrial surface. The spiral arterioles that supply the functionalis (upper layer of the endometrium) have a distinctive coiled appearance that becomes more pronounced during the secretory phase of the menstrual cycle. The spiral arterioles are unusual among resistance arterioles in that they lack significant elastin within the internal lamina (Ferenczy, 1987). Each spiral arteriole provides blood to a uterine luminal surface area estimated at 4 to 9 mm^2 (Bartelmez, 1933), with little or no overlap between these areas. Capillaries branch from the spiral arterioles at all levels supplying the stroma and periglandular networks. Just below the endometrial surface the spiral arterioles break up into the prominent subepithelial capillary plexus. This plexus drains into a number of venules that pass down through the endometrium collecting blood from other capillaries on the way. The numerous venules join to form veins that pass out of the endometrium into the myometrium.

The endometrial microvasculature is formed of capillaries, arterioles, and venules that are indistinguishable at the ultrastructural level from those of most other tissues in the body (Roberts et al, 1992). All endometrial blood vessels are lined by a continuous single layer of endothelial cells arranged on the vascular basement membrane. In mature capillaries the endothelial cells may be surrounded by pericytes, which extend numerous processes that pass through the basement membrane to make direct contact with a number of endothelial cells. Pericytes are believed to play a major role in vessel stability. Still larger vessels are surrounded by one or more layers of smooth muscle cells and these vessels have contractile properties that enable them to regulate blood flow within the endometrium. Microvascular heterogeneity is an important concept that is recognized both between and within organs throughout the body (Auerbach et al, 1985; McCarthy et al, 1991; Garlanda and Dejana, 1997). Heterogeneity between adjacent vessels can be seen at a structural and functional level in most microvascular beds, and is apparent by altered levels of expression of different molecules in different vessels. Examples of this in endometrium include variable endothelial expression of intercellular adhesion molecule-1 (ICAM-1) (Tawia et al, 1993) and basement membrane components such as heparan sulfate proteoglycan (HSPG) (Kelly et al, 1995).

CYCLICAL CHANGES IN UTERINE VASCULAR STRUCTURE AND FUNCTION

Routine histologic dating of endometrial biopsies places little emphasis on the appearance of the capillaries (Noyes et al, 1950). This may be because in hematoxylin and eosin–stained light microscope sections, most of the smaller blood vessels collapse following biopsy and hysterectomy and are thus not visible. It is also important to note, however, that the majority of the endometrial blood vessels show minimal structural changes during the menstrual cycle. The exception to this are the spiral arterioles during the secretory phase, and some of the more overt changes as the vasculature breaks down during the menstrual process. Of microvascular parameters that have been studied, both microvascular density and endothelial cell proliferation indices remain relatively constant throughout the menstrual cycle (Rogers et al, 1993; Goodger and Rogers, 1994). These observations seem surprising given the high level of vascular repair and growth that occurs following menstruation and during the proliferative phase of the endometrial cycle, compared to the absence of growth and repair during the secretory phase of the cycle. However, they do support the concept that endometrial vascular growth is tightly regulated and closely matches tissue growth and metabolic requirements.

Proliferative-Phase Growth and Angiogenesis

The absence of a clear cyclical pattern of endothelial cell proliferation raises key questions about when and how angiogenesis occurs during the menstrual cycle in human endometrium. It would be expected that some significant angiogenesis occurs during postmenstrual repair of the vascular bed (Markee, 1940) and also in conjunction with the rapid endometrial growth that occurs under the influence of estrogen during the proliferative phase of the menstrual cycle (Ferenczy et al, 1979). However, the endothelial cell proliferation data do not support this. Endometrial angiogenesis may occur by a number of different mechanisms (Rogers and Gargett, 1999). These include the conventional sprouting method where endothelial sprouts grow out from existing capillaries and connect with other sprouts to form new vessels, as well as processes such as elongation (Ausprunk et al, 1974), intussusception (Burri and Tadek, 1990), and circulating endothelial progenitor cells (Asahara et al, 1999). There is no definitive evidence for which angiogenic mechanisms occur in human endometrium, although it seems unlikely sprouting plays a major role. Two pieces of evidence support this statement. The first is that proliferating endothelial cells always appear within vessel profiles in human endometrium, and are never associated with structures that could be interpreted as vascular sprouts (for review see Rogers and Gargett, 1999). The second piece of evidence concerns the distribution of the integrin $\alpha_v\beta_3$, which in human endometrium is found on endothelial cells located inside existing vessels and not on sprouts. In contrast, sprouts were observed in angiogenic blood vessels in ovarian tumour tissue, suggesting that although this integrin is associated with sprouting angiogenesis in some tissues, that this is not the case in endometrium (Hii and Rogers, 1998).

In most of the studies to date on endometrial vascular structure and function, no distinction has been made between the different regions of the endometrium. The failure to recognize different endometrial regions is of concern since there is both well-established and emerging evidence of major differences in angiogenesis between the functionalis and the basalis. Ferenczy et al (1979) reported elevated levels of endothelial cell proliferation in the upper third of the functionalis compared to the basalis during the cycle, although statistical analysis of these data was not undertaken. In a subsequent study it has been shown that endothelial cell proliferation levels are significantly higher throughout the cycle in the functionalis compared to the basalis (Rogers et al, 1998). It is also possible that the spiral arterioles and the subepithelial capillary plexus that they supply represent another specialized vascular compartment within the overall endometrial vasculature. The spiral arterioles show major changes in growth and coiling during the secretory phase of the cycle and are known to play a significant role in both implantation and menstruation. These specialized roles may set the spiral arterioles apart, in a functional sense, from the rest of the endometrial vasculature, and should be considered when undertaking structural and functional studies. It may therefore be necessary to consider three functional zones within the human endometrium: the subepithelial capillary plexus and spiral arterioles, the remaining vessels of the functionalis, and the vessels of the basalis. To date almost no studies have considered these functional zones, and many studies have used just functionalis tissue obtained from endometrial biopsy procedures.

Implantation and Placentation

Edema and focal hemorrhage are some of the earliest signs of embryo implantation seen in the mammalian endometrium, and this includes the human (Hertig et al, 1956). Little is known of early trophoblast invasion of the maternal vasculature in the human; however, studies have been made of other primate species such as the macaque, rhesus monkey, and baboon (Enders, 1989; Enders and King, 1991). Based on this work it appears that the trophoblast migrates rapidly into the endometrium, invading small arterioles but not veins directly beneath the implantation site. There is evidence of vascular occlusion, suggesting that trophoblastic lacunae receive their vascular supply by relatively indirect means. During the early stages of invasion, the trophoblast migrates between and along endothelial cells without causing their demise. It seems likely that a sophisticated signaling mechanism exists between the invading trophoblast and the maternal vasculature. For successful implantation to occur, immune and inflammatory responses must be muted and eventually overcome. For successful placentation, the embryonic trophoblast must substantially replace the endothelial lining of the maternal vasculature, and play a major role in remodeling the walls of the spiral arterioles. Inadequate development of this early placental vasculature may lead to major problems with subsequent placental function, including the development of preeclampsia and fetal intrauterine growth retardation.

Secretory-Phase Differentiation and Arteriole Growth

During the secretory phase of the cycle, under the influence of luteal progesterone, the functionalis layer of the endometrium stops proliferating and growing and undergoes differentiation. During the latter half of the secretory phase, after the time when implantation normally occurs, decidual cells first start to develop around the spiral arterioles. Throughout the secretory phase, the spiral arterioles continue to grow, developing a highly tortuous and coiled shape. This is reflected histologically where numerous profiles of the same vessel will appear in a single section due to the heavily coiled nature of the vessel. In a study of vascular smooth muscle cell proliferation in human endometrium across the menstrual cycle, it has been shown that total vascular smooth muscle cell proliferation indices increase from around 2% to between 3.5% and 4% in the midsecretory and late-secretory stages of the cycle. A similar increase occurs when spiral arteriole vascular smooth muscle cells are considered separately, with early secretory proliferation indices of

0.7% rising to approximately 1.6% in the midsecretory and late-secretory stages of the cycle (Abberton et al, 1999). These data suggest that arteriogenesis or growth of arterioles, is under the control of progesterone in the human endometrium. Whether this is a direct or indirect mechanism is unknown; however, it has been reported that endometrial vascular smooth muscle cells contain estrogen and progesterone receptors (Rogers et al, 1996). In this study immunohistochemical techniques demonstrated that not all vascular smooth muscle cells contained estrogen and/or progesterone receptors, thus identifying another parameter where heterogeneity exists between different microvessels.

Menstruation

Most of our current understanding of the vascular events during menstruation is based on the observations of Markee (1940) using intraocular endometrial transplants in the rhesus monkey. In summary, this work shows that 1 to 4 days prior to the onset of bleeding, venous and arterial stasis occurs, sometimes accompanied by vasodilatation; 4 to 24 hours prior to bleeding, vasoconstriction is also seen. About 4 days prior to menstruation, leukocyte infiltration of the endometrium occurs. Menstrual bleeding usually commences from the wall of an arteriole or capillary once a previously constricted spiral arteriole relaxes and blood flow recommences. Approximately 70% of blood is lost in this way. Some blood cells also leave the capillary circulation by diapedesis, as well as by reflux from veins through previously formed breaks in the vasculature. Blood loss from a break in the endometrial vasculature during menstruation normally only lasts for 1 or 2 minutes before ceasing due to spiral arteriole vasoconstriction.

The spontaneous rupture and breakdown of blood vessels during menstruation is a highly unusual vascular event. While the process of menstruation is undoubtedly regulated by a series of coordinated events in a number of different cell types, a number of fundamental rules of vascular function appear to be broken during this unusual event. Despite this, there are a number of strands of evidence that suggest that the endometrial vasculature retains full if somewhat altered function during the menstrual process: ICAM-1 is upregulated on many but not all endometrial endothelial cells prior to menstruation (Tawia et al, 1993); endothelial cell proliferation indices remain high (Goodger and Rogers, 1994); at the ultrastructural level many of the endothelial cells appear relatively normal while surrounding stromal and gland cells degenerate (Roberts et al, 1992); and finally, endothelial cell von Willebrand factor levels are significantly reduced (Au and Rogers, 1993). All of these observations lend support to the hypothesis that the endometrial vasculature, and in particular the endometrial endothelial cells, are specifically adapted for the menstrual process, and help to control this unusual vascular event.

Endometrial Blood Flow and Edema

Studies of endometrial blood flow in the human have shown a rise during the proliferative phase of the cycle with a peak just prior to ovulation (reviewed by Fraser and Peek, 1992). There is a significant positive correlation between unopposed circulating estradiol levels and endometrial blood flow. When progesterone levels rise, this correlation is lost. The mechanism by which estrogen regulates endometrial blood flow is unclear, although since at least some endometrial vascular smooth muscle cells contain estrogen receptors (Rogers et al, 1996), some direct effects of estrogen on blood flow are possible. Other vasoactive compounds that may be under the indirect control of sex steroids in the endometrium include prostaglandins, endothelium-derived relaxing factor or nitric oxide, and endothelin.

Recently, the use of colored Doppler processing in conjunction with a transvaginal ultrasound probe has provided a new mode for investigating uterine blood flow. Using these techniques it has been reported that increased uterine artery pulsatility at the time of embryo transfer during in vitro fertilization may be associated with a reduced implantation rate (Coulam et al, 1994; Steer et al, 1995). The precise vascular events that are responsible for these changes in pulsatility index are unclear; however, it does appear that a direct relationship may exist between uterine blood flow and the success or otherwise of embryo implantation.

It is generally believed that a cessation of endometrial blood flow, or ischemia, just prior to menstruation plays a key role in causing tissue degeneration and vascular rupture. This belief is based on the observation by Markee (1940) that spiral arteriole vasoconstriction occurred in endometrial transplants in the anterior chamber of the eye in rhesus monkeys between 4 and 24 hours before menstrual bleeding commenced. However, similar evidence for a major vasoconstrictive episode prior to or during menstruation does not exist in humans. A review of several studies investigating endometrial blood flow failed to find any evidence for such an ischemic event (Fraser and Peek, 1992). More recently a major study on endometrial perfusion in the human using laser Doppler fluxmetry failed to detect any episodes of transient ischemia in the endometrium during 67 records taken in conscious volunteers, including 13 records obtained during menstruation (Gannon et al, 1997). The same study reported significantly elevated blood flow in the early proliferative and early secretory stages of the cycle.

The blood vessels also play a major role in regulating fluid exchange and balance between the tissues and the blood compartment. There are two occasions during the normal menstrual cycle when stromal edema becomes evident: the midproliferative and the midsecretory phases (Noyes et al, 1950). Tissue edema occurs when there is a shift in net fluid flux from the vasculature to the tissues. Fluid flux is governed by relatively simple physical laws, and an increase will occur if blood pressure rises, if the permeability of the vessel wall increases, or if the osmotic pressure of the tissues rise. Although the times of endometrial edema have been linked to changes of levels of sex steroids, and in particular estrogen, there is little understanding of the mechanisms controlling these events or their physiologic purpose.

Vascular Basement Membrane

Ultrastructural studies show that the endometrial microvascular basement membrane is loosely formed and discontinuous during the early proliferative phase (Roberts et al, 1992). By the late proliferative phase it becomes more distinct, and in the secretory phase develops whorled extensions that include pericytes. Basement membrane degeneration occurs in the late secretory and menstrual phases. By immunohistochemistry the endometrial vascular basement membrane contains collagen IV, laminin, and HSPG (Kelly et al, 1995). Collagen IV and laminin were found in all endometrial vessel basement membanes, while HSPG was seen in only about 55% of vessel walls. During the menstrual phase HSPG, but not collagen IV or laminin, immunostaining decreased significantly in vascular but not glandular basement membranes, suggesting a role for HSPG in vascular remodeling at this time.

FACTORS THAT REGULATE UTERINE VASCULAR GROWTH AND REGRESSION

Sex Steroid Regulation of Endometrial Blood Vessels

It is well accepted that overall control of endometrial growth and regression is primarily regulated by circulating levels of estrogen and progesterone. However, the roles that these steroids play in specifically regulating endometrial vascular growth and regression seem less clear. For example, the highly varied endothelial cell proliferation indices that occur at all stages of the endometrial cycle indicate that estrogen and/or progesterone do not directly regulate this process (Goodger and Rogers, 1994). This statement is supported by all but two of the many immunohistochemical studies of estrogen and progesterone receptor expression in the endometrial endothelial cells. In all cases there has been a failure to demonstrate receptor expression, other than one study that found progesterone receptors in endothelial cells of decidual tissue taken from the fifth to ninth weeks of pregnancy (Wang et al, 1992) and a more recent study suggesting that endometrial endothelial cells do express functional estrogen and progesterone receptors (Iruela-Arispe et al, 1999). Immunohistochemically detectable levels of estrogen and progesterone receptor have been reported in endometrial vascular smooth muscle cells, although interestingly the distribution was heterogeneous, with some cells showing high levels of expression and others appearing negative (Rogers et al, 1996). Estrogen has been shown to have numerous genomic and nongenomic effects on the vasculature in a wide range of studies (Clark, 1997; Nasr and Breckwoldt, 1998). The activity of sex steroid hormones is mediated through two receptor subtypes for both the estrogen and progesterone receptors (ERα and ERβ, PR A and PR B), and there is emerging evidence suggesting that transcriptional activities of these receptors may be regulated differently in different cell types (Graham and Clarke, 1997; Kuiper and Gustafsson, 1997; Keightley, 1998). It is not currently known which receptor subtypes are present in vascular cells. To date over 20 genes with potential vascular actions have been identified that contain consensus repeats for the estrogen receptor, thus identifying them as targets for regulation by the estrogen–estrogen receptor bound complex. Estrogen may also exert effects on gene transcription through other response elements such as AP (activator protein)-1 sites in the promoters of target genes (Paech et al, 1997). These studies have shown that estrogen, and to a lesser extent progesterone, can act on the vasculature both through nongenomic mechanisms and also via receptors in vascular cells or by paracrine action via perivascular cells.

In addition to direct effects of estrogen and progesterone on the vasculature, the endometrium and its resident population of macrophages are capable of producing most, if not all, of the major growth factors and cytokines currently known to play a role in controlling angiogenesis (Giudice, 1994; Sunderkötter et al, 1994; Goodger and Rogers, 1995; Smith, 1995).

Paracrine Factors in Endometrium

Most work to date on endometrial angiogenic factors has focused on vascular endothelial growth factor (VEGF). Attempts have been made to relate VEGF expression in the human endometrium to stages of the menstrual cycle, and while a generalized picture is emerging, there is still some disagreement (Torry and Torry, 1997; Rogers and Gargett, 1999). From messenger RNA (mRNA) studies (in situ hybridization and Northern blots), it appears that VEGF expression in whole endometrium is low during the proliferative phase, increases during the late secretory phase, and reaches a maximum at menses (Charnock-Jones et al, 1993; Shifren et al, 1996; Graubert et al, 1997). However, there is no clear pattern of VEGF protein expression across the menstrual cycle, although expression is greater in the glands than the stroma (Li et al, 1994; Shifren et al, 1996). Most studies of VEGF expression in the human endometrium have been descriptive and only two have had large enough sample numbers for meaningful statistical analysis (Graubert et al, 1997; Lau et al, 1999). More recently an attempt has been made to correlate VEGF production with endothelial cell proliferation in human endometrium throughout the menstrual cycle (Gargett et al, 1999). This study failed to find any correlation between either total glandular or total stromal VEGF production and endothelial cell proliferation. The authors put forward a number of possible explanations for these data including that the majority of human endometrial VEGF has a nonangiogenic role, that focal VEGF associated with individual blood vessels may regulate vessel growth, or that VEGF is not the primary regulator of endothelial cell proliferation in human endometrium. It has recently been shown that the bulk of VEGF produced by glandular cells is secreted apically (Hornung et al, 1998), thus supporting the hypothesis that not all endometrial VEGF is involved in regulating angiogenesis.

Endometrial expression of other members of the VEGF family and their receptors has received little attention to date. VEGF-B, VEGF-C and placenta growth factor (PlGF) are constitutively expressed at low levels throughout the menstrual cycle and do not show variability at different stages. The VEGF receptors, VEGFR-1 and VEGFR-2, are present at low levels throughout the cycle but both show a two- to ninefold increase in mRNA levels in the menstrual phase (Graubert et al, 1997). In contrast, an immunohistochemical study of VEGFR expression has shown that the fraction of VEGFR-2-expressing capillaries was maximal during the proliferative phase, with a second smaller rise in the midsecretory phase, suggesting that significant angiogenesis occurs during the growth phase of the endometrium. Fewer VEGFR-1-expressing capillaries were observed and these showed little variation across the cycle (Meduri et al, 2000).

In human endometrium basic and acidic fibroblast growth factor (bFGF and aFGF) immunoreactivity has been detected in epithelial cells with some pale heterogeneous staining of stromal cells. No alteration in staining across the menstrual cycle was reported (Ferriani et al, 1993). In contrast to this finding, enzyme-linked immunosorbent assay (ELISA) of endometrial extracts and reverse-transcriptase polymerase chain reaction (RT-PCR) have been used to show that endometrial bFGF reaches a peak in the mid- to late proliferative stages of the cycle before reducing in the secretory phase (Fujimoto et al, 1996). In a more detailed recent study, mRNA for bFGF and FGF receptor-1 (FGFR-1) were demonstrated in human endometrium throughout the cycle by RNAse protection assay with a peak in the mid- and late proliferative phases (Sangha et al, 1997). By in situ hybridization bFGF and FGFR-1 gave a strong signal in gland cells throughout the menstrual cycle, but showed reduced expression in stromal cells during the secretory phase. The same pattern of results were obtained in gland and stromal cells using immunohistochemistry for FGFR-1. Interestingly, FGFR-1 was shown to be reduced in the endometrium of women suffering from menorrhagia.

There is high probability that the receptor tyrosine kinases Tie-1 and Tie-2 and that ligands for Tie-2, angiopoietin-1, and angiopoietin-2 may play a major role in regulating growth and regression of the endometrial vasculature (Sato et al, 1995; Hanahan, 1997; Maisonpierre et al, 1997). The Tie-angiopoietin system plays an important role in the maturation of new vessels and the recruitment by endothelial cells of pericytes and vascular smooth muscle cell precursors to the developing vessel. It has been hypothesized that VEGF, angiopoietin-1, and angiopoietin-2 work together to control angiogenesis. In a stable mature vessel, angiopoietin-1 and VEGF alone may not be sufficient to induce angiogenesis. However, angiopoietin-2 may disrupt the cells supporting the blood vessel sufficiently to allow endothelial cell migration and proliferation. Alternatively, in the absence of sufficient amounts of VEGF, it is possible that elevated levels of angiopoietin-2 may cause vascular regression (Hanahan, 1997). This hypothesis predicts a key role for angiopoietin-2 during vascular remodeling. In support of this hypothesis, elevated levels of angiopoietin-2 mRNA are found in ovary, uterus, and placenta but not other adult organs. These tissues undergo constant vascular remodeling with periods of angiogenesis and vascular regression throughout the reproductive cycle (Maisonpierre et al, 1997).

A number of other factors found in endometrium have either direct or indirect angiogenic activity under certain circumstances, including endothelial cell stimulating angiogenic factor, transforming growth factor-α and -β, tumor necrosis factor-α, interleukin-8, and epidermal growth factor (for review see Goodger and Rogers, 1995). Current understanding of the specific roles that these and other factors play in vascular morphogenesis in the endometrium is presently limited.

It is generally thought that angiogenesis is dependent on the balance of angiogenesis promoters and inhibitors. Little information has been published to date on endometrial angiogenesis inhibitors despite the human endometrium being an active site of angiogenesis. Thrombospondin-1 (TSP-1) has significant antiangiogenic effects on endothelial cells including inhibiting proliferation, disrupting focal adhesions, diminishing cell spreading, and inhibiting angiogenesis (Iruela-Arispe et al, 1996). In human endometrium TSP-1 is elevated in the secretory phase of the cycle, and in vitro studies have demonstrated regulation by progesterone.

Several polypeptide fragments that are inhibitory for angiogenesis have now been identified and many of the precursor proteins for these are known to exist in endometrium. Decidualized stromal cells, which appear around the spiral arterioles in the last third of the secretory phase of the nonpregnant menstrual cycle produce prolactin (Narukawa et al, 1994), which may act as a source for the antiangiogenic 16-kd prolactin fragment (Clapp et al, 1993). The endometrium is also a source of plasminogen and plasmin through the actions of tissue-plasminogen activator and urokinase-plasminogen activator (Koh et al, 1992). The antiangiogenic agent angiostatin is a fragment of the plasminogen molecule (O'Reilly et al, 1994). In addition, collagen XVIII is localized to basement membranes of blood vessels, and its proteolysis can result in the production of the angiogenesis inhibitor endostatin (O'Reilly et al, 1997). Canstatin is a 24-kd fragment of the α_2 chain of type IV collagen, which is also localized to basement membranes of blood vessels. Canstatin inhibits endothelial cell proliferation, migration, and tube formation, possibly through an apoptosis-mediated pathway (Kamphaus et al, 2000). Several members of the clotting factor pathway contain antiangiogenic fragments, including the serpin antithrombin (O'Reilly et al, 1999).

Matrix metalloproteinases (MMPs) and their inhibitors, tissue inhibitors of metalloproteinases (TIMPs), play a major role in the breakdown and remodeling of tissues and are found in human endometrium (Salamonsen, 1996). MMP-2, -3, and -9 are expressed in endometrial capillaries and/or arterioles in a cycle-dependent manner, with MMP-2 and -9 in secretory phase capillaries and spiral arterioles, respectively, and all three during menstruation at foci of active tissue breakdown (Freitas et al, 1999). TIMP-1 and -2 are found in endometrial vessels throughout the cycle, with heterogeneous staining at menstruation. The authors of this immunohistochemical study concluded that endometrial MMPs and TIMPs are implicated in vascular remodeling during angiogenesis and menstruation. In addition, MMPs may also play a role in the production of antiangiogenic substances that are fragments of other molecules through their proteolytic activity.

Circulating Leukocytes

In vessels that undergo nonsprouting angiogenesis, where endothelial cells proliferate and migrate within the vessel, it is possible that the source of angiogenic factors for intussusception or elongation is derived from cells from within the vasculature, rather than the tissues. In support of such an hypothesis, it has been demonstrated that some circulating leukocytes, such as neutrophils, constitutively express VEGF in specific granules that are released when these cells become activated (Gaudry et al, 1997; Taichman et al, 1997; Webb et al, 1998). Other leukocytes, such as monocytes and T lymphocytes produce VEGF when activated (Freeman et al, 1995; Harmey et al, 1998). There is evidence that in angiogenesis associated with inflammatory conditions and in atherosclerosis, extravasated leukocytes are a source of VEGF (Edelman et al, 1999; Ramos et al, 1998). It is therefore possible that leukocytes have a role in endometrial angiogenesis by directly delivering VEGF to endothelial cells in vessels undergoing angiogenesis by nonsprouting mechanisms.

REFERENCES

Abberton, K. M., Taylor, N. H., Healy, D. L., and Rogers, P. A. W. 1999. Vascular smooth muscle cell proliferation in arterioles of the human endometrium. *Hum. Reprod.* 14:1072–1079.

Asahara, T., Masuda, H., Takahashi, T., Kalka, C., Pastore, C., Silver, M., Kearne, M., Magner, M., and Isner, J. M. 1999. Bone marrow origin of endothelial progenitor cells responsible for postnatal vasculogenesis in physiological and pathological neovascularization. *Circ. Res.* 85:221–228.

Au, C. L., and Rogers, P. A. W. 1993. Immunohistochemical staining of von Willebrand factor in human endometrium during normal menstrual cycle. *Hum. Reprod.* 8:17–23.

Auerbach, R., Alby, L., and Morrissey, L. W., Tu, M., Joseph, J. 1985. Expression of organ-specific antigens on capillary endothelial cells. *Microvasc. Res.* 29:401–411.

Ausprunk, D. H., Knighton, D. R., and Folkman, J. 1974. Differentiation of vascular endothelium in the chick chorioallantois: a structural and autoradiographic study. *Dev. Biol.* 38:237–248.

Bartelmez, G. W. 1933. Histologic studies on the menstruating mucous membrane during menstruation. *Contrib. Embryo.* 142:142–157.

Burri, P. H., and Tadek, M. R. 1990. A novel mechanism of capillary growth in the rat pulmonary microcirculation. *Anat. Rec.* 228:35–45.

Charnock-Jones, D. S., Sharkey, A. M., Rajput-Williams, J., Burch, D., Schofield, J. P., and Fountain, S. A. 1993. Identification and localization of alternately spliced mRNAs for vascular endothelial growth factor in human uterus and estrogen regulation in endometrial carcinoma cell lines. *Biol. Reprod.* 48:1120–1128.

Clapp, C., Martial, J. A., Guzman, R. C., Rentier-Delure, F., and Weiner, R. I. 1993. The 16-kilodalton N-terminal fragment of human prolactin is a potent inhibitor of angiogenesis. *Endocrinology* 133:1292–1299.

Clark, K. E. 1997. Modulation of vascular responsiveness by steroids. *Curr. Opin. Endocrinol. Diabetes* 4:392–400.

Coulam, C. B., Bustillo, M., Soenksen, D. M., and Britten, S. 1994. Ultrasonographic predictors of implantation after assisted reproduction. *Fertil. Steril.* 62:1004–1010.

Edelman, J. L., Castro, M. R., and Wen, Y. 1999. Correlation of VEGF expression by leukocytes with the growth and regression of blood vessels in the rat cornea. *Invest. Ophthalmol. Vis. Sci.* 40:1112–1123.

Enders, A. C. 1989. Trophoblast differentiation during the transition from trophoblastic plate to lacunar stage of implantation in the rhesus monkey and human. *Am. J. Anat.* 186:85–98.

Enders, A. C., and King, B. F. 1991. Early stages of trophoblastic invasion of the maternal vascular system during implantation in the macaque and baboon. *Am. J. Anat.* 192:329–346.

Ferenczy, A. 1987. Anatomy and histology of the uterine corpus. In: Durman, T. G., ed. *Blaustein's Pathology of the Genital Tract,* 3rd ed. Springer, New York, pp. 257–291.

Ferenczy, A., Bertrand, G., and Gelfand, M. M. 1979. Proliferation kinetics of human endometrium during the normal menstrual cycle. *Am. J. Obstet. Gynecol.* 133:859–867.

Ferriani, R. A., Charnock-Jones, D. S., Prentice, A., Thomas, E. J., and Smith, S. K. 1993. Immunohistochemical localization of acidic and basic fibroblast growth factors in normal human endometrium and endometriosis and the detection of their mRNA by polymerase chain reaction. *Hum. Reprod.* 8:11–16.

Fraser, I. S., and Peek, M. J. 1992. Effects of exogenous hormones on endometrial capillaries. In: Alexander N. J., and d'Arcangues, C., eds. *Steroid Hormones and Uterine Bleeding.* AAAS Press, Washington, pp. 67–79.

Freeman, M. R., Schneck, F. X., Gagnon, M. L., Corless, C., Soker, S., Niknejad, K., Peoples, G. E., and Klagsbrun, M. 1995. Peripheral blood T lymphocytes and lymphocytes infiltrating human cancers express vascular endothelial growth factor: a potential role for T cells in angiogenesis. *Cancer Res.* 55:4140–4145.

Freitas, S., Meduri, G., Le Nestour, E., Bausero, P., and Perrot-Applanat, M. 1999. Expression of metalloproteinases and their inhibitors in blood vessels in human endometrium. *Biol. Reprod.* 61:1070–1082.

Fujimoto, J., Hori, M., Ichigo, S., and Tamaya, T. 1996. Expression of basic fibroblast growth factor and its mRNA in uterine endometrium during the menstrual cycle. *Gynecol. Endocrinol.* 10:193–197.

Gannon, B. J., Carati, C. J., and Verco, C. J. 1997. Endometrial perfusion across the normal human menstrual cycle assessed by laser Doppler fluxmetry. *Hum. Reprod.* 12:132–139.

Gargett, C. E., Lederman, F. L., Lau, T. M., and Rogers, P. A. W. 1999. Lack of correlation between vascular endothelial growth factor production and endothelial cell proliferation in the human endometrium. *Hum. Reprod.* 14:2080–2088.

Garlanda, C., and Dejana, E. 1997. Heterogeneity of endothelial cells. Specific markers. *Arterioscler. Thromb. Vasc. Biol.* 17:1193–1202.
Gaudry, M., Bregerie, O., Andrieu, V., El Benna, J., Pocidalo, M. A., and Hakim, J. 1997. Intracellular pool of vascular endothelial growth factor in human neutrophils. *Blood* 90:4153–4161.
Giudice, L. C. 1994. Growth factors and growth modulators in human uterine endometrium: their potential relevance to reproductive medicine. *Fertil. Steril.* 61:1–17.
Goodger, A. M., and Rogers, P. A. W. 1994. Endometrial endothelial cell proliferation during the menstrual cycle. *Hum. Reprod.* 9:399–405.
Goodger, A. M., and Rogers, P. A. W. 1995. Blood vessel growth in the endometrium. *Microcirculation* 2:329–343.
Graham, J. D., and Clarke, C. L. 1997. Physiological action of progesterone in target tissue. *Endocr. Rev.* 18:502–519.
Graubert, M. D., Mortola, J. F., Kessel, B., and Iruela-Arispe, M. L. 1997. Upregulation of vascular endothelial growth factor in the menstrual phase of normal endometrium may be mediated by hypoxia. Presented at the Society of Gynecology Investigation, San Diego, March 1997.
Hanahan, D. 1997. Signalling vascular morphogenesis and maintenance. *Science* 277:48–50.
Harmey, J. H., Dimitriadis, E., Kay, E., Redmond, H. P., and Bouchierhayes, D. 1998. Regulation of macrophage production of vascular endothelial growth factor (VEGF) by hypoxia and transforming growth factor β-1. *Ann. Surg. Oncol.* 5:271–278.
Hertig, A. T., Rock, J., and Adams, E. C. 1956. A description of 34 human ova within the first 17 days of development. *Am. J. Anat.* 98:435–493.
Hii, L., and Rogers, P. A. W. 1998. Endometrial vascular and glandular expression of integrin $\alpha_v\beta_3$ in women with and without endometriosis. *Hum. Reprod.* 13:1030–1035.
Hornung, D., Lebovic, D. I., and Shifren, J. L. 1998. Vectorial secretion of vascular endothelial growth factor by polarized human endometrial epithelial cells. *Fertil. Steril.* 69:909–915.
Iruela-Arispe, M. L., Porter, P., Bornstein, P., and Sage, E. H. 1996. Thrombospondin-1, an inhibitor of angiogenesis, is regulated by progesterone in the human endometrium. *J. Clin. Invest.* 197:403–412.
Iruela-Arispe, M. L., Rodriguez-Manzaneque, J. C., and Abu-Jawdeh, G. 1999. Endometrial endothelial cells express estrogen and progesterone receptors and exhibit a tissue specific response to angiogenic growth factors. *Microcirculation* 6:127–140.
Kamphaus, G. D., Colorado, P. C., Pandi, D. J., Hopfer, H., Ramchandran, R., and Torre, A. 2000. Canstatin, a novel matrix-derived inhibitor of angiogenesis and tumor growth. *J. Biol. Chem.* 275:1209–1215.
Keightley, M.-C. 1998. Steroid receptor isoforms: exception or rule? *Mol. Cell. Endocrinol.* 137:1–5.
Kelly, F. D., Tawia, S. A., and Rogers, P. A. W. 1995. Immunohistochemical characterisation of human endometrial microvascular basement membrane components during the normal menstrual cycle. *Hum. Reprod.* 10:268–276.
Koh, S. C. L., Wong, P. C., Yuen, R., Chua, S. E., Ng, B. L., and Ratnam, S. S. 1992. Concentration of plasminogen activators and inhibitor in the human endometrium at different phases of the menstrual cycle. *J. Reprod. Fertil.* 96:407–413.
Kuiper, G. G., and Gustafsson, J. A. 1997. The novel estrogen receptor-β subtype: potential role in the cell-and promoter-specific actions of estrogens and anti-estrogens. *FEBS Lett.* 410:87–90.
Lau, T. M., Affandi, B., and Rogers, P. A. 1999. The effects of levonorgestrel implants on vascular endothelial growth factor expression in the endometrium. *Mol. Hum. Reprod.* 5:57–63.
Li, F. X., Gregory, J., and Ahmed, A. 1994. Immunolocalisation of vascular endothelial growth factor in human endometrium. *Growth Factors* 11:277–282.

Maisonpierre, P. C., Suri, C., Jones, P. F., Bartunkova, S., Wiegand, S. J., Radziejewski, C., Compton, D., McClain, J., Aldrich, T. H., Papadopoulos, N., Daly, T. J., Davis, S., Sato, T. N., and Yancopoulos, G. D. 1997. Angiopoietin-2, a natural antagonist for Tie2 that disrupts in vivo angiogenesis. *Science* 277:55–60.

Markee, J. E. 1940. Menstruation in intraocular endometrial transplants in the rhesus monkey. *Contrib. Embryo. Carnegie Inst.* 28:219–308.

McCarthy, S. A., Kuzu, I., and Gatter, K. C. 1991. Heterogeneity of the endothelial cell and its role in organ preference of tumour metastasis. *Trends Pharmacol. Sci.* 12:462–467.

Meduri, G., Bausero, P., and Perrot-Applant, M. 2000. Expression of vascular endothelial growth factor receptors in the human endometrium: modulation during the menstrual cycle. *Biol. Reprod.* 62:439–447.

Narukawa, S., Kanzaki, H., Inoue, T., Imai, K., Higuchi, T., and Hatayama, H. 1994. Androgens induce prolactin production by human endometrial stromal cells in vitro. *J. Clin. Endocrinol. Metab.* 78:165–168.

Nasr, A., and Breckwoldt, M. 1998. Estrogen replacement therapy and cardiovascular protection: lipid mechanisms are the tip of an iceberg. *Gynecol. Endocrinol.* 12:43–59.

Noyes, R. W., Hertig, A. T., and Rock, J. 1950. Dating the endometrial biopsy. *Fertil. Steril.* 1:3–25.

O'Reilly, M. S., Boehm, T., Shing, Y., Fukai, N., Vasios, G., and Lane, S., Flynn E., Birkhead, J. R., Olsen, B. R., and Folkman, J. 1997. Endostatin: an endogenous inhibitor of angiogenesis and tumor growth. *Cell* 88:277–285.

O'Reilly, M. S., Holmgren, L., Shing, Y., Chen, C., Rosenthal, R. A., Moses, M., Lane, W. S., Cao, Y., Sage, E. H., and Folkman, J. 1994. Angiostatin: a novel angiogenesis inhibitor that mediates the supression of metastases by a Lewis lung carcinoma. *Cell* 79:315–328.

O'Reilly, M. S., Pirie-Shepherd, S., Lane, W. S., and Folkman, J. 1999. Antiangiogenic acitivity of cleaved conformation of the serpin antithrombin. *Science*. 285:1926–1928.

Paech, K., Webb, P., Kuiper, G. G. J. M., Nilsson, S., Gustafsson, J. A., Kushner, P. J., and Scanlon, T. S. 1997. Differential ligand activation of estrogen receptors ERα and ERβ at AP1 sites. *Science* 277:1508–1510.

Ramos, M. A., Kuzuya, M., Esaki, T., Miura, S., Satake, S., Asai, T., Kanda, S., Hayashi, T., and Iguchi, A. 1998. Induction of macrophage VEGF in response to oxidized LDL and VEGF accumulation in human atherosclerotic lesions. *Arterioscler.Thromb.Vasc. Biol.* 18:1188–1196.

Ramsey, E. M. 1989. Vascular anatomy. In: Wynn, R. M., and Jollie, W. P., eds. *Biology of the Uterus*, 2nd ed. Plenum Publishing, New York, pp. 57–68.

Roberts, D. K., Parmley, M. C., Walker, N. J., and Horbelt, D. V. 1992. Ultrastructure of the microvasculature in the human endometrium throughout the normal menstrual cycle. *Am. J. Obstet. Gynecol.* 166:1393–1406.

Rogers, P. A. W., Au, C. L., and Affandi, B. 1993. Endometrial microvascular density during the normal menstrual cycle and following exposure to long-term levonorgestrel. *Hum. Reprod.* 8:1396–1404.

Rogers, P. A. W., and Gargett, C. E. 1999. Endometrial angiogenesis. *Angiogenesis* 2:287–294.

Rogers, P. A. W., Lederman, F., Kooy, J., Taylor, N. H., and Healy, D. L. 1996. Endometrial vascular smooth muscle oestrogen and progesterone receptor distribution in women with and without menorrhagia. *Hum. Reprod.* 11:2003–2008.

Rogers, P. A. W., Lederman, F., and Taylor, N. 1998. Endometrial microvascular growth in normal and dysfunctional states. *Hum. Reprod. Update* 4:503–508.

Salamonsen, L. A. 1996. Matrix metalloproteinases and their tissue inhibitors in endometrial remodelling and menstruation. *Reprod. Med. Rev.* 5:185–203.

Sanghah, R. K., Li, X. F., Shams, M., and Ahmed, A. 1997. Fibroblast growth factor receptor-1 is a critical component for endometrial remodeling: localization and expression of basic fibroblast growth factor and FGF-R1 in human endometrium during the menstrual cycle and decreased FGF-R1 expression in menorrhagia. *Lab. Invest.* 77:389–402.

Sato, T. N., Tozawa, Y., Deutsch, U., Wolburg-Buchholz, K., Fujiwara, Y., Gendron-Maguire, M., Gridley, T., Wolburg, H., Risau, W., and Qin, Y. 1995. Distinct roles of the receptor tyrosine kinases Tie-1 and Tie-2 in blood vessel formation. *Nature* 376:70–74.

Shifren, J. L., Tseng, J. F., Zaloudek, C. J., Ryan, I. P., Meng, Y. G., Ferrara, N., Jaffe, R. B., and Taylor, R. N. 1996. Ovarian steroid regulation of vascular endothelial growth factor in the human endometrium: implications for angiogenesis during the menstrual cycle and in the pathogenesis of endometriosis. *J. Clin. Endocrinol. Metab.* 81:3112–3118.

Smith, S. K. 1995. Angiogenic growth factor expression in the uterus. *Hum. Reprod. Update* 1:162–173.

Steer, C. V., Tan, S. L., and Dillon, D. 1995. Vaginal color Doppler assessment of uterine artery impedance correlates with immunohistochemical markers of endometrial receptivity required for the implantation of an embryo. *Fertil. Steril.* 63:101–108.

Sunderkötter, C., Steinbrink, K., Goebeler, M., Bhardwaj, R., and Sorg, C. 1994. Macrophages and angiogenesis. *J. Leukoc. Biol.* 55:410–422.

Taichman, N. S., Young, S., Cruchley, A. T., Taylor, P., and Paleolog, E. 1997. Human neutrophils secrete vascular endothelial growth factor. *J. Leukoc. Biol.* 62:397–400.

Tawia, S. A., Beaton, L. A., and Rogers, P. A. W. 1993. Immunolocalization of the cellular adhesion molecules, intercellular adhesion molecule-1 (ICAM-1) and platelet endothelial cell adhesion molecule (PECAM), in human endometrium throughout the menstrual cycle. *Hum. Reprod.* 8:175–181.

Torry, D. S., and Torry, R. J. 1997. Angiogenesis and the expression of vascular endothelial growth factor in endometrium and placenta. *Am. J. Reprod. Immunol.* 37:21–29.

Wang, J.-D., Fu, Y., Shi, W.-L., Zhu, P. D., Cheng, J., Qiao, G. M., Wang, Y. Q., and Greene, G. L. 1992. Immunohistochemical localisation of progesterone receptor in human decidua of early pregnancy. *Hum. Reprod.* 7:123–127.

Webb, N. J., Myers, C. R., Watson, C. J., Bottomley, M. J., and Brenchley, P. E. 1998. Activated human neutrophils express vascular endothelial growth factor (VEGF). *Cytokine* 10:254–257.

CHAPTER 12

The Clinical Relevance of Disturbances of Uterine Vascular Growth, Remodeling, and Repair

Martha Hickey and Ian S. Fraser

Endometrial blood vessels are unique in the manner in which they undergo monthly cycles of dramatic growth, disruption, remodeling, and repair, rivaled only by the remarkable vascular growth and regression occurring during formation and subsequent demise of the ovarian corpus luteum. In the normal menstrual cycle the regular endometrial vascular and tissue disruption and remodeling is associated with frank bleeding known as menstruation. The volume and duration of this bleeding is generally very precisely controlled by a number of fail-safe mechanisms. The overall cyclical changes occurring in endometrium are regulated by a complex sequence of interacting cellular and molecular processes. Any of these processes may be disturbed under certain circumstances.

The cyclical vascular changes are part of the process of preparing the endometrium for implantation and for supporting a healthy pregnancy, but if pregnancy does not occur, the extensive degree of tissue differentiation that occurs in late secretory, predecidual endometrium has necessitated the unique remodeling process that we call menstruation. This physiologic process includes superficial tissue breakdown, some tissue shedding, tissue remodeling, and then angiogenesis and a new phase of tissue growth.

The morphologic changes are accompanied by changes in endometrial tissue blood flow that are little understood at present. These include cyclical changes in total endometrial flow that are influenced by circulating estrogen levels and other factors (Fraser et al, 1987). Localized fluctuations in flow, a "blush and blanch" phenomenon, also appear to occur (Markee, 1948; Marleschki, 1968) and typical capillary vasomotion has been observed (Hickey et al, 2000a).

Substantial disturbances in vascular morphologic appearance and function may occur during the course of pathologic change in the endometrium or during exposure to exogenous estrogens or progestogens. These disturbances are usually manifest as alterations in the menstrual bleeding pattern, which can vary substantially in frequency, regularity, duration, volume, and overall pattern. Evidence from nearly 30 years ago suggested that endometrial vascular fragility might be increased by combined estrogen-progestogen treatment (Sarfati and Brux, 1973).

Myometrial vessels also show changes in vascular morphology and function, but these are much less obvious and less well understood than those occurring in endometrium. Physiologic changes are most obvious during and after pregnancy, but pathologic disturbances occur in association with uterine myomata and adenomyosis. Alterations in myometrial contractions may also influence endometrial blood flow.

This chapter focuses on those vascular disturbances associated with benign gynecologic conditions, mostly resulting in disturbances of menstrual bleeding, and excludes any consideration of malignant disease or diseases of pregnancy.

PATTERNS OF MENSTRUAL BLEEDING

Menstrual disturbances affect up to 30% of women of reproductive age, and are a major social and medical problem for women, their families, and health providers (Gath et al, 1987; Coulter et al, 1989). Normal menstrual bleeding occurs every 21 to 35 days (Treloar et al, 1970; Belsey and Carlson, 1991; Belsey and Pinol, 1997), between menarche and menopause, and lasts for an average of 5 days, with a total blood loss of less than 60 to 80 mL (Hallberg et al, 1966; Fraser, 1989). Blood loss varies by up to 40% between cycles, especially in menorrhagia (Haynes et al, 1977). Very light intermenstrual bleeding occurs physiologically at ovulation in less than 1% of normal cycles (Vollman, 1977).

Abnormal menstrual bleeding is that which is irregular, excessive, or prolonged. In addition, many women consider their bleeding abnormal if it differs with regard to subtle changes, including appearance, aroma, and pattern of loss, over the duration of the bleeding days, or if they have perceived a substantial change in their pattern.

Approximately half of those women presenting to the gynecologist with abnormal bleeding will have ovulatory or anovulatory dysfunctional uterine bleeding, excessively heavy, prolonged, or frequent bleeding of uterine origin that is not due to pregnancy or to recognizable pelvic or systemic disease (Fraser, 1985). The majority of the rest will have menorrhagia associated with uterine fibroids. In modern Western society, dysfunctional bleeding is regular and ovulatory in approximately 80% of cases (Cameron, 1989).

As well as spontaneous menstrual disorders, bleeding patterns are commonly disturbed in users of sex steroid preparations, including contraceptives and hormone replacement therapy (HRT) (Odlind and Fraser, 1990). This bleeding is characteristically irregular, light, and may be frequent and prolonged. It is particularly common in users of progestogen-only contraceptives and loosely classified as "breakthrough bleeding" (BTB). There is no established long-term treatment for BTB and it commonly leads to discontinuation of contraceptives and HRT when effective alternatives may not be available or utilized (Odlind and Fraser, 1990).

For the endometrium to bleed, endometrial blood vessels and surface epithelium must both break down. Although circulating sex steroids may exert overall control over these vessels, it has conventionally been maintained that endometrial microvessels do not express estrogen (ER) or progestogen receptors (PR), and the influence of exogenous and endogenous sex steroids must therefore be indirect (Perrot-Applanat et al, 1988). However, recent evidence has suggested that there is some endometrial endothelial cell expression of both ER and PR and these cells exhibit a direct tissue specific response to angiogenic growth factors (Iruela-Arispe et al, 1999). There is also increasing evidence that several local mechanisms directly and indirectly determine vascular breakdown and repair (Fraser et al, 1996).

ENDOMETRIAL VASCULAR MORPHOLOGY, GROWTH, BREAKDOWN, AND REMODELING

Normal Anatomy and Morphology of the Endometrial Vasculature

Arterial blood reaches the uterus via the uterine and ovarian arteries and through arcuate arteries in the myometrium (Ramsey, 1982). Radial arteries arise from the arcuate arteries at 90 degrees toward the endometrium. After crossing the endomyometrial junction, small basal arteries arise from the radial arteries and supply the basal endometrium and larger spiral arterioles that run toward the endometrial surface (functional layer). Unlike the microvessels of endometrium, spiral arterioles are acutely hormone sensitive (Perrot-Applanat et al, 1988).

Capillaries arising from the spiral arterioles form a subepithelial plexus that can be visualized hysteroscopically (van Herendael et al, 1987; Hickey et al, 1996). These empty into a venous plexus within the functional layer. These veins subsequently drain into the inner myometrium. Endometrial microvessels consist only of endothelial cells (linked by tight junctions and adhesion molecules), their associated basement membrane, and surrounding pericytes.

Endometrial endothelial basement membrane (BM) is a specialized form of extracellular matrix composed of several heterogeneous macromolecules such as collagen IV, laminin, and heparan sulfate proteoglycan. Vascular BM contributes to vessel growth, differentiation, and cell permeability. The membrane also provides a sheet-like structure to which endothelial cells can closely attach (Bulletti et al, 1988). Because BM is distributed across many cells, changes in its integrity or composition will modify the behavior of other cell groups. Absence or disruption of the BM could thus contribute to endometrial vascular integrity or breakdown.

Pericytes are associated with all vascular capillaries and postcapillary venules. Differences in pericyte morphology and distribution among vascular beds suggests tissue-specific functions. Pericytes may contribute toward the regulation of local blood flow, vessel formation, cellular differentiation (Hirschi and D'Amore, 1996), and angiogenesis. Physical contact between endothelial cells and pericytes, mediated by cell adhesion molecules, integrins, and gap junctions, is likely to influence vascular function in the endometrium.

Much less is known about the structure and role of the endometrial lymphatic system. Lymphatics are thin-walled vessels with a more loosely structured endothelial cell layer than correspondingly sized blood vessels. Unlike endometrial blood vessels, the basement membrane is mostly irregular and fragmented, and may be absent (Casley-Smith and Florey, 1961). Lymphatic vessels have substantially thinner endothelium with fewer nuclei per cross section of the vessel. There are currently few endothelial cell immunohistochemical markers that differentiate endometrial lymphatics from micro–blood vessels. Flt-4 may be one such useful marker.

Approximately two thirds of normal human endometrium has a recognizable lymphatic system, with 1 to 20 vessels per square millimeter, mostly concentrated in the deeper portions of the tissue. There is no obvious correlation between the presence of lymphatics and the stage of the menstrual cycle, or the position of lymphatic vessels and that of other endometrial structures such as glands and blood vessels (Blackwell and Fraser, 1981). Lymphatics may play an important role in the endometrial regression seen preceding normal menstruation (Markee, 1950) and following sex steroid exposure, and in endometrial remodeling (Blackwell and Fraser, 1981).

Vascular Changes During the Normal Menstrual Cycle

Almost all of the data describing vascular changes during the normal menstrual cycle are based on histologic and immunohistochemical observations from human endometrial biopsies and on physiologic data from Markee (1940) observing endometrial explants placed in the anterior chamber of the eye of the rhesus monkey. The physical location of the endometrium makes in vivo studies difficult in the human, and it remains unclear how much information can be accurately extrapolated from these animal and in vitro studies.

The late preovulatory phase is accompanied by a fivefold increase in the length of the spiral arterioles, leading to vascular coiling. Distal spiral arterioles connect to the subepithelial capillary plexus via precapillary arterioles. During the secretory phase, spiral arterioles further increase in size and tortuosity, and the subepithelial capillaries dilate. Premenstrually, the endometrial stroma rapidly regresses by loss of intercellular fluid and spiral arterioles develop even more complex coiling. Menstrual bleeding is preceded by vascular stasis—vasoconstriction with brief relaxation of arterioles leading to bleeding (Markee, 1940). Histologic studies also suggest intense vasoconstriction in human endometrium (Salvatore, 1968). Bleeding also occurs via gaps in vessel walls and epithelium, rupturing hematomas, diapedesis from capillaries, and directly from endometrial veins.

Tissue shed during menstruation is associated with significant apoptosis, disordered expression of adhesion molecules, loss of filamentous actin from cell borders, and fragmentation of endometrial glands (Tabibzadeh, 1996). The normal process of endometrial vascular control involves a delicate balance between the agents of construction and destruction. Interference with this process at any level could potentially precipitate menstrual symptoms such as menorrhagia, irregular and prolonged bleeding, intermenstrual bleeding, and amenorrhea.

MINOR MECHANISMS OF ENDOMETRIAL ANGIOGENESIS, VASCULAR BREAKDOWN, AND REPAIR

Consideration of the mechanisms and potential disturbances of angiogenesis is central to any consideration of disordered uterine vascular structure and function. Vascular density ranges fairly widely from 50 to 1,000 capillaries per mm^2, such that most cells in normal tissues are within 10 to 20 μm of a blood vessel. Disturbances of myometrial and endometrial growth depend on, and may arise from, changes in angiogenesis as well as vascular flow. Angiogenesis is important to cyclic regeneration of the endometrium and to disease processes such as dysfunctional uterine bleeding (DUB), the endometrial response to exogenous sex steroids and to intrauterine contraceptive devices, uterine fibroids, endometriosis, endometrial hyperplasia, and carcinoma (Abulafia and Sherer, 1999).

Conventional (nonendometrial) angiogenesis involves a series of well-defined steps including endothelial cell activation, breakdown of the vascular basement membrane, migration and proliferation of endothelial cells, fusion of sprouts, and tube formation. Angiogenesis may also occur via intussusception of vessel elongation. In the endometrium, conventional markers of angiogenesis have failed to show changes during the normal menstrual cycle despite strong evidence of vascular growth, and there is no correlation between vascular endothelial growth factor (VEGF) production and endothelial cell proliferation in human endometrium (Gargett et al, 1999). This suggests that endometrial vascular growth does not follow the same pattern as that seen in other tissues, and that the stimuli may not be under the overall direct control of estrogen and progesterone (Rogers et al, 1998).

Endometrial angiogenesis is controlled by a variety of angiogenic and antiangiogenic stimuli and angiogenesis may be stimulated by the agents of endometrial breakdown (Tabibzadeh, 1991). Polypeptide growth factors act to regulate the effects of estrogen and progestogens, to control endometrial proliferation and differentiation, and to mediate the effects of sex steroid hormones in an autocrine and paracrine fashion. The expression of VEGF in the glandular epithelium and in the stroma may be increased by estrogen and progesterone (Gordon et al, 1995), suggesting that VEGF may play an important role in the regulation of angiogenesis during the reproductive cycle even if there is no direct correlation between VEGF levels and endothelial cell proliferation (Gargett et al, 1999). Of the growth factors, VEGF has the attribute of being a secreted peptide. Receptors for VEGF were thought to be localized almost exclusively to vascular endothelial cells (Millauer et al, 1993; Neufield et al, 1994), but recent evidence suggests that they may also be found on other cell types such as myometrial smooth muscle, epithelia, spermatozoa, and macrophages. Vascular endothelial growth factor also acts to increase vascular permeability, which may be of importance in the degradation of the extracellular matrix during angiogenesis (Charnock-Jones et al, 1993). Furthermore, VEGF also modulates the expression of a number of proteolytic enzymes involved in angiogenesis (Pepper et al, 1991).

The Ovarian Cycle

Increasing secretion of estrogen from the growing ovarian follicle during the follicular phase of the menstrual cycle eventually stimulates positive feedback from pituitary luteinizing hormone (LH) and ovulation. Progesterone levels rise through the luteal phase due to secretion from the corpus luteum. Progesterone withdrawal from an estrogen-primed endometrium then leads to menstrual bleeding. This bleeding occurs only if the endometrium is primed with estrogen (Corner, 1942; Padwick et al, 1986), and can be prevented by the continued administration of progesterone. Sudden withdrawal of estrogen will often lead to menstrual bleeding, but bleeding will not occur if the withdrawal is gradual (Corner, 1942; Smith, 1990). Different patterns of endometrial steroid exposure have broadly predictable effects on endometrial bleeding patterns (Table 12.1).

The systemic signal that leads to menstruation is the withdrawal of steroid hormones. There is some evidence that tumor necrosis factor-α may serve as one of

TABLE 12.1. Sex steroid hormonal environments associated with predominant patterns of endometrial bleeding.*

Sex steroid environment	Clinical situation
Estrogen and progesterone withdrawal from oestrogen-primed endometrium	Normal menstrual bleeding. Combined oral contraceptive pill (OCP) withdrawal with light bleeding
Withdrawal of unopposed estrogen	Anovulatory dysfunctional uterine bleeding (DUB) with heavy and unpredictable bleeding
High-dose continuous progestogens, e.g., depot medroxyprogesterone acetate	Irregular bleeding and amenorrhea
Low-dose continuous progestogens, e.g., Norplant	Irregular, light bleeding

* Although these may be predominant patterns of bleeding with each type of steroid exposure, small numbers of women may experience a wide range of bleeding patterns or amenorrhea in several of these situations.

the local signals contributing to the process of menstrual shedding and bleeding (Tabibzadeh, 1996). Secretion of metalloproteinases and their subsequent activation induced by plasmin facilitates degradation of extracellular matrices and bleeding. Numerous other molecular systems are also involved. The processes responsible for the cessation of each menstrual episode are not well understood, but it is clear that increasing secretion of estrogen is not essential for this purpose. Locally secreted growth factors are probably of prime importance.

Ovarian steroids induce the activation of a wide range of local factors able to precipitate vascular breakdown. These include prostaglandins (Cameron et al, 1991), macrophages, mast cells, granulated lymphocytes and other migratory leukocytes and their products (Finn, 1986; Bulmer et al, 1987; Clark and Daya, 1990), plasminogen activators (Koh et al, 1992), cytokines (Tabibzadeh and Sun, 1992), lysosomes (Henzl et al, 1972), and matrix metalloproteinases (MMPs) (Hampton and Salamonsen, 1994). These molecules have complex interactions, and the precise sequence of events culminating in endometrial vascular breakdown and repair is still under investigation. Lysosomes release hydrolytic enzymes in premenstrual endometrium that appear to contribute to tissue breakdown, bleeding, remodeling, and subsequent regeneration (Henzl et al, 1972; Christiaens et al, 1982). Lysosomes activate in response to falling progesterone, but it is unclear whether they have a primary or secondary role in the onset of endometrial vascular breakdown (Wang and Fraser, 1990). Increasing evidence suggests that MMPs are the agents responsible for extracellular matrix and vascular breakdown in normal, and perhaps also some types of abnormal, menstrual bleeding (Hampton and Salamonsen, 1994; Vincent et al, 1999). MMPs are a family of at least 24 enzymes capable of degrading all major classes of extracellular matrix proteins, and hence are critical to all tissue remodeling events, as well as the invasion processes involved in cancer and placentation. The control of MMP enzyme activity is complex and takes place at a number of levels.

Endometrial leukocytes are known to be involved in tissue destruction and regeneration. Menstruation is preceded by a marked increase in endometrial stromal granulated lymphocytes, T lymphocytes, granulated mast cells, and macrophages. Polymorphonuclear leukocytes only appear in large numbers in uterine tissues at the onset of menstruation (Kamat and Issacson, 1987; Bulmer et al, 1988) and are often located near blood vessels in the endometrium where they may influence vascular permeability and integrity. Endometrial migratory leukocytes are influenced by changes in ovarian sex steroid production, and small populations of these cells have varying estrogen and/or progesterone receptor positivity (Tabibzadeh and Satyaswaroop, 1989). The adhesion molecules intracellular adhesion molecule-1 (ICAM-1) and vascular cell adhesion molecule (VCAM) are known to be involved in the migration of leukocytes from the circulation, through vessel walls, and into the tissues (Larson and Springer, 1990).

Vasoconstriction of spiral arterioles limits the early phase of normal menstruation. Hemostatic intravascular plug formation occurs in the first 24 hours, but these plugs are small and transient and are rarely seen more than 20 hours after the onset of menstruation (Christiaens et al, 1982). In contrast to hemostatic mechanisms in nonendometrial tissue, menstrual blood does not show persistent clotting (Sixma et al, 1980). There is early coagulation followed by anticoagulation. Coagulation factors are reduced in menstrual blood, and there is increased activity of thrombolytic tissue plasminogen activator (tPA) as well as plasmin and α_2-antiplasmin (Rees et al, 1985). In collected menstrual blood, platelets are dysfunctional (Rees et al, 1984a), suggesting that the normal clotting mechanisms have been "disabled" by menstrual fluid. In the late secretory and menstrual phase there are much reduced levels of von Willebrand factor (vWF), crucial for platelet adherence to damaged vessels, suggesting impaired hemostasis (Au and Rogers, 1993). Gaps can sometimes be seen between endothelial cells without the aggregation of platelets that would normally attach to exposed subendothelial components in other tissues. There is also clinical evidence pointing to increased fragility and rhythmic changes in skin capillaries during menstruation, suggesting a generalized effect of estrogen withdrawal (Brewer, 1938; Clemetson et al, 1962).

Endometrial repair and angiogenesis commence during early menstruation (Christiaens et al, 1982), and epithelial repair is complete by days 4 to 5 of menstruation (Ferenczy et al, 1979).

ABNORMAL VASCULAR GROWTH, FUNCTION, AND BREAKDOWN

Abnormalities of many of the above features of endometrial vascular morphology and function have been demonstrated in a number of clinical situations, but there is still much to be learned about the variations and interactions between different endometrial cell types and molecular systems in abnormal circumstances.

Endometrial Vascular Abnormalities in DUB

There is little evidence of any change in the morphology or density of spiral arterioles in women with ovulatory DUB (Rees et al, 1984b), although spiral arterioles may be markedly suppressed in anovulatory DUB (Farrer-Brown et al, 1970). There is an apparent increase in the numbers of small veins in the deep endometrium and inner myometrium in women with ovulatory and anovulatory DUB (Hourihan et al, 1989; Farrer-Brown et al, 1971).

Ovarian steroids modulate vascular tone and uterine perfusion and hence may indirectly contribute to menstrual disturbance by such mechanisms. Estrogen causes a fall in uterine vascular resistance, and an increase in uterine blood flow, and this effect is lost with the addition of progesterone (Markee, 1948). Progesterone directly modulates calcium-controlled smooth muscle contraction in the uterine artery and stimulates the degradation of endothelin, a potent vasoconstrictor (Casey and MacDonald, 1996). Vasopressin, which acts to decrease endometrial blood flow, is stimulated by estradiol and inhibited by progesterone (Forsling et al, 1982).

Despite normal levels of circulating ovarian steroids, the action of estrogen and progesterone at an endometrial level may be altered in women with DUB. Some authors have reported that expression of endometrial ER and PR is higher in the late secretory phase in women with DUB compared with women with normal menstrual loss (Adelentado et al, 1988), but others have not found this (Critchley et al, 1994). Changes in uterine vascular perfusion have been seen in association with DUB. In women with DUB there is an increase in endometrial flow compared with normal, corresponding with the increasing estradiol secretion in the late follicular phase, and a decrease in flow in the secretory phase (Fraser et al, 1987). In women with anovulatory DUB, endometrial flow rates are variable, but tend to be high.

Ovulatory DUB is associated with a shift in the ratio of endometrial vasoconstricting prostaglandin $F_{2\alpha}$ ($PGF_{2\alpha}$) to vasodilatory prostacyclin and PGE_2 (Smith et al, 1981), and an increase in total endometrial concentration of PG. In anovulatory DUB associated with persistently proliferative endometrium, the availability of arachidonic acid is reduced and PG production is impaired (Smith et al, 1982). Endometrial tissues may be more responsive to the action of vasodilatory PG, PGE_2 and prostacylin via increased receptor concentrations in women with menorrhagia (Adelentado et al, 1988). Successful treatment of ovulatory DUB with prostaglandin synthetase inhibitors (PGSIs) usually leads to a reduction in endometrial concentrations of $PGF_{2\alpha}$ and PGE_2, and inhibits binding of PGE to its receptor (Rees et al, 1988).

Adequate and timely vasoconstriction of the spiral arterioles is an essential component of menstrual hemostasis. Endometrial endothelins (ETs) act as powerful vasoconstrictors, as well as stimulating the release of endothelial relaxing factors such as nitric oxide, atrial naturetic peptide, and prostacyclin. Receptors for endothelins are predominantly located at the endometrial-myometrial junction (Cameron et al, 1993) and are present at increased concentrations just before menstruation. Deficient ET production may prolong or increase menstrual blood loss (MBL) and facilitate menorrhagia and breakthrough bleeding (Marsh et al, 1996). The potent vasorelaxant parathyroid hormone–related protein may also play a role in endometrium (Casey and MacDonald, 1996).

Lysosomes are intracellular membrane-bound vacuoles containing a variety of possible destructive hydrolytic enzymes, and are vital to tissue remodeling and regeneration. Endometrial lysosomes show cyclical variation, with a marked rise in lysosomal enzyme activity in the late secretory phase and during early menstruation (Henzl et al, 1972). Progressive accumulation and release of hydrolytic enzymes in secretory endometrium may be a principal mechanism in cell separation, endometrial bleeding, remodeling, and subsequent regeneration in normal menstruation (Wang and Fraser, 1990). Tissue hypoxia following spiral artery coiling and endometrial regression and vascular stasis stimulates lysosomal activation. Lysosomal enzyme activity in the endometrium is increased in women with ovulatory dysfunctional uterine bleeding and with heavy menstrual bleeding secondary to intrauterine contraceptive device (IUCD) use (Wang et al, 2000).

Macrophages and other migratory leukocytes may be implicated in the control of MBL and contribute to mechanisms of excessive loss. An excessive leukocyte infiltrate has been associated with the copper IUCD. Macrophages, among many other factors, can release platelet activating factor (PAF) and PGE, potent vasodilators that can augment MBL. Endometrial granulated lymphocytes secrete perforins, which may degrade endometrial vascular and other cellular structures and promote bleeding.

Mast cells degranulate premenstrually to secrete heparin, histamine, tryptase, chymase, and probably a range of other substances (Salamonsen and Wooley, 1996). Heparin stimulates endometrial fibrinolysis via the secretion of tPA, and histamine causes endothelial cell contraction creating gaps between vascular endothelial cells. Menstrual bleeding is always accompanied by a component due to endometrial transudation (Fraser et al, 1985). In DUB, endometrial secretion of heparin-like substances is also increased (Foley et al, 1978).

Active fibrinolysis within the uterine cavity prevents organized clot formation and the development of intrauterine adhesions. The defective hemostasis seen in normal menstruation may be exaggerated in ovulatory DUB. Vascular plugs are loose and solely located inside vessels, and reduced fibrin concentrations make them unstable (Sheppard, 1990). Fibrinolysis will occur according to the balance of plasminogen activators, plasminogen inhibitors, and plasmin. Excessive menstrual blood loss is associated with an increase in endometrial fibrinolysis (Kasonde and Bonnar, 1976). Estrogens stimulate and progesterone inhibits the release of plasminogen activators, and progesterone also stimulates the release of fibrinolytic inhibitors (Casslen and Astedt, 1983). An increased concentration of tPA has been observed in women with ovulatory DUB. Overactivation of the fibrinolytic system may "unbalance" the hemostatic system, causing early breakdown of thrombi in the endometrial vessels and excessive blood loss.

Changes in Endometrial Microvessels in Response to Exogenous Sex Steroids

In general, estrogen-containing combined contraceptives produce fewer bleeding complications than progestogen-only methods, and women with anovulatory reactions to progestogen-only methods have more bleeding disturbance than those with some endogenous luteal activity (Landgren et al, 1982). Steroid regimens that suppress ovarian activity produce regular bleeding if the endometrium is repeatedly estrogen-primed, for example, the oral contraceptive pill (OCP), or scant bleeding with spotting when there is no estrogen priming, for example, depomedroxyprogesterone acetate (DMPA). Where ovarian activity is incompletely suppressed (e.g., Norplant, Jadelle, and the progestogen-only pill), fairly troublesome irregular and frequent bleeding patterns are often observed. Bleeding patterns will vary according to the progestogen used, the dose and duration, and the level of endogenous estradiol and progesterone secretion, and will differ markedly between individuals. Bleeding patterns may rapidly alter in response to exogenous sex steroids and persist for long periods in this modified pattern, or they may show change over time. There is much individual variation in bleeding patterns in response to exogenous steroids, and it is not known how much of this variation occurs at a hormonal, vascular, or endometrial tissue or cellular level.

Abnormalities of vascular growth or remodeling may interfere with endometrial vascular stability and lead to unscheduled vascular breakdown, irregular bleeding, or amenorrhea. The pattern, quantity, and appearance of BTB suggest that it may arise from a different vascular source than normal menstrual bleeding (Odlind and Fraser, 1990). Normal menstrual bleeding is thought to arise predominantly from the spiral arterioles (Markee, 1940). Hysteroscopic studies in women using the low-dose levonorgestrel contraceptive implant system Norplant have suggested increased superficial endometrial vascular fragility, and have demonstrated disturbed superficial vascular patterns and small dilated surface veins with bleeding from the endometrial surface in these patients (Hickey et al, 1996).

Mechanisms of Endometrial Vascular Fragility

Changes in the superficial endometrial vascular morphology have been seen in women using low-dose progestogen contraceptives. At hysteroscopy, vascular dilatation (Hickey et al, 1996; Runic et al, 1997), and neovascular formations (Hickey et al, 1996) suggest that vascular development is altered in these women, though conventional markers of angiogenesis have failed to demonstrate this (Subakir et al, 1996). Three-dimensional computer-generated studies of individual endometrial microvessels have demonstrated variable diameter and dilated vessels of abnormal shape and structure (Manconi, Simbar, Markham, and Fraser, unpublished data). These altered vascular appearances may also indicate alterations in local vascular function. For example, vascular endothelial growth factor contributes to vascular permeability and provokes dilatation. Increased endometrial VEGF protein has been observed in progestogen users (Lau et al, 1999).

Low-dose progestogen exposure is associated with a change in the balance between vascular and nonvascular elements in the endometrium, with a relative increase in endometrial microvascular density compared to glandular and stromal elements (Rogers et al, 1993; Hickey et al, 1999a). The implications of this alteration in the capillary to surrounding tissue ratio are unclear.

Extensive subepithelial bleeding has been seen at hysteroscopy in progestogen users (Hickey et al, 1996), and changes in vascular structural integrity that may lead to vascular fragility have also been associated with progestogen use. Endometrial microvascular BM is deficient in progestogen users, particularly during the early months of use when irregular bleeding is most common (Hickey et al, 1999b).

Exposure to progestogens is associated with an increased capacity for endometrial vascular breakdown. An increase in endometrial migratory leukocytes (Clark et al, 1996; Song et al, 1996) and an alteration in the endometrial balance between some MMP and tissue inhibitor metalloproteinase (TIMP) expression (Skinner et al, 1999; Vincent et al, 1999) has been seen in progestogen users. Similarly, ongoing studies in HRT users have shown that the balance between certain endometrial MMPs and their tissue inhibitors (TIMPs) is altered following HRT exposure

(Hickey et al, 2000b). Some MMPs are activated by progestogen withdrawal (Marbaix et al, 1995). Exposure to exogenous progestogens may render the endometrium unresponsive to further progestogen by PR downregulation (Critchley et al, 1998), thus mimicking the progesterone withdrawal in cycling premenstrual endometrium.

Altered endometrial perfusion may contribute to increased endometrial vascular fragility. Pilot studies using laser Doppler in Norplant users show reduced endometrial perfusion in early users and profound alteration of normal vasomotion patterns with an almost total loss of short-term vasomotion (Hickey et al, 2000a). Reduced endometrial perfusion will lead to relative endometrial hypoxia. Hypoxia is a potent vascular destabilizer, compromising endothelial integrity, inducing vascular breakdown, and stimulating angiogenesis via VEGF release (Smith, 1998).

The poor correlation between endometrial vascular function and circulating sex steroid levels may partly be explained by changes in endometrial responsiveness to sex steroids in progestogen users. Some differences have been reported in endometrial sex steroid receptor expression between women using Norplant and those using the higher local dose intrauterine system Mirena (Critchley et al, 1993, 1998; Lau et al, 1999). In women using the 20-μg levonorgestrel-releasing intrauterine system (Mirena), progesterone receptor expression is reduced, rendering the endometrium functionally unresponsive to endogenous and exogenous progestogens (Critchley et al, 1998). A fall in progesterone is known to stimulate MMP activity, and consequent breakdown of the extracellular matrix (ECM) (Marbaix et al, 1995). An endometrium unresponsive to progestogens because of profound receptor downregulation may thus become a fragile tissue with reduced ECM components.

Changes in endometrial vascular hemostasis may also contribute to prolonged bleeding seen in women using steroidal contraceptives, progestogens, and HRT. Spiral arteriole development is markedly curtailed following combined sex steroid exposure (Maqueo, 1980; Hourihan et al, 1986), and hysteroscopic studies suggest that progestogen-induced bleeding may arise from smaller superficial vessels in the endometrium (Hickey et al, 1996). Since spiral arteriole constriction is an essential early mechanism of menstrual hemostasis (Markee, 1940), bleeding arising from other vessels may result in compromised hemostasis and prolonged bleeding. In addition, the powerful vasoconstrictor molecule ETs are reduced in progestogen users, and the ET degrading enzyme neural endopeptidase (NEP) is increased (Casey and MacDonald, 1996; Marsh et al, 1996; Salamonsen et al, 1999). Reduced ET activity may compromise vascular repair as well as hemostasis. This situation could be confused by the role of tissue factor (TF), which is a primary inhibitor of endometrial hemostasis. Messenger RNA (mRNA) for TF in endometrium is reduced in progestogen users compared to secretory controls (Rogers et al, 1997; Runic et al, 1997), which could be expected to reduce the volume and duration of bleeding.

The overall effect of these vascular changes may be to increase endometrial vascular fragility. If the endometrium contains comparatively more microvessels with comparatively less structural support, and greater influx and activity of local molecules able to break down vessels, in conjunction with compromised hemostatic mechanisms, this may induce and perpetuate BTB.

DISTURBANCES OF MYOMETRIAL VASCULAR GROWTH AND REMODELING

The normal vascular supply to the myometrium arises from the uterine and ovarian arteries, which form the arcuate arteries in the myometrium (Ramsey, 1982). Radial arteries arise from the arcuate arteries and pass toward the endometrium, giving off lateral branches to the myometrium as they go. These small lateral arterioles travel in tissue planes between the muscle bundles giving off capillary branches that form a rich interconnecting network. Capillaries drain into an extensive venous network that is not so prominent in the inner myometrium as the outer part (Farrer-Brown et al, 1971).

Uterine Fibroids

Uterine fibroids, or leiomyomas, are the most common tumors in women during the reproductive years. They arise from a proliferation of uterine smooth muscle cells, and their growth appears to be stimulated by both estrogen and progesterone (Rein et al, 1995). The association between fibroids and heavy and/or prolonged menstrual bleeding is well recognized (Rybo et al, 1985; Fraser et al, 1986). The location of fibroids is thought to influence menstrual symptoms, and submucous fibroids (which distort the uterine cavity) are thought to have a greater association with menorrhagia (Fraser et al, 1986; Lumsden, 1998), although the objective evidence for this is limited. Only about 40% of those women having a hysterectomy for subjective menorrhagia associated with a fibroid uterus will be found to have submucous fibroids (Lumsden, 1998).

Clinically, fibroids may be highly vascularized (especially in the pseudocapsule) with large-caliber vessels characteristically arising from the base of the tumor and often visualized at hysteroscopy for submucous myomas. The large-caliber vessels may be quite fragile. Destruction of the fibroid vascular supply, for example by embolization of the local vessels, causes shrinkage (Ravina et al, 1995). Using Doppler ultrasound to observe blood flow through the fibroid, Sosic et al (1996) found that flow was increased and vascular resistance decreased in proportion to fibroid size. Following treatment for subjective menorrhagia with tranexamic acid, Lakhani et al (1998) observed that women without uterine fibroids showed a reduction in uterine artery Doppler pulsatility index and resistance index, and an apparent reduction in the rate of menstrual bleeding. These changes were not seen in subjects with uterine fibroids.

In addition, the myometrial microvasculature is altered in the fibroid uterus (Sapozhnikov, 1987). Perfusion in the vascular plexus around a fibroid increases while perfusion within the myoma decreases. This is associated with a reduced microvascular volume. Myomas are also associated with ultrastructurally altered myometrial vessels. Skopichev and Savitskii (1992) found that endothelial cell tight junctions were disturbed in fibroids, and that this was accompanied by a disturbance of the microvesicular transendothelial transport system. This suggests that there is a dysregulation of angiogenic and other regulatory growth factors associated with myoma growth that may also contribute to menorrhagia. Factors that may prove to be important in this process include basic fibroblast growth factor, VEGF, heparin-binding epidermal growth factor, platelet-derived growth factor, transforming growth factor-β, parathyroid hormone–related protein, and prolactin

(Stewart and Nowak, 1996). Menstrual disorders associated with fibroids are notoriously difficult to manage with medical treatments. With an improved understanding of the pathogenesis of this disease, therapies directed against growth factor abnormalities may result in improved nonsurgical management options.

Adenomyosis

Adenomyosis is a condition in which endometrial glands and stroma are located deep within the myometrium. Most pathologists will make a diagnosis of adenomyosis only when the endometrial glands penetrate more than 2.5 mm into myometrium. The etiology and pathogenic mechanisms responsible for adenomyosis are poorly understood, and the true incidence of this condition is not known, since the only definitive diagnostic technique involves histopathologic examination of the extirpated uterus, though there have been attempts to establish diagnostic criteria using ultrasound and magnetic resonance imaging (MRI) (Brosens and Brosens, 1998). From hysterectomy specimens, it has been estimated that 60% of women with moderate or severe adenomyosis are symptomatic with menorrhagia and dysmenorrhea, and that 80% of adenomyotic cases are associated with fibroids. This suggests a widespread disturbance of myometrial function. Adenomyosis also coexists commonly with endometrial carcinoma (Ferenczy, 1998).

Blood vessels appear to be randomly scattered throughout the adenomyotic uterus (Chiang et al, 1999), and endometrial as well as myometrial vascularization is increased (Ota et al, 1998). The origin of this increased myometrial vascularization is unclear. This ectopic endometrium does not respond to sex steroids in the same manner as eutopic endometrium (Matsumoto et al, 1999), and demonstrates different leukocyte populations (Bulmer et al, 1998). There is evidence that vessels in the ectopic adenomyotic endometrium do not arise from the endometrial vasculature, but may develop de novo within the myometrium (Matsumoto et al, 1999).

ROLE OF THE UTERINE VASCULATURE IN IMPLANTATION AND DISTURBANCES OF IMPLANTATION

Human endometrium undergoes sequential phases of proliferation and secretory changes to support the implantation and growth of an embryo. Survival of the implanting human blastocyst requires that the trophoblast gains access to the maternal circulation. Initially, this occurs by syncytiotrophoblast cells breaching the endometrial capillaries and venules, and subsequently extravillous cytotrophoblast cells penetrate the spiral arteries and change these vessels into high-flow, low-resistance vessels with fibrinoid deposition and medial necrosis. Fetal trophoblast cells gain access to maternal blood within 2 days after the onset of blastocyst implantation in the macaque (Enders and Blankenship, 1997). This process approximates the embryo with the maternal circulation and provides it with ready access to nutrients and oxygen. However, invasion of the vessels may also lead to decidual hemorrhage, which may interfere with implantation and placental development. Decidual cells act to modulate local hemostasis by expression of tissue factor and plasminogen activator type 1 (Lockwood et al, 1997), and hence to maintain pregnancy. Recurrent pregnancy loss is known to be associated with vascular thrombosis (Rai and Regan, 1998). Thrombosis in this syndrome may be the result of the displacement from phospholipid surfaces of annexin-V, an anionic phospholipid-binding protein with potent anticoagulant activity (Rand, 1998). This protein may normally act to prevent thrombosis. Antiphospholipid antibodies reduce the circulating level of annexin-V and may therefore act to promote thrombosis and pregnancy loss in the antiphospholipid antibody syndrome (Rand et al, 1997).

Endometrial capillaries undergo dramatic changes at the time of embryo implantation (Rogers, 1995). VEGF, which acts to promote uterine vascular permeability and dilatation, may also contribute to implantation. Just before implantation and again in early pregnancy, high levels of the VEGF receptor Flk-1 are expressed in the rabbit uterus, in close proximity to endometrial blood vessels. VEGF may thus contribute to the induction of vascular hyperpermeability at implantation and in the subsequent angiogenesis of early pregnancy (Das et al, 1997). In addition, an increase in the expression of inducible nitric oxide (iNOS) and endothelial nitric oxide (eNOS) at the embryonic site following implantation may imply a role in vasodilatation and vascular remodeling (Purcell et al, 1999).

Endometrial thickness and perfusion (as assessed by Doppler blood flow) are positively associated with a good in vitro fertilization (IVF) outcome in women (Yang et al, 1999). This suggests that adequate endometrial proliferation with good secretory change resulting in a thicker endometrium and associated with good levels of tissue perfusion is more likely to successfully support a viable pregnancy.

ARTERIOVENOUS MALFORMATIONS

Arteriovenous malformations (AVMs) of the uterus are rare entities in gynecology, with less than 100 cases reported in the literature. The vascular structure varies considerably but frequently includes a mixture of muscular and thin-walled capillary-like vessels in varying proportions (Fleming et al, 1989). Many muscular vessels have intermediate appearances between veins and arteries, and show prominent intimal fibrous thickening. Most commonly they present with vaginal hemorrhage, but other presentations such as congestive heart failure, postmenopausal bleeding, and an asymptomatic mass have been described. These lesions may be congenital or acquired. Acquired lesions are believed to follow trauma or may arise after choriocarcinoma or other gynecologic malignancies. The mechanism by which congenital lesions develop is unknown, but presumably involves a major local disturbance of angiogenic processes. Diagnosis can be made with color flow Doppler ultrasound or angiography (Hoffman et al, 1997). Additionally, the lesions have been detected using hysteroscopy, hysterosalpingogram, computed tomography, and MRI (Huang et al, 1998).

Acute management consists of hemodynamic stabilization, and may include the placement of a Foley bulb into the uterus or a methylergonovine injection (Flynn and Levine, 1996). Curettage may precipitate life-threatening hemorrhage and is therefore contraindicated when a uterine AVM is suspected (Manolitsas et al, 1994). Ultimate management depends on the patient's desire for fertility. Embolization therapy is variably successful and may allow the preservation of reproductive capacity. If pregnancy is not desired or embolization fails, hysterectomy remains the treatment of choice, as the patient is at risk of sudden and massive hemorrhage.

CONCLUSION

Endometrium is a tissue that exhibits unique cycles of growth, differentiation, coordinated breakdown, repair, and remodeling. Parallel changes in vessels are an integral part of this process, and these vessels show unique features of angiogenesis, growth, function, rupture, and repair, and associated alterations in hemostatic processes. The way in which multiple cellular and molecular systems interact within these carefully coordinated sequences of events is still not well understood. Hence, the many disturbances in cellular and molecular function that have been demonstrated in different endometrial conditions are also poorly understood.

The most obvious disturbances of angiogenesis, vascular structure, and vascular function are seen during prolonged exposure to progestogens, and this situation serves as an important model for the exploration and understanding of how such disturbances may arise and how they may be prevented. Prevention is particularly important because the major clinical symptom of these pathophysiologic processes is erratic, unpredictable, and troublesome breakthrough bleeding that can lead to high discontinuation rates. Spontaneous disorders of menstruation can also be associated with significant disturbances of vascular morphology and function, and sometimes with major pelvic pathology. A better understanding of these underlying molecular processes will lead to improvements in medical management rather than the traditional recourse to surgery such as hysterectomy or endometrial ablation.

REFERENCES

Abulafia, O., and Sherer, D. M. 1999. Angiogenesis of the endometrium. *Obstet. Gynecol.* 94:148–153.

Adelentado, J. M., Rees, M. C. P., Lopez Bernal, A., and Turnbull, A. C. 1988. Increased uterine prostaglandin E receptors in menorrhagic women. *Br. J. Obstet. Gynaecol.* 95:162–165.

Au, C. L., and Rogers, P. A. W. 1993. Immunohistochemical staining of von Willebrand factor in human endometrium during the normal menstrual cycle. *Hum. Reprod.* 8:17–23.

Belsey, E. M., and Carlson, N. 1991. The description of menstrual bleeding patterns: towards fewer measures. *Stat. Med.* 10:267–284.

Belsey, E. M., and Pinol, A. P. Y. 1997. Task Force on Long-Acting Systemic Agents for Fertility Regulation. Menstrual bleeding patterns in untreated women. *Contraception* 55:57–65.

Blackwell, P. M., and Fraser, I. S. 1981. Superficial lymphatics in the functional zone of the normal human endometrium. *Microvasc. Res.* 21:142–152.

Brewer, J. I. 1938. Rhythmic changes in skin capillaries and their relation to menstruation. *Am. J. Obstet. Gynecol.* 36:597–612.

Brosens, J. J., Brosens, I. A. 1998. Adenomyosis uteri. In: Cameron, I. T., Fraser, I. S., and Smith, S. K., eds. *Clinical Disorders of the Endometrium and Menstrual Cycle.* Oxford University Press, Oxford, pp. 297–312.

Bulletti, C., Jansonni, V. M., Polli, V., Cappucini, F., Galassi, A., Flamingni, C. 1988. Basement membrane components in normal, hyperplastic and neoplastic endometrium. *Cancer* 63:142–149.

Bulmer, J. N., Johnson, P. M., and Bulmer, D. 1987. Leukocyte populations in human decidua and endometrium. In: Gill, P. J., and Wegmann, T. J., eds. *Immunoregulation and Fetal Survival.* Oxford University Press, Oxford, pp. 111–134.

Bulmer, J. N., Lunny, D. P., and Hagin, S. V. 1988. Immunohistochemical characterisation of stromal leukocytes in non-pregnant human endometrium. *Am. J. Reprod. Immunol. Microbiol.* 17:83–90.

Bulmer, J. N., Jones, R. K., and Searle, R. F. 1998. Intraepithelial leukocytes in endometriosis and adenomyosis: comparison of eutopic and ectopic endometrium with normal endometrium. *Hum. Reprod.* 13:2910–2915.

Cameron, I. T. 1989. Dysfunctional uterine bleeding. *Baillieres Clin. Obstet. Gynaecol.* 3:315–327.

Cameron, I. T., Davenport, A. P., Brown, M. J., and Smith S. K. 1991. Endothelin-1 stimulates $PGF_{2\alpha}$ release from human endometrium. *Prostaglandins Leukot. Essent. Fatty Acids* 42:155–158.

Cameron, I. T., Plumpton, C., Champeney, R., van Papendorp, C., Ashby, M. J., and Davenport, A. P. 1993. Identification of endothelin-1, endothelin-2 and endothelin-3 in human endometrium. *J. Reprod. Fertil.* 97:251–255.

Casey, M. L., and MacDonald, P. C. 1996. The endothelin-parathyroid hormone-related protein vasoactive peptide system in human endometrium: modulation by transforming growth factor-beta. In: Findlay, J., and Affandi, B., eds. *Menstrual Disorders and Progestin Contraception*. Hum. Reprod. 11(suppl 2):62–82.

Casley-Smith, J. R., and Florey, H. W. 1961. The structure of normal small lymphatics. *Q. J. Exp. Physiol.* 46:101–106.

Casslen, B., and Astedt, B. 1983. Reduced plasminogen activator content of endometrium in oral contraceptive users. *Contraception* 28:181–188.

Charnock-Jones, D. S., Sharkey, A. M., Rajput-Williams, J., Burch, D., Schofield, J. P., Fountain, S. A., Boocock, C. A., and Smith, S. K. 1993. Identification and localisation of alternately spliced mRNAs for vascular endothelial growth factor in human uterus and steroid regulation in endometrial carcinoma cell lines. *Biol. Reprod.* 48:1120–1128.

Chiang, C. H., Chang, M. Y., Hsu, J. J., Chiu. T. H., Lee, K. F., Hsieh, T. T., and Soong, Y. K. 1999. Tumor vascular pattern and blood flow impedance in the differential diagnosis of leiomyoma and adenomyosis by color Doppler sonography. *J. Assist. Reprod. Genet.* 16:268–275.

Christiaens, G. C. M. L., Sixma, J. J., and Haspels, A. A. 1982. Haemostasis in menstrual endometrium: a review. *Am. J. Reprod. Immunol.* 5:78–83.

Clark, D. A., and Daya, S. 1990. Macrophages and other migratory cells in endometrium: relevance to endometrial bleeding. In: d'Arcangues, C., Fraser, I. S., Newton, J. R., and Odlind, V., eds. *Contraception and Mechanisms of Endometrial Bleeding*. Cambridge University Press, Cambridge, pp. 363–379.

Clark, D. A., Wang, S., Rogers, P. A. W., Vince, G., and Affandi, B. 1996. Endometrial lymphomyeloid cells in abnormal uterine bleeding due to levonorgestrel (Norplant). *Hum. Reprod.* 11:1438–1444.

Clemetson, C. A., Blair, L., and Brown, A. B. 1962. Capillary strength and the menstrual cycle. *Ann. NY Acad. Sci.* 93:279–299.

Corner, G. W. 1942. *The Hormones in Human Reproduction.* Princeton University Press, Princeton, NJ.

Coulter, A., Noone, A., and Goldacre, M. 1989. General practitioner's referrals to specialist outpatient clinics. *Br. Med. J.* 299:304–308.

Critchley, H. O. D., Bailey, D. A., Au, C. L., Affandi, B., and Rogers, P. A. 1993. Immunohistochemical sex steroid receptor distibution in endometrium from long-term subdermal levonorgestrel users and during the normal menstrual cycle. *Hum. Reprod.* 8:1632–1639.

Critchley, H. O. D., Abberton, K. M., Taylor, N. H., Healy, D. L., and Rogers, P. A. 1994. Endometrial sex steroid receptor expression in women with menorrhagia. *Br. J. Obstet. Gynaecol.* 101:428–434.

Critchley, H. O. D., Wang, H., Kelly, R. W., Gebbie, A. E., and Glasier, A. F. 1998. Progestin receptor isoforms and prostaglandin dehydrogenase in the endometrium of women using a levonorgestrel-releasing intrauterine system. *Hum. Reprod.* 13:1210–1217.

Das, S. K., Chakraborty, I., Wang, J., Dey, S. K., and Hoffman, L. H. 1997. Expression of vascular endothelial growth factor (VEGF) and VEGF-receptor messenger ribonucleic acids in the peri-implantation rabbit uterus. *Biol. Reprod.* 56:1390–1399.

Enders, A. C., and Blankenship, T. N. 1997. Modification of endometrial arteries during invasion by cytotrophoblast cells in the pregnant macaque. *Acta. Anat. (Basel)* 159:169–193.

Farrer-Brown, G., Beilby, J. O. W., and Tarbit, M. H. 1970. The blood supply of the uterus. 1: arterial vasculature. *J. Obstet. Gynaecol. Br. Commonw.* 77:672–681.

Farrer-Brown, G., Beilby, J. O. W., and Tarbit, M. H. 1971. Venous changes in the endometrium of myomatous uteri. *Obstet. Gynecol.* 38:743–751.

Ferenczy, A. 1998. Pathophysiology of adenomyosis. *Hum. Reprod. Update* 4:313–322.

Ferenczy, A., Bertrand, G., and Gelfand, M. N. 1979. Proliferation kinetics of human endometrium during the normal menstrual cycle. *Am. J. Obstet. Gynecol.* 133:859–867.

Finn, C. A. 1986. Implantation, menstruation and inflammation. *Biol. Rev.* 61:313–328.

Fleming, H., Ostor, A. G., Pickel, H., and Fortune, D. W. 1989. Arteriovenous malformations of the uterus. *Obstet. Gynecol.* 73:209–213.

Flynn, M. K., and Levine, D. 1996. The noninvasive diagnosis and management of a uterine arteriovenous malformation. *Obstet. Gynecol.* 88:650–652.

Foley, M. E., Griffin, B. D., and Zugel, M. 1978. Heparin-like activity in uterine fluid. *Br. Med. J.* 11:322–324.

Forsling, M. L., Stromberg, P., and Akerlund, M. 1982. Effect of ovarian steroids on vasopressin secretion. *J. Endocrinol.* 95:147–151.

Fraser, I. S. 1985. The dysfunctional uterus: dysmenorrhoea and dysfunctional uterine bleeding. In: Shearman, R. P., ed. *Clinical Reproductive Endocrinology.* Churchill Livingstone, Edinburgh, pp. 579–598.

Fraser, I. S. 1989. *Treatment of menorrhagia. Baillieres Clin. Obstet. Gynaecol.* 3:391–402.

Fraser, I. S., Hickey, M., and Song, J. Y. 1996. A comparison of mechanisms underlying disturbances of bleeding caused by spontaneous dysfunctional uterine bleeding or hormonal contraception. *Hum. Reprod.* 11(suppl 2):165–178.

Fraser, I. S., McCarron, G., Hutton, B. F., and Macey, D. 1987. Endometrial blood flow measured by xenon-133 clearance in women with normal menstrual cycles and dysfunctional uterine bleeding. *Am. J. Obstet. Gynecol.* 156:158–166.

Fraser, I. S., McCarron, G., Markham, R., and Resta, T. 1985. Blood and total fluid content of menstrual discharge. *Obstet. Gynecol.* 65:194–198.

Fraser, I. S., McCarron, G., Markham, R., Resta, T., and Watts, A. 1986. Measured menstrual blood loss in women with menorrhagia associated with pelvic disease or coagulation disorder. *Obstet. Gynecol.* 68:630–633.

Gargett, C. E., Lederman, F. L., Lau, T. M., Taylor, N. H., and Rogers, P. A. 1999. Lack of correlation between vascular endothelial growth factor production and endothelial cell proliferation in the human endometrium. *Hum. Reprod.* 14:2080–2088.

Gath, D., Osborn, M., and Bungay, G. 1987. Psychiatric disorder and gynaecological symptoms in middle-aged women: a community survey. *Br. Med. J.* 294:213–218.

Gordon, J. D., Shifren, J. L., Foulk, R. A., Taylor, R. N., and Jaffe, R. B. 1995. Angiogenesis in the female reproductive tract. *Obstet. Gynecol. Surv.* 50:688–696.

Hallberg, L., Hogdahl, A. M., Nilsson, L., and Rybo, G. 1966. Menstrual blood loss—a population study. Variation at different ages and attempts to define normality. *Acta Obstet. Gynecol. Scand.* 45:320–351.

Hampton, A. L., and Salamonsen, L. 1994. Expression of messenger ribonucleic acid encoding matrix metalloproteinases and their tissue inhibitors is related to menstruation. *J. Endocrinol.* 141:R1–R3.
Haynes, P. J., Hodgson, H., Anderson, A. B. M., and Turnbull, A. C. 1977. Measurement of menstrual blood loss in patients complaining of menorrhagia. *Br. J. Obstet. Gynaecol.* 84:763–768.
Henzl, M. R., Smith, R. E., Boost, G., and Tyler, E. T. 1972. Lysosomal concept of menstrual bleeding in humans. *J. Clin. Endocrinol. Metab.* 34:860–875.
Hickey, M., Carati, C., Manconi, F., Gannon, B. J., Dwarte, D., and Fraser, I. S. 2000a. The measurement of endometrial perfusion in Norplant users: a pilot study. *Hum. Reprod.* 15:1086–1091.
Hickey, M., Fraser, I. S., Dwarte, D., and Graham, S. 1996. Endometrial vasculature in Norplant users: preliminary results from a hysteroscopic study. *Hum. Reprod.* 11(suppl 2):35–44.
Hickey, M., Higham, J., Sullivan, M., et al. 2000b. Endometrial bleeding in hormone replacement therapy users: preliminary findings regarding the role of MMP-9. *Fertil. Steril.* in press.
Hickey, M., Simbar, M., Markham, R., Young, L., Manconi, F., Russell, P., and Fraser, I. S. 1999b. Changes in vascular basement membrane in the endometrium of Norplant users. *Hum. Reprod.* 14:716–721.
Hickey, M., Simbar, M., Young, L., Markham, R., Russell, P., and Fraser, I. S. 1999a. A longitudinal study of changes in endometrial microvascular density in Norplant implant users. *Contraception* 59:123–129.
Hirschi, K. K., and D'Amore, P. A. 1996. Pericytes in the microvasculature. *Cardiovasc. Res.* 32:687–698.
Hoffman, M. K., Meilstrup, J. W., Shackelford, D. P., and Kaminski, P. F. 1997. Arteriovenous malformations of the uterus: an uncommon cause of vaginal bleeding. *Obstet. Gynecol. Surv.* 52:736–740.
Hourihan, H. M., Sheppard, B. L., and Bonnar, J. 1986. A morphometric study of the effect of oral norethisterone or levonorgestrel on endometrial blood vessels. *Contraception* 38:603–612.
Hourihan, H. M., Sheppard, B. L., and Bonnar, J. 1989. The morphologic characteristics of menstrual haemostasis in patients with unexplained menorrhagia. *Int. J. Gynecol. Pathol.* 8:221–229.
Huang, M. W., Muradali, D., Thurston, W. A., Burns, P. N., and Wilson, S. R. 1998. Uterine arteriovenous malformations: Gray-scale and Doppler US features with MR imaging correlation. *Radiology* 206:115–123.
Iruela-Arispe, M. L., Rodriguez-Manzaneque, J. C., and Abu-Jawdeh, G. 1999. Endometrial endothelial cells express estrogen and progesterone receptors and exhibit a tissue specific response to angiogenic growth factors. *Microcirculation* 6:127–140.
Kamat, B. R., and Isaacson, P. G. 1987. The immunocytochemical distribution of leukocyte populations in human endometrium. *Am. J. Pathol.* 127:66–73.
Kasonde, J. M., and Bonnar, J. 1976. Plasminogen activators in the endometrium of women using intrauterine contraceptive devices. *Br. J. Obstet. Gynaecol.* 83:315–319.
Koh, S. C. L., Wong, P. C., Yuen, R., Chua, S. E., Ng, B. L., and Ratnam, S. S. 1992. Concentration of plasminogen activators and inhibitors in human endometrium at different stages of the menstrual cycle. *J. Reprod. Fertil.* 96:407–413.
Lakhani, K. P., Marsh, M. S., Purcell, W., and Hardiman, P. 1998. Uterine artery blood flow parameters in women with dysfunctional uterine bleeding and uterine fibroids: the effects of tranexamic acid. *Ultrasound Obstet. Gynecol.* 11:283–285.
Landgren, B. M., Johannisson, E., Masirono, B., and Diczfalusy, E. 1982. Pharmacokinetic and pharmacodynamic investigations with vaginal devices releasing levonorgestrel at a constant, near zero order. *Contraception* 26:567–585.

Larson, R. S., and Springer, T. A. 1990. Structure and function of leukocyte integrins. *Immunol. Rev.* 114:181–217.

Lau, T. M., Affandi, B., and Rogers P. A. W. 1999. The effects of levonorgestrel implants on vascular endothelial growth factor expression in the endometrium. *Mol. Hum. Reprod.* 5:557–563.

Lockwood, C. J., Krikun, G., Hausknecht, V., Wang, E. Y., and Schatz, F. 1997. Decidual cell regulation of hemostasis during implantation and menstruation. *Ann. NY Acad. Sci.* 828:188–193.

Lumsden, M. A. 1998. Clinical presentation of uterine fibroids. *Baillieres Clin. Obstet. Gynaecol.* 12:177–195.

Manolitsas, T., Hurley, V., and Gilford, E. 1994. Uterine arteriovenous malformation—a rare cause of uterine haemorrhage. *Aust. N. Z. J. Obstet. Gynaecol.* 34:197–199.

Maqueo, M. 1980. Vascular and perivascular changes in the endometrium of women using steroidal contraceptives. In: Diczfalusy, E., Fraser, I. S., and Webb, F. T. G., eds. *Endometrial Bleeding and Steroidal Contraception.* Pitman Press, Bath, pp. 138–152.

Marbaix, E., Kokorine, I., Henriet, P., Donnez, J., Courtoy, P. J., and Eeckhout, Y. 1995. The expression of interstitial collagenase in human endometrium is controlled by progesterone and by oestradiol and is related to menstruation. *Biochem. J.* 305:1027–1030.

Markee, J. E. 1940. Menstruation in intraoccular endometrial transplants in the rhesus monkey. *Contrib. Embryol.* 177:220–306.

Markee, J. E. 1948. Morphological basis for menstrual bleeding: relation of regression to the initiation of bleeding. *Bull. NY Acad. Med.* 24:253–268.

Markee, J. E. 1950. The relation of blood flow to endometrial growth and the inception of menstruation. In: Engle, E. T., ed. *Menstruation and Its Disorders.* Charles C. Thomas, Springfield, IL, pp. 165–185.

Marleschki, V. 1968. Hysteroskopische Feststellung der spontanen Perfusionsschwankugen am menschlichen Endometrium. *Zentralbl. Gynakol.* 90:1094–1097.

Marsh, M. M., Findlay, J. K., and Salamonsen, L. A. 1996. Endothelin and menstruation. *Hum. Reprod.* 11(suppl 2):83–89.

Matsumoto, Y., Iwasaka, T., Yamasaki, F., and Sugimori, H. 1999. Apoptosis and Ki-67 expression in adenomyotic lesions and in the corresponding eutopic endometrium. *Obstet. Gynecol.* 94:71–77.

Millauer, B., Wizigmann-Voos, S., Schnürch, H., Martinez, R., Möller, N. P., Risau, W., and Ullrich, A. 1993. High affinity VEGF binding and developmental expression suggests Flk-1 as a major regulator of vasculogenesis and angiogenesis. *Cell* 72:835–846.

Neufield, G., Tessler, G., and Gitay, G. H. 1994. Vascular endothelial growth factor and its receptors. *Prog. Growth Factor Res.* 5:89–97.

Odlind, V., and Fraser, I. S. 1990. Contraception and menstrual bleeding disturbances: a clinical overview. In: d'Arcangues, C., Fraser, I. S., Newton, J. R., and Odlind, V., eds. *Contraception and Mechanisms of Endometrial Bleeding.* Cambridge University Press, Cambridge, pp. 5–31.

Ota, H., Igarashi, S., and Tanaka, T. 1998. Morphometric evaluation of stromal vascularization in the endometrium of adenomyosis. *Hum. Reprod.* 13:715–719.

Padwick, M. L., Pryse-Davies, M. I., and Whitehead, M. 1986. A simple method for determining the optimal dose of progestin in women receiving estrogens. *N. Engl. J. Med.* 315:930–933.

Pepper, M. S., Ferrara, N., Orci, L., and Montesano, R. 1991. Vascular endothelial growth factor (VEGF) induces plasminogen activators and plasminogen activator inhibitor-1 in microvascular endothelial cells. *Biochem. Biophys. Res. Commun.* 181:902–906.

Perrot-Aplanat, M., Groyer-Pickard, M. T., Garcia, E., Lorenzo, F., and Milgrom, E. 1988. Immunocytochemical demonstration of estrogen and progesterone receptors in muscle cells of uterine arteries in rabbits and humans. *Endocrinology* 123:1511–1519.

Purcell, T. L., Given, R., Chwalisz, K., and Garfield, R. E. 1999. Nitric oxide synthetase distribution during implantation in the mouse. *Mol. Hum. Reprod.* 5:467–475.
Rai, R., and Regan, L. 1998. Antiphospholipid syndrome and pregnancy loss. *Hosp. Med.* 59:637–639.
Ramsey, E. M. 1982. Vascular anatomy. In: Wynn, R. M., ed. *Biology of the Uterus*, 2nd ed. Plenum Press, New York, pp. 59–76.
Rand, J. H. 1998. Antiphospholipid antibody syndrome: new insights on thrombogenic mechanisms. *Am. J. Med. Sci.* 316:142–151.
Rand, J. H., Wu, X. X., Andree, H. A., Lockwood, C. J., Guller, S., Scher, J., and Harpel, P. C. 1997. Pregnancy loss in the antiphospholipid-antibody syndrome—a possible thrombogenic mechanism. *N. Engl. J. Med.* 337:154–160.
Ravina, J. A., Herbreteau, D., Ciraru-Vigneron, N., Bouret, J. M., Houdart, E., Aymard, A., and Merland, J. J. 1995. Arterial embolisation to treat uterine myomata. *Lancet* 346:671–672.
Rees, M. C. P., Canete-Soler, R. Lopez-Bernal, A., and Turnbull, A. C. 1988. Effect of fenamates on prostaglandin E receptor binding. *Lancet* 2:541–542.
Rees, M. C. P., Cederholm-Williams, S. A., and Turnbull, A. C. 1985. Coagulation factors and fibrinolytic proteins in menstrual fluid collected from normal and menorrhagic women. *Br. J. Obstet. Gynaecol.* 92:1164–1168.
Rees, M. C. P., Demers L. M., Anderson, A. B. M., and Turnbull, A. C. 1984a. A functional study of platelets in menstrual fluid. *Br. J. Obstet. Gynaecol.* 91:667–672.
Rees, M. C. P., Dunnhill, M. S., Anderson, A. B., and Turnbull, A. C. 1984b. Quantitative uterine histology during the menstrual cycle related to measured blood loss. *Br. J. Obstet. Gynaecol.* 91:662–666.
Rein, M. S., Barbieri, R. L., and Friedman, A. J. 1995. Progesterone: a critical role in the pathogenesis of uterine myomas. *Am. J. Obstet. Gynecol.* 172:14–18.
Rogers, P. A. W. 1995. Current studies on human implantation. *Reprod. Fertil. Dev.* 7:1395–1399.
Rogers, P. A. W., Au, C. L., and Affandi, B. 1993. Endometrial microvascular density during the normal menstrual cycle and following exposure to long-term levonorgestrel. *Hum. Reprod.* 8:1396–1404.
Rogers, P. A. W., Lederman, F., and Taylor, N. 1998. Endometrial microvascular growth in normal and dysfunctional states. *Hum. Reprod. Update* 4:503–508.
Rogers, P. A. W., Salamonsen, L. A., Runic, R., Schata, F., and Krey, L. 1997. Alterations in endometrial stromal cell tissue factor protein and messenger ribonucleic acid expression in patients experiencing abnormal uterine bleeding while using Norplant-2 contraception. *J. Clin. Endocrinol. Metab.* 72:1983–1988.
Runic, R., Schatz, F., Krey, L., Demopoulos, R., Thung, S., Wan, L., and Lockwood, C. J. 1997. Alterations in endometrial stromal cell tissue factor protein and messenger ribonucleic acid expression in patients experiencing abnormal uterine bleeding while using Norplant-2 contraception. J. Clin. Endocrinol. Metab. 82:1983–1988.
Rybo, G., Leman, J., and Tibblin, E. 1985. Epidemiology of menstrual blood loss. In: Baird, D. T., and Michie, E. A., eds. *Mechanism of Menstrual Bleeding.* Raven Press, New York, pp. 181–194.
Salamonsen, L. A., Marsh, M. M., and Findlay, J. K. 1999. Endometrial endothelin: regulator of uterine bleeding and endometrial repair. *Clin. Exp. Pharmacol. Physiol.* 26:154–157.
Salamonsen, L. A., and Woolley, D. E. 1996. Matrix metalloproteinases in normal menstruation. *Hum. Reprod.* 11(suppl 2):124–133.
Salvatore, C. A. 1968. Arteriole alternation in normal and hyperplastic endometrium. *Am. J. Obstet. Gynecol.* 102:493–500.
Sapozhnikov, A. G. 1987. Structure of the blood microcirculatory bed in intramural uterine myomas. *Vopr. Onkol.* 33:43–48.

Sarfati, R., and Brux, J. 1973. Vascular fragility of the endometrium under oestrogen-progestin treatment. *Eur. J. Obstet. Gynecol. Reprod. Biol.* 2:87–94.

Sheppard, B. L. 1990. Coagulation and electron microscopy studies in menorrhagia. In: Shaw, R. W., ed. *Advances in Reproductive Endocrinology*, vol. 2. Parthenon Press, Carnforth, Lancs, UK. pp. 25–42.

Sixma, J., Christiaens G. C. M. L., and Haspels, A. A. 1980. The sequence of haemostatic events in the endometrium during normal menstruation. In: Diczfalusy, E., Fraser, I. S., and Webb, F. T. G., eds. *Endometrial Bleeding and Steroidal Contraception.* Pitman Press, Bath, England, pp. 86–96.

Skinner, J. L., Riley, S. C., Gebbie, A. E., Glasier, A. F., and Critchley, H. O. 1999. Regulation of matrix metalloproteinase-9 in endometrium during the menstrual cycle and following administration of intrauterine levonorgestrel. *Hum. Reprod.* 14:793–799.

Skopichev, V. G., and Savitskii, G. A. 1992. Changes in the human uterine vascular bed in myoma. *Arkh. Patol* 54:27–30.

Song, J. Y., Russell, P., Markham, R., Manconi, F., and Fraser, I. S. 1996. Effects of high dose progestogens on white cells and necrosis in human endometrium. *Hum. Reprod.* 11:1713–1718.

Sosic, A., Skupski, D. W., Streltzoff, J., Yun, H., and Chervenak, F. A. 1996. Vascularity of uterine myomas: assessment by color and pulsed Doppler ultrasound. *Int. J. Gynaecol. Obstet.* 54:245–250.

Smith, S. K. 1990. The physiology of menstruation. In: d'Arcangues, C., Fraser, I. S., Newton, J. R., and Odlind, V., eds. *Contraception and Mechanisms of Endometrial Bleeding.* Cambridge University Press, Cambridge, pp. 33–41.

Smith, S. K. 1998. Angiogenesis, vascular endothelial growth factor and the endometrium. *Hum. Reprod. Update* 4:509–519.

Smith, S. K., Abel, M. H., Kelly, R. W., and Baird, D. T. 1981. A role for prostacyclin (PGI_2) in excessive menstrual bleeding. *Lancet* 1(8219):522–524.

Smith, S. K., Abel, M. H., Kelly, R. W., and Baird, D. T. 1982. The synthesis of prostaglandins from persistent proliferative endometrium. *J. Clin. Endocrinol. Metab.* 55:284–289.

Stewart, E. A., and Nowak, R. A. 1996. Leiomyoma-related bleeding: a classic hypothesis updated for the molecular era. *Hum. Reprod. Update* 2:295–306.

Subakir, S. B., Hadisaputra, W., Handoyo, A. E., and Affandi, B. 1996. Endometrial angiogenic response in Norplant users. *Hum. Reprod.* 11(suppl 2):51–55.

Tabibzadeh, S. 1991. Human endometrium: an active site of cytokine production and action. *Endocr. Rev.* 12:272–290.

Tabibzadeh S. 1996. The signals and molecular pathways involved in human menstruation, a unique process of tissue destruction and remodelling. *Mol. Hum. Reprod.* 2:770–792.

Tabibzadeh, S., and Satyaswaroop, P. G. 1989. Sex steroid receptors in lymphoid cells of human endometrium. *Am. J. Clin. Pathol.* 91:656–663.

Tabibzadeh, S., and Sun, X. Z. 1992. Cytokine expression in human endometrium throughout the menstrual cycle. *Hum. Reprod.* 7:1214–1221.

Treloar, A. E., Boynton, R. E., Behn, B. G., and Brown, B. W. 1970. Variation of the human menstrual cycle throughout reproductive life. *Int. J. Fertil.* 12:77–113.

Van Herendael, B. J., Stevens, M. J., Flakiewicz-Kula, A., and Hansch, C. 1987. Dating of the endometrium by microhysteroscopy. *Gynecol. Obstet. Invest.* 24:114–118.

Vincent, A. J., Malakooti, N., Zhang J., Rogers, P. A., Affandi, B., and Salamonsen, L. A. 1999. Endometrial breakdown in women using Norplant is associated with migratory cells expressing matrix-metalloproteinase-9 (gelatinase B). *Hum. Reprod.* 14:807–814.

Vollman, R. F. 1977. *The Menstrual Cycle.* W. B. Saunders, Philadelphia.

Wang, I. Y. S., and Fraser, I. S. 1990. Lysosomes: an important mediator in the human reproductive tract. *Obstet. Gynecol. Surv.* 45:18–33.

Wang, I. Y., Fraser, I. S., Barsamian, S. P., Manconi, F., Street, D. J., Cornillie, F. J., and Russell, P. 2000. Endometrial lysosomal enzyme activity in ovulatory dysfunctional bleeding, IUCD users and in postpartum women. *Mol. Hum. Reprod.* 6:101–106.

Yang, J. H., Wu, M. Y., Chen, C. D., Jiang, M. C., Ho, H. N., and Yang, Y. S. 1999. Association of endometrial blood flow as determined by a modified colour Doppler technique with subsequent outcome of in-vitro fertilisation. *Hum. Reprod.* 14:1606–1610.

CHAPTER **13**

Angiogenesis in Endometriosis

Michael D. Mueller and Robert N. Taylor

Endometriosis is a common gynecologic disorder defined by the presence, growth, and progression of endometrial tissue outside the uterine cavity. Endometriosis can be asymptomatic, but it typically is associated with a multitude of clinical symptoms, the most common being pelvic pain and impaired fertility. The disease affects about 10% of all reproductive-age women and the prevalence rises to 20% to 50% in infertile women. Genes seem to influence liability to endometriosis (Treloar et al, 1999); however, the mode of hereditary transmission appears to be multifactorial (Malinak et al, 1980). Retrograde menstruation and increased exposure to menstrual flow (reduced parity, longer duration of menstrual periods, or shorter cycle length) have been identified as risk factors for endometriosis. Recently, environmental factors, in particular dioxin (Rier et al, 1993) and other presumed endocrine disrupters have been implicated in its development.

Despite growing numbers of hypotheses, the etiology of the disease still remains an enigma. The implantation hypothesis, proposed by Sampson (1927) in the 1920s is the most widely accepted, albeit with some disagreement (Redwine, 1988). Recent studies have supported the Sampson theory by documenting that retrograde menstruation (Halme et al, 1984) and peritoneal spillage of viable endometrial epithelial cells occur frequently in cycling women (Kruitwagen et al, 1991). Some studies suggest that microscopic peritoneal defects allow the cells to contact the submesothelial matrix (Groothuis et al, 1999) where they proliferate, spread, and grow, sometimes invading deeply into the subperitoneal space. However, other investigators assert that endometrial fragments can adhere directly to intact mesothelium (Witz et al, 1999). Given the ubiquity of retrograde menstruation, it is not clear why endometriosis develops in only a subset of cycling women. Using the analogy of tumor metastasis (Hanahan and Folkman, 1996), we postulated that the angiogenic potential of the intraperitoneal environment may dictate the likelihood that lesions become established (Taylor et al, 1997). Indeed, on visualization, endometriotic implants are often surrounded by exaggerated vascularity, and extrapelvic endometriosis, while rare, typically occurs in well-vascularized organs (Ryan and Taylor, 1997). Microscopic studies have confirmed neovascularization around and within endometriosis implants (Nisolle et al, 1993). Thus, angiogenesis in ectopic implants is likely to represent a critical step in the development of endometriosis.

The sprouting of new blood vessels from preexisting vessels is complex and includes the proteolytic degradation of extracellular matrix, proliferation and migration of endothelial cells, and ultimately the formation of patent capillary tubules supplying the angiogenic stimulus. Many different growth factors and cytokines have been shown to exert chemotactic, mitogenic, modulatory, or inhibitory activities on endothelial cells, smooth muscle cells, and pericytes, and therefore can be expected to participate in angiogenic processes in either direct or indirect ways. As angiogenesis occurs only when the balance of local factors promoting vascular growth exceeds those factors inhibiting angiogenesis, the different paracrine and autocrine regulatory mechanisms promoting or inhibiting angiogenesis are considered separately in this chapter.

PERITONEAL FLUID: A WINDOW ON THE ENVIRONMENT OF ENDOMETRIOSIS

The syndrome of endometriosis is unique to menstruating Old World primate species. In addition to women, spontaneous endometriosis has been observed in female rhesus monkeys and baboons (D'Hooghe, 1997). Some attempts to create small animal models of endometriosis by surgically introducing endometrial fragments into the peritoneal cavity have met with success (Sharpe-Timms, 1997). However, the clinical applicability of these models has been questioned (D'Hooghe, 1997). Due to technical and ethical issues of repeated operative interventions in women and primates, it has been a challenge for investigators to directly observe the natural history and pathogenesis of spontaneous endometriosis.

Fortunately, because of its clinical accessibility and proximity to the site of endometriosis growth, peritoneal fluid has been evaluated as an indicator of the in vivo milieu of endometriosis lesions. Although pelvic fluid is a complex medium containing and influenced by multiple cell types, we (Khorram et al, 1993; Ryan et al, 1995) and other investigators (Oosterlynck et al, 1993) have found this to be a useful biologic fluid to define intercellular mediators of the local reactions associated with endometriosis. Peritoneal fluid seems to play an important role in the process of neovascularization. Oosterlynck et al (1993) used the chick chorioallantoic membrane bioassay to show that pelvic fluid from women with endometriosis had more angiogenic activity than pelvic fluid from normal controls. We confirmed this observation using an in vitro model of endothelial cell proliferation. [^{3}H]thymidine incorporation into nascent cellular DNA demonstrated that peritoneal fluid from women with endometriosis contains significantly more angiogenic activity than pelvic fluid from normal controls (Taylor et al, 1997).

In recent years it has become apparent that a vast array of angiogenic principles are synthesized by the human endometrium. Indeed, as discussed elsewhere in this volume, the human endometrium has evolved a plethora of gene products that serve to promote a carefully regulated regeneration of the endometrial functionalis vascular tree at the beginning of each ovulatory cycle. Borrowing from our growing knowledge of the pro- and antiangiogenic factors synthesized in the normal cycling endometrium and peritoneal fluid, we will consider the potential cytokines, steroids, and eicosanoids that modulate vascularization of endometriosis implants.

ANGIOGENESIS STIMULATORS

Cytokines Influencing Neovascularization in Endometriosis

Cyclical vascular changes within the primate endometrium have been recognized for years. Some of the first observations were the classical experiments of Markee (1940), who transplanted rhesus endometrium into the anterior chamber of the eye, allowing direct visualization of this dynamic tissue. His interpretation that profound endometrial vascular changes paralleled the ovulatory cycle enlightened our understanding of the endocrine regulation of endometrial angiogenesis. In their landmark article, Noyes and colleagues (1950) reported the effects of enhanced vascular permeability and perivascular edema during the secretory phase of the human endometrial cycle, providing early clues that steroids from the corpus luteum orchestrate predictable vascular responses in normal endometrium. In recent years it has become apparent that some steroid hormone actions on uterine tissues are not direct but are effected by paracrine mediators such as cytokines and growth factors (Nelson et al, 1991). Cytokines are key mediators of intercellular communication within the immune system. They have pleiotropic activities on a variety of target cells, exerting proliferative, cytostatic, chemoattractant, or differentiative effects. Specific cytokines that are directly or indirectly involved in the process of angiogenesis in endometriosis are described below.

Interleukin-1

Interleukin-1 (IL-1) is a pleiotropic cytokine playing a key role in the regulation of inflammation and immune responses. Monocytes, activated macrophages from different sources (e.g., peritoneal macrophages), peripheral neutrophil granulocytes, and other cell types produce IL-1. There are two forms of IL-1: IL-1α (17 kd, 159 amino acids) and IL-1β (17 kd, 153 amino acids) encoded by two different genes. At the protein level IL-1α and IL-1β have only approximately 25% homology. However, the three-dimensional structures of the two IL-1 forms are almost identical. Therefore, they bind to the same receptor and show similar biologic activities. Different IL-1 receptors (IL-1RI and II) that bind with different affinities have been described and their densities generally increase following the stimulation of cells by IL-1. Exposure to glucocorticoids increases IL-1 receptor density approximately 10-fold. IL-1β preferentially binds IL-1RII, but the different IL-1 isoforms can bind to both IL-1Rs. Furthermore, IL-1α and IL-1β block the binding of each other to the receptors. IL-1 receptor–mediated signal transduction involves diverse pathways including eicosanoid production and activation of adenylate cyclase, which transiently increases intracellular cyclic adenosine monophosphate (cAMP) levels. Endometrial epithelial cells express IL-1RI, with highest concentrations in the secretory phase (Simón et al, 1993). The IL-1 system has been implicated as an important mediator of embryo implantation, and administration of exogenous IL-1 receptor antagonist prevents successful implantation in mice (Simón et al, 1994), although IL-1 knockout mice appear to have normal fecundity.

IL-1β is the dominant IL-1 secreted by activated peritoneal macrophages and its levels are elevated in peritoneal fluid of women with endometriosis (Mori et al, 1992). There are conflicting data regarding the levels of soluble IL-1 in peritoneal fluid of women with endometriosis. Some studies showed elevated levels in stage I and II endometriosis cases (Fakih et al, 1987; Taketani et al, 1992), whereas others find no significant elevation (Keenan et al, 1995).

We recently showed that the IL-1β production by activated peritoneal macrophages plays an important role in the neovascularization that surrounds endometriotic lesions (Lebovic et al, 2000). By Northern blot analysis we demonstrated a 2.6-fold induction of vascular endothelial growth factor (VEGF) messenger RNA (mRNA) and a 6.3-fold increase in IL-6 mRNA transcripts in endometriosis stromal cells exposed to IL-1β, whereas normal endometrial stromal cells had no significant increase in these transcripts. A similar induction was seen for VEGF and IL-6 protein secretion in the endometriosis-derived cells. Reverse-transcriptase polymerase chain reaction (RT-PCR) for the IL-1RI indicated that the differential effects of IL-1β on normal endometrial and endometriosis cells were associated, in part, with more receptor mRNA in endometriosis-derived cells. Therefore, it appears that IL-1β has the capability of activating an angiogenic phenotype in endometriosis stromal cells but not in normal endometrial cells and that this effect is mediated through IL-1 RI.

Interleukin-6

Interleukin-6 (IL-6) is a multifunctional cytokine that is produced by many different cell types. The main sources in vivo are stimulated monocytes, fibroblasts, and endothelial cells. Endometrial glandular and stromal cells also express IL-6 (Tabibzadeh et al, 1995; Tseng et al, 1996). IL-6 expression varies across the menstrual cycle; secretory-phase endometrium expresses more IL-6 mRNA and IL-6 protein than proliferative-phase endometrium (Vandermolen and Gu, 1996). IL-6 forms a complex with its receptor and associates with a transmembrane glycoprotein, gp130, that is involved in signal transduction. The IL-6 receptor is primarily found in the endometrial glands and to a lesser extent in the stroma (Tabibzadeh et al, 1995). Isolated endometrial stromal cells from the proliferative phase of the cycle have very low levels of IL-6R mRNA (Mueller and Taylor, unpublished observations). IL-6 has pleiotropic effects and acts on a wide range of tissues. In the endometrium and in endometriosis, IL-6 has been suggested to be an important immunomodulator (Tseng et al, 1996). Suggesting a role for IL-6 in angiogenesis, Motro et al (1990) demonstrated that the in vivo expression of IL-6 accompanies vascularization in the development of ovarian follicles and in the maternal decidua following embryonic implantation. Furthermore, IL-6 expression has been noted during wound healing and tumor growth (Mateo et al, 1994), and Cohen et al (1996) provided evidence indicating that IL-6 may promote angiogenesis through the induction of VEGF expression.

Using an in vitro model of endometriosis, established in our laboratory (Ryan et al, 1994), we demonstrated that endometriosis stromal cells derived from ovarian endometriomas express significantly higher basal and cytokine-induced IL-6 than stromal cells derived from normal endometrium (Tseng et al, 1996). Curiously, using this model we found that cells derived from eutopic endometrium of endometriosis patients expressed intermediate levels of IL-6. An attractive interpretation of this finding, provided the Sampson hypothesis is accepted, is that the endometrial lining of women destined to develop endometriosis already manifests an "activated" angiogenic phenotype. As an analogy to Knudson's "two-hit" hypothesis of oncogenesis (Knudson, 1977), women susceptible to the development of endometriosis may carry a "first hit" genetic trait that enhances endometrial cell angiogenic factor production. This proposal is in keeping with the familial trends observed in cases of endometriosis (Kennedy, 1999).

The observation that angiogenic potential is increased in endometriosis implants has been extended by in vivo investigation. IL-6 mRNA was noted to be upregulated significantly in endometriomas relative to normal ovarian tissue (Odukoya et al, 1997). In toto, these findings suggest that the local expression of angiogenic cytokines in endometriosis contributes to the preferential neovascularization of these implants relative to normal tissue.

Interleukin-8

Interleukin-8 (IL-8) is a member of the α or C-X-C chemokine family, a group of small peptide inflammatory mediators characterized by four conserved cysteine residues. In all members of this subfamily the first cysteine residues are separated by a variable amino acid. IL-8 is an 8-kd peptide that is produced by monocytes, lymphocytes, fibroblasts, and epithelial and endothelial cells in response to a variety of proinflammatory stimuli (e.g., IL-1, tumor necrosis factor-α [TNF-α], lipolysaccharide). Biologic activity is exerted by binding to two distinct but homologous receptors, CXCR I and CXCR II, that share 77% amino acid homology. Both of these are members of the G-protein–coupled receptor family. While the type I receptor specifically binds IL-8, the type II receptor also can bind other chemokines (LaRosa et al, 1992). IL-8 recruits and activates polymorphonuclear leukocytes and T lymphocytes and has angiogenic activity (Koch et al, 1992).

In recent years, several authors have suggested a role of IL-8 in endometrial physiology. In vivo, IL-8 protein is selectively localized to the perivascular smooth muscle cells and varies across the menstrual cycle (Milne et al, 1999). In vitro, IL-8 gene expression and protein secretion were documented in cultured endometrial stromal cells stimulated with IL-1β, TNF-α, or serum (Arici et al, 1993). In the latter model, co-incubation with progesterone caused an increase in IL-8 mRNA synthesis and stability (Arici et al, 1996a). Cultured epithelial cells also express IL-8 mRNA, although steady-state concentrations in stromal cells were higher. Therefore, both endometrial stromal and epithelial cells have the potential to secrete IL-8 upon stimulation by hormones and other cytokines, and may contribute to the recruitment of leukocytes (Bulmer et al, 1991) and especially neutrophils into the endometrium. By producing VEGF (Mueller et al, 2000), also known as vascular permeability factor (VPF), neutrophils may modulate endometrial capillary tight junctions, allowing extravasation of lymphomyeloid cells into the stroma. Neutrophil-derived elastase and other proteinases facilitate endothelial cell migration by degrading vascular basement membranes and other subendothelial matrix components of the endometrium and therefore play an indirect role in endometrial angiogenesis.

In endometriosis, we showed that IL-8 protein is present in the peritoneal fluid of patients with endometriosis, and documented a correlation between IL-8 concentration and endometriosis stage (Ryan et al, 1995). We hypothesize that IL-8 is derived from pelvic macrophages and from the endometriosis implants itself, and contributes to the pathogenesis of endometriosis by promoting the neovascularization and hence proliferation of ectopic endometrial implants. Our pelvic fluid findings are supported by studies by Arici et al (1996c) and Rana et al (1996), who showed increased levels of IL-8 secreted by peritoneal macrophages from patients with endometriosis.

Growth Factors Influencing Neovascularization in Endometriosis

Several human endometrial growth factors with known angiogenic and/or endometrial cell mitogenic activity have been identified. Although their precise effects on endometriosis are not as well known at the current time, these are potential regulators of endometriosis implant neovascularization and modulators of the secretion of other angiogenic cytokines and growth factors.

Epidermal Growth Factor

Epidermal growth factor (EGF) is a globular protein of 6 kd and contains three intramolecular disulfide bonds essential for biologic activity. EGF is the prototype

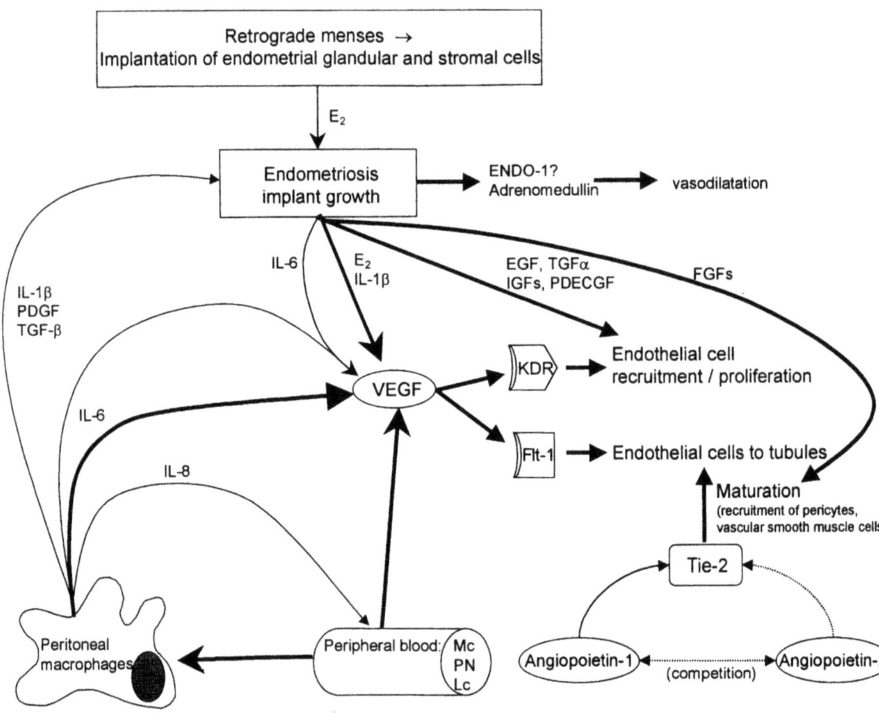

FIGURE 13.1. Growth factor and cytokine networks modulating angiogenesis in endometriosis. Mc, monocytes; PN, polymorphonuclear cells; Lc, lymphocytes; E_2, estradiol. See text for other abbreviations.

of a large family of EGF-like proteins characterized by sequence homologies and the presence of distinct EGF-like repeats containing conserved cysteine residues. Transforming growth factor-α (TGF-α), betacellulin, cripto, epiregulin, and amphiregulin are, among others, also members of this family. These proteins appear to be involved in protein-to-protein and protein-to-cell interactions, presumably mediated by the EGF-like sequence domains. Like EGF, most of these proteins also are involved in embryonic development (Miyauchi et al, 1995). TGF-α shows a strong homology to EGF. Both factors are functionally analogous, bind to the same receptor and show more or less the same biologic activities. However, antibodies specific for EGF do not neutralize TGF-α. Both EGF (Gospodarowicz et al, 1978) and TGF-α have angiogenic activity (Folkman, 1995).

The human EGF receptor-1 (HER-1) is a 170-kd transmembrane glycoprotein that appears to exist in two conformational states binding EGF with either low or high affinity. EGF receptors (EGFRs) are expressed in almost all cell types, including endometriosis epithelial and stromal cells (Ryan and Taylor, unpublished observations). Binding of EGF to its receptor leads to conformational changes, receptor dimerization, and autophosphorylation of receptor tyrosine kinase. In the endometrium EGF and EGFR appear to mediate estrogen-induced proliferative effects through the coupling of signaling pathways (Nelson et al, 1991; Curtis et al, 1996).

In vivo, the synthesis and expression of human endometrial EGF, TGF-α, and EGFR vary with the stage of the menstrual cycle and are increased at the late follicular and luteal stages compared to the early follicular stage (Imai et al, 1995). However, in human endometrial stromal cell monolayer cultures, progesterone alone is sufficient to upregulate EGFR (Singer et al, 1998) and the combination of estradiol and progesterone causes a greater increase in EGFR concentration than does either steroid alone (Watson et al, 1996). EGF and EGFR are expressed in endometriosis implants (Huang and Yeh, 1994). In patients with endometriosis, treatment with goserelin, a gonadotropin-releasing hormone (GnRH) analogue, is associated with a dramatic reduction of EGFR expression in eutopic endometrium and in endometriotic implants (Di Lieto et al, 1997).

EGF is found in varying concentrations in most body fluids. Peritoneal fluid concentrations of EGF are highly variable and there is no significant difference in the concentrations of EGF in the peritoneal fluid of women with or without endometriosis (Huang et al, 1996).

Fibroblast Growth Factors

Fibroblast growth factors (FGFs) constitute a family of nine related 17- to 38-kd proteins playing major roles in development, wound healing, hematopoiesis, and tumorigenesis. Four FGFs (1, 2, 4 and 7) have been identified in human endometrium (Presta, 1988; Ferriani et al, 1993). The two prototype factors—FGF-1 or acidic FGF (aFGF), and FGF-2 or basic FGF (bFGF)—are mitogenic for a variety of cell types, particularly capillary endothelial cells. However, FGF-2 does not induce proliferation in endothelial cells obtained from human decidua, and it is therefore unlikely that in the human endometrium FGF is angiogenic on its own (Grimwood et al, 1995).

In human endometrium FGF-2 has been localized in stromal cells, epithelial glands, and perivascular adventitia (Cordon-Cardo et al, 1990), where it may contribute to angiogenic activity. Whether the levels of endometrial FGF-2 change during the menstrual cycle is controversial. Some authors showed no cyclical influence (Ferriani et al, 1993). Salat-Baroux et al (1994) demonstrated an increase in FGF-1 and -2 during the secretory phase, while others revealed an increase in endometrial basic FGF in the proliferative phase followed by a decrease in the secretory phase (Sangha et al, 1997). There is no difference in the pattern of staining for FGF-1 and FGF-2 when endometrium from normal women is compared to eutopic or ectopic endometrium of women with endometriosis (Ferriani et al, 1993). FGF-2 is present in peritoneal fluid, but its concentration does not change during the menstrual cycle or in the presence of endometriosis (Seli et al, 1998).

FGFR-1 has been identified in human endometrium and markedly decreased FGFR-1 levels were found in patients with menorrhagia, suggesting that FGFR-1 is critical for endometrial "maturation" and regeneration (Sangha et al, 1997).

Insulin-Like Growth Factors

The insulin-like growth factors (IGFs) and their receptors and binding proteins (IGFBPs) represent a complex system of growth regulating factors in human endometrium. The IGFs are nonglycosylated peptides ~7 kd in size and evolutionary relatives of insulin. The IGF type I receptor (IGF-I-R) is structurally similar to the classical insulin receptor. It forms a heterotetrameric transmembrane complex that binds IGF-I and -II ligands with high affinity (1 to 5 nM) and signals cellular mitogenesis via tyrosine kinase domains in its β-subunits (Le Roith, 1996). The IGF type II receptor is a single membrane spanning protein identical to the mannose-6-phosphate receptor that also binds IGFs with high affinity. While it is believed that IGF-II-R does not transduce a mitogenic signal, mice carrying null mutations of a related gene manifest a specific reduction in placental growth (Baker et al, 1993). Six forms of human IGFBPs, ranging in size from 24 to 45 kd, have been identified. These proteins serve as IGF transporters and modulate local bioactive IGF concentrations. The bioactivity of the IGFBPs in turn may be modulated by endogenous serine proteases (Giudice et al, 1994).

In human endometrium IGF-I is expressed predominantly in the proliferative phase of the cycle, while IGF-II is abundant during the secretory phase (Giudice et al, 1993). Some controversy exists as to the localization of these products within the endometrial tissue. In situ hybridization analyses by Zhou et al (1994) suggest that IGF-l and -II transcripts were found predominantly in the stroma, whereas the two receptors were more abundant in the epithelium. Immunoreactive IGF-I and IGF-I-R proteins were reported to be localized predominantly in the glandular epithelium of this tissue (Tang et al, 1994a). Interestingly, the mitogenic effect of IGF-I on endometrial stromal cell cultures is significantly less than that induced by platelet-derived growth factor (PDGF) or EGF (Tang et al, 1994a); however, modulation by IGFBPs might alter these effects in vivo. IGFBPs 1 to 6 mRNA and protein have been identified in human endometrial tissue and cells (Giudice et al, 1994; Zhou et al, 1994).

Direct studies of the IGF system in endometriosis are very limited. Tang et al (1994a) describe more heterogeneous and localized glandular immunostaining of IGF-I-R in endometriosis patients compared to normal subjects. IGF-I and -II and IGFBP-2 and -3 (Giudice et al, 1994) have been identified in pelvic fluid from normal subjects but no differences could be found in IGF-II or IGFBP-3 when compared to fluid from endometriosis patients (Gurgan et al, 1999). Gurgan et al (1999) demonstrated that the serum levels of all three proteins were higher except for a reversed IGF-I pelvic fluid/serum ratio in the early stage of endometriosis. Whether this growth factor system plays a specific role in the development and/or maintenance of endometriosis or progression to late-stage disease remains to be proven.

Platelet-Derived Growth Factor

Platelet-derived growth factor (PDGF) was initially identified in platelets, but macrophages, endothelial cells, fibroblasts, smooth muscle cells, and a number of other cells synthesize PDGF. PDGF is a heat-stable cationic hydrophilic protein of 32 kd. It consists of two related polypeptide chains, PDGF-A and PDGF-B, linked by disulfide bonds. All possible isoforms, i.e., PDGF-AA, PDGF-BB, and PDGF-AB, are biologically active. PDGF is chemotactic and mitogenic for cells of mesenchymal origin, including endothelial cells, although the latter effect appears to be indirect (Nicosia et al, 1994).

Two distinct receptors for PDGF exist and are referred to as PDGFR-α (or CD140a) and PDGFR-β (or CD140b). The synthesis of PDGF receptors is subject to autoregulation by PDGF. The type α receptor binds all three PDGF isoforms with equal and high (dissociation constant [K_d] = 0.1 nM) affinity. The type β receptor binds PDGF-BB with high affinity and PDGF-AB with lower affinity (K_d = 1 nM), but does not bind PDGF-AA (Heldin et al, 1988). The receptors are monomeric transmembrane glycoproteins with an intrinsic intracellular tyrosine kinase domain. Ligand-induced activation of the receptor stimulates receptor phosphorylation, phosphoinositol turnover, Ca^{2+} mobilization, arachidonic acid release, enhanced expression of c-*myc* and c-*fos*, DNA synthesis, and cell mitosis (Williams, 1989).

Normal human endometrium expresses PDGF ligand subunits and receptors (Chegini et al, 1992). Both endometrial epithelial (Munson et al, 1995) and stromal (Surrey and Halme, 1991) cells proliferate in vitro in response to PDGF. Human peritoneal fluid (Halme et al, 1988) and media conditioned by activated peritoneal macrophages (Olive et al, 1991) contain mitogenic activity that has many of the biochemical characteristics of PDGF. Pelvic fluid and purified PDGF were noted to stimulate [^{3}H]thymidine incorporation into the nascent DNA of human endometrial stromal cell cultures (Surrey and Halme, 1991). These findings suggest that PDGF derived from macrophages or from pelvic endometriosis implants themselves might contribute to the growth and progression of endometriosis lesions.

Platelet-Derived Endothelial Cell Growth Factor

Platelet-derived endothelial cell growth factor (PDECGF) is an acidic nonglycosilated cytoplasmic protein of 45 kd. PDECGF encodes the enzyme thymidine phosphorylase, which promotes angiogenesis (Moghaddam et al, 1995). PDECGF is chemotactic for vascular endothelial cells and monocytes and selectively supports the growth of these cells in contrast to many other angiogenesis factors that are also mitogenic for fibroblasts and muscle cells. PDECGF was initially isolated from platelets, but the factor is also synthesized by fibroblasts, vascular smooth muscle cells, and a number of transformed cell lines. PDECGF is not produced by endothelial cells. In the luteal and menstrual phases PDECGF is strongly expressed in endometrial glands. This pattern of expression is reversed in the stroma where the greatest expression is in the proliferative phase of the cycle (Zhang et al, 1997). However, a marked increase in PDECGF expression has also been demonstrated in decidualized endometrial stromal cells (Osuga et al, 1995). In vitro, PDECGF is upregulated following treatment with cytokines (e.g., TNF-α, interferon-γ) (Zhang et al, 1997) or with progesterone (Osuga et al, 1995). PDECGF also has been localized to the epithelial cells lining ovarian endometriomas and in the subepithelial zona associated with angiogenesis in these implants. Modulation of PDECGF during the menstrual cycle is not observed in endometriosis implants, suggesting that PDECGF may be a chronic angiogenic stimulator in these lesions (Fujimoto et al, 1999).

Vascular Endothelial Growth Factor

Vascular endothelial growth factor (VEGF), also known as vascular permeability factor and vasculotropin, is one of the most potent and specific angiogenic factors. The biochemistry and molecular biology of VEGF are well described (Ferrara, 1999a,b). VEGF binds to a family of tyrosine kinase receptors including FIt-1 and kinase domain receptor (KDR), leading to dimer formation, autophosphorylation of the receptors, and, like FGF, activation of mitogen-activated protein kinases (D'Angelo et al, 1995). VEGF receptors are expressed on vascular endothelial cells and human trophoblast cells, which develop an endothelial phenotype during endovascular invasion. The physiologic roles of VEGF in cycling endometrium include neovascularization and implantation. VEGF-induced angiogenesis allows repair of the endometrium following menstruation. It also appears to increase microvascular permeability, permitting the formation of a fibrin matrix that allows endothelial migration and proliferation. VEGF appears to be a central mediator of "cyclic neovascularization" in the human endometrium and it may play an important role in the initiation of different endometrial pathologies (Lebovic et al, 1999). In the human endometrium VEGF mRNA begins to increase in the midproliferative phase and reaches a maximum in late proliferative and secretory endometrium (Shifren et al, 1996).

We and others have demonstrated that VEGF is an estrogen- and progestin-responsive angiogenic factor (Cullinan-Bove and Koos, 1993; Shifren et al, 1996). The mechanism of VEGF regulation by estrogen, progesterone, or their various antagonists is still under debate (Greb et al, 1997). It is not known if ovarian steroids increase VEGF mRNA stability, transcriptional activity, or both. VEGF mRNA and protein are expressed in different compartments of the endometrial tissue during the ovulatory cycle and there is a directional VEGF secretion by polarized human endometrial epithelial cells in vitro (Hornung et al, 1998). During the estrous cycle of the rat, VEGF mRNA is predominantly expressed in the luminal endometrial epithelium during proestrus (when the estradiol/progesterone ratio is highest) but it also is expressed in the stromal compartment during diestrus when progesterone dominates (Karuri et al, 1998). In primate menstrual and proliferative phases, VEGF mRNA and protein are most concentrated in the epithelium (Shifren et al, 1996; Greb et al, 1997; Zhang et al, 1998), whereas stromal cells become more diffusely positive for VEGF in the secretory phase (Shifren et al, 1996; Zhang et al, 1998). Besides the ovarian steroids, other factors are known to upregulate VEGF expression including hypoxia (Liu et al, 1995), IL-1β (Lebovic et al, 2000), PDGF and TGF-β (Brogi et al, 1994), EGF (Goldman et al, 1993), and PGE_2 (Ben-Av et al, 1995).

VEGF has been postulated to play a role in angiogenesis of endometriosis. Donnez et al (1998) demonstrated that eutopic glandular epithelium in endometriosis patients had significantly higher VEGF content during the late secretory phase compared to the endometrium of patients without endometriosis, implying that shed endometrial fragments in women with endometriosis have greater potential to recruit new vessels than women without endometriosis. The authors suggest that endometriosis arises from the peritoneal seeding of viable endometrial cells during retrograde menstruation and that red lesions be considered the initial angiogenic stage of implantation. These findings correlate with earlier morphologic and morphometric studies, demonstrating prominent vascularity in eutopic endometrium and red peritoneal lesions (Nisolle et al, 1993). During their evolution, the red lesions induce an inflammatory reaction and provoke bleeding and hemosiderin deposition, and the lesions become black. Indeed, lower VEGF levels were found in black lesions, explaining the decrease in vascularization, followed by fibrosis and white opacification at these inactive sites (Nisolle et al, 1993; Donnez et al, 1998). Furthermore, peritoneal fluid concentrations of soluble VEGF from patients with endometriosis (McLaren et al, 1996a; Shifren et al, 1996) correlate significantly with the stage of disease. The cellular sources of VEGF in peritoneal fluid have not been precisely defined. While evidence exists for production of this factor by endometriosis lesions themselves (Shifren et al, 1996), activated peritoneal macrophages and other leukocytes also have the capacity to synthesize and secrete VEGF (McLaren et al, 1996b). These combined data suggest that VEGF expression influences the vascularization and progression of endometriotic lesions.

Transforming Growth Factor-β

Transforming growth factor-β (TGF-β) is the prototype of a family of proteins known as the TGF-β superfamily where inhibins, activin A, and müllerian inhibiting substance (MIS) are also included. TGF-β exists in at least five isoforms, known as TGF-β-1 to -5, that are not related to TGF-α. TGF-β-1 is the prevalent form and is found almost ubiquitously, while the other isoforms are expressed in a more limited spectrum of cells and tissues. The biologically active forms of all isoforms are disulfide-linked dimers. The different isoforms of TGF-β are encoded by different genes and arise by proteolytic cleavage of longer precursors. Virtually all cell types possess TGF-β receptors of at least three general classes (types 1, 2, and 3), which are incompletely characterized (Massagué et al, 1990). As a result, TGF-β has a variety of biologic activities that include both growth stimulatory and inhibitory effects. It appears to play important roles in tissue morphogenesis and development, and thus is likely to be important in the cyclical regeneration of the human endometrium. It also is an important modulator of macrophage and natural killer (NK) cell function (Hooper, 1991). TGF-β appears to have an indirect effect on angiogenesis and inconsistent actions on endothelial cell proliferation (Folkman, 1995).

Human epithelial and stromal endometrial cells express TGF-β mRNA transcripts corresponding to three of the isoforms (Tang et al, 1994b) and secrete immunoreactive TGF-β protein into their conditioned media. In keeping with an increased level of TGF-β mRNA in proliferative endometrium, the addition of estradiol to endometrial stromal cell cultures increased the steady-state concentration of TGF-β mRNA in these cells (Arici et al, 1996b). Human endometrial cells also express TGF-β receptor mRNAs (Dumont et al, 1995) and binding proteins (Tang et al, 1994b).

In a cell growth inhibition bioassay, Oosterlynck et al (1994) demonstrated that TGF-β concentrations are elevated in the pelvic fluid of patients with endometriosis and they speculated that at the mean concentration observed (11.4 ± 3.3 ng/mL) TGF-β might act as a mitogen for endometriosis implants. At low concentrations (<12.5 ng/mL) TGF-β increases endometrial stromal cell proliferation, although this effect is lost at higher ligand concentrations (Tang et al, 1994b).

Angiopoietins

The angiopoietins are novel ligands that specifically activate Tie-2, a member of a family of receptor tyrosine kinases predominantly expressed in endothelial cells and upregulated in microvessels of malignant tumors (Davis et al, 1996; Peters et al, 1998). Tie receptors, angiopoietin-1 (Ang-1), and its naturally occurring antagonist, angiopoietin-2 (Ang-2) are essential for the normal development of the embryonic vasculature (Maisonpierre et al, 1997). Addition of Ang-1 to endothelial cell cultures does not promote proliferation (Davis et al, 1996), but it is chemotactic for endothelial cells and can induce sprouting and branching of new vessels in vitro (Koblizek et al, 1998). The VEGFs and the angiopoietins play complementary and coordinated roles in vascular development. Whereas VEGF promotes endothelial cell differentiation, proliferation, and primitive vessel formation, Ang-1 acts via the Tie-2 receptor to remodel these primitive vessels and helps to maintain and stabilize the mature vasculature (Peters, 1998). In adults, Ang-2 is expressed primarily at sites of vascular remodeling, where it counteracts blood vessel maturation mediated by Ang-1.

Ang-1, Ang-2, and Tie-2 expression have been demonstrated in the endometrium, and it seems that Ang-1 and Tie-2 are downregulated in dysfunctional uterine bleeding (Ahmed et al, 2000). Although angiopoietins have not yet been described in endometriotic lesions, it is tempting, by analogy to tumor neovascularization, to speculate about a possible role in endometriosis. It has been proposed that destabilization by Ang-2 in the absence of VEGF leads to frank vessel regression, whereas such destabilization in the presence of high VEGF levels facilitates the angiogenic response (Maisonpierre et al, 1997).

Endometriosis Proteins-I and -II

Endometriosis protein-I (ENDO-I) and endometriosis protein-II (ENDO-II) are two proteins that were identified as uniquely synthesized and secreted in ectopic endometrial tissues (Sharpe and Vernon, 1993; Sharpe et al, 1993). By sequence analysis ENDO-II has been identified as tissue inhibitor of metalloproteinases-1 (Sharpe-Timms et al, 1995). ENDO-I, a complex acidic glycoprotein, is a haptoglobin (Hp)-like protein (Piva and Sharpe-Timms, 1999). Independently of the stage of the reproductive cycle ENDO-I transcript levels are significantly higher in ectopic pelvic endometrium than in eutopic endometrium of the same individuals (Piva and Sharpe-Timms, 1999). Local peritoneal production of ENDO-I may be associated with localized angiogenesis and altered immune phenomena related to endometriosis.

Adrenomedullin

Adrenomedullin is a 52-amino-acid peptide that is structurally related to the calcitonin gene-related peptide and amylin (Kitamura et al, 1995). Adrenomedullin has numerous physiologic roles including vasodilatation. In the rat uterus, adrenomedullin attenuates galanin-stimulated smooth muscle contraction (Upton et al, 1997). Similar to other members of the calcitonin gene-related peptide family, adrenomedullin increases intracellular cAMP through a specific seven-transmembrane-spanning G-protein–associated receptor (Shimekake et al, 1995). Recently, prominent expression of adrenomedullin has been documented in normal and cancerous human endometrial epithelial cells (Michishita et al, 1999). No differences could be demonstrated between the abundance of adrenomedullin mRNA in normal and in carcinomatous endometrium (Michishita et al, 1999). The precise roles of adrenomedullin in the human uterus remain to be identified. However, production and secretion of vasodilator substances such as adrenomedullin to enhance the availability of nutrient factors to small endometriosis implants may be of critical importance.

Ovarian Steroids

It is unlikely that ovarian steroids are direct mediators of endometrial or endometriosis angiogenesis. Most investigators have failed to identify detectable levels of classical estrogen (α) or progesterone (A and B) receptors in endometrial vascular endothelial cells (Perrot-Applanat et al, 1994; Taylor et al, 1999). However, by immunohistochemistry and RT-PCR, some recent reports demonstrated estrogen and progesterone receptors expression in human endometrial endothelial cells (Iruela-Arispe et al, 1999; Vazquez et al, 1999). In these cells, progesterone receptors appear to inhibit mitotic activity (Vazquez et al, 1999).

Indirect angiogenic effects of these steroids were shown in the rat uterus (Cullinan-Bove and Koos, 1993) and human endometrium (Charnock-Jones et al, 1993; Shifren et al, 1996; Lebovic et al, 2000). Estrogens and progestins upregulate VEGF mRNA and protein production in uterine tissues and cells. If similar results are confirmed in endometriosis tissues, these studies support the obligatory role of ovarian steroids on endometriosis implants (DiZerega et al, 1980) and the observation that downsregulating doses of GnRH analogues decrease the size and vascularity of endometriosis lesions (Schriock et al, 1985).

Eicosanoids (PGE_2)

Many of the arachidonic acid metabolites have potent vasoactive effects; among these, there are reports of angiogenesis. Prostaglandin E_2 (PGE_2), which is synthesized by endometrial and endometriosis stromal cells, has been shown to possess angiogenic activity (Ben-Av et al, 1995). Recently, thromboxane A_2 has been shown to have mitogenic and chemotactic effects on endothelial cells (Daniel et al, 1999) and antagonists of this eicosanoid inhibit angiogenesis.

ANTIANGIOGENIC FACTORS IN ENDOMETRIOSIS

The regulation of angiogenesis is tightly controlled in adults. Indeed, the only normal adult tissues that undergo substantial physiologic angiogenesis are restricted to the female—namely, the endometrium and ovary. A delicate balance of angiogenic and antiangiogenic principles accomplishes the maintenance of a stable vasculature. In the preceding sections we have described the major proangiogenic proteins and lipids expressed in normal endometrium and endometriosis. Here we will present the limited, but growing evidence that antiangiogenic molecules also are expressed in these tissues.

Interferon-γ

Interferon (IFN)-γ is a pleiotropic, proinflammatory cytokine mainly produced by T cells and activated NK cells. Two molecular forms of the biologically active protein of 20 and 25 kd have been described. IFN-γ has antiviral, antiparasitic, and antiproliferative activities for a number of normal and transformed cells. IFN-γ synergizes with TNF-α and TNF-β in inhibiting the proliferation of various cell types. The growth inhibitory activities of IFN-γ are more pronounced than those of the other interferons. However, the main biologic activity of IFN-γ appears to be immunomodulatory in contrast to the other interferons, which are mainly antiviral. The IFN-γ receptor is expressed on the surfaces of all cell types except erythrocytes and shows homology to the superfamily of cytokine receptors (Lundell and Narula, 1994).

In human endometrium interferon receptors do not change during the menstrual cycle (Tabibzadeh, 1990) and IFN-γ is secreted by human endometrium explant cultures (Tabibzadeh, 1991). In the endometrium it is postulated that IFN-γ opposes the mitogenic effect of estrogen via a paracrine, antiproliferative action (Tabibzadeh et al, 1988). Lymphoid cells may increase the local concentration of IFN-γ, exerting a cytostatic effect on normal epithelial cells and inducing human leukocyte antigen (HLA)-DR expression (Tabibzadeh et al, 1986). Consequently, IFN-γ might limit the cellular growth of the endometrium throughout the menstrual cycle, particularly in the basal layer where a greater density of lymphoid aggregates has been noted. This hypothesis is supported by the low proliferative activity and the presence of HLA-DR expression in the basal layer of the human endometrium.

Peritoneal fluid concentrations of IFN-γ are either unaffected (Khorram et al, 1993; Keenan et al, 1994) or even lower (Wu et al, 1998) in patients with endometriosis. But soluble IFN-γ concentrations in peritoneal fluid samples may underestimate the actual concentrations in vivo, due to the ubiquity of its receptor and a high rate of cellular uptake. Within endometriosis implants themselves, more T lymphocytes and macrophages have been documented than the number found in eutopic endometrium (Klein et al, 1994). Additionally, in situ hybridization studies reveal that mRNA expression for IFN-γ is significantly greater in leukocytes of ectopic implants than of eutopic tissues (Klein et al, 1994). Protein localization studies reveal that the receptor for IFN-γ is present in glandular epithelium of ectopic endometrium, suggesting a possible paracrine role for resident leukocytes and IFN-γ in regulating cell proliferation in endometriosis. In an in vitro angiogenesis model Tsuruoka et al (1988) showed that IFN-γ inhibits capillary formation in a dose-dependent manner. Other effects of IFN-γ that may contribute to infertility include embryotoxicity and adverse affects on sperm motility (Hill, 1992).

Angiostatin

Angiostatin was the first molecule specifically isolated as a potential tumor-derived angiogenesis inhibitor (O'Reilly et al, 1994). Angiostatin is a 38-kd fragment of the clotting cascade precursor plasminogen. Tumor cells do not produce angiostatin, but certain tumors can produce or activate proteases capable of generating angiostatin from circulating plasminogen. Angiostatin specifically inhibits endothelial cell proliferation and migration. High concentrations of plasmin and serine proteases are present in ectopic endometrial implants (Fernandez-Shaw et al, 1995; Sillem et al, 1998) and provide a potential source of angiostatin (Folkman, 1997).

Endostatin

Endostatin, a carboxy-terminal fragment of collagen XVIII, also is a potent angiogenesis inhibitor (O'Reilly et al, 1997). Fragments of 16 to 21 kd in size are found in human plasma (John et al, 1999). Endostatin inhibits microvascular endothelial cell proliferation at doses of 100 to 500 ng/mL (Zetter, 1998). Comparative analyses of collagen expression in normal and ectopic endometrium have not been performed systematically. However, Harrington et al (1999), using immunohistochemistry and Western blotting techniques, failed to observe differences in most extracellular matrix components including collagen IV, laminin, vitronectin, and fibronectin. These findings argue against the hypothesis that decreased substrate for the generation of endostatin (i.e., collagen) might contribute to the pathogenesis of endometriosis.

Thrombospondin

Thrombospondin-1 (TSP-1), a 450-kd glycoprotein, has been shown to suppress angiogenesis in vivo and in vitro. TSP-1 is released from platelet α granules by thrombin and is deposited in the extracellular matrix (Adams, 1997). In the endometrium, TSP-1 is cycle dependent and regulated by progesterone (Iruela-Arispe et al, 1996). Transcripts for TSP-1 were observed predominantly in stromal cells, but mRNA also was detected in some vascular cells. Since progesterone action appears to be attenuated in endometriosis lesions (Osteen et al, 1996), the production of this antiangiogenic protein may be reduced in ectopic implants relative to normal endometrium.

16K Prolactin

16K prolactin is a 16-kd amino-terminal fragment of prolactin formed by enzymatic cleavage of intact 23-kd prolactin in the pituitary gland and in target tissues for prolactin (Ferrara et al, 1991). This fragment has been shown to be antiangiogenic in vivo and in vitro, while intact prolactin is inactive (Ferrara et al, 1991; D'Angelo et al, 1999). This activity is mediated by specific, high-affinity, saturable binding sites (52 kd, 32 kd) for the 16-kd protein that are distinct from the normal prolactin receptor. Decidualized human endometrium expresses prolactin in vivo and in vitro. The prolactin gene can be induced in endometrial stromal cells in vitro by progestins, cAMP analogues, and ligands that elevate cellular cAMP levels (Telgmann and Gellersen, 1998). Luteal phase intraperitoneal concentrations of prolactin were significantly elevated in women with endometriosis; however, whether the derivation of this hormone was pituitary or endometrial was not ascertained (Chew et al, 1990). Cleavage of the peritoneal fluid prolactin to generate the antiangiogenic 16-kd fragment has not been demonstrated, but production of active protein from human decidual prolactin has been reported (Bentzien et al, 1997).

Tumor Necrosis Factor-α

TNF-α has been shown to both promote and inhibit angiogenesis. In breast carcinoma, TNF-α can upregulate PDECGF, leading to enhanced endothelial mitogenesis and increased vascular density (Leek et al, 1998). By contrast, it has been shown that TNF-α can prevent VEGF signaling by downregulating VEGF-receptor gene expression (Patterson et al, 1996). These antiangiogenic properties may be of therapeutic use (Shimazawa et al, 1993).

TNF-α levels are increased in peritoneal fluid of endometriosis patients (Keenan et al, 1995) and its production by peritoneal macrophages also is increased in this condition (Rana et al, 1996). We have described the many possible roles of TNF-α in endometriosis (see above).

Steroids

We have discussed that ovarian steroids may have indirect angiogenic effects. Conversely, angiostatic steroids of the pregnane (Crum et al, 1985) and estratropone (Miller et al, 1997) classes have been described. There is no evidence that concentrations of these steroids are decreased in endometriosis; however, this hypothesis is testable.

ANTIANGIOGENESIS THERAPY FOR ENDOMETRIOSIS

Antiangiogenic therapies are being aggressively evaluated for the treatment of cancer. Unfortunately for those of us who seek novel and efficacious medical therapies for the treatment of endometriosis, the therapeutic window of the currently available agents may be too narrow. A specific and problematic dilemma in the field of infertility in endometriosis patients is that our therapeutic task is to inhibit fertility-impairing influences of ectopic endometrium while preserving normal eutopic endometrial function. Although certain differences between ectopic and eutopic tissues have been described (e.g., increased complement 3 protein [Tao et al, 1997], increased IL-6 [Tseng et al, 1996], increased tenascin [Harrington et al, 1999], and increased IL-1β induced VEGF expression [Lebovic et al, 2000] in ectopic sites), many of the biochemical characteristics of cells isolated from endometriosis and normal endometrium are remarkably similar (Ryan et al, 1994). Thus, if physicians wish to specifically target endometriosis cells, while sparing eutopic endometrium from untoward effects of medical therapy, highly specific and selective drugs will need to be developed.

CONCLUSION

Endometriosis is a fascinating and complex disease resulting from a dysregulation between exfoliated menstrual endometrium and the intraabdominal environment. Increased concentrations of activated pelvic macrophages and lymphocytes and elevated levels of specific cytokines and growth factors, reviewed above, support this hypothesis. The precise roles of these soluble factors are currently unknown, but we propose that a complex network of locally produced cytokines, growth factors, steroids, and eicosanoids modulate the growth and inflammatory behavior of ectopic endometrial implants via neovascularization. The enhanced secretion of local proangiogenic proteins by endometriosis lesions and associated immune cells (and the concomitant reduction of antiangiogenic principles) promotes the recruitment of capillaries toward the growing lesions. Ultimately, a cascade of effects on the peritoneal microenvironment results in implant proliferation and invasion. Future therapeutic strategies to target these angiogenic stimuli have the potential to block the vascular pathogenesis of endometriosis. However, we must not ignore the likely physiologic actions of these same antiangiogenic molecules in eutopic endometrial function. Indeed, unless these agents are selectively or preferentially reversible in the eutopic endometrium, angiosuppressive therapy may be more amenable for the treatment of endometriosis-associated pain than for infertility.

ACKNOWLEDGMENTS

Many of the studies summarized in this chapter were supported by National Institutes of Health grant HD33238 (R.N.T.) and by a grant from the Swiss National Science Foundation (M.D.M.).

REFERENCES

Adams, J. 1997. Thrombospondin-1. *Int. J. Biochem. Cell. Biol.* 29:861–865.

Ahmed, A., Acevedo, C., Nijjar, S., Aplin, J., Yancopoulos, G., and Shams, M. 2000. Angiopoietin-1 and Tie-2 are downregulated in dysfunctional uterine bleeding. *J. Soc. Gynecol. Invest.* 7(suppl 1):57A.

Arici, A., Head, J., MacDonald, P., and Casey, M. 1993. Regulation of interleukin-8 gene expression in human endometrial cells in culture. *Mol. Cell. Endocrinol.* 94:195–204.

Arici, A., MacDonald, P., and Casey, M. 1996a. Progestin regulation of interleukin-8 mRNA levels and protein synthesis in human endometrial stromal cells. *J. Steroid. Biochem. Mol. Biol.* 58:71–76.

Arici, A., MacDonald, P. C., and Casey, M. L. 1996b. Modulation of the levels of transforming growth factor beta messenger ribonucleic acids in human endometrial stromal cells. *Biol. Reprod.* 54:463–469.

Arici, A., Tazuke, S. I., Attar, E., Kliman, H. J., and Olive, D. L. 1996c. Interleukin-8 concentration in peritoneal fluid of patients with endometriosis and modulation of interleukin-8 expression in human mesothelial cells. *Mol. Hum. Reprod.* 2:40–45.

Baker, J., Liu, J., Robertson, E., and Efstratiadis, A. 1993. Role of insulin-like growth factors in embryonic and postnatal growth. *Cell* 75:73–82.

Ben-Av, P., Crofford, L. J., Wilder, R. L., and Hla, T. 1995. Induction of vascular endothelial growth factor expression in synovial fibroblasts by prostaglandin E and interleukin-1: a potential mechanism for inflammatory angiogenesis. *FEBS Lett.* 372:83–87.

Bentzien, F., Vigne, J.-L., Weiner, R., and Taylor, R. 1997. Potential role of cathepsin D in the formation of antiangiogenic fragment of prolactin in the human endometrium. *Endocrine Society Abstracts* P1–33.

Brogi, E., Wu, T., Namiki, A., and Isner, J. 1994. Indirect angiogenic cytokines upregulate VEGF and bFGF gene expression in vascular smooth muscle cells, whereas hypoxia upregulates VEGF expression only. *Circulation* 90:649–652.

Bulmer, J., Longfellow, M., and Ritson, A. 1991. Leukocytes and resident blood cells in endometrium. *Ann. NY Acad. Sci.* 622:57–68.

Charnock-Jones, D., Sharkey, A., Rajput-Williams, J., Burch, D., Schofield, J., and Fountain, S. 1993. Identification and localization of alternately spliced mRNAs for vascular endothelial growth factor in human uterus and estrogen regulation in endometrial carcinoma cell lines. *Biol. Reprod.* 48:1120–1128.

Chegini, N., Rossi, M., and Masterson, B. 1992. Platelet-derived growth factor (PDGF), epidermal growth factor (EGF), and EGF and PDGF beta-receptors in human endometrial tissue: localization and in vitro action. *Endocrinology* 130:2373–2385.

Chew, P., Peh, K., Loganath, A., Gunasegaram, R., and Ratnam, S. 1990. Elevated peritoneal fluid luteinizing hormone and prolactin concentrations in infertile women with endometriosis. *Int. J. Gynaecol. Obstet.* 33:35–39.

Cohen, T., Nahari, D., Cerem, L. W., Neufeld, G., and Levi, B. Z. 1996. Interleukin 6 induces the expression of vascular endothelial growth factor. *J. Biol. Chem.* 271:736–741.

Cordon-Cardo, C., Vlodavsky, I., Haimovitz-Friedman, A., Hicklin, D., and Fuks, Z. 1990. Expression of basic fibroblast growth factor in normal human tissues. *Lab. Invest.* 63:832–840.

Crum, R., Szabo, S., and Folkman, J. 1985. A new class of steroids inhibits angiogenesis in the presence of heparin or a heparin fragment. *Science* 230:1375–1378.

Cullinan-Bove, K., and Koos, R. 1993. Vascular endothelial growth factor/vascular permeability factor expression in the rat uterus: rapid stimulation by estrogen correlates with estrogen-induced increases in uterine capillary permeability and growth. *Endocrinology* 133:829–837.

Curtis, S., Washburn, T., Sewall, C., DiAugustine, R., Lindzey, J., and Couse, J. 1996. Physiological coupling of growth factor and steroid receptor signaling pathways: estrogen receptor knockout mice lack estrogen-like response to epidermal growth factor. *Proc. Natl. Acad. Sci. USA* 93:12626–12630.

D'Angelo, G., Martini, J. F., Iiri, T., Fantl, W. J., Martial, J., and Weiner, R. I. 1999. 16K human prolactin inhibits vascular endothelial growth factor-induced activation of Ras in capillary endothelial cells. *Mol. Endocrinol.* 13:692–704.

D'Angelo, G., Struman, I., Martial, J., and Weiner, R. 1995. Activation of mitogen-activated protein kinases by vascular endothelial growth factor and basic fibroblast growth factor in capillary endothelial cells is inhibited by the antiangiogenic factor 16-kDa-terminal fragment of prolactin. *Proc. Natl. Acad. Sci. USA* 92:6374–6378.

Daniel, T., Liu, H., Morrow, J., Crews, B., and Marnett, L. 1999. Thromboxane A2 is a mediator of cyclooxygenase-2-dependent endothelial migration and angiogenesis. *Cancer Res.* 59:4574–4577.

Davis, S., Aldrich, T., Jones, P., Acheson, A., Compton, L., and Jain, V. 1996. Isolation of angiopoietin-1, a ligand for the TIE2 receptor, by secretion-trap expression cloning. *Cell* 87:1161–1169.

D'Hooghe, T. M. 1997. Clinical relevance of the baboon as a model for the study of endometriosis. *Fertil. Steril.* 68:613–625.

Di Lieto, A., De Rosa, G., Albano, G., Staibano, S., Campanile, M., and Pontillo, M. 1997. Influence of goserelin depot therapy on epidermal growth factor receptor expression in pelvic endometriosis. *Gynecol. Endocrinol.* 11:17–20.

DiZerega, G., Barber, D., and Hodgen, G. 1980. Endometriosis: role of ovarian steroids in intiation, maintenance and suppression. *Fertil. Steril.* 33:649–653.

Donnez, J., Smoes, P., Gillerot, S., Casanas-Roux, F., and Nisolle, M. 1998. Vascular endothelial growth factor (VEGF) in endometriosis. *Hum. Reprod.* 13:1686–1690.

Dumont, N., O'Connor-McCourt, M., and Philip, A. 1995. Transforming growth factor-beta receptors on human endometrial cells: identification of the type I, II, and III receptors and glycosyl-phosphatidylinositol anchored TGF-beta binding proteins. *Mol. Cell. Endocrinol.* 111:57–66.

Fakih, H., Baggett, B., Holtz, G., Tsang, K., Lee, J., and Williamson, H. 1987. Interleukin-1: a possible role in the infertility associated with endometriosis. *Fertil. Steril.* 47:213–217.

Fernandez-Shaw, S., Marshall, J., Hicks, B., Barlow, D., and Starkey, P. 1995. Plasminogen activators in ectopic and uterine endometrium. *Fertil. Steril.* 63:45–51.

Ferrara, N. 1999a. Role of vascular endothelial growth factor in the regulation of angiogenesis. *Kidney Int.* 56:794–814.

Ferrara, N. 1999b. Vascular endothelial growth factor: molecular and biological aspects. *Curr. Top. Microbiol. Immunol.* 237:1–30.

Ferrara, N., Clapp, C., and Weiner, R. 1991. The 16K fragment of prolactin specifically inhibits basal or fibroblast growth factor stimulated growth of capillary endothelial cells. *Endocrinology* 129:896–900.

Ferriani, R. A., Charnock-Jones, D. S., Prentice, A., Thomas, E. J., and Smith, S. K. 1993. Immunohistochemical localization of acidic and basic fibroblast growth factors in normal human endometrium and endometriosis and the detection of their mRNA by polymerase chain reaction. *Hum. Reprod.* 8:11–16.

Folkman, J. 1995. Clinical applications of research on angiogenesis. *N. Engl. J. Med.* 333:1757–1764.

Folkman, J. 1997. Addressing tumor blood vessels. *Nat. Biotechnol.* 15:510.

Fujimoto, J., Sakaguchi, H., Hirose, R., and Tamaya, T. 1999. Expression of platelet-derived endothelial cell growth factor (PD-ECGF) related to angiogenesis in ovarian endometriosis. *J. Clin. Endocrinol. Metab.* 84:359–362.

Giudice, L., Dsupin, B., Gargosky, S., Rosenfeld, R., and Irwin, J. 1994. The insulin-like growth factor system in human peritoneal fluid: its effects on endometrial stromal cells and its potential relevance to endometriosis. *J. Clin. Endocrinol. Metab.* 79:1284–1293.

Giudice, L., Dsupin, B., Jin, I., Vu, T., and Hoffman, A. 1993. Differential expression of messenger ribonucleic acids encoding insulin-like growth factors and their receptors in human uterine endometrium and decidua. *J. Clin. Endocrinol. Metab.* 76:1115–1122.

Goldman, C., Kim, J., Wong, W., King, V., Brock, T., and Gillespie, G. Y. 1993. Epidermal growth factor stimulates vascular endothelial growth factor production by human malignant glioma cells: a model of glioblastoma multiforme pathophysiology. *Mol. Biol. Cell* 4:121–133.

Gospodarowicz, D., Brown, K., Birdwell, C., and Zetter, B. 1978. Control of proliferation of human vascular endothelial cells. Characterization of the response of human umbilical vein endothelial cells to fibroblast growth factor, epidermal growth factor, and thrombin. *J. Cell Biol.* 77:774–788.

Greb, R., Heikinheimo, O., Williams, R., Hodgen, G., and Goodman, A. 1997. Vascular endothelial growth factor in primate endometrium is regulated by oestrogen-receptor and progesterone-receptor ligands in vivo. *Hum. Reprod.* 12:1280–1292.

Grimwood, J., Bicknell, R., and Rees, M. 1995. The isolation, characterization and culture of human decidual endothelium. *Hum. Reprod.* 10:2142–2148.

Groothuis, P., Koks, C., de Goeij, A., Dunselman, G., Arends, J., and Evers, J. 1999. Adhesion of human endometrial fragments to peritoneum in vitro. *Fertil. Steril.* 71:1119–1124.

Gurgan, T., Bukulmez, O., Yarali, H., Tanir, M., and Akyildiz, S. 1999. Serum and peritoneal fluid levels of IGF I and II and insulinlike growth binding protein-3 in endometriosis. *J. Reprod. Med.* 44:450–454.

Halme, J., Hammond, M., Hulka, J., Raj, S., and Talbert, L. 1984. Retrograde menstruation in healthy women and in patients with endometriosis. *Obstet. Gynecol.* 64:151–154.

Halme, J., White, C., Kauma, S., Estes, J., and Haskill, S. 1988. Peritoneal macrophages from patients with endometriosis release growth factor activity in vitro. *J. Clin. Endocrinol. Metab.* 66:1044–1049.

Hanahan, D., and Folkman, J. 1996. Patterns and emerging mechanisms of the angiogenic switch during tumorigenesis. *Cell* 86:353–364.

Harrington, D., Lessey, B., Rai, V., Bergqvist, A., Kennedy, S., and Manek, S. 1999. Tenascin is differentially expressed in endometrium and endometriosis. *J. Pathol.* 187:242–248.

Heldin, C., Backstrom, G., Ostman, A., Hammacher, A., Ronnstrand, L., and Rubin, K. 1988. Binding of different dimeric forms of PDGF to human fibroblasts: evidence for two separate receptor types. *EMBO* J. 7:1387–1393.

Hill, J. 1992. Immunological factors in endometriosis and endometriosis-associated reproductive failure. *Med. Clin. North Am.* 3:583–596.

Hooper, W. 1991. The role of transforming growth factor-beta in hematopoiesis. A review. *Leuk. Res.* 15:179–184.

Hornung, D., Lebovic, D., Shifren, J., Vigne, J., and Taylor, R. 1998. Vectorial secretion of vascular endothelial growth factor by polarized human endometrial epithelial cells. *Fertil. Steril.* 69:909–915.

Huang, J. C., Papasakelariou, C., Dawood, M. Y. 1996. Epidermal growth factor and basic fibroblast growth factor in peritoneal fluid of women with endometriosis. *Fertil. Steril.* 65:931–944.

Huang, J., and Yeh, J. 1994. Quantitative analysis of epidermal growth factor receptor gene expression in endometriosis. *J. Clin. Endocrinol. Metab.* 79:1097–1101.

Imai, T., Kurachi, H., Adachi, K., Adachi, H., Yoshimoto, Y., and Homma, H. 1995. Changes in epidermal growth factor receptor and the levels of its ligands during menstrual cycle in human endometrium. *Biol. Reprod.* 52:928–938.

Iruela-Arispe, M., Porter, P., Bornstein, P., and Sage, E. 1996. Thrombospondin-1, an inhibitor of angiogenesis, is regulated by progesterone in the human endometrium. *J. Clin. Invest.* 97:403–412.

Iruela-Arispe, M. L., Rodriguez-Manzaneque, J. C., and Abu-Jawdeh, G. 1999. Endometrial endothelial cells express estrogen and progesterone receptors and exhibit a tissue specific response to angiogenic growth factors. *Microcirculation* 6:127–140.

John, H., Preissner, K. T., Forssmann, W. G., and Ständker, L. 1999. Novel glycosylated forms of human plasma endostatin and circulating endostatin-related fragments of collagen XV. *Biochemistry* 38:10217–10224.

Karuri, A., Kumar, A., and Mukhopadhyay, D. 1998. Differential expression and selective localization of vascular permeability factor/vascular endothelial growth factor in the rat uterus during the estrous cycle. *J. Endocrinol.* 159:488–499.

Keenan, J., Chen, T. T., Chadwell, N. L., Torry, D. S., and Caudle, M. R. 1994. Interferon-gamma (IFN-gamma) and interleukin-6 (IL-6) in peritoneal fluid and macrophage-conditioned media of women with endometriosis. *Am. J. Reprod. Immunol.* 32:180–183.

Keenan, J. A., Chen, T. T., Chadwell, N. L., Torry, D. S., and Caudle, M. R. 1995. IL-1 beta, TNF-alpha, and IL-2 in peritoneal fluid and macrophage-conditioned media of women with endometriosis. *Am. J. Reprod. Immunol.* 34:381–385.

Kennedy, S. 1999. The genetics of endometriosis. *Eur. J. Obstet. Gynecol. Reprod. Biol.* 82:129–133.

Khorram, O., Taylor, R., Ryan, I., Schall, T., and Landers, D. 1993. Peritoneal fluid concentrations of the cytokine RANTES correlate with the severity of endometriosis. *Am. J. Obstet. Gynecol.* 169:1545–1549.

Kitamura, K., Kangawa, K., Matsuo, H., and Eto, T. 1995. Adrenomedullin. Implications for hypertension research. *Drugs* 49:485–495.

Klein, N., Pergola, G., Tekmal, R., Montoya, I., Dey, T., and Schenken, R. 1994. Cytokine regulation of cellular proliferation in endometriosis. *Ann. NY Acad. Sci.* 734:322–332.

Knudson, A. G. Jr. 1977. Genetics and etiology of human cancer. *Adv. Hum. Genet.* 8:1–66.

Koblizek, T. I., Weiss, C., Yancopoulos, G. D., Deutsch, U., and Risau, W. 1998. Angiopoietin-1 induces sprouting angiogenesis in vitro. *Curr. Biol.* 8:529–532.

Koch, A., Polverini, P., Kunkel, S., Harlow, L., Di Pietro, L., and Elner, V. 1992. Interleukin-8 as a macrophage-derived mediator of angiogenesis. *Science* 258:1798–1801.

Kruitwagen, R., Poels, L., Willemsen, W., de Ronde, I. J., Jap, P. H., and Rolland, R. 1991. Endometrial epithelial cells in peritoneal fluid during the early follicular phase. *Fertil. Steril.* 55:297–303.

LaRosa, G., Thomas, K., Kaufmann, M., Mark, R., White, M., and Taylor, L. 1992. Amino terminus of the interleukin-8 receptor is a major determinant of receptor subtype specificity. *J. Biol. Chem.* 267:25402–25406.

Le Roith, D. 1996. Insulin-like growth factor receptors and binding proteins. *Baillieres Clin. Endocrinol. Metab.* 10:49–73.

Lebovic, D., Bentzien, F., Chao, V., Garrett, E., Meng, Y., and Taylor, R. 2000. Induction of an angiogenic phenotype in endometriosis stromal cell cultures by interleukin-1β. *Mol. Hum. Reprod.* 6:269–275.

Lebovic, D. I., Mueller, M. D., and Taylor, R. N. 1999. Vascular endothelial growth factor in reproductive biology. *Curr. Opin. Obstet. Gynecol.* 11:255–260.

Leek, R., Landers, R., Fox, S., Ng, F., Harris, A., and Lewis, C. 1998. Association of tumour necrosis factor alpha and its receptors with thymidine phosphorylase expression in invasive breast carcinoma. *Br. J. Cancer* 77:2246–2251.

Liu, Y., Cox, S., Morita, T., and Kourembanas, S. 1995. Hypoxia regulates vascular endothelial growth factor gene expression in endothelial cells. Identification of a 5′ enhancer. *Circ. Res.* 77:638–643.

Lundell, D., and Narula, S. 1994. Structural elements required for receptor recognition of human interferon-gamma. *Pharm. Ther.* 64:1–21.

Maisonpierre, P., Suri, C., Jones, P., Bartunkova, S., Wiegand, S., and Radziejewski, C. 1997. Angiopoietin-2, a natural antagonist for Tie2 that disrupts in vivo angiogenesis. *Science* 277:55–60.

Malinak, L. R., Buttram, V. C., Jr., Elias, S., and Simpson, J. L. 1980. Heritage aspects of endometriosis. II. Clinical characteristics of familial endometriosis. *Am. J. Obstet. Gynecol.* 137:332–337.

Markee, J. 1940. Menstruation in intraocular endometrial transplants in the rhesus monkey. *Contrib. Embryol.* 177:220–230.

Massagué, J., Cheifetz, S., Boyd, F., Boyd, F., and Andres, J. 1990. TGF-beta receptors and TGF-beta binding proteoglycans: recent progress in identifying their functional properties. *Ann. NY Acad. Sci.* 593:59–72.

Mateo, R. B., Reichner, J. S., and Albina, J. E. 1994. Interleukin-6 activity in wounds. *Am. J. Physiol.* 266(6 pt 2): R1840–1844.

McLaren, J., Prentice, A., Charnock-Jones, D., and Smith, S. 1996a. Vascular endothelial growth factor (VEGF) concentrations are elevated in peritoneal fluid of women with endometriosis. *Hum. Reprod.* 11:220–223.

McLaren, J., Prentice, A., Charnock-Jones, D., Millican, S., Muller, K., and Sharkey, A. 1996b. Vascular endothelial growth factor is produced by peritoneal fluid macrophages in endometriosis and is regulated by ovarian steroids. *J. Clin. Invest.* 98:482–489.

Michishita, M., Minegishi, T., Abe, K., Kangawa, K., Kojima, M., and Ibuki, Y. 1999. Expression of adrenomedullin in the endometrium of the human uterus. *Obstet. Gynecol.* 93:66–70.

Miller, T. A., Bulman, A. L., Thompson, C. D., Garst, M. E., and Macdonald, T. L. 1997. Synthesis and structure-activity profiles of A-homoestranes, the estratropones. *J. Med. Chem.* 40:3836–3841.

Milne, S., Critchley, H., Drudy, T., Kelly, R., and Baird, D. 1999. Perivascular interleukin-8 messenger ribonucleic acid expression in human endometrium varies across the menstrual cycle and in early pregnancy decidua. *J. Clin. Endocrinol. Metab.* 84:2563–2567.

Miyauchi, A., Momoeda, M., Nakabayashi, M., Osuga, Y., Yano, T., and Tsutsumi, O., et al. 1995. Regulation of the plasminogen activator/plasmin system by epidermal growth factor in cultured human endometrial cells. *Hum. Reprod.* 10:3284–3288.

Moghaddam, A., Zhang, H. T., Fan, T. P., Hu, D. E., Lees, V. C., and Turley, H. 1995. Thymidine phosphorylase is angiogenic and promotes tumor growth. *Proc. Natl. Acad. Sci. USA* 92:998–1002.

Mori, H., Sawairi, M., Nakagawa, M., Itoh, N., Wada, K., and Tamaya, T. 1992. Expression of interleukin-1 (IL-1) beta messenger ribonucleic acid (mRNA) and IL-1 receptor antagonist mRNA in peritoneal macrophages from patients with endometriosis. *Fertil. Steril.* 57:535–542.

Motro, B., Itin, A., Sachs, L., and Keshet, E. 1990. Pattern of interleukin 6 gene expression in vivo suggests a role for this cytokine in angiogenesis. *Proc. Natl. Acad. Sci. USA* 87:3092–3096.

Mueller, M. D., Lebovic, D. I., Garrett, E., and Taylor, R. N. 2000. Neutrophils infiltrating the endometrium express vascular endothelial growth factor (VEGF): potential role in endometrial angiogenesis. *Fertil. Steril.* 74:107–112.

Munson, L., Upadhyaya, N., and Van Meter, S. 1995. Platelet-derived growth factor promotes endometrial epithelial cell proliferation. *Am. J. Obstet. Gynecol.* 173:1820–1825.

Nelson, K., Takahashi, T., Bossert, N., Walmer, D., and McLachlan, J. 1991. Epidermal growth factor replaces estrogen in the stimulation of female genital-tract growth and differentiation. *Proc. Natl. Acad. Sci. USA* 88:21–25.

Nicosia, R., Nicosia, S., and Smith, M. 1994. Vascular endothelial growth factor, platelet-derived growth factor, and insulin-like growth factor-1 promote rat aortic angiogenesis in vitro. *Am. J. Pathol.* 145:1023–1029.

Nisolle, M., Casanas Roux, B., Anaf, V., Mine, J., and Donnez, J. 1993. Morphometric study of the stromal vascularization in peritoneal endometriosis. *Fertil. Steril.* 59:681–684.

Noyes, R., Hertig, A., and Rock, J. 1950. Dating the endometrial biopsy. *Fertil. Steril.* 1:3–25.

Odukoya, O. A., Ajjan, R., Lim, K., Watson, P. F., Weetman, A. P., and Cooke, I. D. 1997. The pattern of cytokine mRNA expression in ovarian endometriomata. *Mol. Hum. Reprod.* 3:393–397.

Olive, D., Montoya, I., Riehl, R., and Schenken, R. 1991. Macrophage-conditioned media enhance endometrial stromal cell proliferation in vitro. *Am. J. Obstet. Gynecol.* 164:953–958.

Oosterlynck, D. J., Meuleman, C., Sobis, H., Vandeputte, M., and Koninckx, P. R. 1993. Angiogenic activity of peritoneal fluid from women with endometriosis. *Fertil. Steril.* 59:778–782.

Oosterlynck, D., Meuleman, C., Waer, M., and Koninckx, P. 1994. Transforming growth factor-beta activity is increased in peritoneal fluid from women with endometriosis. *Obstet. Gynecol.* 83:287–292.

O'Reilly, M. S., Boehm, T., Shing, Y., Fukai, N., Vasios, G., and Lane, W. S. 1997. Endostatin: an endogenous inhibitor of angiogenesis and tumor growth. *Cell* 88:277–285.

O'Reilly, M. S., Homgren, L., Shing, Y., Chen, C., Rosenthal, R., and Moses, M. 1994. Angiostatin: a novel angiogenesis inhibitor that mediates the suppression of metastases by a Lewis lung carcinoma. *Cell* 88:1–20.

Osteen, K., Bruner, K., and Sharpe-Timms, K. 1996. Steroid and growth factor regulation of matrix metalloproteinase expression and endometriosis. *Semin. Reprod. Endocrinol.* 14:247–255.

Osuga, Y., Toyoshima, H., Mitsuhashi, N., and Taketani, Y. 1995. The presence of platelet-derived endothelial cell growth factor in human endometrium and its characteristic expression during the menstrual cycle and early gestational period. *Hum. Reprod.* 10:989–993.

Patterson, C., Perrella, M., Endege, W., Yoshizumi, M., Lee, M., and Haber, E. 1996. Down-regulation of vascular endothelial growth factor receptors by tumor necrosis factor-alpha in cultured human vascular endothelial cells. *J. Clin. Invest.* 98:490–496.

Perrot-Applanat, M., Deng, M., Fernandez, H., Lelaidier, C., Meduri, G., and Bouchard, P. 1994. Immunohistochemical localization of estradiol and progesterone receptors in human uterus throughout pregnancy: expression in endometrial blood vessels. *J. Clin. Endocrinol. Metab.* 78:216–224.

Peters, K. G. 1998. Vascular endothelial growth factor and the angiopoietins: working together to build a better blood vessel. *Circ. Res.* 83:342–343.

Peters, K., Coogan, A., Berry, D., Marks, J., Iglehart, J., and Kontos, C. 1998. Expression of Tie2/Tek in breast tumour vasculature provides a new marker for evaluation of tumour angiogenesis. *Br. J. Cancer* 77:51–56.

Piva, M., and Sharpe-Timms, K. L. 1999. Peritoneal endometriotic lesions differentially express a haptoglobin-like gene. *Mol. Hum. Reprod.* 5:71–78.

Presta, M. 1988. Sex hormones modulate the synthesis of basic fibroblast growth factor in human endometrial adenocarcinoma cells: implications for the neovascularization of normal and neoplastic endometrium. *J. Cell Physiol.* 137:593–597.

Rana, N., Braun, D. P., House, R., Gebel, H., Rotman, C., and Dmowski, W. P. 1996. Basal and stimulated secretion of cytokines by peritoneal macrophages in women with endometriosis. *Fertil. Steril.* 65:925–930.

Redwine, D. 1988. Mülleriosis: the single best-fit model of the origin of endometriosis. *J. Reprod. Med.* 33:915–920.

Rier, S. E., Martin, D. C., Bowman, R. E., Dmowski, W. P., and Becker, J. L. 1993. Endometriosis in rhesus monkeys (*Macaca mulatta*) following chronic exposure to 2,3,7,8-tetrachlorodibenzo-p-dioxin. *Fundam. Appl. Toxicol.* 21:433–441.

Ryan, I. P., Schriock, E. D., and Taylor, R. N. 1994. Isolation, characterization, and comparison of human endometrial and endometriosis cells in vitro. *J. Clin. Endocrinol. Metab.* 78:642–649.

Ryan, I., and Taylor, R. 1997. Endometriosis and infertility: new concepts. *Obstet. Gynecol. Surv.* 52:365–371.

Ryan, I., Tseng, J., Schriock, E., Khorram, O., Landers, D., and Taylor, R. 1995. Interleukin-8 concentrations are elevated in peritoneal fluid of women with endometriosis. *Fertil. Steril.* 63:929–932.

Salat-Baroux, J., Romain, S., Alvarez, S., Antoine, J., Kopp, K., and Raulais, D. 1994. Biochemical and immunohistochemical multiparametric analysis of steroid receptors and growth factor receptors in human normal endometrium in spontaneous cycles and after the induction of ovulation. *Hum. Reprod.* 9:200–208.

Sampson, J. 1927. Peritoneal endometriosis due to the menstrual dissemination of endometrial tissue into the peritoneal cavity. *Am. J. Obstet. Gynecol.* 14:422–469.

Sangha, R., Li, X., Shams, M., and Ahmed, A. 1997. Fibroblast growth factor receptor-1 is a critical component for endometrial remodeling: localization and expression of basic fibroblast growth factor and FGF-R1 in human endometrium during the menstrual cycle and decreased FGF-R1 expression in menorrhagia. *Lab. Invest.* 77:389–402.

Schriock, E., Monroe, S., Henzl, M., and Jaffe, R. 1985. Treatment of endometriosis with a potent agonist of gonadotropin-releasing hormone (nafarelin). *Fertil. Steril.* 44:583–588.

Seli, E., Zeyneloglu, H. B., Senturk, L. M., Bahtiyar, O. M., Olive, D. L., and Arici, A. 1998. Basic fibroblast growth factor: peritoneal and follicular fluid levels and its effect on early embryonic development. *Fertil. Steril.* 69:1145–1148.

Sharpe, K. L., and Vernon, M. W. 1993. Polypeptides synthesized and released by rat ectopic uterine implants differ from those of the uterus in culture. *Biol. Reprod.* 48:1334–1340.

Sharpe, K., Zimmer, R., Griffin, W., and Penney, L. 1993. Polypeptides synthesized and released by human endometriosis differ from those of the uterine endometrium in cell and tissue explant culture. *Fertil. Steril.* 60:839–851.

Sharpe-Timms, K. L. 1997. Basic research in endometriosis. *Obstet. Gynecol. Clin. North Am.* 24:269–290.

Sharpe-Timms, K. L., Penney, L. L., Zimmer, R. L., Wright, J. A., Zhang, Y., and Surewicz, K. 1995. Partial purification and amino acid sequence analysis of endometriosis protein-II (ENDO-II) reveals homology with tissue inhibitor of metalloproteinases-1 (TIMP-1). *J. Clin. Endocrinol. Metab.* 80:3784–3787.

Shifren, J., Tseng, J., Zaloudek, C., Ryan, I., Meng, Y., and Ferrara, N. 1996. Ovarian steroid regulation of vascular endothelial growth factor in the human endometrium: implications for angiogenesis during the menstrual cycle and in the pathogenesis of endometriosis. *J. Clin. Endocrinol. Metab.* 81:3112–3118.

Shimazawa, R., Miyachi, H., Takayama, H., Kuroda, K., Kato, F., and Kato, M. 1993. Antiangiogenic activity of tumor necrosis factor-alpha production regulators derived from thalidomide. *Biol. Pharmaceut. Bull.* 22:224–226.

Shimekake, Y., Nagata, K., Ohta, S., Kambayashi, Y., Teraoka, H., and Kitamura, K. 1995. Adrenomedullin stimulates two signal transduction pathways, cAMP accumulation and Ca^{2+} mobilization, in bovine aortic endothelial cells. *J. Biol. Chem.* 270:4412–4417.

Sillem, M., Prifti, S., Neher, M., and Runnebaum, B. 1998. Extracellular matrix remodelling in the endometrium and its possible relevance to the pathogenesis of endometriosis. *Hum. Reprod. Update* 4:730–735.

Simón, C., Frances, A., and Piquette, G. 1994. The immune mediator interleukin-1 receptor antagonist (IL-1ra) preevents embryo implantation. *Endocrinology* 134:521–528.

Simón, C., Piquette, G., Frances, A., Westphal, L., Heinrichs, W., and Polan, M. 1993. Interleukin-1 type I receptor messenger ribonucleic acid expression in human endometrium throughout the menstrual cycle. *Fertil. Steril.* 59:791–796.

Singer, G. A., Strowitzki, T., Rettig, I., and Kimmig, R. 1998. Flow cytometric detection and binding studies of human endometrial stromal cell epidermal growth factor receptor in monolayer culture: influence of progesterone. *Mol. Hum. Reprod.* 4:577–583.

Surrey, E., and Halme, J. 1991. Effect of platelet-derived growth factor on endometrial stromal cell proliferation in vitro: a model for endometriosis? *Fertil. Steril.* 56:672–679.

Tabibzadeh, S. 1990. Evidence of T cell activation and potential cytokine action in human endometrium. *J. Clin. Endocrinol. Metab.* 71:645–653.

Tabibzadeh, S. 1991. Human endometrium: an active site of cytokine production and action. *Endocr. Rev.* 1991:272–290.

Tabibzadeh, S., Gerber, M., and Satyaswaroop, P. 1986. Induction of HLA-DR antigen expression in human endometrial epithelial cells in vitro by recombinant gamma-interferon. *Am. J. Pathol.* 125:90–96.

Tabibzadeh, S., Kong, Q. F., Babaknia, A., and May, L. T. 1995. Progressive rise in the expression of interleukin-6 in human endometrium during menstrual cycle is initiated during the implantation window. *Hum. Reprod.* 10:2793–2799.

Tabibzadeh, S., Satyaswaroop, P., and Rao, P. 1988. Antiproliferative effect of interferon-g in human endometrial epithelial cell in vitro: potential local growth modulatory role in endometrium. *J. Clin. Endocrinol. Metab.* 67:131–138.

Taketani, Y., Kuo, T. M., and Mizuno, M. 1992. Comparison of cytokine levels and embryo toxicity in peritoneal fluid in infertile women with untreated or treated endometriosis. *Am. J. Obstet. Gynecol.* 167:265–270.

Tang, X., Rossi, M., Masterson, B., and Chegini, N. 1994a. Insulin-like growth factor I (IGF-I), IGF-I receptors, and IGF binding proteins 1–4 in human uterine tissue: tissue localization and IGF-I action in endometrial stromal and myometrial smooth muscle cells in vitro. *Biol. Reprod.* 50:1113–1125.

Tang, X., Zhao, Y., Rossi, M., Abu-Rustum, R., Ksander, G., and Chegini, N. 1994b. Expression of transforming growth factor-beta (TGF beta) isoforms and TGF beta type II receptor messenger ribonucleic acid and protein, and the effect of TGF betas on endometrial stromal cell growth and protein degradation in vitro. *Endocrinology* 135:4450–4459.

Tao, X. J., Sayegh, R. A., and Isaacson, K. B. 1997. Increased expression of complement component 3 in human ectopic endometrium compared with the matched eutopic endometrium. *Fertil. Steril.* 68:460–467.

Taylor, R., Ryan, I., Moore, E., Hornung, D., Shifren, J., and Tseng, J. 1997. Angiogenesis and macrophage activation in endometriosis. *Ann. NY Acad. Sci.* 828:194–207.

Taylor, R., Ryan, I., Tseng, J., Shifren, J., Hornung, D., and Lebovic, D. 1999. Ovarian steroids and angiogenesis. In: Lemay, A., and Maheux, R., eds. *Understanding Endometriosis: Advances in Research and Practice*. Parthenon, Carnforth., UK pp. 131–137.

Telgmann, R., and Gellersen, B. 1998. Marker genes of decidualization: activation of the decidual prolactin gene. *Hum. Reprod. Update* 4:472–479.

Treloar, S., O'Connor, D., O'Connor, V., and Martin, N. 1999. Genetic influences on endometriosis in an Australian twin sample. *Fertil. Steril.* 71:701–710.

Tseng, J., Ryan, I., Milam, T., Murai, J., Schriock, E., and Landers, D. 1996. Interleukin-6 secretion in vitro is up-regulated in ectopic and eutopic endometrial stromal cells from women with endometriosis. *J. Clin. Endocrinol. Metab.* 81:1118–1122.

Tsuruoka, N., Sugiyama, M., Tawaragi, Y., Tsujimoto, M., Nishihara, T., and Goto, T. 1988. Inhibition of in vitro angiogenesis by lymphotoxin and interferon-gamma. *Biochem. Biophys. Res. Commun.* 155:429–435.

Upton, P., Austin, C., Taylor, G., Nandha, K., Clark, A., and Ghatei, M. 1997. Expression of adrenomedullin (ADM) and its binding sites in the rat uterus: Increased number of binding sites and ADM messenger ribonucleic acid in 20-day pregnant rats compared with nonpregnant rats. *Endocrinology* 138:2508–2514.

Vandermolen, D. T., and Gu, Y. 1996. Human endometrial interleukin-6 (IL-6): in vivo messenger ribonucleic acid expression, in vitro protein production, and stimulation thereof by IL-1 beta. *Fertil. Steril.* 66:741–747.

Vazquez, F., Rodraiguez-Manzaneque, J. C., Lydon, J. P., Edwards, D. P., O'Malley, B. W., and Iruela-Arispe, M. L. 1999. Progesterone regulates proliferation of endothelial cells. *J. Biol. Chem.* 274:2185–2192.

Watson, H., Franks, S., and Bonney, R. C. 1996. Regulation of epidermal growth factor receptor synthesis by ovarian steroids in human endometrial cells in culture. *J. Reprod. Fertil.* 107:199–205.

Williams, L. 1989. Signal transduction by the platelet-derived growth factor receptor. *Science* 243:1564–1570.

Witz, C. A., Monotoya-Rodriguez, I. A., and Schenken, R. S. 1999. Whole explants of peritoneum and endometrium: a novel model of the early endometriosis lesion. *Fertil. Steril.* 71:56–60.

Wu, M. H., Yang, B. C., Hsu, C. C., Lee, Y. C., and Huang, K. E. 1998. The expression of soluble intercellular adhesion molecule-1 in endometriosis. *Fertil. Steril.* 70:1139–1142.

Zetter, B. R. 1998. Angiogenesis and tumor metastasis. *Annu. Rev. Med.* 49:407–424.

Zhang, L., Mackenzie, I., Rees, M., and Bicknell, R. 1997. Regulation of the expression of the angiogenic enzyme platelet-derived endothelial cell growth factor/thymidine phosphorylase in endometrial isolates by ovarian steroids and cytokines. *Endocrinology* 138:4921–4930.

Zhang, L., Scott, P., Turley, H., Leek, R., Lewis, C., and Gatter, K. 1998. Validation of anti-vascular endothelial growth factor (anti-VEGF) antibodies for immunohistochemical localization of VEGF in tissue sections: expression of VEGF in the human endometrium. *J. Pathol.* 185:402–408.

Zhou, J., Dsupin, B., Giudice, L., and Bondy, C. 1994. Insulin-like growth factor system gene expression in human endometrium during the menstrual cycle. *J. Clin. Endocrinol. Metab.* 79:1723–1734.

PART IV

Vascular Morphogenesis in the Placenta

CHAPTER 14

Placental Vascular Morphogenesis: Introduction and Overview

D. Stephen Charnock-Jones and Stephen K. Smith

While it is self-evident that the placenta is essential for the mammalian reproductive process, historically its role has been confused. The precise architectural arrangement of the blood vessels has been a matter of dispute, and this in part can be explained by the diversity of structures observed in common domestic, laboratory, and wild species (for a review of this subject area, see Kaufmann and Burton, 1994). However, by the beginning of the last century it had been established that the placenta served as an organ for fetal nutrition, respiration, and as an endocrine organ. All of the different placental types show specialized adaptations to facilitate exchange between the fetal and maternal circulatory systems. Both the human and murine placentae are of the hemochorial type, and although there are marked differences between them, knockout and transgenic studies in the mouse have been extremely informative. It is a reasonable assumption that the requirements for exchange play the most significant role in determining placental morphology, and therefore the adaptations of the fetal vasculature to facilitate this process are fundamental. In a range of species as gestation advances and fetal demand increases, the capacity of the placenta for exchange increases. In the human, there is a continual elaboration of the principal functional units of the placenta, the terminal villi, and a progressive reduction in the mean thickness of the villous membrane separating the two circulations. These changes, and the changes seen in other species, are dependent on continued angiogenesis, and therefore placental angiogenesis is central for the success of pregnancy.

CLINICAL CONSEQUENCES OF LACK OF ADAPTATION

The failure of the normal development of the placenta from the early trophectoderm have several consequences for women. At its simplest, the outcome is a failed pregnancy. Miscarriage is associated with a wide range of associative causes including chromosomal abnormalities and uterine abnormalities but also altered adaptation in the placenta. This is manifested by increased perfusion of the early placenta as shown by increased Doppler flow in the body of the placenta. Rather than there being a need for the rapid establishment of the hemochorial placenta, it seems that the trophoblast needs to reduce the exposure of the fetal circulation to that of the mother in the early stages of pregnancy (Jauniaux et al, 1994).

Perhaps the best manifestation of a failed adaptation of the placenta to a clinical consequence is the situation that arises with preeclampsia. The studies of Robertson and Colleagues (1976) suggested that trophoblast invasion of the spiral arterioles was inhibited and trophoblast cells were not found in the vessels at the level of endometrial/myometrial junction. This was associated with a failure of the maternal spiral arterioles to lose their muscular coat, resulting in vessels that can continue to respond to pressor agents. Such changes are less well characterized for intrauterine growth retarded infants, as only about 50% show such changes. These clinical consequences of failed maternal fetal adaptation are discussed in greater detail in Chapters 15, 16, and 17.

While the importance of the vascular adaptation to pregnancy at the implantation site cannot be overstated, it should be remembered that where this goes wrong, systemic manifestations may arise as in preeclampsia. The increased endothelial cell activation, platelet aggregation, and edema are all systemic consequences of the disease (Roberts, 1998). As such they constitute an important element of the role of pathologic angiogenesis in women's health.

DEVELOPMENT OF THE HUMAN PLACENTA

The development of the placenta is complex and dynamic and our understanding of how this organ adapts to the changing demands during pregnancy is undergoing significant revision. The previous assumption that maternal blood bathes and flows around the early villi would appear to be erroneous. The timing and role of regulation by oxygen in villous development and function is described in detail in Chapter 15.

The vasculature of the placenta is dynamically regulated; as fetal demand rises, the diffusional capacity of the placenta rises and this is accomplished by remodeling of the villous capillaries and the overlying trophoblast. These structural changes may be regulated by both physical and chemical signals such as flow and stretch or local po_2. How these alter the cellular behavior is not clear but a number of polypeptide growth factors are known to profoundly affect endothelial cells and to be regulated by the some of the same physical forces and chemical stimuli. In addition, studies in mice have revealed some insights into the complex interaction between endothelial cells (or their progenitors) and other cell types.

The ability to selectively inactivate specific genes in the mouse, so-called knockout technology, has led to a much improved understanding of embryonic and placental development. Two families of such molecules that have been studied in this way and shown to be of central importance are the vascular endothelial growth factor (VEGF) family and the angiopoietin (Ang) family. Knockout of the receptors and the ligands has shown that various members of these families play a central role and are required for placental vascularization.

THE VEGF SYSTEM AND PLACENTAL GROWTH

The VEGF family currently comprises five members. VEGF-A is a homodimeric glycoprotein of between 21 and 46 kd and was first isolated using two bioassays. One of these was dependent on the stimulation of growth of endothelial cells and the other on stimulation of vascular permeability (Ferrara et al, 1992; Ferrara and Davis-Smyth, 1997). The VEGF-A exists in five alternatively spliced forms ranging in size from 121 to 206 amino acids in length (Houck et al, 1992; Charnock-Jones et al, 1993). The different splice variants show varying affinities for heparan sulfate, and in fact the shorter form ($VEGF_{121}$) does not bind to heparin but is freely soluble. The longer forms, while secreted, bind heparin strongly and are generally cell-surface associated. Additionally portions of the longer forms can be released as active fragments by plasmin digestion (Keyt et al, 1996).

VEGF-A immunoreactivity and the messenger RNA (mRNA) encoding it have been localized in the human placenta (Sharkey et al, 1993; Shiraishi et al, 1996; Vuckoviv et al, 1996; Vuorela et al, 1997; Clark et al, 1998b). While there are some discrepancies in the relative amount produced by each cell type, a consensus view would suggest that it is produced by both maternal and fetal macrophages, villus trophoblast, and some villous fibroblasts.

VEGF-A has many direct actions on endothelial cells and a few actions on nonendothelial cells. The endothelial-specific actions are all in some way linked with the process of angiogenesis. VEGF-A is able to stimulate vasodilatation, an increase in microvascular permeability, protease release, migration and proliferation of the endothelial cells, and lumen formation. Each of these processes is required for angiogenesis. For example, nitric oxide is released by endothelial cell following stimulation with VEGF-A, as are a variety of proteases, including collaginase IV, which is able to degrade basement membrane, and protease inhibitors plasminogen activator inhibitor-1 (PAI-1) and PAI-2 (Pepper and Montesano, 1990; Pepper et al, 1991, 1992; Papapetropoulos et al, 1997; Murohara et al, 1998). VEGF-A is also chemotactic for endothelial cells and a potent mitogen for this cell type. This wide range of important endothelial effects would suggest that VEGF-A plays an essential role in both physiologic and pathologic angiogenesis. This indeed is the case, and knockout studies have shown that two copies of this gene are required for embryonic development. In these studies the surprising observation was made that even in the heterozygous state the mutant embryos died in utero (Carmeliet et al, 1996; Ferrara et al, 1996). This suggests that at critical points during development an adequate level of VEGF-A is required to support angiogenesis and organogenesis. At these times half the normal VEGF-A is insufficient. In addition, the vasculature within the yolk cycle in such embryos was defective with an absence of larger collecting vessels. This in it self would have serious consequences on the development of the embryo. Physiologic angiogenesis also occurs in the developing corpus luteum, and inactivation of VEGF-A completely inhibits this process (Ferrara et al, 1998).

VEGF-A has also been shown to be essential for the angiogenesis that either is the cause of or is associated with a variety of pathologic conditions. For example, solid tumor growth requires new blood vessel growth. Blocking the actions of VEGF-A using neutralizing antibodies, a soluble receptor or a dominant negative receptor inhibits the tumor growth (Kim et al, 1993; Millauer et al, 1994; Kong et al, 1998). Indeed there are anti-VEGF-A therapeutic agents currently undergoing clinical trails in cancer patients based on these strategies. Other examples of pathologic angiogenesis are diabetic retinopathy, retinopathy of prematurity, and rheumatoid arthritis (Sato, 1996).

VEGF-A acts via two tyrosine kinase receptors—Flt-1 and KDR (or in the mouse flk-1). Both these receptors have a similar structure comprising an extracellular domain (itself containing seven immunoglobulin-like loops), a transmem-

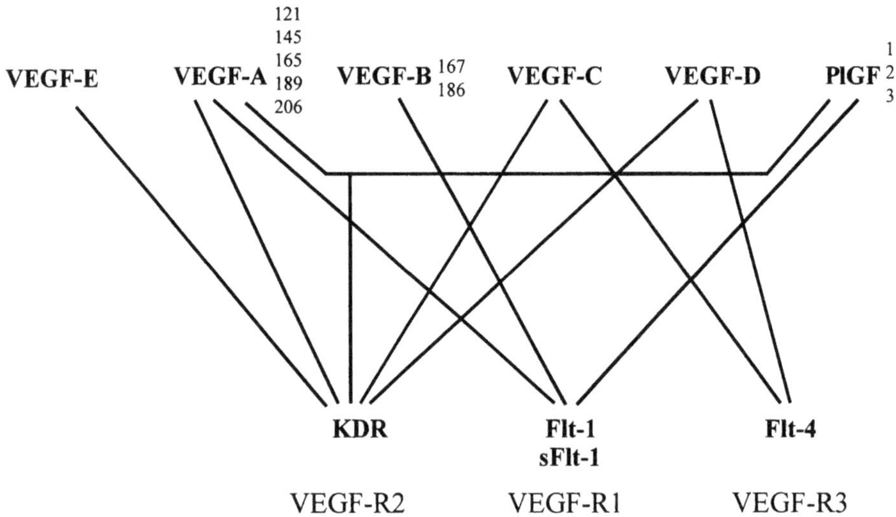

FIGURE 14.1. Schematic summary of the interactions between the various members of the vascular endothelial growth factor (VEGF) family of growth factors and their receptors.

brane domain, and a split intracellular tyrosine kinase domain (Shibuya et al, 1990; Terman et al, 1991). Ligand-binding specificity is conferred by immunoglobulin-like domain 2, and both receptors are present on endothelial cells (Davis-Smyth et al, 1996). Figure 14.1 summarizes the binding of the VEGF family members to the three VEGF tyrosine kinase receptors. Knockout studies have shown that complete ablation of either receptor causes embryonic lethality though the phenotypes do show some differences. In the Flk-1 knockout angioblast differentiation is blocked and both vasculogenesis and hematopoiesis fail in these animals (Shalaby et al, 1995). In the Flt-1 animals angioblast differentiation occurs normally but the structure of the vessels is severely disrupted, so much so that this mutation is also lethal (Fong et al, 1995). The vessels in the yolk sac at around embryo day E8.5 lack an organized vascular structure, although there are many endothelial cells present. Some of these endothelial cells appeared to be within the lumen of the vessels as well as lining them. The embryonic vasculature was similarly disorganized and these embryos died by embryo day E9.0, probably due to a failure in the yolk sac circulation.

The actions of VEGF-A are generally restricted to the endothelial cells, and the pattern of receptor localization agrees with this. However, one notable exception to this is the trophoblast. We have shown by in situ hybridization that the human villous and extravillous trophoblast cells contain large amounts of the mRNA apparently encoding Flt-1. Furthermore, the choriocarcinoma cell line BeWo, which is a commonly used model for the trophoblast, responded to exogenous VEGF-A. At the time these data were published, we suggested that VEGF-A may have direct action on the trophoblast (Charnock-Jones et al, 1994).

THE ROLE OF SOLUBLE FLT-1 IN REGULATING VEGF ACTION

However, understanding of the role of VEGF-A is complicated by the fact that there is a soluble form of the Flt-1 receptor that acts as an antagonist of VEGF-A. This receptor was first described in a human umbilical vein endothelical cell (HUVEC) cDNA library and subsequently in the conditioned media of these cells by Kendall and Thomas (Kendall and Thomas, 1993; Kendall et al, 1996). This soluble receptor is formed by alternative splicing of the Flt-1 pre-mRNA and leads to a premature termination of translation after immunoglobulin domain 6. Other natural forms of soluble VEGF-A receptors exist containing four or five of the extracellular immunoglobulin-like loops (Boocock et al, 1995). Therefore these receptors are able to bind VEGF-A at high affinity but unable to signal since the tyrosine kinase domain is missing.

When we performed ligand-binding studies with ^{125}I-VEGF-A to localize functional receptors in the placenta, we were surprised to find that the VEGF-A binding was confined exclusively to the endothelial cells. There was no detectable binding to the trophoblast cells. The explanation for this discrepancy between the in situ hybridization and the ligand-binding results is that the trophoblast actually expresses the soluble form of the receptor. We confirmed this by in situ hybridization with a probe specific for the mRNA encoding the sFlt-1. The secreted sflt-1 protein was readily detectable in villus conditioned media and in maternal serum. It was not detectable in the serum of nonpregnant women or males, and we have made similar observation in mice (Clark et al, 1998a; He et al, 1999).

Due to the initial in situ hybridization results, we and many others have sought to identify direct actions of VEGF-A on trophoblast. This has not been straightforward due to the problems of culturing primary trophoblast. Several authors have demonstrated convincingly that trophoblast-derived cell lines respond in a variety of ways to VEGF-A or placenta growth factor (PlGF), which also binds to the Flt-1 receptor. For example, the line HTR-8 shows increased [^{3}H]thymidine incorporation when treated with PlGF (in the presence of heparan sulfate proteoglycan) (Athanassiades and Lala, 1998). The cell line used by Ahmed et al (1997) releases NO following VEGF treatment and this is both time and dose dependent (Ahmed et al, 1997; Diss et al, 1992). However, there is considerable debate concerning the true identity and usefulness of these and other trophoblast cell lines. A recent publication describes the validation of trophoblast-derived cell lines (King et al, 2000). It is clear that none of the lines has all the features of fresh trophoblasts; indeed, many produce proteins known not to be produced by cytotrophoblasts. Therefore, it remains an open question as to whether VEGF-A or PlGF do indeed act directly on trophoblasts. Given that these cells produce large amounts of sFlt-1, which is a potent VEGF-A antagonist, this would seem unlikely.

The mRNA encoding this soluble form of the receptor is produced by many endothelial cells and as previously mentioned is highly expressed by trophoblast. Thus, the central question remains as to the role of membrane bound and soluble forms. This has been addressed by Hiratsuka et al (1998), who performed a selective deletion of the tyrosine kinase portion of the flt-1 receptor. This mutation was not lethal and the animals survived and reproduced normally. The only detectable difference between these and normal animals was an alteration in monocyte migration in response to VEGF-A. This is consistent with the previously published data that showed that the monocyte chemotactic response to VEGF-A was mediated by Flt-1 and that the signaling portion of the receptor is not necessary for placental or embryonic development. However, the mutation introduced in this study did not disrupt the production of soluble Flt-1, and therefore it is possible that it is the soluble form of the receptor that is actually required for normal vasculature development. Indeed Fong et al (1999) have shown that ablation of Flt-1 (both soluble and membrane bound) leads to aberrant angioblast commitment and inappropriate vascular channel development in chimeric embryos.

It is therefore probable that the sFlt-1, produced by the trophoblast cells, is the essential protein produced by this gene. The VEGF knockout studies, already described, show that too little VEGF is lethal but too much may also be dangerous. This suggestion is supported by the studies of Fong et al (1995) and also directly, by those of He et al (1999) in which recombinant VEGF was administered to pregnant mice. This caused an increase in resorption frequency, a decrease in embryo weight and placental hemorrhage. Thus, the level of bioactive VEGF would seem to be of critical importance in regulating vessel growth and function during embryonic and placental development, and can be regulated both positively and by a natural antagonist. This is similar to the "angiogenic switch" described by Hanahan and Folkman 1996) in which it is the balance of pro- and antiangiogenic factors (of which there are many) that determines whether the endothelial cells switch from one state to the other. Different tissues have different set positions for this switch and the placenta seems to be very sensitive to excess VEGF-A.

PLACENTA GROWTH FACTOR

Another member of the VEGF family that is highly expressed in the placenta is placenta growth factor (PlGF) (Maglione et al, 1991) (see Chapter 16). This factor also binds the Flt-1 receptor but does not bind KDR (or the third VEGF receptor Flt-4) (Park et al, 1994). However, the precise role of this factor in placental development is at present unclear. Some studies have suggested that PlGF is at best a weak mitogen for endothelial cells and therefore is not likely to be of great physiologic importance. However, others have shown in a range of in vivo model systems that PlGF is as potent as VEGF in stimulating new vessel growth (Ziche et al, 1997). This factor is able to form heterodimers with VEGF when expressed by the same cell and these heterodimers would appear to have intermediate potency (Cao et al, 1996). In situ hybridization studies of the human placenta show that PlGF is predominantly produced by extravillous trophoblast, whereas VEGF is produced by macrophages and to a lesser extent by villous trophoblast. These data suggest that the level of heterodimerization in vivo will be low (Clark et al, 1998b). The predominant form would be a PlGF/PlGF homodimer, and indeed this is readily detectable in the serum of pregnant women. To date no studies have been reported in which endogenous PlGF activity has been blocked in vivo; thus, at present no physiologic process has been shown to be PlGF dependent. Interestingly in the context of placental function plasma PlGF levels are reduced in patients who have preeclampsia (Torry et al, 1999).

The studies published to date are all cross-sectional in nature, although a longitudinal study has been performed in which plasma PlGF levels were shown to be low in patients prior to the onset of preeclampsia and therefore this may be a worthwhile clinical test.

Other members of the VEGF family (VEGF-B, VEGF-C, VEGF-D and VEGF-E) have been shown to have potent effects on endothelial cells in a variety of systems but at present little information is available about their role in placental development.

THE ANGIOPOIETIN FAMILY

Another important family of angiogenic growth factors recently discovered is that of the angiopoietins. Angiopoietin-1 and angiopoietin-2 both bind to the tyrosine kinase receptor tie-2. While current data would suggest that Ang-2 is a competitive antagonist of Ang-1, as more data are accumulated in different endothelial systems this may be seen as a simplification. Nonetheless, knockout studies have shown that ablation of Ang-1 or tie-2 leads to embryonic death with the vascular development in the embryo. Similarly, overexpression of the presumed antagonist Ang-2 also leads to embryonic death (Davis et al, 1996; Maisonpierre et al, 1997; Takakura et al, 1998).

The role of this family of proteins is quite distinct from that of the VEGF family and mediates the interaction between endothelial cells and the closely associated pericytes or vascular muscle cells. It would appear that endothelial cells are sensitive to growth factor withdrawal if they do not have pericytes closely associated with them to provide trophoblast support. Thus, in endothelial cells that are growing rapidly in a growth factor–rich environment pericytes may well be absent, but if for any reason the growth factors decline, then in the absence of pericytes the endothelial cells will die back. This has been elegantly shown by several authors (Alon et al, 1995; Benjamin and Keshet, 1997; Benjamin et al, 1998). These data suggest that the presence of the pericyte to support the endothelial cell is fundamental to the maintenance of a stable vessel. It seems likely that the angiopoietins are the mediators of this interaction (Hanahan, 1997).

While the function of the angiopoietins have been best described in knockout mice, the presence of these factors have been described in the human placenta. In situ hybridization studies show that the mRNAs encoding Ang-2 mRNA is present in the villous trophoblast and readily detectable in the first trimester, and the level drops as gestation progresses. The Ang-1 mRNA is not detectable at any time by in situ hybridization. However, both mRNAs can be detected by reverse-transcriptase polymerase chain reaction (RT-PCR) and their ratio determined. This ratio changes in favor of Ang-1 as gestation proceeds. Of greater interest is the finding that this ratio is oxygen sensitive. The level of the two mRNAs is regulated differently and Ang-1 mRNA becoming increasingly unstable as the po_2 is reduced. Thus, under reduced oxygen the Ang1/Ang-2 ratio is shifted toward Ang-2. This together with a rise in VEGF-A would be expected to lead to new vessel growth.

In all of these studies, the analytic emphasis is generally on the embryo and the possible placental effects are often overlooked. A clear example of this can be seen in the analysis of the Hepatocyte Growth Factor/Scatter Factor (HGF)/SF knockout mice. The two papers that describe this analysis attribute the cause of embryonic death to a failure in placental development or embryonic liver failure (Schmidt et al, 1995; Uehara et al, 1995). In the human it would appear that HGF/SF may be a trophoblast growth factor as well as affecting villous endothelial cells so a placental phenotype is not unexpected (Clark et al, 1996).

THE ROLE OF OXYGEN IN REGULATING PLACENTAL ANGIOGENESIS

One of the most important functions of the placenta is the transfer of oxygen between the two circulatory systems. It therefore would not be surprising if the local po_2 in some way regulated placental development. Our understanding of placental oxygenation during early pregnancy has undergone radical revision over the last decade (see Chapter 15). It would appear that before approximately 12 weeks' gestation aggregates of cytotrophoblast cells virtually occlude the mouths of the maternal spiral arteries, ensuring that any flow into the intervillous space is a slow seepage of blood or even a plasma filtrate (Hustin and Schaaps, 1987; Burton et al, 1999). Thus, the early placenta develops in an environment with a low po_2, which rises rapidly at the start of the second trimester.

Several of the growth factors described above are acutely regulated by the local po_2. Indeed VEGF is regulated transcriptionally and at the level of message stability by decreasing po_2. This has been well studied and is an important mechanism of regulation on VEGF produced by tumors (Levy et al, 1996). However, it is also of great relevance physiologically since, in the endometrium, for example, VEGF is regulated in a similar way. Placental fibroblast VEGF production is similarly oxygen sensitive (Wheeler et al, 1995). PlGF appears also to be regulated by oxygen but in the opposite direction, i.e., PlGF levels are suppressed by low po_2 (Ahmed et al, 2000). The absolute values of, and, more importantly, the ratio of Ang-1 and Ang-2 are altered dramatically in vitro by manipulating the po_2. Thus, many of the polypeptide growth factors that have been shown to have specific effects on endothelial cells and are present in the placenta are oxygen sensitive. It is not surprising, therefore, that in women living at high altitude, where the prevailing po_2 is significantly lower the architecture of the placental villi is altered (Ali et al, 1996; Burton et al, 1996). The changes that have been described would be expected to increase the diffusional capacity of the placenta and include increased branching and an increase in the formation of terminal villus formation.

These interactions are complex and involve the interplay of differentially regulated factors that act on a variety of cell types. Furthermore, since they occur at the scale of a few cells within an individual villus, precise definition of the importance of each will require considerable research. Nonetheless, the rapid advances using knockout animals and in vitro modeling have shown which are the most important molecules. The challenge now is to relate these findings to pathologic conditions and to identify therapeutic strategies to intervene clinically.

REFERENCES

Ahmed, A., Dunk, C., Ahmad, S., and Khaliq, A. 2000. Regulation of placental vascular endothelial growth factor (VEGF) and placenta growth factor (PlGF) and soluble Flt-1 by oxygen—a review. *Placenta* 21(suppl A):S16–24.

Ahmed, A., Dunk, C., Kniss, D., and Wilkes, M. 1997. Role of VEGF receptor-1 (Flt-1) in mediating calcium-dependent nitric oxide release and limiting DNA synthesis in human trophoblast cells. *Lab. Invest.* 76:779–791.

Ali, K. Z. M., Burton, G. J., Morad, N., and Ali, M. E. 1996. Does hypercapillarization influence the branching pattern of terminal villi in the human placenta at high altitude? *Placenta* 17:677–682.

Alon, T., Hemo, I., Itin, A., Pe'er, J., Stone, J., and Keshet, E. 1995. Vascular endothelial growth factor acts as a survival factor for newly formed retinal vessels and has implications for retinopathy of prematurity. *Nat. Med.* 1:1024–1028.

Athanassiades, A., and Lala, P. K. 1998. Role of placenta growth factor (PIGF) in human extravillous trophoblast proliferation, migration and invasiveness. *Placenta* 19:465–473.

Benjamin, L. E., Hemo, I., and Keshet, E. 1998. A plasticity window for blood vessel remodelling is defined by pericyte coverage of the preformed endothelial network and is regulated by PDGF-B and VEGF. *Development* 125:1591–1598.

Benjamin, L. E., and Keshet, E. 1997. Conditional switching of vascular endothelial growth factor (VEGF) expression in tumors: induction of endothelial cell shedding and regression of hemangioblastoma-like vessels by VEGF withdrawal. *Proc. Natl. Acad. Sci. USA* 94:8761–8766.

Boocock, C. A., Charnock-Jones, D. S., Sharkey, A. M., McLaren, J., Barker, P. J., Wright, K. A., Twentyman, P. R., and Smith, S. K. 1995. Expression of vascular endothelial growth factor and its receptors flt and KDR in ovarian carcinoma. *J. Natl. Cancer Inst.* 87:506–516.

Burton, G. J., Jauniaux, E., and Watson, A. L. 1999. Maternal arterial connections to the placental intervillous space during the first trimester of human pregnancy: the Boyd collection revisited. *Am. J. Obstet. Gynecol.* 181:718–724.

Burton, G. J., Reshetnikova, O. S., Milovanov, A. P., and Teleshova, O. V. 1996. Stereological evaluation of vascular adaptations in human placental villi to differing forms of hypoxic stress. *Placenta* 17:49–55.

Cao, Y., Chen, H., Zhou, L., Chiang, M. K., Anand-Apte, B., Weatherbee, J. A., Wang, Y., Fang, F., Flanagan, J. G., and Tsang, M.-S. 1996. Heterodimers of placenta growth factor/vascular endothelial growth factor: Endothelial activity, tumor cell expression, and high affinity binding to Flk-1/KDR. *J. Biol. Chem.* 271:3154–3162.

Carmeliet, P., Ferreira, V., Breier, G., Pollefeyt, S., Kieckens, L., Gertenstein, M., Fahrig, M., Vandenhoeck, A., Harpal, K., Eberhardt, C., Decleucq, C., Pawling, J., Moons, L., Collen, D., Risau, W., and Nagy, A. 1996. Abnormal blood vessel development and lethality in embryos lacking a single VEGF allele. *Nature* 380:435–439.

Charnock-Jones, D. S., Sharkey, A. M., Boocock, C. A., Ahmed, A., Plevin, R., Ferrara, N., and Smith, S. K. 1994. Vascular endothelial growth factor receptor localization and activation in human trophoblast and choriocarcinoma cells. *Biol. Reprod.* 51:524–530.

Charnock-Jones, D. S., Sharkey, A. M., Rajput-Williams, J., Burch, D., Schofield, J. P., Fountain, S. A., Boocock, C. A., and Smith, S. K. 1993. Identification and localization of alternately spliced mRNAs for vascular endothelial growth factor in human uterus and estrogen regulation in endometrial carcinoma cell lines. *Biol. Reprod.* 48:1120–1128.

Clark, D. E., Smith, S. K., He, Y., Day, K. A., Licence, D. R., Corps, A. C., Lammoglia, R., and Charnock-Jones, D. S. 1998a. A vascular endothelial growth factor antagonist is produced by the human placenta and released into the maternal circulation. *Biol. Reprod.* 59:1540–1548.

Clark, D. E., Smith, S. K., Licence, D., Evans, A. L., and Charnock-Jones, D. S. 1998b. Comparison of expression patterns for placenta growth factor, vascular endothelial growth factor (VEGF), VEGF-B and VEGF-C in the human placenta throughout gestation. *J. Endocrinol.* 159:459–467.

Clark, D. E., Smith, S. K., Sharkey, A. M., Sowter, H. M., and Charnock-Jones, D. S. 1996. Hepatocyte growth factor/scatter factor and its receptor c-met: localisation and expression in the human placenta throughout pregnancy. *J. Endocrinol.* 151:459–467.

Davis, S., Aldrich, T. H., Jones, P. F., Acheson, A., Compston, D. L., Jain, V., Ryan, T. E., Bruno, J., Radziejewski, C., Maisonpierre, P. C., and Yancopoulos, G. D. 1996. Isolation of angiopoietin-1, a ligand for the TIE2 receptor, by secretion-trap expression cloning. *Cell* 87:1161–1169.

Davis-Smyth, T., Chen, H., Park, J., Presta, L. G., and Ferrara, N. 1996. The second immunoglobulin-like domain of the VEGF tyrosine kinase receptor Flt-1 determines ligand binding and may initiate a signal transduction cascade. *EMBO J.* 15:4919–4927.

Diss, E. M., Gabbe, S. G., Moore, J. W., and Kniss, D. A. 1992. Study of thromboxane and prostacyclin metabolism in an in vitro model of first-trimester human trophoblast. *Am. J. Obstet. Gynecol.* 167:1046–1052.

Ferrara, N., Carver-Moore, K., Chen, H., Dowd, M., Lu, L., O'Shea, H. S., Powell-Braxton, L., Hillan, K. J., and Moore, M. W. 1996. Heterozygous embryonic lethality induced by targeted inactivation of the VEGF gene. *Nature* 380:439–442.

Ferrara, N., Chen, H., Davis-Smith, T., Gerber, H. P., Nguyen, T. N., Peers, D., Chisholm, V., Hillan, K. J., and Schwall, R. H. 1998. Vascular endothelial growth factor is essential for corpus luteum angiogenesis. *Nat. Med.* 4:336–340.

Ferrara, N., and Davis-Smyth, T. 1997. The biology of vascular endothelial growth factor. *Endocr. Rev.* 18:4–25.

Ferrara, N., Houck, K., Jakeman, L., and Leung, D. W. 1992. Molecular and biological properties of the vascular endothelial growth factor family of proteins. *Endocr. Rev.* 13:18–32.

Fong, G. H., Rossant, J., Gertsenstein, M., and Breitman, M. L. 1995. Role of the Flt-1 receptor tyrosine kinase in regulating the assembly of vascular endothelium. *Nature* 376:66–70.

Fong, G. H., Zhang, L., Bryce, D. M., and Peng, J. 1999. Increased hemangioblast commitment, not vascular disorganization, is the primary defect in flt-1 knock-out mice. *Development* 126:3015–3025.

Hanahan, D. 1997. Signaling vascular morphogenesis and maintenance. *Science* 277:48–50.

Hanahan, D., and Folkman, J. 1996. Patterns and emerging mechanisms of the angiogenic switch during tumorigenesis. *Cell* 86:353–364.

He, Y., Smith, S. K., Day, K. A., Clark, D. E., Licence, D. R., and Charnock-Jones, D. S. 1999. Alternative splicing of vascular endothelial growth factor (VEGF)-R1 (FLT-1) pre-mRNA is important for the regulation of VEGF activity. *Mol. Endocrinol.* 13:537–545.

Hiratsuka, S., Minowa, O., Kuno, J., Noda, T., and Shibuya, M. 1998. Flt-1 lacking the tyrosine kinase domain is sufficient for normal development and angiogenesis in mice. *Proc. Natl. Acad. Sci. USA* 95:9349–9354.

Houck, K. A., Leung, D. W., Rowland, A. M., Winer, J., and Ferrara, N. 1992. Dual regulation of vascular endothelial growth factor bioavailability by genetic and proteolytic mechanisms. *J. Biol. Chem.* 267:26031–26037.

Hustin, J., and Schaaps, J. P. 1987. Echographic [corrected] and anatomic studies of the maternotrophoblastic border during the first trimester of pregnancy. *Am. J. Obstet. Gynecol.* 157:162–168.

Jauniaux, E., Zaidi, J., Jurkovic, D., Campbell, S., and Hustin, J. 1994. Comparison of colour Doppler features and pathological findings in complicated early pregnancy. *Hum. Reprod.* 9:2432–2437.

Kaufmann, P., and Burton, G. J. 1994. Anatomy and genesis of the placenta. In: Knobil, E., and Neill, J. D., eds. *The Physiology of Reproduction.* Raven Press, New York, pp. 441–484.

Kendall, R. L., and Thomas, K. A. 1993. Inhibition of vascular endothelial cell growth factor activity by an endogenously encoded soluble receptor. *Proc. Natl. Acad. Sci. USA* 90:10705–10709.

Kendall, R. L., Wang, G., and Thomas, K. A. 1996. Identification of a natural soluble form of the vascular endothelial growth factor receptor, FLT-1, and its heterodimerization with KDR. *Biochem. Biophys. Res. Commun.* 226:324–328.

Keyt, B. A., Berleau, L. T., Nguyen, H. V., Chen, H., Heinsohn, H., Vandlen, R., and Ferrara, N. 1996. The carboxyl-terminal domain (111–165) of vascular endothelial growth factor is critical for its mitogenic potency. *J. Biol. Chem.* 271:7788–7795.

Kim, K. J., Li, B., Winer, J., Armanini, M., Gillett, N., Phillips, H. S., and Ferrara, N. 1993. Inhibition of vascular endothelial growth factor-induced angiogenesis suppresses tumour growth in vivo. *Nature* 362:841–844.

King, A., Thomas, L., and Bischof, P. 2000. Cell culture models of trophoblast II: trophoblast cell lines—a workshop report. *Placenta* 21(suppl A):S113–S119.

Kong, H. L., Hecht, D., Song, W., Kovesdi, I., Hackett, N. R., Yayon, A., and Crystal, R. G. 1998. Regional suppression of tumor growth by in vivo transfer of a cDNA encoding a secreted form of the extracellular domain of the flt-1 vascular endothelial growth factor receptor. *Hum. Gene Ther.* 9:823–833.

Levy, A. P., Levy, N. S., and Goldberg, M. A. 1996. Post-transcriptional regulation of vascular endothelial growth factor by hypoxia. *J. Biol. Chem.* 271:2746–2753.

Maglione, D., Guerriero, V., Viglietto, G., Delli-Bovi, P., and Persico, M. G. 1991. Isolation of a human placenta cDNA coding for a protein related to the vascular permeability factor. *Proc. Natl. Acad. Sci. USA* 88:9267–9271.

Maisonpierre, P. C., Suri, C., Jones, P. F., Bartunkova, S., Wiegand, S. J., Radziejewski, C., Compton, D., McClain, J., Aldrich, T. H., Papadopoulos, N., Daly, T. J., Davis, S., Sato, T. N., and Yancopoulos, G. D. 1997. Angiopoietin-2, a natural antagonist for Tie2 that disrupts in vivo angiogenesis. *Science* 277:55–60.

Millauer, B., Shawver, L. K., Plate, K. H., Risau, W., and Ullrich, A. 1994. Glioblastoma growth inhibited in vivo by a dominant-negative Flk-1 mutant. *Nature* 367:576–578.

Murohara, T., Horowitz, J. R., Silver, M., Tsurumi, Y., Chen, D., Sullivan, A., and Isner, J. M. 1998. Vascular endothelial growth factor/vascular permeability factor enhances vascular permeability via nitric oxide and prostacyclin. *Circulation* 97:99–107.

Papapetropoulos, A., Garcia-Cardena, G., Madri, J. A., and Sessa, W. C. 1997. Nitric oxide production contributes to the angiogenic properties of vascular endothelial growth factor in human endothelial cells. *J. Clin. Invest.* 100:3131–3139.

Park, J. E., Chen, H. H., Winer, J., Houck, K. A., and Ferrara, N. 1994. Placenta growth factor. Potentiation of vascular endothelial growth factor bioactivity, in vitro and in vivo, and high affinity binding to Flt-1 but not to Flk-1/KDR. *J. Biol. Chem.* 269:25646–25654.

Pepper, M. S., Ferrara, N., Orci, L., and Montesano, R. 1991. Vascular endothelial growth factor (VEGF) induces plasminogen activators and plasminogen activator inhibitor-1 in microvascular endothelial cells. *Biochem. Biophys. Res. Commun.* 181:902–906.

Pepper, M. S., and Montesano, R. 1990. Proteolytic balance and capillary morphogenesis. *Cell. Differ. Dev.* 32:319–328.

Pepper, M. S., Sappino, A. P., Montesano, R., Orci, L., and Vassalli, J. D. 1992. Plasminogen activator inhibitor-1 is induced in migrating endothelial cells. *J. Cell. Physiol.* 153:129–139.

Roberts, J. M. 1998. Endothelial dysfunction in preeclampsia. *Semin. Reprod. Endocrinol.* 16:5–15.

Robertson, W. B., Brosens, I., and Dixon, G. 1976. Maternal uterine vascular lesions in the hypertensive complications of pregnancy. *Perspect. Nephrol. Hypertens.* 5:115–127.

Sato, Y. 1996. Current state of anti-angiogenic therapy. *Biotherapy* 10:972–978.

Schmidt, C., Bladt, F., Goedecke, S., Brinkmann, V., Zschiesche, W., Sharpe, M., Gherardi, E., and Birchmeier, C. 1995. Scatter factor/hepatocyte growth factor is essential for liver development. *Nature* 373:699–702.

Shalaby, F., Rossant, J., Yamaguchi, T. P., Gertsenstein, M., Wu, X. F., Breitman, M. L., and Schuh, A. C. 1995. Failure of blood-island formation and vasculogenesis in Flk-1 deficient mice. *Nature* 376:62–66.

Sharkey, A. M., Charnock-Jones, D. S., Boocock, C. A., Brown, K. D., and Smith, S. K. 1993. Expression of mRNA for vascular endothelial growth factor in human placenta. *J. Reprod. Fertil.* 99:609–615.

Shibuya, M., Yamaguchi, S., Yamane, A., Ikeda, T., Tojo, A., Matsushime, H., and Sato, M. 1990. Nucleotide sequence and expression of a novel human receptor-type tyrosine kinase gene (flt) closely related to the fms family. *Oncogene* 5:519–524.

Shiraishi, S., Nakagawa, K., Kinukawa, N., Nakano, H., and Sueishi, K. 1996. Immunohistochemical localization of vascular endothelial growth factor in the human placenta. *Placenta* 17:111–121.

Takakura, N., Huang, X. L., Naruse, T., Hamaguchi, I., Dumont, D. J., Yancopoulos, G. D., and Suda, T. 1998. Critical role of the TIE2 endothelial cell receptor in the development of definitive hematopoiesis. *Immunity* 9:677–686.

Terman, B. I., Carrion, M. E., Kovacs, E., Rasmussen, B. A., Eddy, R. L., and Shows, T. B. 1991. Identification of a new endothelial cell growth factor receptor tyrosine kinase. *Oncogene* 6:1677–1683.

Torry, D. S., Wang, H. S., Wang, T. H., Caudle, M. R., and Torry, R. J. 1999. Preeclampsia is associated with reduced serum levels of placenta growth factor. *Am. J. Obstet. Gynecol.* 179:1539–1544.

Uehara, Y., Minowa, O., Mori, C., Shiota, K., Kuno, J., Noda, T., and Kitamura, N. 1995. Placental defect and embryonic lethality in mice lacking hepatocyte growth factor/scatter factor. *Nature* 373:702–705.

Vuckoviv, M., Ponting, J., Terman, B. I., Niketic, V., Seif, M. W., and Kumar, S. 1996. Expression of the vascular endothelial growth factor receptor, KDR, in human placenta. *J. Anat.* 188:361–366.

Vuorela, P., Hatva, E., Lymboussaki, A., Kaipainen, A., Joukov, V., Persico, M. G., Alitalo, K., and Halmesmaki, E. 1997. Expression of vascular endothelial growth factor and placenta growth factor in human placenta. *Biol. Reprod.* 56:489–494.

Wheeler, T., Elcock, C. L., and Anthony, F. W. 1995. Angiogenesis and the placental environment. *Placenta* 16:289–296.

Ziche, M., Maglione, D., Ribatti, D., Morbidelli, L., Lago, C. T., Battisti, M., Paoletti, I., Barra, A., Tucci, M., Parise, G., Vincenti, V., Granger, H. J., Viglietto, G., Persico, M. G. 1997. Placenta growth factor-1 is chemotactic, mitogenic, and angiogenic. *Lab Invest* 76:517–531

CHAPTER 15

Placental Functional Morphology

Graham J. Burton

The placenta performs a remarkable variety of functions, acting as the fetal lung, liver, and kidneys in addition to serving as a physical and immunologic barrier separating the maternal and fetal circulations. Each of these different functions places its own special demands on the structure of the organ, some of which are potentially in conflict. However, it is a reasonable assumption that the requirements for diffusional exchange play the most significant role in determining placental morphology. The rate of diffusion of a gas across a membrane is governed by the Fick equation, and so the structural determinants are the surface area available for exchange and the thickness of the membrane. As gestation advances there is a continual elaboration of the principal functional units of the placenta, the terminal villi, and a progressive reduction in the mean thickness of the villous membrane separating the two circulations. Both these changes facilitate gaseous exchange, and are dependent on continued angiogenesis within the villi.

PLACENTAL VASCULOGENESIS

The human placenta is a discoid organ comprising an elaborately branched villous tree that is bathed directly by the maternal blood. Development begins at the time of implantation when the trophectodermal layer of the blastocyst wall proliferates and differentiates into an outer multinucleated layer, the syncytiotrophoblast, and an underlying population of uninucleate progenitor cytotrophoblast cells. Tongues of trophoblast penetrate the maternal endometrium, and soon acquire a mesodermal core (Kaufmann and Burton, 1994). For many years the origin of the mesoderm was debated, but it now seems clear that it is derived by precocious activity at the caudal end of the future primitive streak (Luckett, 1978). The mesoderm is therefore closely associated with the allantoic diverticulum, but while this sac remains rudimentary in the human, the mesoderm migrates over the entire inner surface of the trophectoderm, penetrating the trophoblastic projections to form villous precursors. Hemangioblastic cells are first recognizable within these mesodermal cores between days 18 and 20 postfertilization as clusters or cords of closely approximated cells, often lying just beneath the trophoblastic basement membrane (Demir et al, 1989). These clusters are conspicuous by the extent and

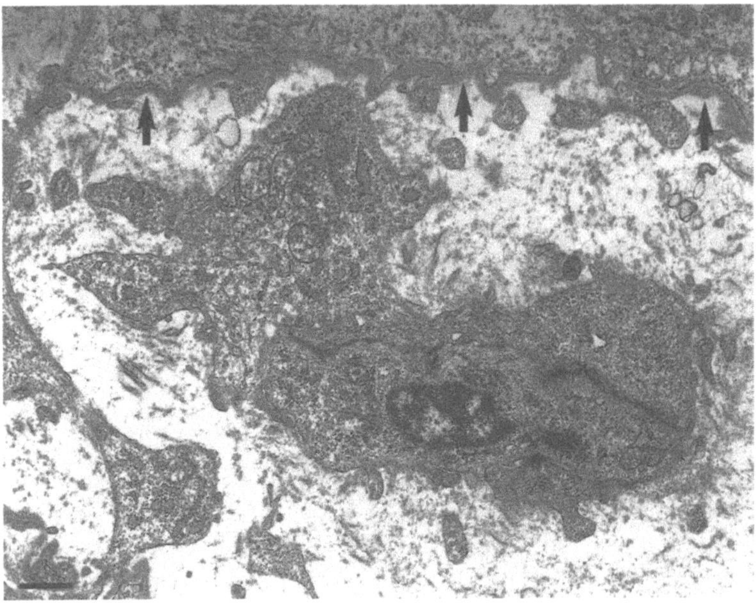

FIGURE 15.1. Electron micrograph of the villous core from an 18-week placenta. The early stages of capillary development are represented by aggregates of mesodermal cells deep to the trophoblast basement membrane (arrow). The cells are linked by conspicuous electron-dense junctional complexes and no intercellular space is evident. Scale bar = 1 μm.

electron density of the junctional complexes linking the cells. The membranes of adjacent cells are closely juxtaposed over a large area, and an intercellular space is barely visible (Figure 15.1).

Other examples are observed in which the cluster has a more circular profile, and in which the junctional complexes are more restricted and localized toward the periphery of the cells. A distinct intercellular space separates the cells and appears to be continuous within the junctional complexes (Figure 15.2). Further examples are seen in which the intercellular space has enlarged to form a distinct lumen, and the whole structure is clearly recognizable as an early capillary (Figure 15.3). It is notable that additional cells within the villous mesodermal core make extensive and intimate contact with the presumptive endothelial cells, suggesting that a recruitment process takes place. At this stage there is no basement membrane surrounding the vessel. After viewing specimens of 7 to 11 weeks menstrual age, Dempsey (1972) concluded that mitotic figures are not seen in the early placental endothelial cells, and was of the opinion that capillary enlargement at this stage is due to proliferation and incorporation of pericytes alone. By contrast, other authors have observed and illustrated dividing endothelial cells during the first trimester (Castellucci et al, 1987; Jones and Fox, 1991), and so it seems probable that both processes are involved.

The factors regulating placental angiogenesis at this stage of pregnancy are far from clear. The prevailing oxygen tension is an important factor to consider, and our understanding of the situation during early pregnancy has undergone radical revision over the last decade. Prior to this it was held that maternal blood flows

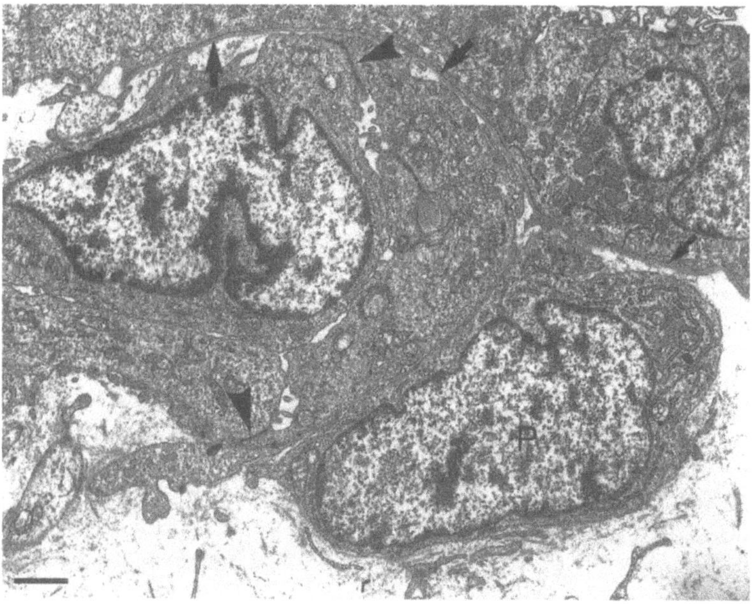

FIGURE 15.2. Other aggregates are seen in which the junctional complexes (arrowheads) between the cells are more restricted, and a clear intercellular space is visible in the center. Note the proximity to the trophoblast basement membrane (arrow), and the presence of closely associated mesodermal cells (P) from the villous core. Scale bar = 1 μm.

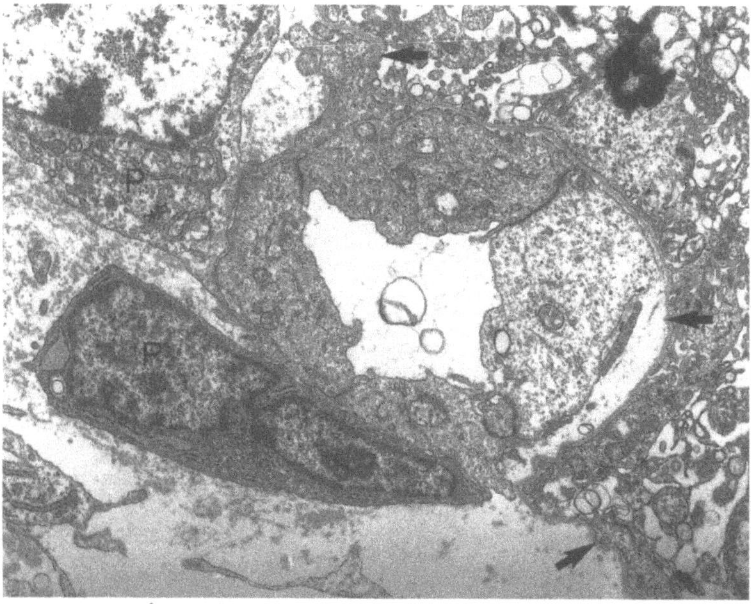

FIGURE 15.3. A later stage of capillary development in which a lumen is now clearly present. Again, note the close proximity to the trophoblast basement membrane (arrow), and the recruitment of mesodermal cells as pericytes (P). Scale bar = 1 μm.

freely through the developing placenta from shortly after implantation onward. This supposition was first challenged by ultrasonic and anatomic studies that indicated that a continuous maternal circulation is not fully established until around 12 weeks of pregnancy (Hustin et al, 1988; Jaffe et al, 1997). Before this stage aggregates of cytotrophoblast cells virtually occlude the mouths of the maternal spiral arteries, ensuring that any flow into the intervillous space is a slow seepage of blood or even a plasma filtrate (Hustin and Schaaps, 1987; Burton et al, 1999a). Direct measurements of the oxygen tension within the placenta have confirmed a mean tension of 17.9 mm Hg at 8 to 10 weeks of pregnancy, which rises to 60.7 mm Hg at 12 to 13 weeks (Rodesch et al, 1992; Jauniaux et al, 2000). The placental villi therefore develop initially in a low oxygen environment, but at the start of the second trimester they are exposed to rapidly rising concentrations. In an attempt to quantify the effect of this change on vascular development, villi from 6 to 10 weeks were compared morphometrically with others from 11 to 15 weeks (Jauniaux et al, 1991). A steady increase in the number of capillary profiles within each villous profile was seen over this period, and the percentage of the villous volume occupied by capillaries rose from 2.7% (SD ± 0.8) to 4.1% (SD ± 0.9). These data suggest that capillary development proceeds independently of the prevailing oxygen tension, although they provide no information on the differentiation of hemangioblast precursor cells.

By contrast, a higher rate of capillary proliferation has been reported in placentas at 10 to 12 weeks gestational age from women with iron-deficiency anemia compared to controls (Kadyrov et al, 1998). The mean number of capillary profiles per villous profile increased with the severity of the anemia, rising from 4.14 in controls (Hb > 110 g/L) to 11.7 in severely anemic women (Hb < 90 g/L). Despite this large increase, there was no change in the percentage volume of the villous stromal core occupied by the capillaries, for the mean capillary diameter tended to be smaller. The authors attribute their findings to excessive branching angiogenesis in response to the maternal anemia, although predicting the impact of a reduced hemoglobin concentration on the already low oxygen tension within the intervillous space in the absence of significant circulation of maternal blood is difficult. A further note of caution should be added when interpreting the data from both studies. Comparisons based on planar morphometric data, such as the number of capillaries per villous profile, are only meaningful if there has been no simultaneous change in overall villous morphology (Mayhew and Burton, 1988). This applies to fractional volumes as well, for if the villi are larger or more edematous in one condition than another, the reference space will alter. Under ideal conditions it is the total number of endothelial cells or the total capillary length that is significant, but with first-trimester material the complete placenta is rarely obtainable.

PLACENTAL MATURATION

During early pregnancy the villi are large, and display a low surface area to volume ratio (Figure 15.4). Their diffusional characteristics are limited, but the favorable fetoplacental weight ratio at this stage of gestation undoubtedly ensures that the placenta is capable of meeting fetal demands. Fetal weight increases exponentially during the third trimester, slowing down only in the last few weeks before term, whereas placental weight shows only a gradual increment. The placental villous tree undergoes considerable transformation during this period, aimed at meeting

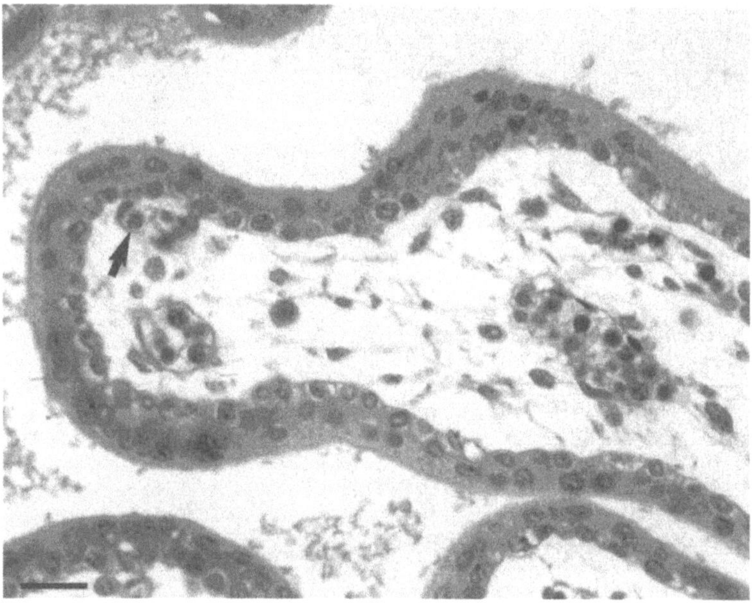

FIGURE 15.4. Low-power view of villi at approximately 7 weeks of gestation. The diffusional capacity of such villi will be low due to their small surface area to volume ratio and limited vascularization. Small-caliber capillaries (arrow) can be seen close to the trophoblast basement membrane. Scale bar = 20 μm.

these new demands. In general the changes can be summarized as an elaboration of smaller terminal villi with a high surface area to volume ratio, an increased vascularization, and a reduction in the thickness of the villous membrane. Angiogenesis and vascular remodeling are key to all three processes.

Terminal villi are the functional units of the placenta, particularly for diffusional exchange. They are first elaborated at around 20 weeks of gestation and the process continues to term, providing a surface area of 12 to 14 m^2 (Jackson et al, 1992). Characteristically terminal villi are short, stubby protuberances, up to 100 μm in length and 80 μm in diameter, that arise from the surfaces of larger branches of the villous tree, the intermediate villi. The formation of new terminal villi does not appear to be due to local proliferation in either the trophoblast or mesodermal compartments. Rather, it is thought to be due to the capillaries within an intermediate villus elongating at a faster rate than the villus itself (Kaufmann et al, 1985). As a result the capillaries become coiled, and when loops obtrude from the villous surface they raise the trophoblast covering in a blister-like fashion, thus forming a new terminal villus. It is therefore to be expected that terminal villi will have a high vascular complement, and this is immediately apparent when viewing cross sections (Figure 15.5). Assessing the true size and proportion of the capillaries in vivo is difficult, however, for they rapidly collapse postpartum. Studies based on villous samples aspirated in situ at the time of cesarean section indicate that they occupy approximately 35% of the overall villous volume (Sen et al, 1979), and a similar figure is obtained if placentas are perfusion-fixed at physiologic pressures after delivery (Burton et al, 1987).

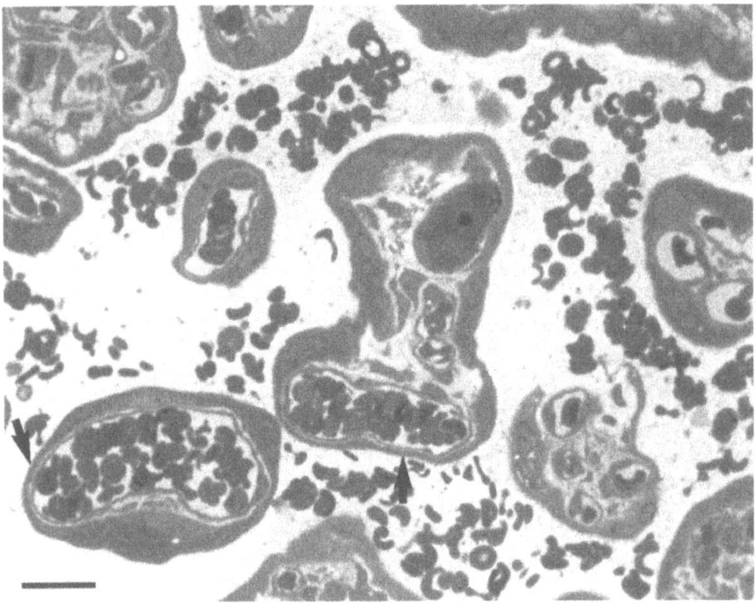

FIGURE 15.5. Terminal villi from a mature placenta demonstrating the higher surface area to volume ratio, and increased vascularity at this stage of pregnancy. Notably, the fetal capillaries are closely approximated to the trophoblast surface, which is locally thinned to form vasculosyncytial membranes (arrow). These are considered to be the principal sites of diffusional exchange. Scale bar = 20 μm.

CAPILLARY STRUCTURE

Throughout gestation the placental capillary endothelial cells are of the continuous nonfenestrated type, and are joined by tight junctions (Figure 15.6) (Heinrich et al, 1976; Jones and Fox, 1991). In the past, two types of endothelial cell have been described on the basis of their cross-sectional appearances (Nikolov and Schiebler, 1981). Both types contain a moderate number of pinocytotic vesicles, but while one set possesses an abundance of intermediate filaments, the other is characterized by the presence of numerous secretory granules and rough endoplasmic reticulum. Morphometric studies have revealed, however, that an unequal distribution of organelles exists between the perinuclear and peripheral parts of placental endothelial cells (Karimu and Burton, 1995). The contrasting appearances described can be accounted for by differing planes of section through individual cells, casting doubt on whether two types of cell genuinely exist.

The stage of gestation at which a basement membrane can be seen surrounding the capillaries has been variously reported (Jones and Fox, 1991), but most sources agree that one is present by the start of the third trimester (Figure 15.6). It contains laminin, fibronectin, and type IV collagen, and is secreted by both the endothelial cells and their pericytes (Yamada et al, 1987). Pericytes are seen in close approximation to capillaries, and their cell processes ramify over the outer surface of the endothelial cells (Figure 15.6).

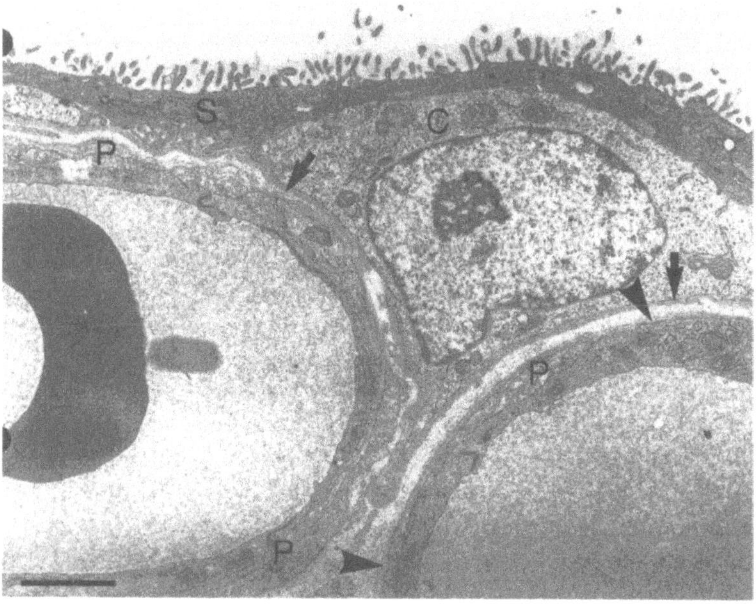

FIGURE 15.6. Electron micrograph of two capillary profiles in a term placenta. They are closely approximated to the trophoblastic basement membrane (arrow) with only a few scanty collagen fibers intervening. The endothelial cells are nonfenestrated, and are linked by tight junctions. Pericytes (P) are present, along with a basement membrane (arrowheads). S, syncytiotrophoblast; C, cytotrophoblast. Scale bar = 2 μm.

FORMATION OF VASCULOSYNCYTIAL MEMBRANES

It is notable that the outer walls of the capillaries are often closely approximated to the overlying trophoblastic basement membrane (Figure 15.5). At these points the syncytiotrophoblast is locally thinned, and appears to be devoid of mitochondria and many other organelles. The thickness of the villous membrane separating the maternal and fetal circulations may be reduced to as little as 1 to 2 μm at these sites, which are termed vasculosyncytial membranes. Because their structure is similar in many respects to the alveolus of the lung, these sites are considered to be the principal sites of diffusional exchange. A reduction in their incidence is frequently associated with fetal distress (Fox, 1967).

The presence of vasculosyncytial membranes facilitates gaseous exchange in two ways. First, mathematical models reveal that 90% of the total resistance to oxygen diffusion between the maternal and fetal hemoglobins is offered by the villous membrane (Laga et al, 1973; Mayhew et al, 1986). Reducing the thickness of the epithelium at a vasculosyncytial membrane will therefore have a beneficial effect on membrane conductance. Second, the placenta has a high metabolic rate, and it is considered that 30% to 40% of the total oxygen consumed by the fetoplacental unit is utilized by the organ. Since the trophoblast is responsible for much of the protein synthesis and active transport, it is likely that it accounts for a high percentage of this uptake. Hence, it must be expected that if the trophoblast were distributed uniformly in thickness over the villous surface, a steep oxygen concentration gradient would exist between the intervillous space and the fetal capillaries in the villous core. By analogy with electrical circuits in this situation, the placental and fetal demands are placed in series. However, an uneven distribution of the trophoblast over the villous surface, with reduction in thickness over the capillaries and accumulations for metabolic purposes elsewhere on the villous surface, reduces the amount of oxygen that is abstracted before the fetal hemoglobin is reached. In this situation the placental and fetal demands are placed more in parallel, and a higher oxygen tension can be achieved in the fetal circulation (Steven and Samuel, 1976).

Given the key role of vasculosyncytial membranes in placental diffusional exchange, it is important to consider how they are formed, and the factors that may regulate this process. These are difficult questions to answer, since longitudinal studies cannot be performed in the human. However, two contrasting theories have been proposed that place the motive force within the trophoblast and the fetal capillaries, respectively. First, early in vitro studies suggested that under low oxygen conditions the syncytiotrophoblast may stream away from vasculosyncytial membranes, causing a reduction in membrane thickness (Tominaga and Page, 1966). More stringent analyses using stereologic techniques to measure membrane thickness in villi maintained under different oxygen concentrations did not support this claim, and indeed cast doubt on the suitability of villous organ culture as a suitable model system (Burton et al, 1989). An insuperable problem with this technique is that the fetal capillaries are no longer perfused, and their consequent partial collapse has a major confounding impact on the values of many parameters measured. Second, the close morphologic relationship that exists between the capillaries and the trophoblast basement membrane suggested to many observers that vasculosyncytial membranes are formed by the obtrusion of fetal capillaries into the trophoblast layer (Pisarski and Topilko, 1966; Bouw et al, 1976). Support for this theory was provided by the demonstration that a precise and consistent relationship exists between the volume of the fetal capillaries and the mean thickness of the villous membrane (Burton and Feneley, 1992). The strength of this relationship indicates that significant redistribution of the trophoblast over the villous surface does not take place independently of a change in capillary diameter.

CAPILLARY SINUSOIDS

The obtrusion of the capillaries into the trophoblast layer is facilitated by another aspect of their gross morphology, in that the caliber of the villous capillaries is not uniform along their length. Localized dilatations, achieving diameters of 20 to 40 μm, occur toward the end of gestation (Boyd and Hamilton, 1970; Kaufmann et al, 1985), and are commonly referred to as sinusoids. Use of the microvascular corrosion-casting technique has revealed that these dilatations are almost invariably located at the tips of terminal villi, and on the outer wall of tight capillary bends (Figure 15.7). This pattern strongly suggests that mechanical factors such as wall tension or shear stress may be important in their formation. Wall tension will be greatest on the outer surface of a bend, and consequently the capillaries will dilate preferentially at this point under the influence of the increasing transmural pressure toward term. Ballooning of the capillary outward will bring the wall into contact with the trophoblastic basement membrane, and the pressure exerted from

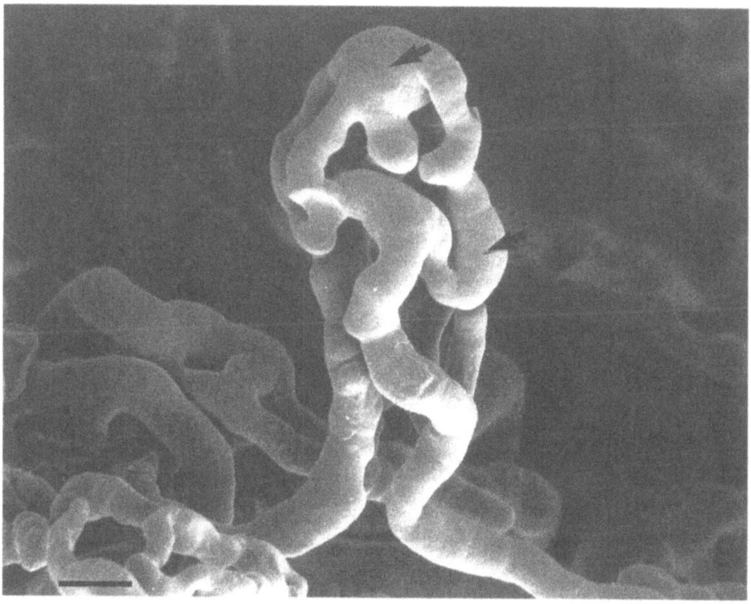

FIGURE 15.7. A microvascular corrosion cast of a villous capillary network from a term placenta. Sinusoidal dilatations (arrow) of the capillaries occur at the tip of the villus and at the sites of tight bends. Several capillaries may radiate from one sinusoid, suggesting these are points of fusion or sprouting angiogenesis. (From Burton, 1987, with permission.) Scale bar = 20 μm.

the undersurface may lead to trophoblast remodeling and local attenuation in thickness. As a result a vasculosyncytial membrane is formed (Burton and Tham, 1992).

Evidence that this may be the case was provided by experiments in which the fetal placental circulation was perfused in vitro at different arterial pressures. Increasing the pressure resulted in an asymptotic rise in the fractional volume of the villi occupied by the capillaries, with most of the change occurring within the range of 40 to 60 mm Hg (Karimu and Burton, 1994a). This rise was accompanied by a similar increase in the amount of capillary surface area per unit volume of villous tissue, and a reduction in the thickness of the villous membrane. If the pressure in the intervillous space was raised simultaneously, then these effects were blocked, confirming that the pressure differential between the two circulations plays a key role in determining the size of the sinusoids (Karimu and Burton, 1994b). Thus, over the physiologic range the sinusoid walls display elastic properties, and this may be facilitated by the fact that whereas for much of their length the villous capillaries are closely encircled by collagen fibers, these are sparse in the vicinity of the sinusoids (Kaufmann et al, 1977). Further distention is restrained at higher pressures, possibly due to the large numbers of vimentin microfilaments contained within the endothelial cells, and/or the complete basement membrane that surrounds the capillaries at this stage of gestation (Jones and Fox, 1991). However, the possibility remains that the diffusional characteristics of the placenta can be significantly altered by relatively small changes in fetal blood pressure. Peripheral vasoconstriction during fetal hypoxia results in increased fetal blood pressure and diversion of much of the blood flow in the descending aorta to the umbilical circulation. On the basis of the results above, it is to be expected that the sinusoids will dilate, and consequently both the rate of placental blood flow and the diffusion capacity of the villous membrane will adapt so as to improve fetal oxygenation. The restriction of capillary dilation to a preferential orientation at the site of sinusoids ensures that maximal changes in membrane thickness can be achieved with minimal changes in the fetal extracorporeal blood volume (Karimu and Burton, 1994a).

Sinusoids are not totally dependent on a distending pressure for their existence, however, for they remain present, albeit to a reduced extent, after cessation of the fetal circulation at delivery. Given that cyclic strain and shear stress are powerful mitogenic stimuli for endothelial cells (Resnick and Gimbrone, 1995), it is likely that the sinusoids will be sites of cell proliferation. Some evidence for this come from the observation that frequently several capillaries appear to connect at a single sinusoid, suggesting these may be sites of sprouting angiogenesis (Figure 15.7). Thus, as gestation advances and the dynamics of the fetal circulation increase, cell recruitment into the capillary wall will lead to permanent enlargement.

The formation of the sinusoids and vasculosyncytial membranes therefore seems to be closely interlinked (Figure 15.8). Besides reducing the diffusion distance between the two circulations, with all the benefits that this confers, the sinusoids further facilitate placental exchange by their effects on the local blood flow. The increase in vessel diameter will result in a reduction in the rate of flow past the thinned villous membrane, allowing a greater time for transfer to take place.

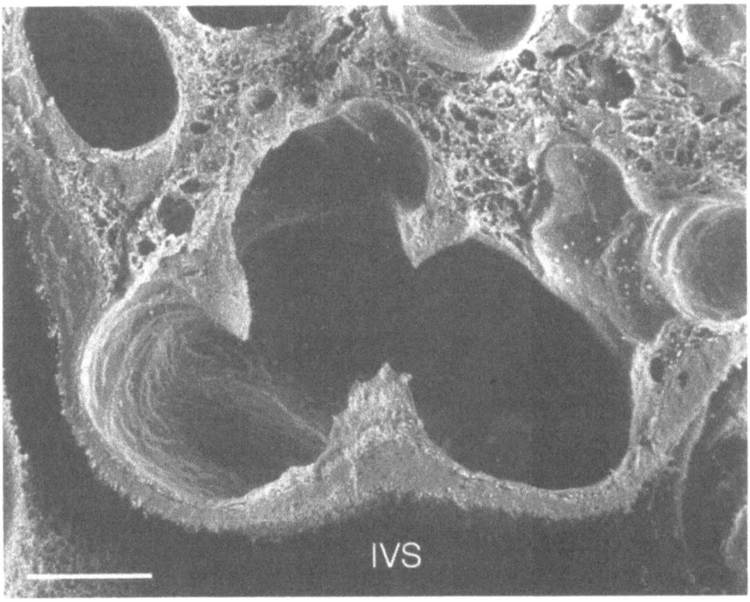

FIGURE 15.8. Scanning electron micrograph of a cryofractured perfusion-fixed terminal villus from a term placenta illustrating the manner by which the capillary sinusoids bring the fetal blood into close approximation to the maternal blood in the intervillous space (IVS). The vasculosyncytial membrane can be seen as a dome-like elevation on the villous surface. (From Burton, 1987, with permission.) Scale bar = 10μm.

In addition, it is likely that the often abrupt transition in diameter will cause the flow through the sinusoid to become complex, although not turbulent. Such a pattern would bring a greater number of fetal erythrocytes into close proximity to the maternal circulation than would linear flow. Finally, it has also been suggested that the presence of sinusoids reduces the vascular resistance within the fetal capillary network, and that they may assist in ensuring an even distribution of blood flow through the various parts of the villous tree (Kaufmann et al, 1985, 1988).

MORPHOLOGIC DETERMINANTS OF PLACENTAL VASCULAR RESISTANCE

Many attempts have been made to correlate alterations in placental vascular resistance as assessed by Doppler waveform patterns with morphologic changes in the villous vasculature (Jauniaux and Burton, 1993). During development there is a profound reduction in resistance that correlates with the proliferation of terminal villi and their vascular beds (Jauniaux et al, 1992). However, changes associated with more subtle degrees of placental maturation at term have not proved possible to detect in this way. No differences in waveform were observed between normal term placentas that were graded completely immature or mature on the Grannum system, although there was a tendency for a greater capillary length and surface area in the latter group (Degani et al, 1991; Burton and Jauniaux, 1995).

There are two major difficulties with this approach. First, according to Poiseuille's law, the resistance of a vessel is related to the fourth power of the radius, and so even the most minor luminal collapse postpartum can have a profound impact on calculations. Second, the relative contributions of the various elements of the placental vascular tree to overall resistance are not yet known. The umbilical arteries ramify over the chorionic plate, branching as they do so, before entering a stem villus. Here they are surrounded by smooth muscle cells and by an adventitia. As the stem villi branch, so do the arteries, until the level of arterioles is reached within the intermediate villi. It is these that feed the capillary beds within the terminal villi, but unlike resistance vessels in many vascular beds, these seem to be maximally dilated under normal conditions (Boura and Walter, 1991). There is no autonomic innervation of the placental vasculature, and so the role they play in determining resistance is uncertain. However, vascular resistance will depend not only on the cross-sectional area of these arteries but also on how they are arranged geometrically.

The branching pattern of vessels can be encoded in techniques such as the Strahler system, which has been used to quantify fluvial networks, but this is a time-consuming activity and has not been applied to the placenta. Instead, various mathematical models have been devised with differing degrees of sophistication (Thompson and Trudinger, 1990; Guiot et al, 1992). These models predict that the pulsatility index will rise if areas of the villous tree are occluded, or if the caliber of the vessels is reduced. Both these predictions have been confirmed by clinical studies. For example, Jackson et al (1995) found a reduction in the volumes of intermediate and terminal villi in small-for-gestational-age pregnancies associated with reduced diastolic flow velocity, but the villi present had the normal vascular complement. On the other hand, several studies have reported that in more severe cases of absent or reversed end-diastolic flow the fetal capillaries are more slender, less branched, and straighter than normal (Hitschold et al, 1993; Krebs et al, 1996; Todros et al, 1999). The cause for this maldevelopment is not yet known, but on the basis of comparisons with other clinical conditions it has been suggested that it may be due to too much oxygen reaching the placenta rather than too little (Kingdom and Kaufmann, 1997).

EFFECT OF OXYGEN SUPPLY ON PLACENTAL VASCULATURE

The effects of oxygen supply on placental structure have recently been reviewed (Burton et al, 1999b). Attention has focused in the past on clinical conditions in which the placenta has been considered hypoxic, but the morphologic findings have often been inconsistent or even contradictory between studies. There are a number of reasons why this might be the case. First, the criteria for claiming intrauterine hypoxia often relate to clinical signs at the time of delivery, and so their relevance to placental development during the second and early third trimesters is questionable. Second, hypoxia is a generic term embracing heterogeneous conditions that may elicit different responses within the placenta (Burton et al, 1996). Third, predicting the oxygen tension within the intervillous space on the basis of maternal blood flow rate alone is hazardous, for in reality the tension will always be a balance between maternal supply and fetoplacental extraction.

One of the few conditions in which the maternal arterial oxygen tension is universally lowered is pregnancy at high altitude, for atmospheric oxygen content falls exponentially with ascent. Up to 2,800 m the placenta appears grossly normal, and total villous volume and surface area remain constant (Reshetnikova et al, 1994). At higher altitudes there is impoverished growth of the villous tree, and a compensatory increase in the volume of the intervillous space (Jackson et al, 1987). Within the villi there are significant changes in the fetal vasculature, and again these appear to be altitude-dependent. Up to 2,800 m the fetal capillaries occupy a greater percentage volume of the terminal villi than at sea level, and there is a greater capillary surface area per unit volume of villous tissue. By contrast, there is no significant increase in overall capillary length per unit volume, and so mean capillary diameter tends to increase, although the differences fail to reach statistical significance (Burton et al, 1996). At altitudes up to 3,600 m mean capillary diameter appears to be reduced, but there is a tendency for a greater capillary/villus length ratio, and the capillaries lie more peripherally within the villi (Jackson et al, 1988). These differences may reflect changes in the expression of angiogenic factors, for the higher incidence of chorangioma beyond 3,600 m suggests that normal capillary development may be disrupted by overstimulation (Soma et al, 1995; Reshetnikova et al, 1996).

These measurements provide an accurate assessment of the total villous or capillary volume and surface area, which as discussed are important determinants of the placenta's diffusing capacity. They do not provide any indication of how that volume is distributed in three-dimensional space, however. Studies on animal placentas have indicated that hypoxia may stimulate capillary branching as well as proliferation (Krebs et al, 1997). Branching is a difficult property to quantify, but on the basis of the model of terminal villus formation discussed previously one would predict that increased capillary branching would be reflected by denser clustering of shorter, more stubby terminal villi. Scanning electron microscopy has confirmed that this is indeed the case in placentas from an altitude of 3,000 m (Ali et al, 1996). More branching, and the establishment of a greater number of short capillary loops, may confer physiologic benefits to the fetus by shortening the transit time for fetal blood through the villous tree.

Since the oxygen tension within the intervillous space falls from 16 weeks until term (Soothill et al, 1986), a progressive increase in capillary branching might therefore be expected under normal conditions. The findings of slender, straight capillaries in cases of intrauterine growth retardation with absent end-diastolic flow are clearly contrary to these expectations. Consequently, it has been proposed that these placentas are exposed to an hyperoxic environment, resulting from reduced oxygen extraction by the fetus (Kingdom and Kaufmann, 1997). In many ways this is an attractive hypothesis, although there is a circularity in the argument as currently presented (Burton, 1997). Furthermore, differences have been described in the villous vasculature of normal placentas between the well-oxygenated zones close to the sites of maternal arterial inflow, and the more peripheral venous areas (Critchley and Burton, 1987). These changes are considerably more subtle than those reported above in intrauterine growth retardation, casting doubt as to whether oxygen tension alone is a sufficient explanation for the latter's causation.

CONCLUSION

For effective exchange the placenta needs to approximate the fetal and maternal circulations as closely as possible over a large area. Angiogenesis and capillary remodeling are key processes in achieving both these requirements through their effects on terminal villus and vasculosyncytial membrane formation. There is much evidence to support a role for the prevailing oxygen tension in regulating placental vascular development, but given the delicacy of the structures concerned and their susceptibility to pressure differentials, it is likely that mechanical stimuli are also important. Only by exploring the interaction between these will a more complete understanding of the control of placental angiogensis be obtained.

REFERENCES

Ali, K. Z. M., Burton, G. J., Morad, N., and Ali, M. E. 1996. Does hypercapillarization influence the branching pattern of terminal villi in the human placenta at high altitude? *Placenta* 17:677–682.

Boura, A. L. A., and Walter, W. A. W. 1991. Autocoids and the control of vascular tone in the human umbilical-placental circulation. *Placenta* 12:453–477.

Bouw, G. M., Stolte, L. A. M., Baak, J. P. A., and Oort, J. 1976. Quantitative morphology of the placenta. I. Standardization of sampling. *Eur. J. Obstet. Gynecol. Reprod. Biol.* 6:325–331.

Boyd, J. D., and Hamilton, W. J. 1970. *The Human Placenta.* Heffer and Sons, Cambridge.

Burton, G. J. 1987. The time structure of the human placental villus as revealed by scanning electron microscopy. *Scanning Micross.* 1:1811–1828.

Burton, G. J. 1997. On "Oxygen and placental villous development: origins of fetal hypoxia." *Placenta* 18:625–626.

Burton, G. J., and Feneley, M. R. 1992. Capillary volume fraction is the principal determinant of villous membrane thickness in the human placenta at term. *J. Dev. Physiol.* 17:39–45.
Burton, G. J., Ingram, S. C., and Palmer, M. E. 1987. The influence of the mode of fixation on morphometrical data derived from terminal villi in the human placenta at term: a comparison of immersion and perfusion fixation. *Placenta* 8:37–51.
Burton, G. J., and Jauniaux, E. 1995. Sonographic, stereological and Doppler flow velocimetric assessments of placental maturity. *Br. J. Obstet. Gynaecol.* 102:818–825.
Burton, G. J., Jauniaux, E., and Watson, A. L. 1999a. Maternal arterial connections to the placental intervillous space during the first trimester of human pregnancy: the Boyd Collection revisited. *Am. J. Obstet. Gynecol.* 181:718–724.
Burton, G. J., Jauniaux, E., and Watson, A. L. 1999b. Influence of oxygen supply on placental structure. In: O'Brien, P. M. S., Wheeler, T., and Barker, D. J. P., eds. *Fetal Programming: Influences on Development and Disease in Later Life. RCOG Press*, London, pp. 326–341.
Burton, G. J., Mayhew, T. M., and Robertson, L. A. 1989. Stereological re-examination of the effects of varying oxygen tensions on human placental villi maintained in organ culture for up to 12h. *Placenta* 10:263–273.
Burton, G. J., Reshetnikova, O. S., Milovanov, A. P., and Teleshova, O. V. 1996. Stereological evaluation of vascular adaptations of human placental villi to differing forms of hypoxic stress. *Placenta* 17:49–55.
Burton, G. J., and Tham, S. W. 1992. The formation of vasculo-syncytial membranes in the human placenta. *J. Dev. Physiol.* 18:43–47.
Castellucci, M., Celona, A., Bartels, H., Steininger, B., Benedetto, V., and Kaufmann, P. 1987. Mitosis of the Hofbauer cell: possible implications for a fetal macrophage. *Placenta* 8:65–76.
Critchley, G. R., and Burton, G. J. 1987. Intralobular variations in barrier thichkness in the mature human placenta. *Placenta* 8:185–194.
Degani, S., Paltieli, Y., Lewinsky, R., Abramovici, D., and Sharf, M. 1991. Fetal and utero-placental flow velocity waveforms and placental grading. *Am. J. Perinatol.* 8:47–49.
Demir, R., Kaufmann, P., Castellucci, M., Erbengi, T., and Kotowski, A. 1989. Fetal vasculogenesis and angiogenesis in human placental villi. *Acta Anat.* 136:190–203.
Dempsey, E. W. 1972. The development of capillaries in the villi of early human placentas. *Am. J. Anat.* 134:221–238.
Fox, H. 1967. The incidence and significance of vasculo-syncytial membranes in the human placenta. *J. Obstet. Gynaecol. Br. Commonw.* 47:28–33.
Guiot, C., Pianta, P. G., and Todros, T. 1992. Modelling the feto-placental circulation: 1. A distributed network predicting umbilical haemodynamics throughout pregnancy. *Ultrasound Med. Biol.* 18:535–544.
Heinrich, D., Metz, J., Raviola, E., and Forssmann, W. G. 1976. Ultrastructure of perfusion-fixed fetal capillaries in the human placenta. *Cell Tissue Res.* 172:157–169.
Hitschold, T., Weiss, E., Beck, T., Huntefering, H., and Berle, P. 1993. Low target birth weight or growth retardation? Umbilical Doppler flow velocity waveforms and histometric analysis of fetoplacental vascular tree. *Am. J. Obstet. Gynecol.* 168:1260–1264.
Hustin, J., and Schaaps, J. P. 1987. Echographic and anatomic studies of the maternotrophoblastic border during the first trimester of pregnancy. *Am. J. Obstet. Gynecol.* 157:162–168.
Hustin, J., Schaaps, J. P., and Lambotte, R. 1988. Anatomical studies of the utero-placental vascularisation in the first trimester of pregnancy. *Trophoblast Res.* 3:49–60.
Jackson, M. R., Mayhew, T. M., and Boyd, P. A. 1992. Quantitative description of the elaboration and maturation of villi from 10 weeks of gestation to term. *Placenta* 13:357–370.

Jackson, M. R., Mayhew, T. M., and Haas, J. D. 1987. Morphometric studies on villi in human term placentae and the effects of altitude, ethnic grouping and sex of newborn. *Placenta* 8:487–495.

Jackson, M. R., Mayhew, T. M., and Haas, J. D. 1988. On the factors which contribute to thinning of the villous membrane at high altitude. II. An increase in the degree of peripheralization of fetal capillaries. *Placenta* 9:9–18.

Jackson, M. R., Walsh, A. J., Morrow, R. J., Mullen, J. B. M., Lye, S. J., and Ritchie, J. W. K. 1995. Reduced placental villous tree elaboration in small-for-gestational-age pregnancies: relationship with umbilical artery Doppler waveforms. *Am. J. Obstet. Gynecol.* 172:518–525.

Jaffe, R., Jauniaux, E., and Hustin, J. 1997. Maternal circulation in the first-trimester human placenta—myth or reality? *Am. J. Obstet. Gynecol.* 176:695–705.

Jauniaux, E., and Burton, G. J. 1993. Correlation of umbilical Doppler features and placental morphometry: the need for uniform methodology. *Ultrasound Obstet. Gynecol.* 3:233–235.

Jauniaux, E., Burton, G. J., Moscosco, G. J., and Hustin, J. 1991. Development of the early placenta: a morphometric study. *Placenta* 12:269–276.

Jauniaux, E., Jurkovic, D., Campbell, S., and Hustin, J. 1992. Doppler ultrasound features of the developing placental circulations: correlation with anatomic findings. *Am. J. Obstet. Gynecol.* 166:585–587.

Jauniaux, E., Watson, A. L., Hempstock, J., Bao, Y.-P., Skepper, J. N., and Burton, G. J. 2000. Onset of maternal arterial bloodflow and placental oxidative stress; a possible factor in human early pregnancy failure. *Am. J. Path.* 157:2111–2122.

Jones, C. J. P., and Fox, H. 1991. Ultrastructure of the normal human placenta. *Electron Microsc. Rev.* 4:129–178.

Kadyrov, M., Kosanke, G., Kingdom, J., and Kaufmann, P. 1998. Increased fetoplacental angiogenesis during first trimester in anaemic women. *Lancet* 352:1747–1749.

Karimu, A. L., and Burton, G. J. 1994a. Compliance of the human placental villous membrane at term: the concept of the feto-placental unit as an autoregulating gas exchange system. *Trophoblast Res.* 8:541–558.

Karimu, A. L., and Burton, G. J. 1994b. The effects of maternal vascular pressure on the dimensions of the placental capillaries. *Br. J. Obstet. Gynaecol.* 101:57–63.

Karimu, A. L., and Burton, G. J. 1995. Human term placental capillary endothelial cell specialization: a morphometric study. *Placenta* 16:93–99.

Kaufmann, P., Bruns, U., Leiser, R., Luckhardt, M., and Winterhager, E. 1985. The fetal vascularisation of term placental villi. II. Intermediate and terminal villi. *Anat. Embryol.* 173:203–214.

Kaufmann, P., and Burton, G. J. 1994. Anatomy and genesis of the placenta. In: Knobil, E., and Neill, J. D., eds. *The Physiology of Reproduction. Raven Press*, New York, pp. 441–484.

Kaufmann, P., Luckhardt, M., and Leiser, R. 1988. Three-dimensional representation of the fetal vessel system in the human placenta. *Trophoblast Res.* 3:113–137.

Kaufmann, P., Stark, J., and Stegner, H. E. 1977. The villous stroma of the human placenta. 1. The ultrastructure of fixed connective tissue cells. *Cell Tissue Res.* 177:105–121.

Kingdom, J. C. P., and Kaufmann, P. 1997. Oxygen and placental villous development: origins of fetal hypoxia. *Placenta* 18:613–621.

Krebs, C., Longo, L. D., and Leiser, R. 1997. Term ovine placental vasculature: comparison of sea level and high altitude conditions by corrosion cast and histomorphometry. *Placenta* 18:43–51.

Krebs, C., Macara, L. M., Leiser, R., Bowman, A. W., Greer, I. A., and Kingdom, J. C. P. 1996. Intrauterine growth restriction with absent end-diastolic flow velocity in the umbilical artery is associated with maldevelopment of the placental terminal villous tree. *Am. J. Obstet. Gynecol.* 175:1534–1542.

Laga, E. M., Driscoll, S. G., and Munro, H. N. 1973. Quantitative studies of human placenta. *I. Morphometry. Biol. Neonate* 23:231–259.
Luckett, W. P. 1978. Origin and differentiation of the yolk sac and extraembryonic mesoderm in presomite human and rhesus monkey embryos. *Am. J. Anat.* 152:59–97.
Mayhew, T. M., and Burton, G. J. 1988. Methodological problems in placental morphometry: apologia for the use of stereology based on sound sampling practice. *Placenta* 9:565–581.
Mayhew, T. M., Jackson, M. R., and Haas, J. D. 1986. Microscopical morphology of the human placenta and its effects on oxygen diffusion: a morphometric study. *Placenta* 7:121–131.
Nikolov, S. D., and Schiebler, T. H. 1981. Über Endothelzellen in Zottengefäβen der reifen menschlichen placenta. *Acta Anat.* 110:338–344.
Pisarski, T., and Topilko, A. 1966. Comparative study of the vascular syncytial membranes of the human placenta in light and electron microscopy. *Pol. Med. J.* 5:630–638.
Reshetnikova, O. S., Burton, G. J., and Milovanov, A. P. 1994. Effects of hypobaric hypoxia on the feto-placental unit: the morphometric diffusing capacity of the villous membrane at high altitude. *Am. J. Obstet. Gynecol.* 171:1560–1565.
Reshetnikova, O. S., Burton, G. J., Milovanov, A. P., and Fokin, E. I. 1996. Increased incidence of placental chorangioma in high altitude pregnancies: hypobaric hypoxia as a possible aetiological factor. *Am. J. Obstet. Gynecol.* 174:557–561.
Resnick, N., and Gimbrone, M. A. 1995. Hemodynamic forces are complex regulators of endothelial gene expression. *FASEB J.* 9:874–882.
Rodesch, F., Simon, P., Donner, C., and Jauniaux, E. 1992. Oxygen measurements in endometrial and trophoblastic tissues during early pregnancy. *Obstet. Gynecol.* 80:283–285.
Sen, D. K., Kaufmann, P., and Schweikhart, G. 1979. Classification of placental villi. II. Morphometry. *Cell Tissue Res.* 200:425–434.
Soma, H., Watanabe, Y., and Hata, T. 1995. Chorangiosis and chorangioma in three cohorts of placentas from Nepal, Tibet and Japan. *Reprod. Fertil. Dev.* 7:1533–1538.
Soothill, P. W., Nicolaides, K. H., Rodeck, C. H., and Campbell, S. 1986. Effect of gestational age on fetal and intervillous blood gas and acid-base values in human pregnancy. *Fetal Ther.* 1:168–175.
Steven, D. H., and Samuel, C. A. 1976. Morphological basis of placental transport. In: Wilkinson, A. W., ed. *Early Nutrition and Later Development. Pitman Medical*, London, pp. 18–30.
Thompson, R. S., and Trudinger, B. J. 1990. Doppler waveform pulsatility index and resistance, pressure and flow in the umbilical placental circulation; an investigation using a mathematical model. *Ultrasound Med. Biol.* 16:449–458.
Todros, T., Sciarrone, A., Piccoli, E., Guiot, C., Kaufmann, P., and Kingdom, J. 1999. Umbilical Doppler waveforms and placental villous angiogenesis in pregnancies complicated by fetal growth restriction. *Obstet. Gynecol.* 93:499–503.
Tominaga, T., and Page, E. W. 1966. Accommodation of the human placenta to hypoxia. *Am. J. Obstet. Gynecol.* 94:679–691.
Yamada, T., Isemura, M., Yamaguchi, Y., Munakata, H., Hayashi, N., and Kyogoku, M. 1987. Immunochemical localization of fibronectin in the human placentas in their different stages of maturation. *Histochemistry* 86:579–584.

CHAPTER 16

Placenta Growth Factor

M. Graziella Persico

The search for regulators of angiogenesis has led to the identification of a number of genes that, when tested in vitro or in vivo (in animal models), help to elucidate the complex phenomenon of vascular formation during development and pathologic conditions. Many new growth factors and receptors and other molecules previously thought to be specific for other systems (e.g., the nervous system) have now been found to play a role in angiogenesis and are currently under analysis. One such example is *neuropilin-1*, which is expressed predominantly in developing neurons (for a review see Fujisawa and Kitsukawa, 1998). *Neuropilin-1* null mutants die during embryonic development due to disorganization of nerve pathways (Kitsukawa et al, 1997) as well as several vascular defects (Kawasaki et al, 1999).

Still, the greater part of research is focused on the vascular endothelial growth factor (VEGF) family of structurally related growth factors (for a review see Bussolino et al, 1997; Persico et al, 1999), of which VEGF is the first identified member. This family includes placenta growth factor (PlGF, Maglione et al, 1991), VEGF-A (Ferrara and Henzel, 1989; Gospodarowicz et al, 1989; Keck et al, 1989; Levy et al, 1989; Conn et al, 1990), VEGF-B (Olofsson et al, 1996), VEGF-C (Joukov et al, 1996), Fos-induced growth factor (FIGF also named VEGF-D) (Orlandini et al, 1996), and VEGF-E (Wise et al, 1999). All members of this family are able to stimulate the proliferation of endothelial cells as well as neoangiogenesis. In all these proteins, intra- and interchain disulfide bonds are involved in the formation of the dimeric active proteins. In addition, PlGF, VEGF-A, and VEGF-B can form heterodimeric molecules (Figure 16.1) (DiSalvo et al, 1995; Cao et al, 1996a; Olofsson et al, 1996).

Alternative splicing of the *PlGF* primary transcript leads to two forms of the mature human PlGF protein (Figure 16.2) (Maglione et al, 1993a). The two forms, named PlGF-1 and PlGF-2 (or $PlGF_{131}$ and $PlGF_{152}$), differ by the insertion of a highly basic 21-amino-acid stretch at the carboxyl-end of the protein. This additional basic region confers on PlGF-2 the ability to bind to heparin (Hauser and Weich, 1993; Maglione et al, 1993a; Barillari et al, 1998), and, in vivo, PlGF-2 is suggested to bind to the heparan sulfates of the extracellular matrix.

The *PlGF* gene has been assigned to two syntenic regions, chromosome 14q24 in humans and chromosome 12qD in mice (DiPalma et al, 1996), in an attempt to

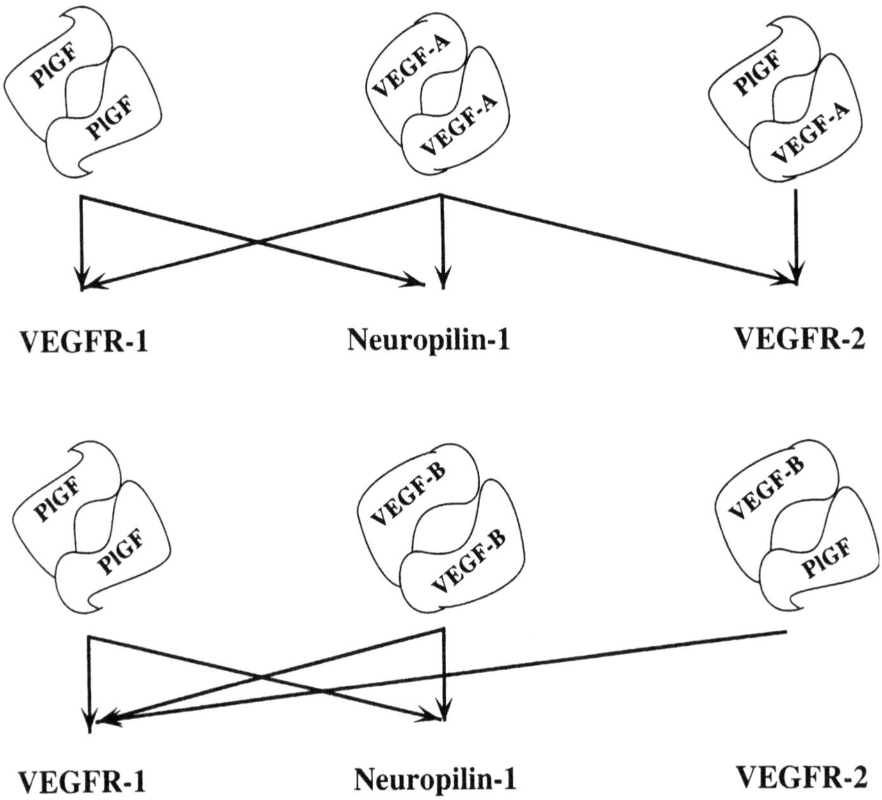

FIGURE 16.1. Schematic representation of the binding of homodimers and heterodimers of placenta growth factor (PlGF), vasculor endothelial growth factor (VEGF)-A and VEGF-B to receptors VEGFR-1, VEGFR-2, and neuropilin-1.

correlate diseases and mouse mutants located on these chromosome regions and to provide additional information on PlGF function.

PlGF RECEPTORS

A series of apparently discordant results has accumulated over the years on the receptor-binding effects of PlGF and on its effect on the cell under test. The use of primary cultures of endothelial cells obtained from different tissues or of transfected cells, overexpressing the receptor of interest, has rendered the comparison from different laboratories very difficult. Therefore, the data reviewed here have to be considered as a contribution to the understanding of the action of PlGF, keeping in mind that they are obtained in vitro on cells that could have lost some of the characteristics they have in vivo.

Until recently, the endothelial cell-specific VEGF receptor-1 (VEGFR-1, also named Flt-1), VEGFR-2 (also named Flk-1/KDR), and VEGFR-3 (also named Flt-4) were considered the only three receptor-type tyrosine kinases interacting

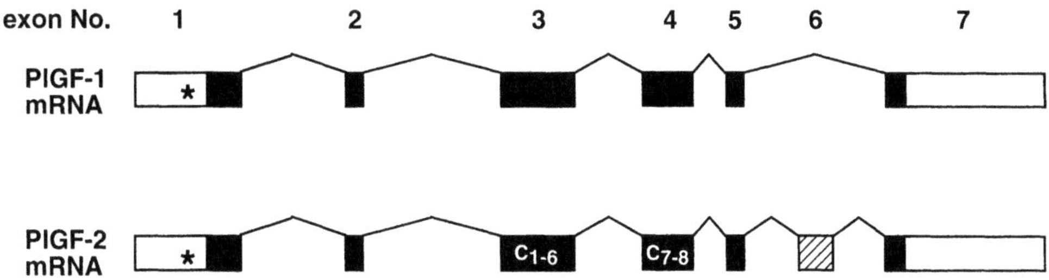

FIGURE 16.2. Exon-intron organization of PlGF transcripts. The asterisk indicates the open reading frame (ORF) present in the 5′ region. Hatched boxes represent the exon containing the basic aminoacid region responsible for heparin binding.

with the VEGF family growth factors (for a review see Shibuya et al, 1999; Taipale et al, 1999). In particular, VEGFR-1 was considered to be the only receptor through which the homodimeric PlGF could act on the endothelial cells. With the recent identification of neuropilin-1 as an additional receptor for PlGF and the finding that this receptor is also present on endothelial cells (Migdal et al, 1998), all previous data should be reassessed.

Homodimeric PlGF binds to the VEGFR-1 receptor (Figure 16.1) (Terman et al, 1994; Cao et al, 1996a; Sawano et al, 1996; Landgren et al, 1998) and the immunoglobulin G (IgG)-like domain 2 of the VEGFR-1 is responsible for the binding specificity of PlGF (Davis-Smyth et al, 1996). Iodinated PlGF-2, but not PlGF-1, can bind to neuropilin-1 receptor on human umbilical vein endothelial cells (HUVECs) (Migdal et al, 1998). Still, both isoforms can induce migration of these cells in the Boyden chamber assay but not proliferation (Migdal et al, 1998).

Since PlGF has been shown to bind and to induce autophosphorylation of the VEGFR-1 tyrosine kinase receptor (Sawano et al, 1996; Landgren et al, 1998) but not of VEGFR-2 (Park et al, 1994; Terman et al, 1994), it appears that PlGF exerts its mitogenic and chemotactic effects on endothelial cells through the activation of the VEGFR-1 intracellular signaling pathway. PlGF induces DNA synthesis but not migration of porcine aortic endothelial (PAE) cells overexpressing VEGFR-1 receptor (Landgren et al, 1998). Our findings that PlGF is mitogenic and chemotactic for bovine coronary postcapillary venule endothelial cells (CVEC) and HUVEC in vitro (Maglione et al, 1991; Ziche et al, 1997) are in contrast to the data by Migdal et al (1998), who reported that PlGF is unable to stimulate proliferation of HUVEC. The recent observation, however, that VEGFR-1 is able to mediate signaling in HUVEC in response to both PlGF and VEGF, leading to distinct biologic responses, suggests that VEGFR-1 does not act as a decoy receptor but is indeed able to signal intracellularly (Clauss et al, 1996). In addition, both PlGF and VEGF are able to induce migration of monocytes through activation of VEGFR-1 (Barleon et al, 1996; Clauss et al, 1996).

PlGF is the most abundant member of the VEGF family in the human dermal microvascular endothelial cells in culture (Yonekura et al, 1999). In normoxic conditions the ratio between PlGF and VEGF messenger RNAs (mRNAs) is about 160:1. Inhibition of PlGF translation by antisense results also in the inhibition of cell proliferation under hypoxic conditions (Yonekura et al, 1999). These new findings assign a role to PlGF in the direct control of endothelial cell proliferation either competing with VEGF for the binding to VEGFR-1, therefore forcing the binding of VEGF to the VEGFR-2 or regulating the amount of heterodimeric PlGF/VEGF that can then bind to VEGFR-2 and activate cell proliferation.

Purified heterodimeric VEGF/PlGF has been shown to bind also to the VEGFR-2 receptor (Cao et al, 1996a) but at a much lower affinity than VEGF (Cunningam et al, 1999). VEGFR-2 being the sole mediator of calcium mobilization in HUVEC, the PlGF/VEGF heterodimer is still able to produce a small rise in intracellular calcium through VEGFR-2 (Cunningam et al, 1999).

PlGF IN PREGNANCY

PlGF is abundantly expressed in both human and mouse placenta throughout gestation as well as in invasive and noninvasive hydatiform moles, as determined by Northern blot and semiquantitative polymerase chain reaction (PCR) analyses (Hauser and Weich, 1993; Maglione et al, 1993a; DiPalma et al, 1996). Immunohistochemistry reveals the presence of PlGF in the vasculosyncytial membrane and in the media of large blood vessels of the placenta villi (Khaliq et al, 1996). By in situ hybridization analysis, *PlGF* has been found to be expressed in the villous trophoblast, syncytiotrophoblast, and extravillous trophoblast, while *VEGF* is expressed in cells of mesenchymal origin within the chorionic plate, decidual macrophages, and decidual glands (Vuorela et al, 1997; Clark et al, 1998). Only homodimeric PlGF is produced in the human placenta cells. Mouse *Plgf* and *Vegf* mRNAs are abundant in trophoblastic giant cells associated with the parietal yolk sac during early embryogenesis (Achen et al, 1997). Therefore, these cells can produce heterodimeric *Plgf/Vegf*, which could be responsible for vascularization in the decidua and placenta at the early stages of embryogenesis. The different expression pattern between mice and humans introduces complications in the use of the mouse as a model system to study the role of these factors during pregnancy or in pathologic states such as preeclampsia.

PlGF has been detected in the serum of pregnant women at a level ranging from about 50 ng/mL in early pregnancy to about 400 ng/mL in term pregnancy; PlGF has also been detected in the amniotic fluid throughout pregnancy (Vuorela-Vepsäläinen et al, 1999). A comparative study on a large number of healthy pregnant women throughout gestation and at delivery in normal pregnancy and preeclampsia has shown that PlGF is present in the blood serum during gestation (Torry et al, 1998). In particular, the serum PlGF level during normal pregnancy increases from the first to the late second trimester, and subsequently declines from 30 weeks' gestation to delivery. Much lower levels of serum PlGF are found in preeclampsia pregnancy and during labor. Combining these data with the data on the expression pattern of PlGF in human placenta (Khaliq et al, 1996; Vuorela et al, 1997; Clark et al, 1998) suggests that a lower level of serum PlGF during preeclampsia indicates a malfunction of trophoblasts, which contribute to the pathogenesis of the disease.

Plgf EXPRESSION IN NORMAL TISSUES AND TUMORS

In rodents, *Plgf* gene expression has been studied in vivo and during embryo development. Mouse *Plgf* mRNA is present at a high level in the placenta and at a lower level in the lung and thyroid (DiPalma et al, 1996). *Plgf* mRNA is absent during

TABLE 16.1. Placenta growth factor (PlGF) and vascular endothelial growth factor (VEGF) expression in human tumors.

	PlGF	VEGF
Hypervascular renal carcinoma[1]	21/23	26/27
Thyroid carcinomas[2]	12/25	2/25
Hemangioblastomas[3]	+	+
Brain tumors[4]	25/39	34/39
Ovarian epithelial tumors[5]	–	|
Cervical squamous carcinomas[6]	15/29	ND
Nonseminomas[7]	8/26	23/26
Seminomas[7]	5/18	13/18
Breast carcinomas[8]	0/39	20/39
Colon carcinomas[8]	1/7	6/7
Kaposi sarcoma[8]	10/10	10/10
Lung carcinomas[8]	0/8	6/8
Neuroblastomas[8]	0/5	5/5
Esophageal carcinomas[8]	0/8	7/8
Ovaric adenocarcinomas[8]	10/39	30/39
Meningiomas[9]	4/16	3/16

+/–, presence/absence of PlGF messenger RNA (mRNA); ND, not done; [1]Takahashi et al, 1994; [2]Viglietto et al, 1995; [3]Hatva et al, 1996; [4]Nomura et al, 1998; [5]Sowter et al, 1997; [6]Kodama et al, 1997; [7]Viglietto et al, 1996; [8]Our unpublished results; [9]Donnini et al, 1999.

development of the mouse embryo, at least as analyzed by in situ hybridization (our unpublished results). Low levels of *PlGF* mRNA are also detected by Northern blot in human thyroid, heart, brain, lung, and skeletal muscle, but not in kidney and pancreas (Table 16.1) (Maglione et al, 1993a; Viglietto et al, 1995). *PlGF* gene expression has been analyzed in a series of different tumors (Table 16.1) parallel to the analysis of *VEGF* gene expression. Human thyroid carcinoma cell lines with different tumorigenic potentials as well as primary thyroid tumors show an elevated *VEGF* expression at the same time that *PlGF* expression is downregulated (Viglietto et al, 1995). Takahashi and coworkers (1994) found that *PlGF* is expressed in 21 out of 23 hypervascular renal cell carcinomas but not in the adjacent normal kidney tissue.

VEGF is dramatically upregulated in the majority of hypervascular germ cell tumors (Viglietto et al, 1996). In contrast, the level of *PlGF* transcript in germ cell tumors is generally very low (13/44). In hemangioblastoma tumors, high levels of *PlGF* are detected only in a small number of cases, while *VEGF* is always upregulated (Hatva et al, 1996). In both thyroid follicular cells and NTERA2/D1 cells, the *PlGF* mRNA expression appears to be associated with the presence of differentiation markers (Viglietto et al, 1995, 1996). The analysis of 36 primary and three metastatic brain tumors has revealed that *PlGF* mRNA is present in 64% of cases. All the hypervascular tumors (20 out of 20) expressed *PlGF* (Nomura et al, 1998). The analysis of 16 meningiomas by Northern blot has detected the presence of *PlGF* mRNA in four tumors and of VEGF-A in three tumors (Donnini et al, 1999).

PlGF EXPRESSION IN PATHOLOGIC CONDITIONS

Compared to normal thyroid, the expression of VEGF and PlGF and their receptors is markedly enhanced in goiters from patients with Graves' disease, suggesting that hypervascularity in human goiters may result from coordinate PlGF and VEGF secretion by thyrocytes and from the enhanced expression of receptors VEGFR-1 and VEGFR-2 by thyroid capillaries (Viglietto et al, 1997). This hypothesis is supported by the data obtained in an animal model of goiter formation. The mRNAs for *PlGF* and *VEGF* are coordinately upregulated in a time-dependent manner by increased levels of thyroid-stimulating hormone (TSH) in the thyroid of propylthiouracil-fed rats compared to the thyroid gland of control rats (Viglietto et al, 1997).

PlGF may represent an important element of hormone-dependent angiogenesis that occurs in the thyroid gland. In fact, TSH treatment of cultured thyroid cells (PC Cl 3 cell line) induces a time- and dose-dependent upregulation of both *PlGF* and *VEGF* mRNA expression. Altogether these results strongly suggest a key role for PlGF in the hypervascularization of thyroid goiters. Although previous work has suggested that VEGF produced by thyroid cells in response to TSH and/or the IgGs of Graves' disease might represent the most likely candidate for the angiogenesis factor produced in the thyroid gland (Sato et al, 1995), later studies have suggested the existence of a complex interplay between VEGF and PlGF in the stimulation of goiter vascularization (Viglietto et al, 1997).

A poor placental vascular development is present in intrauterine growth restriction. A 2.3-fold higher level of *PlGF* has been found in intrauterine growth restriction placentas as compared to gestationally matched normal placentas (Khaliq et al, 1999). Also, another report has shown *PlGF* gene upregulation due to hyperoxia in a model where cats are exposed to hyperoxia for 5 days (Simpson et al, 1999).

Immunohistochemistry analysis has been used to assign a role to PlGF in diabetic retinopathy (Khaliq et al, 1998); in this test, PlGF has been detected in the endothelial cells and perivascular regions of newly formed blood vessels of excised fibrovascular preretinal membranes, in superficial retinal vessels in diabetic retinas adjacent to neovascular preretinal membranes. PlGF was absent in normal retinas, in diabetic retinas that had received extensive treatment, and in nonvascular epiretinal membranes.

POSTTRANSCRIPTIONAL REGULATION OF THE *PlGF* GENE

Differential regulation of the expression of *VEGF* and *PlGF* genes is observed in normal and in pathologic conditions. Here we review data obtained in several in vitro systems aimed at unraveling the mechanisms of such regulation. While *VEGF* expression appears mostly to be regulated at the transcriptional level, *PlGF* expression undergoes transcriptional and posttranscriptional regulation. Hypoxia has been shown to be a potent inducer of angiogenesis (for a review see Guillemin and Krasnow, 1997); however, no change in the *PlGF* mRNA level is detected in response to hypoxia in the teratocarcinoma cell line NTERA2/D1, in the choriocarcinoma cell lines JAR and JEG-3, and the hepatocarcinoma HepG2 cell line (Cao et al, 1996b; Viglietto et al, 1996). The concentration of homodimeric and heterodimeric VEGF and PlGF proteins changes in JEG-3 cell-conditioned media as detected by enzyme-linked immunosorbent assay (ELISA) in hypoxic condition (Cao et al, 1996b). In particular, VEGF and PlGF homodimers and VEGF/PlGF heterodimers vary from 0.25, 9.3, and 0.15 ng/mL to 0.8, 18, and 0.5 ng/mL, respectively. These changes correspond to a fourfold increase of the VEGF homodimer and VEGF/PlGF heterodimer and a twofold increase of the PlGF homodimer in hypoxic condition. The increase in PlGF protein can be explained by a posttranscriptional regulation that also controls the amount of heterodimeric and homodimeric VEGF.

An interesting posttranscriptional control mechanism of *PlGF* expression has been proposed (Maglione et al, 1993b). The elimination of the 5′ untranslated region of *PlGF* mRNA increases PlGF synthesis. The *cis*-acting element responsible for this effect corresponds to a small open reading frame (ORF) positioned upstream of the PlGF coding sequence (Maglione et al, 1993b). Mutation of the starting AUG codon of this small ORF is sufficient to eliminate the posttranscriptional control. This small ORF potentially codes for a 15-amino-acid-long peptide and is present in human and bovine *PlGF* mRNAs but is absent in the mouse.

IN VITRO AND IN VIVO ACTIVITY OF HOMODIMERIC AND HETERODIMERIC PlGF

PlGF-1 and PlGF-2 have been purified in several laboratories. Despite the contrasting results obtained, some properties of PlGF activity have now been clarified (Table 16.2). PlGF-1 and PlGF-2, purified from the conditioned media of transfected cells, are unable to stimulate proliferation of adrenal cortex-derived capillary endothelial (ACCE) cells even at concentrations as high as 530 ng/mL (Park et al, 1994). Purified recombinant PlGF-2, obtained from baculovirus

TABLE 16.2. Known biologic activity of PlGF.

	Cell proliferation	Cell migration
Endothelial cells		
PAE (VEGFR-1 overexpression)[1]	+	–
CVEC[2]	+	+
HUVEC[2,5]	+	+
HUVEC[3]	–	+
ACCE[4]	–	ND
BAEC[5]	+	ND
Other cell types		
Monocytes[6]	ND	+
Extravillous trophoblasts[7]	+	–

+, activity; –, no activity; ND, not done; [1]Landgreen et al, 1998; [2]Ziche et al, 1997; Maglione et al, 1991; [3]Migdal et al, 1998; [4]Park et al, 1994; [5]Hauser and Weich, 1993; [6]Barleon et al, 1996; [7]Athanassiades and Lala, 1998. ACCE, adrenal cortex-derived capillary endothelial; BAEC, bovine aortic endothelial cell; CVEC, coronary postcapillary venule cell; HUVEC, human umbilical vein endothelial cell.

infected cells, is able to stimulate, proliferation of bovine aortic endothelial cells (BAECs) (Hauser and Weich, 1993) and HUVECs but not the endothelial cells from hepatic sinusoids (Sawano et al, 1996). We have demonstrated that recombinant PlGF-1, purified from stably transfected human fetal kidney cells, triggers the proliferation of cultured CVECs and HUVECs in a dose-dependent manner (Ziche et al, 1997). The mitogenic effects of PlGF exerted on HUVECs (10% over control at 10 ng/mL) is significantly lower than the effect exerted on CVECs (40% over control at 10 ng/mL). However, the mitogenic effects of PlGF exerted on HUVECs is also significantly lower than that exerted by VEGF, suggesting that PlGF is a weaker mitogen for HUVECs compared to VEGF (see Sawano et al, 1996). The effect exerted by $PlGF_{131}$ on CVECs is comparable to that elicited by VEGF (Ziche et al, 1997).

These data raise interesting questions about the heterogeneous response of endothelial cells to growth factors. The preferential target for the angiogenic effects of PlGF appears to be the endothelium of postcapillary venules, which represent the natural site for new capillary growth and which are likely to be more prone to respond to mitogenic stimuli than the umbilical cord vein endothelium. These results confirm a similar finding by Sawano and coworkers (1996), who raised the possibility that PlGF may stimulate the growth of only certain types of endothelial cells derived from specific tissues (e.g., PlGF is able to stimulate CVECs, coronary porcine aorta (CPA) BAECs, and less potent HUVECs, but not endothelial cells derived from liver sinusoids).

We have demonstrated that human PlGF-1 induces a dose-dependent angiogenic response in the rabbit cornea assay and in the chorioallantoic membrane assay (Ziche et al, 1997). The comparison of the relative potency in inducing neoangiogenesis in vivo of PlGF with the potency of VEGF and basic fibroblast growth factor (bFGF) shows that at a dose of 100 ng/pellet the extent of corneal vascularization induced by the three growth factors is substantially similar. At 200 ng/pellet, PlGF-1 is clearly more effective in promoting and supporting neovascular growth than are bFGF and VEGF. The kinetics of vascular growth and morphogenesis in the neoangiogenesis elicited by PlGF-1, bFGF, and VEGF are slightly different as shown by the score of vessel density. PlGF and VEGF appear to be more efficient in the early stages of cornea neovascularization than bFGF. PlGF-1 and $VEGF_{165}$ are able to induce the formation of new capillaries approximately 36 hours earlier than bFGF, suggesting a faster recruitment of quiescent cells from capillary endothelium.

In vitro assays have shown that PlGF can induce, in primary term trophoblasts, specific activation of the stress-activated protein kinase (SAPK) pathways, c-Jun-N-terminal kinase (JNK), and p38 kinases, as well as protect the trophoblasts from apoptosis induced by growth factor withdrawal (Desai et al, 1999). Therefore, PlGF can be considered an autocrine modulator of trophoblasts (Desai et al, 1999). Exogenous PlGF-1 stimulates proliferation, in the presence of excess heparan sulfate proteoglycans (HSPGs), of the first trimester extravillous trophoblasts (EVTs), which express the VEGFR-1 receptor gene (Athanassiades and Lala, 1998). However, PlGF-1 in either the presence or absence of HSPGs has no effect on EVT migration or invasiveness as measured in the transwell Boyden chamber in the presence or absence of Matrigel, respectively (Athanassiades and Lala, 1998). VEGFR-1, but not VEGFR-2, is detected in isolated cytotrophoblasts and in in vitro differentiated syncytiotrophoblasts (Shore et al, 1997).

Two different research groups have analyzed the vascular permeability activity of PlGF and shown that both $PlGF_{131}$ and $PlGF_{152}$ have very low activity as tested in the Miles assay (Park et al, 1994; Sawano et al, 1996). In the Miles assay, as well as in the proliferation assay on ACCE cells, PlGF is able to potentiate the activity of minimally effective doses of VEGF (Park et al, 1994). These data allowed Park and coworkers (1994) to speculate on the possibility that the receptor VEGFR-1 behaves as a "decoy" having little or no transducing activity. According to this hypothesis, PlGF would release VEGF from the VEGFR-1 receptor and then increase its availability to the other receptor VEGFR-2.

PlGF and VEGF can also form heterodimeric molecules in cells where both genes are expressed (DiSalvo et al, 1995; Cao et al, 1996a). Comparative analysis of the mitogenic activity of the homodimeric and heterodimeric forms of these two factors on HUVECs shows that the PlGF/VEGF heterodimer is approximately 20- to 50-fold less potent than the homodimeric VEGF (Cao et al, 1996a). Comparable activity of VEGF and PlGF/VEGF is observed in the ability of HUVECs to induce motility and stimulate the phosphorylation of the VEGFR-2 receptor (Cao et al, 1996a). PlGF, VEGF, and PlGF/VEGF heterodimer are unable to enhance survival or cloning efficiency of human hematopoietic progenitors isolated from adult bone marrows, fetal liver, or cord blood samples (Ratajczak et al, 1998).

IN VIVO MODEL FOR THE STUDY OF *Plgf* AND ITS RECEPTORS

The generation of transgenic animals in which a gene can be deleted or mutated is helpful in the understanding of the role of VEGF growth factor family and their receptors for both embryo development and adult life. *VEGFR-1* and *VEGFR-2* deficiencies cause in utero death of mouse embryos between 8.5 and 9.5 days postcoitum (Fong et al, 1995; Shalaby et al, 1995). The *neuropilin-1* null mutation causes defects in the vascularization of the central and peripheral nervous system as well as agenesis of branchial arch-related great vessels and dorsal aorta (Kawasaki et al, 1999). Mice lacking one *VEGF* allele (and presumably producing half the amount of VEGF) die in utero at day 9.5 postcoitum with abnormal blood vessel formation (Carmeliet et al, 1996; Ferrara et al, 1996). To study the involvement of Plgf during embryo development and in the physiology of the endothelial cells of the adult, mouse models were generated where exons 2, 3, and 4 of the *Plgf* gene have been substituted with a cassette containing the *Neo* resistance gene. This mutation removes seven of the eight cysteines of the Plgf protein. $Plgf^{-/-}$ mutant mice obtained are healthy and fertile, and do not show apparent gross phenotypic alteration (Carmeliet et al, 2000). Consequently, *Plgf*, even though highly expressed in the placenta throughout gestation, does not seem to interfere with embryo development, at least in mice.

However, the *Plgf* null mice model has revealed that PlGF activity appears to be required during pathologic conditions to stimulate neoangiogenesis and arteriogenesis. PlGF is required for correct wound healing, growth of collateral vessels after occlusion of supply vessels, vessel maturation and remodeling by stimulating growth and recruitment of smooth muscle cells around the newly formed vessels, recruitment of macrophages, increase in vascular permeability, leakage of fibronectin, and finally tumor growth (Carmeliet et al, 2000).

The mechanism by which PlGF acts in pathologic conditions still remains to be fully elucidate. Several hypothesis can be raised that have to be tested: (1) PlGF enhances the activity of VEGF, (2) the PlGF/VEGF heterodimer is the molecule responsible for the activities described above, and (3) PlGF undergoes a different (probably faster) upregulation that could be required in pathologic conditions.

CONCLUSION

The new findings on the requirement for PlGF only in pathologic angiogenesis and arteriogenesis open a new view of the experimental approach to follow in the future also for the other molecules implicated in the complex phenomenon of vascular growth and physiology. The *Plgf*$^{-/-}$ mouse model is an ideal system to test other functions of PlGF coupled with the administration of recombinant PlGF protein as well as in combination with other mutations in the genes involved in the angiogenesis control.

ACKNOWLEDGMENTS

We thank A. Secondulfo for correcting and typing the manuscript. This work was supported by a grant from the Associazione Italiana Ricerca sul Cancro (AIRC).

REFERENCES

Achen, M. G., Gad, J. M., Stacker, S. A., and Wilks, A. F. 1997. Placenta growth factor and vascular endothelial growth factor are co-expressed during early embryonic development. *Growth Factors* 15:69–80.

Athanassiades, A., and Lala, P. K. 1998. Role of placenta growth factor (PlGF) in human extravillous trophoblast proliferation, migration and invasiveness. *Placenta* 19:465–473.

Barillari, G., Albonici, L., Franzese, O., Modesti, A., Liberati, F., Barillari, P., Ensoli, B., Manzari, V., and Santeusanio, G. 1998. The basic residues of placenta growth factor type 2 retrieve sequestered angiogenic factors into a soluble form. *Am. J. Pathol.* 152:1161–1166.

Barleon, B., Sozzani, S., Zhou, F., Weich, H. A., Mantovani, A., and Marmé, D. 1996. Migration of human monocytes in response to vascular endothelial growth factor (VEGF) is mediated via the VEGF receptor flt-1. *Blood* 87:3336–3343.

Bussolino, F., Mantovani, A., and Persico, G. 1997. Molecular mechanisms of blood vessel formation. *Trends Biochem. Sci.* 22:251–256.

Cao, Y., Chen, H., Zhou, L., Chiang, M.-K. Anand-Apte, B., Weatherbee, J. A., Wang, Y., Fang, F., Flanagan, J. G., and Lik-Shing-Tsang, M. 1996a. Heterodimers of placenta growth factor/vascular endothelial growth factor. Endothelial activity, tumor cell expression, and high affinity binding to Flk-1/KDR. *J. Biol. Chem.* 271:3154–3162.

Cao, Y., Linden, P., Shima, D., Brone, F., and Folkman, J. 1996b. In vivo angiogenic activity and hypoxia induction of heterodimers of placenta growth factor/vascular endothelial growth factor. *J. Clin. Invest.* 98:2507–2511.

Carmeliet, P., Ferreira, V., Breier, G., Pollefeyt, S., Kieckens, L., Gertsenstein, M., Fahrig, M., Vandenhoeck, A., Harpal, K., Eberhardt, C., Declercq, C., Pawling, J., Moons, L., Collen, D., Risau, W., and Nagy, A. 1996. Role of tissue factor in embryonic blood vessel development. *Nature* 383:73–75.

Carmeliet, P., Vincenti, V., Moons, L., Luttun, A., Devy, L., Scholz, D., Brusselmans, K., Bono, F., Dipalma, T., Stalmans, I., Fukao, H., Dewerchin, M., Noel, M., Lutgens, E., Barra, A., Blacher, S., Laux, V., Detmar, M., Herbert, J.-M., Foidart, J.-M., Schaper, W.,

Collen, D., Persico, M. G. 2000. Placental growth factor (PLGF): a specific stimulator of VEGF in pathological angiogenesis and arteriogenesis. Submitted.

Clark, D. E., Smith, S. K., Licence, D., Evans, A. L., and Charnock-Jones, D. S. 1998. Comparison of expression patterns for placenta growth factor, vascular endothelial growth factor (VEGF), VEGF-B and VEGF-C in the human placenta throughout gestation. *J. Endocrinol.* 159:459–467.

Clauss, M., Weich, H. A., Breier, G., Knies, U., Röckl, W., Waltenberger, J., and Risau, W. 1996. The vascular endothelial growth factor receptor Flt-1 mediates biological activities. *J. Biol. Chem.* 271:17629–17634.

Conn, G., Bayne, M. L., Soderman, D. D., Kwok, P. W., Sullivan, K. A., Palisi, T. M., Hope, D. A., and Thomas, K. A. 1990. Amino acid and cDNA sequences of a vascular endothelial cell mitogen that is homologous to platelet-derived growth factor. *Proc. Natl. Acad. Sci. USA* 87:2628–2632.

Cunningham, S. A., Tran, T. M., Arrate, M. P., and Brock, T. A. 1999. Characterization of vascular endothelial cell growth factor interactions with the kinase insert domain-containing receptor tyrosine kinase. *J. Biol. Chem.* 274:18421–18427.

Davis-Smyth, T., Chen, H., Park, J., Presta, L. G., and Ferrara, N. 1996. The second immunoglobulin-like domain of the VEGF tyrosine kinase receptor Flt-1 determines ligand binding and may initiate a signal transduction cascade. *EMBO J.* 15:4919–41927.

Desai, J., Holt-Shore, V., Torry, R. J., Caudle, M. R., and Torry, D. S. 1999. Signal transduction of biological function of placenta growth factor in primary human trophoblast. *Biol. Reprod.* 60:887–892.

DiPalma, T., Tucci, M., Russo, G., Maglione, D., Lago, C. T., Romano, A., Saccone, S., DellaValle, G., De Gregorio, L., Dragani, T. A., Viglietto, G., and Persico, M. G. 1996. The placenta growth factor gene of the mouse. *Mamm. Genome* 7:6–12.

DiSalvo, J., Bayne, M. L., Conn, G., Kwock, P. W., Trivedi, P. G., Soderman, D. D., Palisi, T. M., Sullivan, K. A., and Thomas, K. A. 1995. Purification and characterization of a naturally occurring vascular endothelial growth factor—placenta growth factor. *J. Biol. Chem.* 270:7717–7723.

Donnini, S., Machein, M. R., Plate, K. H., and Weich, H. A. 1999. Expression and localization of placenta growth factor and PlGF receptors in human meningiomas. *J. Pathol.* 189:66–71.

Ferrara, N., Carver-Moore, K., Chen, H., Dowd, M., Lu, L., O'Shea, K. S., Powell-Braxton, L., Hillan, K. J., and Moore, M. W. 1996. Heterozygous embryonic lethality induced by targeted inactivation of the VEGF gene. *Nature* 380:439–442.

Ferrara, N., and Henzel, W. J. 1989. Pituitary follicular cells secrete a novel heparin-binding growth factor specific for vascular endothelial cells. *Biochem. Biophys. Res. Commun.* 161:851–858.

Fong, G. H., Rossant, J., Gertsenstein, M., Breitman, M. L. 1995. Role of the Flt-1 receptor tyrosine kinase in regulating the assembly of vascular endothelium. *Nature* 376:66–70.

Fujisawa, H., and Kitsukawa, T. 1998. Receptors for collapsin/semaphorins. *Curr. Opin. Neurobiol.* 8:587–592.

Gospodarowicz, D., Abraham, J. A., and Schilling, J. 1989. Isolation and characterization of a vascular endothelial mitogen produced by pituitary-derived folliculo stellate cells. *Proc. Natl. Acad. Sci. USA* 86:7311–7315.

Guillemin, K., and Krasnow, M. A. 1997. The hypoxic response: Huffing and HIFing. *Cell* 89:9–12.

Hatva, E., Bohling, T., Jaaskelainen, J., Persico, M. G., Haltia, M., and Alitalo, K. 1996. Vascular growth factors and receptors in capillary hemangioblastomas and hemangiopericytomas. *Am. J. Pathol.* 148:763–775.

Hauser, S., and Weich, H. A. 1993. A heparin-binding form of placenta growth factor (PlGF-2) is expressed in human umbilical vein endothelial cells and placenta. *Growth Factors* 9:259–268.

Joukov, V., Pajusola, K., Kaipainen, A., Chilov, D., Lahtinen, I., Kukk, E., Saksela, O., Kalkkinen, N., and Alitalo, K. 1996. A novel vascular endothelial growth factor, VEGF-C, is a ligand for the Flt4 (VEGFR-3) and KDR (VEGFR-2) receptor tyrosine kinases. *EMBO J.* 15:290–298.

Kawasaki, T., Kitsukawa, T., Bekku, Y., Matsuda, Y., Sanbo, M., Yagi, T., and Fujisawa, H. 1999. A requirement for neuropilin-1 in embryonic vessel formation. *Development* 126:4895–4902.

Keck, P. J., Hauser, S. D., Krivi, G., Sanzo, K., Warren, T., Feder, J., and Connolly, D. T. 1989. Vascular permeability factor, an endothelial cell mitogen related to PDGF. *Science* 246:1309–1312.

Khaliq, A., Dunk, C., Jiang, J., Shams, M., Li, X. F., Acevedo, C., Weich, H., Whittle, M., and Ahmed, A. 1999. Hypoxia down-regulates placenta growth factor, whereas fetal growth restriction up-regulates placenta growth factor expression: molecular evidence for "placental hyperoxia" in intrauterine growth restriction. *Lab. Invest.* 79:151–170.

Khaliq, A., Foreman, D., Ahmed, A., Weich, H., Gregor, Z., McLeod, D., and Boulton, M. 1998. Increased expression of placenta growth factor in proliferative diabetic retinopathy. *Lab. Invest.* 78:109–116.

Khaliq, A., Li, X. F., Shams, M., Sisi, P., Acevedo, C. A., Whittle, M. J., Weich, H. A., and Asmed, A. 1996. Localization of placenta growth factor (PlGF) in human term placenta. *Growth Factors* 13:243–250.

Kitsukawa, T., Shimiizu, M., Sanbo, M., Hirata, T., Taniguchi, M., Bekku, Y., Yagi, T., and Fujisawa, H. 1997. Neuropilin-semaphorin III/D-mediated chemorepulsive signals play a crucial role in peripheral nerve projection in mice. *Neuron* 19:995–1005.

Landgren, E., Schiller, P., Cao, Y., and Claesson-Welsh, L. 1998. Placenta growth factor stimulates MAP kinase and mitogenicity but not phospholipase C-gamma and migration of endothelial cells expressing Flt 1. *Oncogene* 16:359–367.

Levy, A. P., Tamargo, R., Brem, H., and Nathans, D. 1989. An endothelial cell growth factor from the mouse neuroblastoma cell line NB 41. *Growth Factors* 2:9–19.

Maglione, D., Guerriero, V., Rambaldi, M., Russo, G., Persico, M. G. 1993b. Translation of the placenta growth factor mRNA is severely affected by a small open reading frame localized in the 5′ untranslated region. *Growth Factors* 8:141–152.

Maglione, D., Guerriero, V., Viglietto, G., Ferraro, M. G., Aprelikova, O., Alitalo, K., Del Vecchio, S., Lei, K. J., Chou, J. Y., and Persice, M. G. 1993a. Two alternative mRNAs coding for the angiogenic factor, placenta growth factor (PlGF), are transcribed from a single gene of chromosome 14. *Oncogene* 8:925–931.

Maglione, D., Guerriero, V., Viglietto, G., Delli-Bovi, P., and Persico, M. G. 1991. Isolation of a human placenta cDNA coding for a protein related to the vascular permeability factor. *Proc. Natl. Acad. Sci. USA* 88:9267–9271.

Migdal, M., Huppertz, B., Tessler, S., Comforti A., Shibuya, M., Reich, R., Baumann, H., and Neufeld, G. 1998. Neuropilin-1 is a placenta growth factor-2 receptor. *J. Biol. Chem.* 273:22272–22278.

Nomura, M., Yamagishi, S., Harada, S., Yamashima, T., Yamashita, J., and Yamamoto, H. 1998. Placenta growth factor (PlGF) mRNA expression in brain tumors. *J. Neurooncol.* 40:123–130.

Olofsson, B., Pajusola, K., von Euler, G., Chilov, D., Alitalo, K., and Eriksson, U. 1996. Genomic organization of the mouse and human genes for vascular endothelial growth factor B (VEGF-B) and characterization of a second splice isoform. *J. Biol. Chem.* 271:19310–19317.

Orlandini, M., Marconcini, L., Ferruzzi, R., and Oliviero, S. 1996. Identification of a c-fos-induced gene that is related to the platelet-derived growth factor/vascular endothelial growth factor family. *Proc. Natl. Acad. Sci. USA* 93:11675–11680.

Park, J. E., Chen, H. H., Winer, J., Houck, K. A., and Ferrara, N. 1994. Placenta growth factor. Potentiation of vascular endothelial growth factor bioactivity, *in vitro* and *in vivo*, and high affinity binding to Flt-1 but not to VEGFR-2. *J. Biol. Chem.* 269:25646–25654.

Persico, M. G., Vincenti, V., and DiPalma, T. 1999. Structure, expression and receptor-binding properties of placenta growth factor (PlGF). *Curr. Top. Microbiol. Immunol.* 237:31–40.

Ratajczak, M. Z., Ratajczak, J., Machalinski, B., Majka, M., Marlicz, W., Carter, A., Pietrzkowski, Z., and Gewirtz, A. M. 1998. Role of vascular endothelial growth factor (VEGF) and placenta-derived growth factor (PlGF) in regulating human haemopoietic cell growth. *Br. J. Haematol.* 103:969–979.

Sato, K., Yamazaki, K., Shizume, K., Kanaji, Y., Obara, T., Ohsumi, K., Demura, H., Yamaguchi, S., and Shibuya, M. 1995. Stimulation by thyroid-stimulating hormone and Graves' immunoglobulin G of vascular endothelial growth factor mRNA expression in human thyroid follicles in vitro and flt mRNA expression in the rat thyroid in vivo. *J. Clin. Invest.* 96:1295–1302.

Sawano, A., Takahashi, T., Yamaguchi, S., Aonuma, M., and Shibuya, M. 1996. Flt-1 but not KDR/Flk-1 tyrosine kinase is a receptor for placenta growth factor, which is releated to vascular endothelial growth factor. *Cell Growth Differ.* 7:213–221.

Shalaby, F. Rossant, J., Yamaguchi, T. P., Gertsenstein, M., Wu, X. F., Breitman, M. L., and Schuh, A. C. 1995. Failure of blood-island formation and vasculogenesis in Flk-1-deficient mice. *Nature* 376:62–66.

Shibuya, M., Ito, N., and Claesson-Welsh, L. 1999. Structure and function of vascular endothelial growth factor receptor-1 and -2. *Curr. Top. Microbiol. Immunol.* 237:59–84.

Shore, V. H., Wang, T. H., Wang, C. L., Torry, R. J., Caudle, M. T., and Torry, D. S. 1997. Vascular endothelial growth factor, placenta growth factor and their receptors in isolated human trophoblast. *Placenta* 18:657–665.

Simpson, D. A., Murphy, G. M., Bhaduri, T., Gardiner, T. A., Archer, D. B., and Stitt, A. W. 1999. Expression of the VEGF gene family during retinal vaso-obliteration and hypoxia. *Biochem. Biophys. Res. Commun.* 262:333–340.

Sowter, H. M., Corps, A. N., Evans, A. L., Clark, D. E., Charnock-Jones, D. S., and Smith, S. K. 1997. Expression and localization of the vascular endothelial growth factor family in ovarian epithelial tumors. *Lab Investigation* 77:607–614.

Taipale, J., Makinen, T., Arighi, E., Kukk, E., Karkkainen, M., and Alitalo, K. 1999. Vascular endothelial growth factor receptor-3. *Curr. Top. Microbiol. Immunol.* 237:85–96.

Takahashi, A., Sasaki, H., Kim, S. J., Tobisu, K., Kakizoe, T., Tsukamoto, T., Kumamoto, Y., Sugimura, T., and Terada, M. 1994. Markedly increased amount of messenger RNAs for vascular endothelial growth factor and placenta growth factor in renal cell carcinoma associated with angiogenesis. *Cancer Res.* 54:4233–4237.

Terman, B. I., Khandke, L., Dougher-Vermazan, M., Maglione, D., Lassam, N. J., Gospodarowicz, D., Persico, M. G., Bohlen, P., and Eisinger, M. 1994. VEGF receptor subtypes KDR and FLT1 show different sensitivities to heparin and placenta growth factor. *Growth Factors* 11:187–195.

Torry, D. S., Wang, H. S., Wang, T. H. , Caudle, M. R., and Torry, R. J. 1998. Preeclampsia is associated with reduced serum levels of placenta growth factor. *Am. J. Obstet. Gynecol.* 179:1539–1544.

Viglietto, G., Maglione, D., Rambaldi, M., Cerutti, J., Romano, A., Trapasso, F., Fedele, M., Ippolito, P., Chiappetta, G., Botti, G., Fusco, A., and Persico, M. G. 1995. Upregulation of vascular endothelial growth factor (VEGF) and downregulation of placenta growth factor (PlGF) associated with malignancy in human thyroid tumors and cell lines. *Oncogene* 11:1569–1579.

Viglietto, G., Romano, A., Maglione, D., Rambaldi, M., Paoletti, I., Lago, C. T., Califano, D., Monaco, C., Mineo, A., Santelli, G., Manzo, G., Botti, G., Chiappetta, G., and Persico, M. G. 1996. Neovascularization in human germ cell tumors correlates with a marked increase in the expression of the vascular endothelial growth factor but not the placenta growth factor. *Oncogene* 13:577–587.

Viglietto, G., Romano, A., Manzo, G., Chiappetta, G., Paoletti, I., Califano, D., Galati, M. G., Mauriello, V., Bruni, P., Lago, C. T., Fusco, A., and Persico, M. G. 1997. Upregulation of the angiogenic factors PlGF, VEGF and their receptors (Flt-1, Flk-1/KDR) by TSH in cultured thyrocytes and in the thyroid gland of thiouracil-fed rats suggest a TSH-dependent paracrine mechanism for goiter hypervascularization. *Oncogene* 15:2687–2698.

Vuorela, P., Hatva, E., Lymboussaki, A., Kaipainen, A., Joukov, V., Persico, M. G., Alitalo, K., and Halmesmäki, E. 1997. Expression of vascular endothelial growth factor and placenta growth factor in human placenta. *Biol. Reprod.* 56:489–494.

Vuorela-Vepsäläinen, P., Alfthan, H., Orpana, A., Alitalo, K., Stenman, U.-H., and Halmesmäki, E. 1999. Vascular endothelial growth factor is bound in amniotic fluid and maternal serum. *Hum. Reprod.* 14:1346–1351.

Wise, L. M., Veikkola, T., Mercer, A. A., Savory, L. J., Fleming, S. B., Caesar, C., Vitali, A., Makinen, Y., Alitalo, K., and Stacker, S. A. 1999. Vascular endothelial growth factor (VEGF)-like protein from orf virus NZ2 binds to VEGFR2 and neuropilin-1. *Proc. Natl. Acad. Sci. USA* 96:3071–3076.

Yonekura, H., Sakurai, S., Liu, X., Migita, H., Wang, H., Yamagishi, S., Nomura, M., Abedin, Md. J., Unoki, H., Yamamoto, Y., and Yamamoto, H. 1999. Placenta growth factor and vascular endothelial growth factor B and C expression in microvascular endothelial cells and pericytes. *J. Biol. Chem.* 274:35172–35178.

Ziche, M., Maglione, D., Ribatti, D., Morbidelli, L., Lago, C. T., Battisti, M., Paoletti, I., Barra, A., Tucci, M., Parise, G., Vincenti, V., Granger, H. J., Viglietto, G., and Persico, M. G. 1997. Placenta growth factor-1 is chemotactic, mitogenic and angiogenic. *Lab. Invest.* 76:517–531.

CHAPTER 17

Early Placental Vascular Morphogenesis

Caroline H. Damsky and Susan J. Fisher

Survival of the eutherian embryo/fetus depends on formation of a transient but vital organ, the placenta. This process, termed placentation, is the first test of the embryo's differentiative and organogenesis capacity. It has two goals: attaching the conceptus to the uterus, and bringing the fetal and maternal circulations into close proximity to facilitate effective gas and nutrient/waste exchange. These functions require that fetal placental cells (trophoblasts) acquire an invasive phenotype. In mammals that form a hemochorial placenta (e.g., humans and mice), fetal trophoblasts come in direct contact with maternal blood. Thus, placentation also entails the unique requirement for close cooperation and direct cellular contact between two immunologically distinct organisms.

During implantation and subsequent placental development, fetal trophoblasts and the uterine cells they encounter intricately regulate expression of several classes of molecules. These include adhesion receptors and their ligands, proteinases and their inhibitors, growth factors/cytokines and their receptors, immunomodulators, and transcription factors (for review see Cross et al, 1994; Rinkenberger et al, 1997). This chapter focuses on the profound changes in trophoblast adhesion molecule expression that occur during uterine invasion, with emphasis on the human cells, termed cytotrophoblasts (CTBs). We include information about the functional relevance of this transformation that has been gained from studying the pregnancy syndrome preeclampsia, in which CTBs display specific defects in adhesion molecule switching. Data gathered from the studies we review (emphasizing information from 1996 to the present) point to the central importance of the intimate relationship that develops between fetal trophoblasts and the maternal vasculature, as well as to strategies for uncovering the regulatory factors that control formation of these highly unusual connections.

PLACENTAL CYTOTROPHOBLASTS INVADE THE UTERUS, WHERE THEY ENGINEER A HYBRID FETAL/MATERNAL VASCULATURE

Figure 17.1A (Zhou et al, 1997b) diagrams the human fetal-maternal interface at 10 weeks of gestation. The fetal portion of the placenta is highly branched,

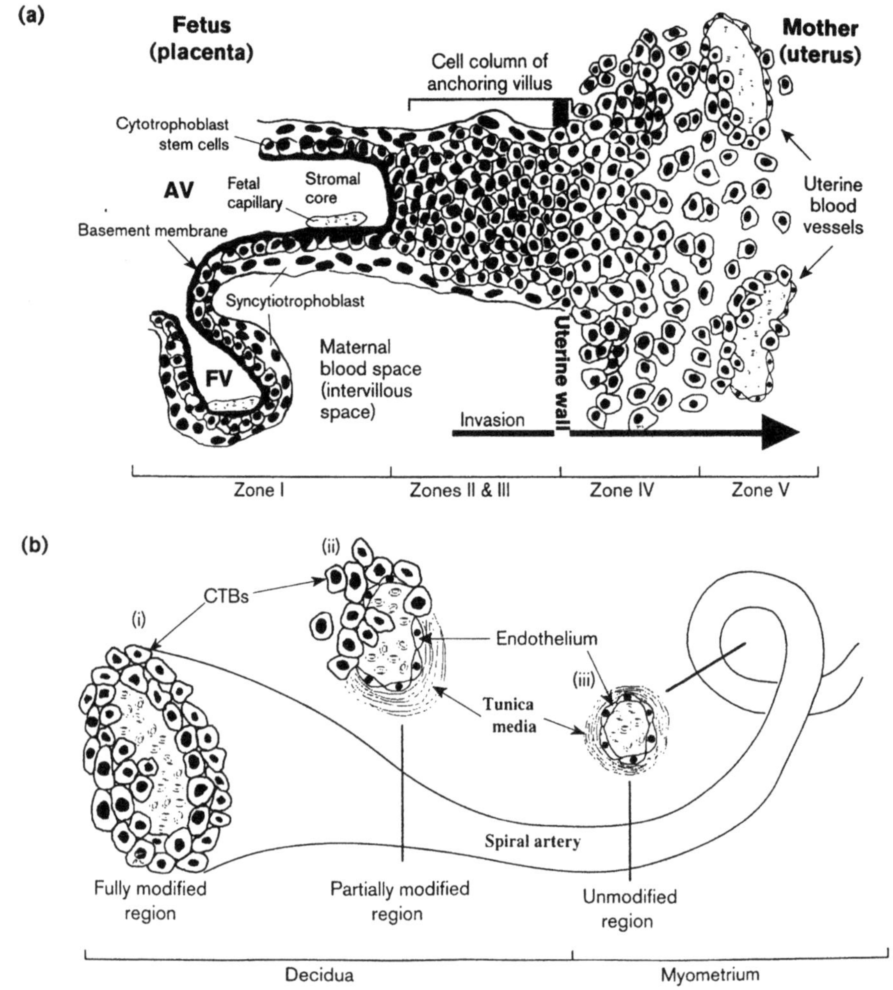

FIGURE 17.1. (A) Diagram of a longitudinal section of an anchoring chorionic villus (AV) at the fetal-maternal interface at about 10 weeks of gestational age. The AV functions as a bridge between the fetal and maternal compartments, whereas floating villi (FV) are suspended in the intervillous space and are bathed by maternal blood. Cytotrophoblasts (CTBs) in AV (zone I) form cell columns (zones II and III). CTBs then invade the uterine interstitium (decidua and first third of the myometrium; zone IV) and maternal vasculature (zone V), thereby anchoring the fetus to the mother and accessing the maternal circulation. Zone designations mark areas in which CTBs have distinct patterns of adhesion molecule expression. (B) Diagram of a uterine (spiral) artery in which endovascular invasion is in progress (10 to 18 weeks of gestation). Endometrial and then myometrial segments of spiral arteries are modified progressively. In fully modified regions (i) the vessel diameter is large. CTBs are present in the lumen and occupy the entire surface of the vessel wall. A discrete muscular layer (tunica media) is not evident. (ii) Partially modified vessel segments. CTBs and maternal endothelium occupy discrete regions of the vessel wall. In areas of intersection, CTBs appear to lie deep to the endothelium and in contact with the vessel wall. (iii) Unmodified vessel segments in the myometrium. Vessel segments in the superficial third of the myometrium will become modified when endovascular invasion reaches its fullest extent (around mid-gestation), while deeper segments of the same artery will retain their normal structure. (Modified from Zhou et al, 1997b, with permission.)

consisting of numerous *floating chorionic villi* that end blindly in the intervillous space. These villi have a syncytiotrophoblast covering that arises by fusion of the underlying polarized CTB stem cells, which in turn are attached to a basement membrane. The fetal vasculature develops within the villus stromal core. The syncytium is bathed by maternal blood. Thus, transport/exchange between the fetal and maternal circulations takes place across a trophoblast barrier. Before this can occur, the embryo must anchor itself to the uterine wall and release maternal blood from its arterial network into the intervillous space. This requires that a subset of CTBs invade the maternal compartment. *Anchoring chorionic villi* have evolved to fulfill this function. Aggregates of CTBs, termed columns, emanate from the tips of these villi and attach to the uterine wall. These CTBs go on to invade the uterine stroma (interstitial invasion) and its vasculature (endovascular invasion).

The microscopic anatomy of endovascular invasion has been described in humans and other primates (Hamilton and Boyd, 1966; Ramsey et al, 1976; Blankenship et al, 1993; Blankenship and Enders, 1997b). In a process that has been likened to dripping candle wax, masses of invasive CTBs (also called intermediate trophoblasts; Shih and Kurman, 1996) breach the termini of spiral arterioles. As these fetal cells move in a retrograde direction, they entirely replace the resident maternal endothelium and a portion of the muscular tunica media (Figure 17.1B). The result is uterine arterioles that are hybrid structures composed of fetal and maternal cells. Why endovascular invasion stops at the myometrial segments of these vessels is unknown. However, the functional consequences of this highly unusual process are clear: CTB invasion transforms the maternal arterioles from small-bore, high-resistance to large-bore, low-resistance vessels capable of meeting the ever-increasing fetal demands for maternal blood. Endovascular invasion of uterine veins, which is comparatively limited, allows the venous return of blood that has circulated through the intervillous space.

INVASIVE CYTOTROPHOBLASTS UNDERGO AN EPITHELIAL-TO-ENDOTHELIAL TRANSFORMATION IN THEIR ADHESION MOLECULE PHENOTYPE

These microscopic features of endovascular invasion trigger a myriad of mechanistic questions: What signals cause a CTB subpopulation to leave the fetal compartment? Do changes in their adhesive phenotype help them acquire an invasive phenotype? How do they execute the takeover of maternal blood vessels within the uterus?

Little is known about the selection process that regulates formation of anchoring villi. The trophoblast basement membrane and subjacent stroma are potential sources of signals (Figure 17.1A, zone I). Interestingly, the counteradhesive extracellular matrix (ECM) molecule tenascin is expressed at the stromal surface of the trophoblast basement membrane only at sites where anchoring villi form (Damsky et al, 1992). One of its receptors, $\alpha_v\beta_6$, is selectively expressed by the first one to two layers of CTBs at these same sites (Zhou et al, 1997b), which suggests that this interaction is important in column initiation. Unfortunately, mechanistic insights beyond these and other descriptive studies of trophoblast basement membrane composition are lacking.

Once CTBs commit to differentiation along the invasive pathway, their adhesion receptor phenotype is exquisitely modulated in a stepwise fashion. Particularly striking is the reduced staining for adhesion receptors characteristic of polarized CTB epithelial stem cells—integrin $\alpha_6\beta_4$ and E-cadherin—and the onset of expression of adhesion receptors characteristic of endothelium—VE-(vascular) cadherin, the immunoglobulin G (IgG)-family receptors vascular cell adhesion molecule (VCAM)-1 and platelet-endothelial cell adhesion molecule (PECAM)-1, and integrins $\alpha_v\beta_3$ and $\alpha_1\beta_1$ (Damsky et al, 1992; Vicovac et al, 1995; Blankenship and Enders, 1997a; Zhou et al, 1997b). All CTBs in this pathway, regardless of location, stain for Mel-CAM, a melanoma-associated IgG-family receptor also expressed on endothelium (Shih and Kurman, 1996). Finally, CTBs within the maternal blood vessels turn on neural cell adhesion molecule (NCAM; CD56) (Burrows et al, 1994; Blankenship and King, 1996; Proll et al, 1996; Zhou et al, unpublished observations), an adhesion receptor that is also expressed by maternal natural killer cells that home to the pregnant uterus (Starkey et al, 1988).

What are the functional consequences of this amazing transformation? One of the end points we have analyzed is invasiveness. Since CTB stem cells isolated from the fetal compartment replicate the adhesion receptor switching observed in situ as they invade Matrigel in vitro, we have used function-blocking antibodies to understand the role of individual molecules. These studies show that integrins $\alpha_v\beta_3$ and $\alpha_1\beta_1$ promote CTB invasion, but $\alpha_5\beta_1$ restrains it. Thus, as CTBs invade, they have a compensatory mechanism for titrating their tumor-like behavior (Damsky et al, 1994). Their regulation of cadherin expression reinforces this concept. E-cadherin staining, which is high on CTB stem cells, is greatly reduced on CTBs once they start to invade, although the protein continues to be produced. Its ability to mediate tight adhesion and restrain invasion, however, may be reduced by the upregulation of VE-cadherin expression by invasive CTBs (Zhou et al, 1997b). Supporting this idea is the observation that VE-cadherin can displace N-cadherin from cell-cell contact sites (Navarro et al, 1998). Cadherin-11 is also upregulated on extravillous CTBs and on decidualizing endometrial stroma (MacCalman et al, 1996). Whether this interaction promotes or restrains interstitial CTB invasion is not yet clear. Another functional end point of this adhesion molecule switch that we are currently investigating is the ability of endovascular CTBs to adopt at least some endothelial functions.

Changes in adhesion phenotype are accompanied by equally significant changes in CTB expression of ECM-degrading proteinases (Queenan et al, 1987; Fisher et al, 1989). Of particular functional importance is the ability of invasive CTBs to express and activate matrix metalloproteinase (MMP)-9, another switch that enhances CTB invasiveness in vitro (Librach et al, 1991). The coordinate upregulation of tissue inhibitor of metalloproteinase (TIMP)-3 expression is additional evidence for balancing mechanisms that hold CTB invasion in check (Bass et al, 1997). Importantly, these regulatory networks also involve maternal components. For example, studies in genetically engineered mice indicate that decidual expression of MMPs and TIMPs also affects the extent of trophoblast invasion (Alexander et al, 1996; Rinkenberger et al, 1997).

THE FAILURE OF INVASIVE CYTOTROPHOBLASTS TO SWITCH THEIR ADHESION MOLECULE PHENOTYPE IS ASSOCIATED WITH THE SERIOUS PREGNANCY DISORDER PREECLAMPSIA

Preeclampsia is a serious pregnancy complication that affects 7% of first-time pregnancies in the United States (Berg et al, 1996; Levine et al, 1997). The mother shows signs of widespread alterations in endothelial function (e.g., high blood pressure, proteinuria, and edema) (Roberts et al, 1989; Ness and Roberts, 1996). In some cases the fetus stops growing, resulting in intrauterine growth retardation. Compounding the dangers of this condition is the fact that the maternal and fetal signs can appear suddenly at any time from mid–second trimester until term—hence, the name preeclampsia (derived from the Greek *eklampsis*, sudden flash or development).

The preeclampsia syndrome reveals the significance of the epithelial-to-endothelial transformation that invasive CTBs normally undergo. Floating chorionic villi in preeclamptic placentas are relatively unaffected. However, anchoring villi and the invasive CTBs they spawn show distinct anomalies. The extent of interstitial invasion by CTBs is variable, but frequently shallow. Endovascular invasion is consistently rudimentary, making it extremely difficult to find any maternal vessels that contain CTBs (Brosens et al, 1972; Zhou et al, 1993, 1997a).

These anatomic defects suggest that in preeclampsia, CTB differentiation along the invasive pathway is abnormal. Biopsies of the uterine wall of women with this syndrome show that invasive CTBs retain expression of adhesion receptors characteristic of stem cells ($\alpha_6\beta_4$, E-cadherin) and fail to turn on receptors that promote invasion ($\alpha_1\beta_1$, $\alpha_v\beta_3$) and/or assumption of an endothelial phenotype (e.g., VE-cadherin, $\alpha_v\beta_3$, VCAM-1, PECAM-1) (Zhou et al, 1993, 1997a). A notable exception is the integrin $\alpha_5\beta_1$, which is expressed relatively normally; however, consistent with the placental pathology, our functional studies show that this adhesion molecule is actually antiinvasive (Zhou et al, 1993; Damsky et al, 1994). Finally, the cells' aberrant phenotype cannot be reversed by removing them from the maternal environment. CTBs isolated from the placentas of preeclamptic patients fail to upregulate $\alpha_1\beta_1$, express lower levels of MMP-9, and invade poorly compared to CTBs from control placentas of similar gestational age (Lim et al, 1997).

The association between failed CTB differentiation and preeclampsia suggests that this syndrome is fundamentally a trophoblast disorder (Redman, 1990). The fact that removal of the placenta is still the only known cure further substantiates this idea. The current challenge is to unravel the pathways that connect abnormal CTB differentiation/invasion to the maternal vascular disease. This task is complicated by evidence suggesting that several routes lead to the same end point. For example, placental mosaicism can produce aspects of preeclampsia, which suggests that genetically altered trophoblasts can initiate this syndrome (Kalousek and Vekemans, 1996). Conversely, women with altered endothelial function (e.g., patients with diabetes or lupus erythematosus) are more prone to develop preeclampsia, suggesting that maternal vascular defects can predate the placental abnormalities (Roberts and Redman, 1993; Ness and Roberts, 1996). We theorize that these seemingly disparate findings can be explained by a unifying hypothesis:

faulty vascular connections between invasive CTBs and uterine vessels, whatever the cause, are the initiating event in preeclampsia.

Will information about the disease process translate into improved methods for detecting women destined to develop this potentially life-threatening condition? In this regard, ECM proteins and adhesion receptors that are indicative of maternal vascular injury/activation have been evaluated as potential markers. For example, elevated cellular fibronectin in maternal plasma midway through gestation predates the onset of symptoms (Taylor et al, 1991). Increased levels of soluble adhesion receptors, such as VCAM-1, may also have predictive value (Lyall et al, 1994; Austgulen et al, 1997; Krauss et al, 1997). Consistent with these findings, platelets in women with preeclampsia circulate in a more activated state, as measured by enhanced binding of pituitary adenylate cyclase–activating polypeptide (PAC)-1 antibody, which recognizes activated $\alpha_{IIb}\beta_3$ integrin fibrinogen receptor, and elevated levels of cell surface P-selectin and PECAM-1 (Konijnenberg et al, 1997). To date, however, no marker of sufficient prognostic value to gain widespread use has been discovered.

THE NEW FRONTIER: REGULATING CYTOTROPHOBLAST PSEUDOVASCULOGENESIS

Our work has uncovered a seemingly unique process in which human CTBs undergo pseudovasculogenesis—the process whereby these epithelial cells of ectodermal origin acquire an adhesion receptor repertoire resembling that of endothelial cells. We theorize that this switch permits the heterotypic adhesive interactions that allow fetal and maternal cells to cohabit the uterine vasculature during normal pregnancy. The observation that this switching process is defective in preeclampsia further suggests its functional importance.

These unexpected findings highlight the potential importance of factors that govern vasculogenesis/angiogenesis in the regulation of placental development. Undoubtedly, this complex differentiation program is controlled by a hierarchy of regulatory factors. At a physiologic level, we already know that oxygen tension, which has profound effects on blood vessels, controls the switch between CTB proliferation and differentiation (Genbacev et al, 1996, 1997). In these experiments, control (6 to 8 weeks of gestation) anchoring villus explants were maintained in either a 20% or an 8% O_2 atmosphere, mimicking standard culture conditions or the environment within the uterine interstitium (Figure 17.1A, zone IV), respectively. Other villi were cultured in 2% O_2, mimicking the hypoxic conditions in the fetal compartment near the uterine lumen at this time (Figure 17.1A, zone I; intervillous space). In 20% and 8% O_2, CTBs exit the cell cycle, upregulate $\alpha_1\beta_1$, and become highly invasive. In hypoxia, CTBs continue to proliferate, but they fail to express $\alpha_1\beta_1$ integrin and do not invade. These observations suggest the following model: Before CTBs access the maternal blood supply (e.g., 10 weeks of gestation) (Rodesch et al, 1992; Jaffe et al, 1997), the hypoxic environment near the uterine lumen in which early placental development occurs favors CTB proliferation, a phenomenon observed in situ. As interstitial invasion proceeds, invasive CTBs encounter a positive oxygen gradient that favors differentiation/invasion. Interestingly, hypoxia in vitro mimics some of the effects seen in preeclampsia, suggesting the possible consequences of failed endovascular invasion in vivo.

We envision that oxygen acts through a variety of downstream effectors that include molecular families known to play an important role in vasculogenesis/angiogenesis (recently reviewed in Yancopoulos et al, 1998). In this regard vascular endothelial growth factor (VEGF) family members and their receptors are attractive candidates. There are several reports that CTBs, as well as fetal and maternal macrophages, express VEGF (Charnock-Jones et al, 1994; Vuorela et al, 1997). CTBs also express placenta growth factor (PlGF)—a unique placental form of VEGF. Unlike VEGF, PlGF is not responsive to hypoxia, its homodimers are not angiogenic, and they bind VEGF receptor (VEGFR)-1 (Flt-1), but not VEGFR-2 (kinase–containing receptor, domain KDR). PlGF also forms heterodimers with VEGF that bind both receptors (Cao et al, 1996). VEGF upregulates expression of $\alpha_1\beta_1$ and $\alpha_v\beta_3$ integrins in endothelial cells (Senger et al, 1997)—the same proinvasive integrins that are upregulated during CTB differentiation, tempting speculation that VEGF-family factors play analogous roles in invasive CTBs and angiogenic endothelial cells. CTBs also express Flt-1 and KDR, although their spatial and temporal patterns of expression are still unclear (Cheung, 1997; Shore et al, 1997). Other factors and their cognate receptors that regulate vasculogenesis/angiogenesis, including fibroblast growth factors/receptors (Friesel and Maciag, 1995), Ties/angiopoietins (Davis et al, 1996; Suri et al, 1996), and thrombospondins (Iruela-Arispe and Dvorak, 1997; Kyriakides et al, 1998; Volpert et al, 1998), are also expressed in the placenta (Dumont et al, 1995; Izumi et al, 1996).

Given this complexity, it is likely that the regulated balance of pro- and antiangiogenic mechanisms is critical to coordinating the development of the hybrid vascular structures, composed of CTBs and maternal cells, that control blood flow to the fetus. Placental development actually requires the coordination of three different vasculogenesis/angiogenesis-like programs. First, fetal blood vessels develop in the stroma of floating chorionic villi (Figure 17.1A, fetal capillary). This process probably employs a vasculogenesis/angiogenesis program similar to that taking place within the embryo proper (Risau, 1997). Second, differentiating/invading CTBs undergo pseudovasculogenesis, a novel autocrine differentiation program that allows them to take on an endothelial adhesion phenotype (Figure 17.1A, zones II to V). Third, endovascular invasion by CTBs occurs without triggering significant angiogenesis in the maternal vessels, which suggests that CTBs also have novel paracrine mechanisms for rendering quiescent the resident uterine vasculature that they remodel (Figure 17.1A, zone V, uterine blood vessels). How are these different processes coordinated within the same microenvironment? The novel requirements of CTB pseudovasculogenesis suggest that known regulatory molecules may have novel actions; alternatively, novel regulatory molecules may control the behavior of these unusual cells.

The importance of trophoblast differentiation to embryonic/fetal development is also revealed by the numerous genetically engineered mice with placental defects (for listings of genes affecting implantation/placentation see Cross et al, 1994; Rinkenberger et al, 1997). In the context of normal and abnormal human CTB differentiation/invasion, some of the most interesting include hepatocyte growth factor (HGF)/scatter factor and its c-Met receptor (Uehara et al, 1995; Birchmeier et al, 1997), VCAM-1 and the integrin α_4 subunit (Gurtner et al, 1995; Kwee et al, 1995; Yang et al, 1995), and the hypoxia-inducible transcription factor HIF-1β (aryl hydrocarbon receptor nuclear translocator ARNT) (Kozak et al, 1997). Although the discovery that these and other molecules—e.g., mammaliar achaete-scute homologue (Mash-2) (Guillemot et al, 1994; Nakayama et al, 1997), Ets-2 (Yamamoto et al, 1998), wingless (Wnt2) (Monkley et al, 1996), and interleukin-11R (IL-11R) (Robb et al, 1998)—play important roles in placentation has for the most part been unexpected, the trophoblast phenotypes of genetically engineered mice, which like humans form a hemochorial placenta, will continue to be extremely informative regarding mechanisms of human placental development and CTB pseudovasculogenesis.

CONCLUSION

Recent advances have uncovered the unexpected finding that specialized fetal cells of the human placenta—CTBs—undergo a novel pseudovasculogenesis differentiation program that enables them to masquerade as the endothelial and smooth muscle components of maternal uterine vessels. Future research into the regulatory mechanisms that govern this unusual transformation will offer fascinating insights about how the placenta forms. Will the results of these studies be applicable to other normal and abnormal processes that require similar sorts of plasticity? One interesting possibility is that tumor cells, which like CTBs are marauders in search of a blood supply, might co-opt portions of the differentiation program described here. Whether lessons learned from placental development could be used to develop additional antiangiogenesis cancer therapies, analogous to those described by Folkman and coworkers (O'Reilly et al, 1996, 1997), remains to be investigated.

ACKNOWLEDGMENTS

Work from the authors' laboratories was supported by grants from the National Institutes of Health (HD26732; HD30367; HL64597) and grant 6RT-30-40 from the California Tobacco-Related Disease Research Program.

REFERENCES

Alexander, C. M., Hansell, E. J., Behrendtsen, O., Flannery, M. L., Kishnani, N. S., Hawkes, S. P., and Werb, Z. 1996. Expression and function of matrix metalloproteinases and their inhibitors at the maternal-embryonic boundary during mouse embryo implantation. *Development* 122:1723–1736.

Austgulen, R., Lien, E., Vince, G., and Redman, C. W. 1997. Increased maternal plasma levels of soluble adhesion molecules ICAM-1, VCAM-1, E-selectin in preeclampsia. *Eur. J. Obstet. Gynecol. Reprod. Biol.* 71:53–58.

Bass, K. E., Li, H., Hawkes, S. P., Howard, E., Bullen, E., Vu, T. K., McMaster, M., Janatpour, M., and Fisher, S. J. 1997. Tissue inhibitor of metalloproteinase-3 expression is upregulated during human cytotrophoblast invasion in vitro. *Dev. Genet.* 21:61–67.

Berg, C. J., Atrash, H. K., Koonin, L. M., and Tucker, M. 1996. Pregnancy-related mortality in the United States, 1987–1990. *Obstet. Gynecol.* 88:161–167.

Birchmeier, C., Bladt, F., and Yamaai, T. 1997. The functions of HGF/SF and its receptor, the c-Met tyrosine kinase, in mammalian development. *Ciba Found. Symp.* 212:169–182.

Blankenship, T. N., and Enders, A. C. 1997a. Expression of platelet-endothelial cell adhesion molecule-1 PECAM by macaque trophoblast cells during invasion of the spiral arteries. *Anat. Rec.* 247:413–419.

Blankenship, T. N., and Enders, A. C. 1997b. Trophoblast cell-mediated modifications to uterine spiral arteries during early gestation in the macaque. *Acta Anat. Basel* 158:227–236.

Blankenship, T. N., Enders, A. C., and King, B. F. 1993. Trophoblastic invasion and the development of uteroplacental arteries in the macaque: immunohistochemical localization of cytokeratins, desmin, type IV collagen, laminin, and fibronectin. *Cell Tissue Res.* 272:227–236.

Blankenship, T. N., and King, B. F. 1996. Macaque intra-arterial trophoblast and extravillous trophoblast of the cell columns and cytotrophoblastic shell express neural cell adhesion molecule NCAM. *Anat. Rec.* 245:525–531.

Brosens, I. A., Robertson, W. B., and Dixon, H. G. 1972. The role of the spiral arteries in the pathogenesis of preeclampsia. *Obstet. Gynecol. Annu.* 1:177–191.

Burrows, T. D., King, A., and Loke, Y. W. 1994. Expression of adhesion molecules by endovascular trophoblast and decidual endothelial cells: implications for vascular invasion during implantation. *Placenta* 15:21–33.

Cao, Y., Linden, P., Shima, D., Browne, F., and Folkman, J. 1996. In vivo angiogenic activity and hypoxia induction of heterodimers of placenta growth factor/vascular endothelial growth factor. *J. Clin. Invest.* 98:2507–2511.

Charnock-Jones, D. S., Sharkey, A. M., Boocock, C. A., Ahmed, A., Plevin, R., Ferrara, N., and Smith, S. K. 1994. Vascular endothelial growth factor receptor localization and activation in human trophoblast and choriocarcinoma cells. *Biol. Reprod.* 51:524–530.

Cheung, C. Y. 1997. Vascular endothelial growth factor: possible role in fetal development and placental function. *J. Soc. Gynecol. Invest.* 4:169–177.

Cross, J. C., Werb, Z., and Fisher, S. J. 1994. Implantation and the placenta: key pieces of the development puzzle. *Science* 266:1508–1518.

Damsky, C. H., Fitzgerald, M. L., and Fisher, S. J. 1992. Distribution patterns of extracellular matrix components and adhesion receptors are intricately modulated during first trimester cytotrophoblast differentiation along the invasive pathway, in vivo. *J. Clin. Invest.* 89:210–222.

Damsky, C. H., Librach, C., Lim, K. H., Fitzgerald, M. L., McMaster, M. T., Janatpour, M., Zhou, Y., Logan, S. K., and Fisher, S. J. 1994. Integrin switching regulates normal trophoblast invasion. *Development* 120:3657–3666.

Davis, S., Aldrich, T. H., Jones, P. F., Acheson, A., Compton, D. L., Jain, V., Ryan, T. E., Bruno, J., Radziejewski, C., Maisonpierre, P. C., and Yancopoulos, G. D. 1996. Isolation of angiopoietin-1, a ligand for the TIE2 receptor, by secretion-trap expression cloning. *Cell* 87:1161–1169.

Dumont, D. J., Fong, G. H., Puri, M. C., Gradwohl, G., Alitalo, K., and Breitman, M. L. 1995. Vascularization of the mouse embryo: a study of flk-1, tek, tie, and vascular endothelial growth factor expression during development. *Dev. Dyn.* 203:80–92.

Fisher, S. J., Cui, T. Y., Zhang, L., Hartman, L., Grahl, K., Zhang, G. Y., Tarpey, J., and Damsky, C. H. 1989. Adhesive and degradative properties of human placental cytotrophoblast cells in vitro. *J. Cell Biol.* 109:891–902.

Friesel, R. E., and Maciag, T. 1995. Molecular mechanisms of angiogenesis: fibroblast growth factor signal transduction. *FASEB J.* 9:919–925.

Genbacev, O., Joslin, R., Damsky, C. H., Polliotti, B. M., and Fisher, S. J. 1996. Hypoxia alters early gestation human cytotrophoblast differentiation/invasion in vitro and models the placental defects that occur in preeclampsia. *J. Clin. Invest.* 97:540–550.

Genbacev, O., Zhou, Y., Ludlow, J. W., and Fisher, S. J. 1997. Regulation of human placental development by oxygen tension. *Science* 277:1669–1672.

Guillemot, F., Nagy, A., Auerbach, A., Rossant, J., and Joyner, A. L. 1994. Essential role of Mash-2 in extraembryonic development. *Nature* 371:333–336.

Gurtner, G. C., Davis, V., Li, H., McCoy, M. J., Sharpe, A., and Cybulsky, M. I. 1995. Targeted disruption of the murine VCAM1 gene: essential role of VCAM-1 in chorioallantoic fusion and placentation. *Genes Dev.* 9:1–14.

Hamilton, W. J., and Boyd, J. D. 1966. Trophoblast in human utero-placental arteries. *Nature* 212:906–908.

Iruela-Arispe, M. L., and Dvorak, H. F. 1997. Angiogenesis: a dynamic balance of stimulators and inhibitors. *Thromb. Haemost.* 78:672–677.

Izumi, S., Slayden, O. D., Rubin, J. S., and Brenner, R. M. 1996. Keratinocyte growth factor and its receptor in the rhesus macaque placenta during the course of gestation. *Placenta* 17:123–135.

Jaffe, R., Jauniaux, E., and Hustin, J. 1997. Maternal circulation in the first-trimester human placenta—myth or reality? *Am. J. Obstet. Gynecol.* 176:695–705.

Kalousek, D. K., and Vekemans, M. 1996. Confined placental mosaicism. *J. Med. Genet.* 33:529–533.

Konijnenberg, A., Stokkers, E. W., van der Post, J. A., Schaap, M. C., Boer, K., Bleker, O. P., and Sturk, A. 1997. Extensive platelet activation in preeclampsia compared with normal pregnancy: enhanced expression of cell adhesion molecules. *Am. J. Obstet. Gynecol.* 176:461–469.

Kozak, K. R., Abbott, B., and Hankinson, O. 1997. ARNT-deficient mice and placental differentiation. *Dev. Biol.* 191:297–305.

Krauss, T., Kuhn, W., Lakoma, C., and Augustin, H. G. 1997. Circulating endothelial cell adhesion molecules as diagnostic markers for the early identification of pregnant women at risk for development of preeclampsia. *Am. J. Obstet. Gynecol.* 177:443–449.

Kwee, L., Baldwin, H. S., Shen, H. M., Stewart, C. L., Buck, C., Buck, C. A., and Labow, M. A. 1995. Defective development of the embryonic and extraembryonic circulatory systems in vascular cell adhesion molecule. *Development* 121:489–503.

Kyriakides, T. R., Zhu, Y. H., Smith, L. T., Bain, S. D., Yang, Z., Lin, M. T., Danielson, K. G., Iozzo, R. V., LaMarca, M., McKinney, C. E., Ginns, E. I., and Bornstein, P. 1998. Mice that lack thrombospondin 2 display connective tissue abnormalities that are associated with disordered collagen fibrillogenesis, an increased vascular density, and a bleeding diathesis. *J. Cell Biol.* 140:419–430.

Levine, R. J., Hauth, J. C., Curet, L. B., Sibai, B. M., Catalano, P. M., Morris, C. D., DerSimonian, R., Esterlitz, J. R., Raymond, E. G., Bild, D. E., Clemens, J. D., and Cutler, J. A. 1997. Trial of calcium to prevent preeclampsia. *N. Engl. J. Med.* 337:69–76.

Librach, C. L., Werb, Z., Fitzgerald, M. L., Chiu, K., Corwin, N. M., Esteves, R. A., Grobelny, D., Galardy, R., Damsky, C. H., and Fisher, S. J. 1991. 92-kD type IV collagenase mediates invasion of human cytotrophoblasts. *J. Cell Biol.* 113:437–449.

Lim, K. H., Zhou, Y., Janatpour, M., McMaster, M., Bass, K., Chun, S. H., and Fisher, S. J. 1997. Human cytotrophoblast differentiation/invasion is abnormal in preeclampsia. *Am. J. Pathol.* 151:1809–1818.

Lyall, F., Greer, I. A., Boswell, F., Macara, L. M., Walker, J. J., and Kingdom, J. C. 1994. The cell adhesion molecule, VCAM-1, is selectively elevated in serum in preeclampsia: does this indicate the mechanism of leucocyte activation? *Br. J. Obstet. Gynaecol.* 101:485–487.

MacCalman, C. D., Furth, E. E., Omigbodun, A., Bronner, M., Coutifaris, C., and Strauss, J. F. R. 1996. Regulated expression of cadherin-11 in human epithelial cells: a role for cadherin-11 in trophoblast-endometrium interactions? *Dev. Dyn.* 206:201–211.

Monkley, S. J., Delaney, S. J., Pennisi, D. J., Christiansen, J. H., and Wainwright, B. J. 1996. Targeted disruption of the Wnt2 gene results in placentation defects. *Development* 122:3343–3353.

Nakayama, H., Liu, Y., Stifani, S., and Cross, J. C. 1997. Developmental restriction of Mash-2 expression in trophoblast correlates with potential activation of the notch-2 pathway. *Dev. Genet.* 21:21–30.

Navarro, P., Ruco, L., and Dejana, E. 1998. Differential localization of VE- and N-cadherins in human endothelial cells: VE-cadherin competes with N-cadherin for junctional localization. *J. Cell Biol.* 140:1475–1484.

Ness, R. B., and Roberts, J. M. 1996. Heterogeneous causes constituting the single syndrome of preeclampsia: a hypothesis and its implications. *Am. J. Obstet. Gynecol.* 175:1365–1370.

O'Reilly, M. S., Boehm, T., Shing, Y., Fukai, N., Vasios, G., Lane, W. S., Flynn, E., Birkhead, J. R., Olsen, B. R., and Folkman, J. 1997. Endostatin: an endogenous inhibitor of angiogenesis and tumor growth. *Cell* 88:277–285.

O'Reilly, M. S., Holmgren, L., Chen, C., and Folkman, J. 1996. Angiostatin induces and sustains dormancy of human primary tumors in mice. *Nat. Med.* 2:689–692.

Proll, J., Blaschitz, A., Hartmann, M., Thalhamer, J., and Dohr, G. 1996. Human first-trimester placenta intra-arterial trophoblast cells express the neural cell adhesion molecule. *Early Pregnancy* 2:271–275.

Queenan, J. T., Jr., Kao, L. C., Arboleda, C. E., Ulloa-Aguirre, A., Golos, T. G., Cines, D. B., and Strauss, J. F. D. 1987. Regulation of urokinase-type plasminogen activator production by cultured human cytotrophoblasts. *J. Biol. Chem.* 262:10903–10906.

Ramsey, E. M., Houston, M. L., and Harris, J. W. 1976. Interactions of the trophoblast and maternal tissues in three closely related primate species. *Am. J. Obstet. Gynecol.* 124:647–652.

Redman, C. W. 1990. Platelets and the beginnings of preeclampsia. *N. Engl. J. Med.* 323:478–480.

Rinkenberger, J. L., Cross, J. C., and Werb, Z. 1997. Molecular genetics of implantation in the mouse. *Dev. Genet.* 21:6–20.

Risau, W. 1997. Mechanisms of angiogenesis. *Nature* 386:671–674.

Robb, L., Li, R., Hartley, L., Nandurkar, H. H., Koentgen, F., and Begley, C. G. 1998. Infertility in female mice lacking the receptor for interleukin 11 is due to a defective uterine response to implantation. *Nat. Med.* 4:303–308.

Roberts, J. M., and Redman, C. W. 1993. Pre-eclampsia: more than pregnancy-induced hypertension. *Lancet* 341:1447–1451.

Roberts, J. M., Taylor, R. N., Musci, T. J., Rodgers, G. M., Hubel, C. A., and McLaughlin, M. K. 1989. Preeclampsia: an endothelial cell disorder. *Am. J. Obstet. Gynecol.* 161:1200–1204.

Rodesch, F., Simon, P., Donner, C., and Jauniaux, E. 1992. Oxygen measurements in endometrial and trophoblastic tissues during early pregnancy. *Obstet. Gynecol.* 80:283–285.

Senger, D. R., Claffey, K. P., Benes, J. E., Perruzzi, C. A., Sergiou, A. P., and Detmar, M. 1997. Angiogenesis promoted by vascular endothelial growth factor: regulation through alpha1beta1 and alpha2beta1 integrins. *Proc. Natl. Acad. Sci. USA* 94:13612–13617.

Shih, I. M., and Kurman, R. J. 1996. Expression of melanoma cell adhesion molecule in intermediate trophoblast. *Lab. Invest.* 75:377–388.

Shore, V. H., Wang, T. H., Wang, C. L., Torry, R. J., Caudle, M. R., and Torry, D. S. 1997. Vascular endothelial growth factor, placenta growth factor and their receptors in isolated human trophoblast. *Placenta* 18:657–665.

Starkey, P. M., Sargent, I. L., and Redman, C. W. 1988. Cell populations in human early pregnancy decidua: characterization and isolation of large granular lymphocytes by flow cytometry. *Immunology* 65:129–134.

Suri, C., Jones, P. F., Patan, S., Bartunkova, S., Maisonpierre, P. C., Davis, S., Sato, T. N., and Yancopoulos, G. D. 1996. Requisite role of angiopoietin-1, a ligand for the TIE2 receptor, during embryonic angiogenesis. *Cell* 87:1171–1180.

Taylor, R. N., Crombleholme, W. R., Friedman, S. A., Jones, L. A., Casal, D. C., and Roberts, J. M. 1991. High plasma cellular fibronectin levels correlate with biochemical and clinical features of preeclampsia but cannot be attributed to hypertension alone. *Am. J. Obstet. Gynecol.* 165:895–901.

Uehara, Y., Minowa, O., Mori, C., Shiota, K., Kuno, J., Noda, T., and Kitamura, N. 1995. Placental defect and embryonic lethality in mice lacking hepatocyte growth factor/scatter factor. *Nature* 6516:702–705.

Vicovac, L., Jones, C. J., and Aplin, J. D. 1995. Trophoblast differentiation during formation of anchoring villi in a model of the early human placenta in vitro. *Placenta* 16:41–56.
Volpert, O. V., Lawler, J., and Bouck, N. P. 1998. A human fibrosarcoma inhibits systemic angiogenesis and the growth of experimental metastases via thrombospondin-1. *Proc. Natl. Acad. Sci. USA* 95:6343–6348.
Vuorela, P., Hatva, E., Lymboussaki, A., Kaipainen, A., Joukov, V., Persico, M. G., Alitalo, K., and Halmesmeaki, E. 1997. Expression of vascular endothelial growth factor and placenta growth factor in human placenta. *Biol. Reprod.* 56:489–494.
Yamamoto, H., Flannery, M. L., Kupriyanov, S., Pearce, J., McKercher, S. R., Henkel, G. W., Maki, R. A., Werb, Z., and Oshima, R. G. 1998. Defective trophoblast function in mice with a targeted mutation of Ets2. *Genes Dev.* 12:1315–1326.
Yancopoulos, G. D., Klagsbrun, M., and Folkman, J. 1998. Vasculogenesis, angiogenesis, and growth factors: ephrins enter the fray at the border. *Cell* 93:661–664.
Yang, J. T., Rayburn, H., and Hynes, R. O. 1995. Cell adhesion events mediated by alpha 4 integrins are essential in placental and cardiac development. *Development* 121:549–560.
Zhou, Y., Damsky, C. H., Chiu, K., Roberts, J. M., and Fisher, S. J. 1993. Preeclampsia is associated with abnormal expression of adhesion molecules by invasive cytotrophoblasts. *J. Clin. Invest.* 91:950–960.
Zhou, Y., Damsky, C. H., and Fisher, S. J. 1997a. Preeclampsia is associated with failure of human cytotrophoblasts to mimic a vascular adhesion phenotype. One cause of defective endovascular invasion in this syndrome? *J. Clin. Invest.* 99:2152–2164.
Zhou, Y., Fisher, S. J., Janatpour, M., Genbacev, O., Dejana, E., Wheelock, M., and Damsky, C. H. 1997b. Human cytotrophoblasts adopt a vascular phenotype as they differentiate. A strategy for successful endovascular invasion? *J. Clin. Invest.* 99:2139–2151.

Index